Venomous Reptiles
of the United States, Canada,
and Northern Mexico

Venomous Reptiles
of the United States, Canada,
and Northern Mexico

VOLUME 2   *Crotalus*

Carl H. Ernst and Evelyn M. Ernst

THE JOHNS HOPKINS UNIVERSITY PRESS
Baltimore

© 2012 The Johns Hopkins University Press
All rights reserved. Published 2012
Printed in the United States of America on acid-free paper
9 8 7 6 5 4 3 2 1

The Johns Hopkins University Press
2715 North Charles Street
Baltimore, Maryland 21218-4363
www.press.jhu.edu

Library of Congress Cataloging-in-Publication Data

Ernst, Carl H.
    Venomous reptiles of the United States, Canada, and northern Mexico / Carl H. Ernst and
Evelyn M. Ernst.
        p.   cm.
    Includes bibliographical references and index.
    ISBN-13: 978-0-8018-9875-4 (v. 1 : alk. paper)
    ISBN-10: 0-8018-9875-7 (v. 1 : alk. paper)
1. Poisonous snakes—North America.   2. Heloderma—North America.   I. Ernst, Evelyn M.
II. Title.
    QL666.O6E773 2011
    597.9′165097—dc22   2010036966

    ISBN-13: 978-0-8018-9876-1 (v. 2 : alk. paper)
    ISBN-10: 0-8018-9876-5 (v. 2 : alk. paper)

A catalog record for this book is available from the British Library.

*Special discounts are available for bulk purchases of this book. For more information, please contact
Special Sales at 410-516-6936 or specialsales@press.jhu.edu.*

The Johns Hopkins University Press uses environmentally friendly book materials, including
recycled text paper that is composed of at least 30 percent post-consumer waste, whenever
possible.

In loving memory of our parents,

George Henry Ernst and Evelyn Mae Schlotzhauer
Willis Raymond Chasteen and Maudie Belle Parker

Contents

Preface

This volume is devoted to the 21 rattlesnakes belonging to the genus *Crotalus* of the family Viperidae.

Each species has its own account, which includes a detailed description, geographic variation and confusing species, karyotype and fossil record (if known), distribution, and habitat preferences. These are followed by a detailed life history, a description of its venom delivery system, and discussions of its venom and envenomation symptoms and severity. Literature cited is through June 30, 2010.

The species accounts are followed by a glossary of scientific names, a bibliography, and an index to common and scientific names.

The first volume of the book includes chapters on venoms, envenomation and its treatment, conservation of venomous reptiles, and the family chapters and species accounts of the Helodermatidae (beaded lizards, two species), Elapidae (elapid snakes, four coralsnakes, one seasnake), and a subset of Viperidae (four American moccasins and two pygmy rattlesnakes).

Many people have contributed to publication of this book by giving advice, data, encouragement, materials, or specimens through the years: Daniel Beck, William Brown, Vincent J. Burke, Roger Conant, Terry Creque, Bela Demeter, Henry Fitch, Oscar Flores-Villela, J. Whitfield Gibbons, Howard Gloyd, Steve Gotte, Carl Kauffeld, Jeffrey Lovich, Roy McDiarmid, John Orr, Robert Reynolds, Howard Reinert, Richard Seigel, Addison Wynn, and George Zug. William Brown graciously introduced CHE to his timber rattlesnake research dens. Over the years these students of CHE helped with data, specimen collections, and field studies: Thomas Akre, Timothy Boucher, Timothy Brophy, Christopher Brown, Terry Creque, Sandra D'Alessandro, Lee French, Dale Fuller, Steve Gotte, Kerry Hansknecht, Traci Hartsell, Blair Hedges, Arndt Laemmerzahl, Jeffrey Lovich, Peter May, John McBreen, John Orr, Carol Robertson, Steven Sekscienski, James Snyder, Sheila Tuttle, John Wilder, James Wilgenbusch, Gordon Wilson, and Thomas Wilson. The following persons supplied specimens or photographs: the family of the late Roger W. Barbour, Richard Bartlett, Ted Borg, Jeff Boyd, Eric Dugan, Dale Fuller, J. Whitfield Gibbons, Steve Gotte, Paul Hampton, James Harding, Blair Hedges, Alex Henderson, Jr., Jeffrey Lovich, Robert Lovich, Barry Mansell, Peter May, Liam McGranaghan, Bradley Moon, Ali Rabatsky, and John Tashjian. Special thanks go to Richard Greene, Polly Lasker, Martha Rosen, Courtney Shaw, and David Steere, of the Natural History Library of the Smithsonian Institution, and to the library staffs at Franklin and Marshall College and Millersville University of Pennsylvania for their help in acquiring the necessary literature.

Introduction

Several aspects of the biology of our venomous reptiles need to be explained so that the reader will have a better understanding of these when they appear in the individual species accounts. The common names used for the amphibians and reptiles mentioned in the text are adapted from Crother (2000) and Liner (1994).

Karyotype: The karyotype consists of the number and physical appearance of its diploid chromosome pairs, and its terminology needs some explanation. The karyotypical morphology of a species gives evidence of its evolution and systematic position, as well as relationships with other species within its genus. Each chromosome has a small, unstaining, constricted region called a centromere, which may be located at different positions in different pairs of chromosomes. Its position is used to designate the chromosome as either metacentric, having the centromere located near the middle so the two arms are of equal or near equal length; acrocentric, having the centromere near one end of the chromosome so that the two arms are of unequal length (one long, one short); or telocentric, where the centromere is very near one end of the chromosome. Snakes possess sex chromosomes: in females the sex chromosomes are of unequal length (heteromorphic) and are termed ZW; in males the sex chromosomes are of equal or near equal length (homomorphic) and are termed ZZ. A nucleolus organizer region (NOR) is present on a pair of chromosomes.

Thermal ecology: The behavior and the physiology of our venomous reptiles are basically driven by the ambient or environmental temperature. Most are active when the temperatures of their surroundings are between 15 and 30°C. Because our reptiles are conformers, with a body temperature generally matching that of their external surroundings, they are referred to as *ectotherms*; their body temperatures vary, so they may also be termed *poikilotherms*. In the text, where possible, a discussion of the thermal ecology of each species is presented. In these discussions, the following abbreviations are used for the sake of brevity to denote various temperatures: environmental temperature (ET), air temperature (AT), surface or soil temperature (ST), body temperature (BT), cloacal temperature (CT), critical thermal maximum (CT_{max}), critical thermal minimum (CT_{min}), and incubation temperature (IT).

Movement and space relationships: Venomous reptiles, like other vertebrates, are normally active in only a relatively small portion of the available habitat. This area is termed their *home range*, or *activity range*, and all life behaviors are normally carried out within it. However, some do not have suitable overwintering habitat within their home range, and so must make a fall movement beyond the limit of their summer home range to a distant hibernaculum, and return from it the following spring. Such one-way movements may be relatively short, 100–300 m, but sometimes they are of several kilometers.

Reproduction: All reptiles have internal fertilization at the upper end of the oviduct and direct development (no metamorphic stage occurs). The North American viperid snakes covered in this volume are all *ovoviparous* species, with young that are born alive after a *gestation period* (GP). In ovoviviparous snakes, the embryos (covered with a membranous amniotic sac) develop within the female's reproductive tract, then pass out of her cloacal vent (*parturition*). No placental connection is formed between the embryo and the female, so no nourishment, other than the yolk within the original yolk sac, is provided.

In the text, when possible, the following reproductive parameters are provided for each species: size and age at maturity; gametic and hormonal cycles of both sexes; the mating season(s); courtship and mating behaviors; the season of birth (parturition) of ovoviparous species; and descriptions of eggs, hatchlings, or neonates. The *relative clutch mass* (RCM), the total clutch mass divided by the female postparturient mass (Barron 1997; Seigel and Fitch 1984), is presented where known.

Diet and feeding behavior: Viperid snakes consume other animals. With so many different species adapted to various habitats, these North American species take a wide variety of prey, and a prey list is presented for each species. They inject venomous saliva into the prey. The snakes usually release the prey after they are envenomated and later follow their odor trail; they finally swallow their prey when they are dead or incapacitated. A description of the venom delivery system is included in each species account, and notes are presented on the feeding behavior of each reptile.

Predation: Snakes, like all other vertebrates, have their own predators. Those animals that prey on rattlesnakes are listed, and their means of defense are described.

Venoms: A venom is a highly toxic poison that one animal injects into another. The venom of snakes is produced in modified salivary glands (Duvernoy's glands, or venom glands) located toward the rear of the upper jaw. These glands are ducted to either the base of a grooved tooth or into the hollow center of a special injecting fang. The chemistry of such venoms is complicated, but, overall, venoms are about 90% protein (by dry weight), and most of the proteins are enzymes. About 25 different enzymes have been isolated from snake venoms. Ten of these occur in the venom of most snakes. Proteolytic enzymes (involved in the breakdown of tissue proteins), phospholipases (either mildly or highly toxic to muscles and nerves), and hyaluronidases (dissolve intercellular materials and speed the spread of venom through the prey's tissue) are the most common types. Other enzymes are collagenases (which break down connective tissues), ribonucleases, deoxyribonucleases, nucleotides, amino acid oxidases, lactate dehydrogenases, and acidic or basic phosphatases, all of which disrupt normal cellular function. (See the chapter "Venom" in volume 1; Ernst and Zug 1996; Russell 1983; Tu 1977; and Zug and Ernst 2004 for more complete discussions of venom chemistry.) Venoms evolved first as a prey-capture device, and second as a defensive tool.

Because North American snakes of the family Viperidae possess dangerous venoms, where information is available, the toxicity and symptoms in human envenomations are presented for each species.

Populations: The population dynamics, including sizes, sex (male to female) ratio, juvenile to adult ratio, size and age classes, conservation issues, and survival status are given, as far as are known, for each species.

Abbreviations

AT	air temperature
BM	body mass
BT	body temperature
CITES	Convention on International Trade in Endangered Species
CT	cloacal temperature
CT_{max}	critical thermal maximum
CT_{min}	critical thermal minimum
DNA	deoxyribonucleic acid
DOR	dead on road (road-killed)
ESA	Endangered Species Act (United States)
ET	environmental (ambient) temperature
FL	fang length
FL_{max}	maximum recorded fang length
FMR	field metabolic rate
GP	gestation period
HD	head depth
HL	head length
HW	head width
IP	incubation period
IT	incubation temperature
IUCN	International Union for Conservation of Nature
LD_{25}	venom dosage at which 25% of the envenomated animals die
LD_{50}	venom dosage at which 50% of the envenomated animals die
mtDNA	mitochondrial DNA
MYP	million years before present
nDNA	nuclear DNA
NOR	nucleolus organizer region
Q_{10}	temperature coefficient
RCM	relative clutch/litter mass
RNA	ribonucleic acid
rRNA	ribosomal RNA

SMR	standard metabolic rate
ST	surface temperature (soil temperature)
SVL	snout-vent length (measured from tip of snout to anal vent)
TBL	total body length (measured from tip of snout to tip of tail; from tip of snout to base of rattle of rattlesnakes)
TBL_{max}	maximum recorded total body length
TL	tail length
USFWS	U.S. Fish and Wildlife Service
WT	water temperature
YBP	years before present

Venomous Reptiles
of the United States, Canada,
and Northern Mexico

Viperidae
Viperid Snakes

The vipers are venomous snakes that evolved from colubrid ancestors, probably different from those that may have given rise to the elapids. Differences between the vertebrae of colubrids and viperids are greater than those between elapids and viperids (Johnson 1956). The fossil history of snakes dates from the early Cretaceous (Barremian) more than 120 million YBP; that of true vipers (Viperinae) dates from the early Miocene about 23.8–22.8 YBP (Szyndlar and Rage 2002). The earliest fossil examples of North American Viperidae are from early Miocene (Arikareean) deposits in Nebraska; pitvipers (Crotalinae, see below) date from the early Miocene (Hemingfordian) of Delaware (Holman 1979, 1995).

The family is found in Asia, Europe, Africa, and the Americas. McDiarmid et al. (1999) proposed that it consists of about 32 genera and 223 species; however, David and Ineich (1999) recognized 33 genera and 236 species. Lawson et al. (2005), based on the study of mitochondrial and nuclear DNA, listed 34 genera. The species are assigned to 4 subfamilies: Azemiopinae, Causiinae, Crotalinae, and Viperinae. Only pitvipers of the subfamily Crotalinae (about 19 genera, 158 species; David and Ineich 1999) occur in the Americas, where, according to Campbell and Lamar (2004), 11 genera and 115 species occur. Species richness is greatest in Central America (Reed 2003a).

American pitviper species are extremely variable in both TBL and geographic range size, and range size is positively correlated with TBL. The TBL of rattlesnakes is measured from the tip of the rostrum to the base of the tail; the rattle length is excluded. Available continental area strongly influences individual species' range size, and trends in range sizes may have been structured more by historical biogeography than by macroecological biotic factors. Little support exists for Bergmann's Rule, as body size does not increase significantly with either latitude or elevation. Range area and median range latitude are positively correlated above 15°N, indicating a possible Rapoport Rule effect at high northern latitudes (Reed 2003a).

Pitvipers evolved from true vipers (Viperinae) in the Old World (Brattstrom 1964a; Darlington 1957), and these ancestral pitvipers apparently reached the Americas by crossing the Bering Land Bridge from Asia. However, the lineage of Old World pitvipers most closely related to American ones has not been determined (Castoe and Parkinson 2006). Today, 27 species in the genera, *Agkistrodon* (4), *Crotalus* (21), and *Sistrurus* (2), occur in the portion of North America covered in this book.

The families of venomous snakes are differentiated by their venom delivery systems. The Viperidae has evolved an advanced solenoglyphous dentition (fig. 1) that allows the snake to strike, envenomate, and withdraw from struggling prey to avoid injury.

The maxillae have become shortened horizontally while becoming deep vertically, and are capable of movement on the prefrontal and ectopterygoid bones, thus allowing the fangs on the maxillae to rotate until they lie against the palate when not in use

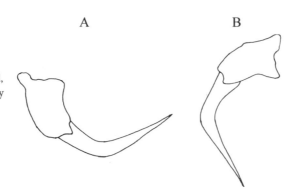

Figure 1. The solenoglyphous fang arrangement of viperid snakes; note the shortened, rotational maxillary bone: A, fang at rest against roof of mouth; B, fang rotated into striking (stabbing) position.

and the mouth is closed (Cundall 2002; Klauber 1972). During the strike the maxillae of adults rotate 70–110° (Cundall 2002), causing the fangs to move downward in a vertical plane and forward into stabbing positions. The fang itself is not capable of movement, but is rigidly fixed within an individual socket located in the maxillary bone; no other teeth are present on the bone. The combination of relatively long, curved fangs (in regard to the shorter, rigid, proteroglyphous fangs of the family Elapidae; see individual species accounts in both families for comparisons) and the ability to rotate them forward constitute the basis of a very efficient biting mechanism (Kardong and Bels 1998). The opposite fangs can be rotated separately. The ability to rotate the fangs present in vipers is almost entirely absent in all other snakes; Fairley (1929) noted that some Australian elapid death adders (*Acanthophis* sp.) also possess to a variable degree the power of elevating and rotating forward the fangs, but the mechanism of rotation differs essentially from that of the vipers. The African *Atractaspis* sp. of the moleviper family Atractaspidae are also solenoglyphous and have the ability to rotate their fangs.

To understand this process it is necessary to remember that the cranial bones of nearly all snakes are loosely joined together and allow a great amount of movement and distention that enables them to swallow thicker-bodied prey. Elastic ligaments connect the bones and are responsible for the great ability of distention.

Several other bones help to transmit the tilting motion to the maxillary bone. The quadrate bone drops posteriorly and causes the pterygoid bone to move forward. This in turn causes an anterior motion of the palatine bone which, being connected to the maxilla below the lacrimal hinge, pushes it forward.

Contraction of several muscles actually puts the mechanism into motion. The sphenopterygoid muscles are the elevator muscles of the fangs. They arise along the median ridge of the base of the skull, and, extending backward, are inserted upon the enlarged posterior terminus of the pterygoid bone. When these muscles contract, the pterygoids are pulled anteriorly. This pushes the ventral side of the maxillary bone forward, while its dorsum is held in place by the lacrimal hinge, resulting in the tip of the fang pointing downward instead of backward. The external pterygoid muscle opposes the sphenopterygoid muscle. It originates from a joint between the quadrate bone and the lower jaw, and extends anteriorly to insert on the outer surface of the maxillary bone. Contraction by it pulls the maxilla rearward, resulting in a backward and upward movement of the point of the fang. During the strike, the mouth opens nearly 180°, and the fangs point forward in a stabbing position. Vipers do not bite, but instead thrust the fangs like a sword into their prey (Kardong and Bels 1998).

Groombridge (1986) and Young and Jackson (2008) reported that in the genus *Crotalus*, the connective tissue link between the fascia of the pterygoideus and the anterior body of the M. compressor glandulae muscle is developed into a prominent tendinous band. A more or less prominent longitudinal plane of separation within the pterygoideus forms a distinct glandulae portion. Few or many lateral fibers of the pterygoideus do not pass posteriorly with the main body of the muscle, but divert laterally toward the venom gland, where they attach to the tendinous band that frequently extends to form an additional aponeurosis (a thicker and denser, deep fascia that covers, invests, and forms the terminus and attachment point of muscles) on the lateral surface of the glandulae portion. This band becomes confluent with the capsule of the venom gland and also, to a variable extent, with the transversoglandulare ligament. The arrangement of the M. pterygoideus glandulae, therefore, is a specialization that increases venom expulsion.

Normally in the pitvipers (Crotalinae), the pterygoideus glandulae muscle passes between the ectopterygoid bone and the venom gland; however, in the species of *Agkistrodon*, the connection between the venom gland and this unmodified, poorly developed muscle is either absent or consists of only a few fibers (Groombridge 1986; Young and Jackson 2008). In addition, the skull of these snakes has a single, robust medioventral basioccipital process consisting of contributions from both the basioccipital and basisphenoid bones. This process serves as a major site for insertion of the rectus capitus muscles. Ruben and Geddes (1983) thought that *Agkistrodon* fang length may be correlated with this structural arrangement, and that it facilitates prey envenomation. Once the fangs engage the prey, the cranium and upper jaw move forward and downward on the prey as the posterior end of the cranium is elevated, pushing the fangs deeper into the prey (Kardong 1975). The rectus capitus are the major muscles involved in these movements. Other myology associated with the jaw and strike apparatus is reviewed in Kardong (1990).

The viper fang is an elongated, very pointed, curved tooth (Ernst 1982a; Klauber 1939, 1972). It contains two cavities: the pulp cavity, which is located on the concave side, and the venom canal, which lies on the convex side of the tooth. The venom canal has an opening at either end. At its beginning is a short, relatively wide, anterior slit, the entrance lumen, which receives venom from the venom duct. Another narrower, more elongated, slit-like opening, the discharge orifice, through which venom leaves the fang to flow into the object bitten, is located on the anterior side of the fang near its tip. Both openings are produced from gaps in the suture formed when the embryonic tooth folded from a flat one to a narrow, tubular one. It is often possible to trace a more or less evident depressed line representing the suture between the two openings.

Microscopic inspection of a cross section of a fang reveals that the venom canal is only a deep groove that became enclosed when the walls crossed over it anteriorly. Some opisthoglyphous (fixed rear-fanged) colubrid snakes still portray this primitive (less derived) condition. The depressed line indicates where the walls met. Because the surface of the venom canal is composed of enamel, as is also the outer surface of the fang, the inner lining is actually the anterior surface and the outer layer is the original posterior surface of the embryonic flat tooth.

The functional fangs of all venomous snakes are replaced periodically (Burkett 1966; Ernst 1962, 1982a; Fitch 1960; Klauber 1939, 1972; Smith 1952; Tomes 1877). This happens whether or not the fang is damaged or worn, but it is not known if the replacement process is more rapid when the fang has been broken before a change would normally occur. Because the fangs are rather delicate, such a provision for replacement appears necessary. It also provides for longer fangs as the snake grows. Some reported average replacement rates for vipers are *Agkistrodon bilineatus*, 20.7%; *A. contortrix*, 19.7%, 20.1%; *A. piscivorus*, 19%, 33.3%; *Calloselasma rhodostoma*, 12.8%; *Daboia russelii*, 10.1%; and *Deinagkistrodon acutus*, 12.3% (Burkett 1966; Ernst 1982a, 1982b; Fitch 1960).

On each maxillary bone is a pair of fang sockets positioned side by side with the inner a little forward of the outer. These sockets are used alternately to hold the functional fang. A section through the jaw shows that the future fangs lie in two alternating rows of sockets behind the fangs, a series leading to each of the two anterior sockets. These future fangs are not fully formed, but are in various stages of development from back to front. The most mature of the 5–6 reserve fangs always lies behind the vacant socket (Ernst 1982a; Klauber 1939, 1972). The first replacement fang of *Daboia russelii* is only 0.1–0.3 mm distant from the functional fang (Ernst 1982a). When the time for a change arrives, this reserve fang moves into the vacant socket anterior to it and becomes fastened in it and attached to the venom canal. The old fang drops out, leaving a vacant socket, which will be occupied, in due course, by the next replacement fang. The active fangs, therefore, alternate between the inner and outer positional sockets.

Separating the two rows of sockets is a heavy, protecting wall of connective tissue. This wall is of greater thickness than that which separates each succeeding fang from its fellow anterior fang on the same side. It is impossible for a fang of the inner row to move into the outer row, and vice versa.

The reserve fangs can be found in all degrees of advancement. The fangs are replaced so frequently that measurements do not ordinarily indicate a larger size of the first reserve fang as compared with the functional fang that it will succeed, but this is known in some rattlesnakes of the genus *Crotalus*. Ernst (1982a) reported the first reserve fang of *Daboia russelii* was only slightly longer than that of the functional one, but never more than 0.1 mm longer. He also reported that the teeth in the replacement series in that snake range in graduated lengths from that of the most anterior reserve fang backward to a short posterior spike about 0.2 mm long.

During embryonic development, the growing fangs are developed not as a unit, but from the point upward (Ernst 1962, 1982a). The point is fully formed and hardened long before the upper part takes shape. In addition, while the morphology of the fang reveals derivation from a flat plate rolled into a tube, the actual growth does not proceed in this manner. Instead, the tubular shape is evident from the earliest period of development. Klauber (1939, 1972) described a typical series in *Crotalus atrox,* which proceeded from a functional fang through six reserve fangs; the last (distal) reserve fang was only a small conical object attached to the connective tissue wall.

Barton (1950) reported that neonate *Crotalus horridus* are fully equipped with replacement fangs, as were several newborn *Agkistrodon contortrix* examined by Ernst (1962). The fact that the reserve fang in one of Barton's rattlesnakes was already assuming the functional position suggests that the first fang-shedding must take place at a very early age.

In the United States, 99% of envenomations by snakes are caused by pitvipers. Venoms of viperid snakes are predominantly hemotoxic, but neurotoxic components are present in many species. The venom is produced in a pair of serous, tubular, glands, one located on each upper jaw below and behind the orbit and above the corner of the mouth. The gland is elongated and typically almond- or pear-shaped, with its broadest portion posterior. When filled with venom, it bends downward posteriorly around the corner of the mouth.

The venom gland in species of *Crotalus* is composed of two discrete secretory regions: a small anterior accessory gland and a large posterior main gland that are joined by a short duct. Within the main gland, the venom is produced in several, normally 4–5, lobes of secretory cells that may make up as much as 80% of the gland's total cell count. These cells are similar to those found in the Duvernoy's gland of the Colubridae. The main gland has at least four distinct cell-types: secretory cells (the dominant cell-type), mitochondria-rich cells, horizontal (secretory stem) cells, and "dark" (myoepithelial) cells. The anterior accessory gland contains six cell-types, including mucosecretory cells and several types of mitochondria-rich cells. Release of venom into the tubules draining into the lumen of the main gland is by exocytosis of granules and by release of intact membrane-bound vesicles (Mackessy 1991). At least in the South American pitviper *Bothrops jararaca,* α_1-adrenoreceptors trigger the production of venom in the secretory cells by activating phosphatidylinositol 4, 5-biphosphate hydrolysis and extracellular signal-related kinase phosphorylation (Kerchove et al. 2008).

The venom is drained from the cells through small tubules into a hollow central lumen. The lumen in turn joins the venom duct, through which the venom flows anteriorly to the base of the fang. The venom duct is completely surrounded, at least proximally, by masses of secreting cells that may act as valves to regulate the flow of venom to the fang. Venom drawn from the lumen of the venom gland is less toxic than that drawn from the fang, so the accessory gland's secretions, although nontoxic, may activate some venom components. The venom duct does not enter the fang, but instead opens adjacent to it within a sheath of connective tissue surrounding the base of the fang. The tissue sheath acts as a seal around the fang that directs the flow of venom into the fang's central venom canal and outward into the object bitten.

The muscle M. levator angulioris (or M. adductor externus superficialis; McDowell 1986) surrounds the venom gland dorsally and posteriorly, and is the only muscle involved in the discharge of venom from the gland. The dorsal portion of the muscle originates from the skull or dorsal head muscles via fasciae. The most important fibers originate in the parietal portion of the skull and insert in the posteroventral bend of the venom gland. A large ligament is also present that attaches the anterior-dorsal portion of the gland to the postorbital-parietal region of the skull. When emptying the venom gland, the muscle compresses and pulls it dorsally and medially and creates lateral and posterior pressure to force the venom through tubules into the central lumen and then anteriorly into the venom duct.

In both the Viperidae and Elapidae, fangs work in conjunction with muscles and other structures to form a complete venom delivery system that functions like a hypodermic syringe and needle (a case of human ingenuity mimicking nature). The body of the syringe represents the venom gland and venom duct to its throat; the plunger, the muscles surrounding the venom gland; and the needle, the fang.

Postfrontal, coronoid, and pelvic bones are absent, as also are teeth on the premaxillae. The prefrontal bone does not contact the nasal bone, and the ectopterygoid is elongated. A scale-like supratemporal bone suspends the quadrate. The hyoid is either Y-shaped or U-shaped, with two long superficially placed, parallel arms. All body vertebrae contain elongated hypapophyses. The left lung is absent or vestigial; a tracheal lung is usually present (Wallach 1998). The hemipenis is bilobed or double, with proximal spines and distal calyces, and a bifurcate or semicentrifugal sulcus spermaticus. Both left and right oviducts are well developed. Head scalation is similar to that of colubrids, but some, like most *Crotalus* rattlesnakes, have the dorsal surface covered with small scales instead of enlarged plates. Body scales are keeled. Pupils are vertical slits (Allen and Neill 1950b). Reproduction is either oviparous or ovoviviparous (ovoviviparous in North American species). Detailed discussions of the various characteristics of vipers are presented in Ineich et al. (2006), Marx et al. (1988), and Marx and Rabb (1972); we refer the reader to these publications.

The species in Crotalinae are called pitvipers because of the pair of small holes in their face that open on each side between the eye and nostril. The maxilla is hollowed out dorsally to accommodate the pit, and the membrane at the base of the hole is extremely sensitive to infrared radiations. It allows the snake to detect modest temperature fluctuations within its surroundings, especially those emitted from warm-blooded prey, but it is also used to direct behavioral thermoregulation. This suggests that the pits might be general purpose organs used to drive a suite of behaviors (Krochmal et al. 2004).

The pit organ is a sensitive infrared receptor that records changes from the normal background heat radiation, detects warm prey, and helps direct the snake's strike. It is usually located in the loreal scale of those crotalids having such a scale, or, in those without, in the position on the side of the face corresponding to the region of this scale. Because of this location, the cavities are sometimes referred to as loreal pits. In rattlesnakes (*Crotalus, Sistrurus*) the opening lies below a line from the nostril to the orbit and slightly closer to the former. Each crotaline pit is subdivided into an outer chamber and a smaller inner chamber by a cornified epidermal membrane about 0.025 mm thick. The principal component of the membrane is a single layer of specialized parenchyma cells with osmiophil reticular cytoplasm that lies between two layers of extremely attenuated epidermis (Bullock and Fox 1957; Lynn 1931; Noble and Schmidt 1937). The membrane is innervated by the trigeminal cranial nerve (V), particularly by its ophthalmic and maxillary branches. These pits are similar to, but not identical to, the elaborate lip pits present between the labial scales of pythons and some boas, but 5 to 10 times more sensitive. Noble and Schmidt (1937) thought that crotaline pit organs may have evolved from those on the upper jaw of boids, but this has not been proven. Instead, it is more probable the two types of snakes evolved the pit organs via convergence. Many axons enter the membrane and the innervation is rather dense. The axons lose their myelin, taper to about 1 µm, then expand into flattened palmate structures that bear many branched processes terminating freely over an average area of about 1,500 µm², overlapping only slightly with adjacent units but leaving virtually no area unsupplied (Bullock and Fox 1957). *Crotalus* have about 500 to 1,500 axon endings per mm².

Bullock and Fox (1957) reported the transmission spectrum of the fresh membrane in *Crotalus* rattlesnakes shares broad absorption peaks at 3 and 6 μm and about 50% is transmitted in other regions out to 16 μm. The visible light spectrum is at least 50% transmitted; however, much is probably lost through reflection. The strongest absorption takes place at wavelengths shorter than 490 μm. A continual transmission of impulses occurs from the pit membrane to the brain (Bullock and Cowles 1952). The rate of this continual message is independent of the snake's BT, but is dependent upon the average radiation from all objects in the receptive field. The membranes are highly sensitive to infrared environmental wavelengths of 15,000–40,000 Å, and any warm or cold object causes a temporary change in the rate of impulse transmission, the response being correlated to sudden ET changes. So the pit organ serves to recognize the presence of any object that is warmer or colder than its surroundings; during tests by Roelke and Childress (2007), three species of pitvipers distinguished between warm and cool targets, while four species of true vipers and two species of colubrid snakes did not exhibit this ability. The field is determined to include a cone extending horizontally from 10° across the midline to a point approximately at right angles to the rattlesnake's body from 45° above to 35° below the horizontal. This allows the receptive fields of the two pits to overlap in front of the snake, and together they survey a 180° field anterior to it. Sensitivity varies with the wavelength, but is generally greater to infrared emissions in the range of 2–3 μm than to shorter or longer wavelengths. The snakes seem able to detect and respond to temperature contrasts of as little as 0.001°C or less (Bakken and Krochmal 2007). Possibly the snakes can detect an ET change of about 0.003°C in 0.1 second, but this estimation may be false, as, according to Krochmal and Bakken (2005), it has been wrongly applied. Either way, the pit membrane is not as sensitive as it might be; its sensitivity is not higher than that calculated for human thermal receptors (Bullock and Fox 1957). Still, Bullock and Diecke (1956) have shown it is certainly sensitive enough to detect objects with STs differing by only 0.1°C, and there is probably no adaptive advantage to having a more sensitive receptor. Several data inputs are needed, like the signal strength (prey size and distance are important) and the quantity and quality of the background "noise" (which must be filtered out). Ideally, prey should be within 0.75 to 1.0 m, while predators should be within 0.75 to 1.5 m away (Krochmal and Bakken 2003).

By visualizing the temperature contrast images formed on the facial pit membrane using optical and heat transfer analysis (including heat transfer through the air in the pit chambers as well as by thermal infrared radiation), Bakken and Krochmal (2007) found the image on the membrane to be poorly focused and of very low temperature contrast. Heat flow through air in the pit chambers severely retards sensitivity, particularly for small snakes with small facial pit chambers. The opening of the facial pit seems to be larger than optimal for detecting small prey at 0.5 m. Angular sharpness (resolution) and image strength and contrast vary greatly with the size of the pit opening, resulting in the patterns of natural background temperatures obscuring prey and other environmental characteristics, creating false patterns. It appears important for snakes to select ambush sites with uniform, noncomplex backgrounds and strong thermal contrasts.

Interestingly, certain other membranes within the mouths of pitvipers may also be sensitive to thermal stimulation (Chiszar et al. 1986a; Dickman et al. 1987). The infrared data received via the pit organs are integrated with visual data in the optic tectum of rattlesnakes (Hartline et al. 1978; Newman and Hartline 1981). De Cock Bunning (1983) thought that, depending on the influence of ecological demands, visual or chemical cues are the main information in the behavioral phases before the strike, but in situations with little input (i.e., at night or in a rodent's burrow, etc.), hunting behavior is guided primarily by radiation of warm objects (see Ernst and Zug 1996; Ford and Burghardt 1993; Krochmal et al. 2004; Molenaar 1992, Newman and Hartline 1982; or Zug and Ernst 2004 for additional information on pit organs).

KEY TO NORTHERN AMERICAN GENERA OF VIPERIDAE

1a. No scaly rattle or button at end of tail . . . . *Agkistrodon* (vol. 1)

1b. Scaly rattle or button present at tip of tail 2

2a. Dorsal surface of head covered with nine enlarged plate-like scales . . . .
. . . .*Sistrurus* (vol. 1)

2b. Dorsal surface of head covered with small scales, or less than nine enlarged plates . . . . *Crotalus* (vol. 2)

Crotalus Linnaeus, 1758
Rattlesnakes
Cascabeles

Campbell and Lamar (2004) recognize 30 species of *Crotalus*, and a later checklist by Beaman and Hayes (2008) includes 35 species. Twenty-one species (some considered only subspecies by Beaman and Hayes and Campbell and Lamar) occur in areas of North America covered in this book.

KEY TO THE NORTHERN AMERICAN SPECIES OF *CROTALUS*

This key is based on adult snakes, and is adapted from those of Campbell and Lamar (2004), Ernst and Ernst (2003), Klauber (1972), and Wright and Wright (1957), and from characteristics listed in Gloyd (1940).

1a. Supraocular scales raised or horn-like. . . . . *cerastes*

1b. Supraocular scales not raised or horn-like 2

2a. Spinal ridge on anterior dorsum . . . . *basiliscus*

2b. No spinal ridge present 3

3a. Tail unicolored black or dark brown, or dark gray with only faint banded rings
. . . . 4

3b. Tail not unicolored black, dark brown, or dark gray; tail rings usually distinct, but may be restricted to anterior portion of tail 6

4a. Usually six or fewer scales in the internasal-prefrontal region 5

4b. Usually more than seven scales in the internasal-prefrontal region . . . . *horridus*

5a. A pair of irregular dark brown or black stripes extending backward from the parietal region onto the neck; ventrals total 184–195 . . . . *totonacus*

5b. No irregular dark stripes extending from the parietal region onto the neck; ventrals total 164–199. . . . . *molossus*

6a. Tip of rostrum sharply raised; two pale stripes extend backward from nostril and mental areas to corner of mouth . . . . *willardi*

6b. Tip of rostrum not sharply raised; no pale stripes on side of face, or two pale stripes extending diagonally backward from orbit to corner of mouth 7

7a. Upper preocular scale subdivided horizontally, vertically, or both 8

7b. Upper preocular scale not divided, or only occasionally divided 10

8a. No small scales separate the rostral scale from the prenasal scale 9

8b. A series of small scales separates the rostral scale from the prenasal scale . . . . .
. . . . *mitchellii*

9a. Supraocular scales pitted, sutured, or with the outer edges broken (ragged) . . .
. . . . *stephensi*

9b. Supraocular scales not pitted or sutured, or with the outer edges broken (ragged)
. . . . *lepidus*

10a Dorsal body pattern consists of dark cross bands, not blotches . . . . *tigris*

Geographic area of
North America
covered in this
book.

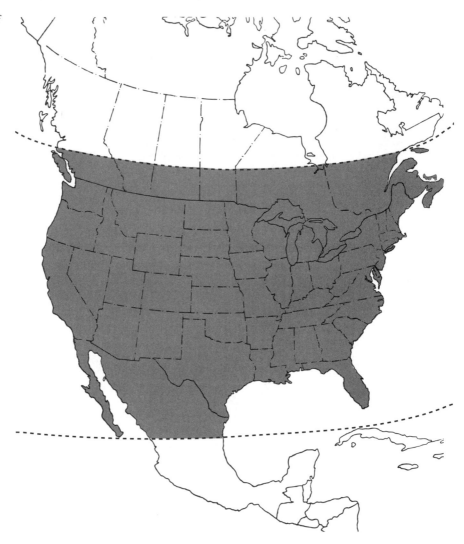

10b. Dorsal body pattern consists of dark blotches or spots, not cross bands 11

11a. Tail rattle vestigial, normally consisting of only a "button" segment; endemic on
      Isla de Santa Catalina in the Gulf of Mexico . . . . *catalinensis*

11b. Tail rattle not vestigial and consisting of several segments; not confined to Isla
      de Santa Catalina in the Gulf of Mexico. 12

12a. Dorsal body pattern consists of a series of paired, small, circular or elliptical dark
      spots. . . . . *pricei*

12b. Dorsal body pattern consists of large, square, diamond-shaped, dark blotches;
      light tail bands larger than dark ones 13

13a. Dorsal body pattern of large, square, or diamond-shaped, dark blotches; dark tail
      bands decidedly narrower than the light bands; normally only two rows of
      rather large intersupraocular scales present . . . . *scutulatus*

13b. Dark tail bands broader than light tail bands; normally more than two rows of
      intersuprascapular scales 14

14a. Tail bands well developed, not faded, usually only two internasal scales. 15

14b. Tail bands poorly developed or faded; usually more than two internasal scales (taxa of the *viridis* complex). 19

15a. Light tail bands white or cream-colored. . . . . *adamanteus*

15b. Light tail bands gray. 16

16a. Body pink, red, or reddish-brown; usually no dark speckles in dorsal body blotches; anterior infralabials usually transversely divided . . . . *ruber*

16b. Body gray, brown, or orangish; dark speckles may be present in the dorsal blotches; anterior infralabials not normally transversely divided 17

17a. Well-defined dorsal body blotches 18

17b. Dorsal body blotches somewhat faded, not well-developed; endemic on Isla Tortuga in the Gulf of California . . . . *tortugensis*

18a. Dark speckles present in dorsal blotches; usually contact between the postnasal and preocular scales; supraocular scales do not bend upward anteriorly; TBL > 95 cm; only reaches extreme northeastern Baja California Norte . . . . *atrox*

18b. Dark speckles absent from dorsal blotches; contact rarely occurs between the postnasal and preocular scales; supraocular scales bend upward anteriorly; TBL < 90 cm; absent from extreme northern Baja California Norte, main range is south on peninsular Baja California . . . . *enyo*

19a. Normally only one loreal scale present. 20

19b. Normally at least two loreal scales present. 22

20a. Dark dorsal blotches are narrow and angular or diamond-shaped; San Luis Obispo and Kern counties southward in Southern California to northern Baja California Norte and the islands of Santa Catalina and Coronado del Sur . . . 21

20b. Dark dorsal body blotches are hexagonal, oval, or almost circular; southern Canada south to San Louis Obispo and Kern counties, California . . . .
. . . . *oreganus oreganus*

21a. TBL_{max} > 70 cm; endemic to Coronado del Sur . . . . *helleri caliginis*

21b. TBL_{max} < 75 cm; Southern California, northern Baja California Norte, and Santa Catalina. . . . . *helleri helleri*

22a. TBL_{max} < 75 cm. 23

22b. TBL_{max} > 75 cm. 24

23a. Body color pink, red, or reddish-brown; dorsal body blotches irregularly oval to rectangular-shaped; northeastern and north-central Arizona and adjacent northwestern New Mexico and southwestern Colorado . . . . *viridis nuntius*

23b. Body color cream, yellowish-brown, or tan; dorsal body blotches rectangular to oval, often faint or absent and barely darker than the body color; southwestern Wyoming, eastern Utah, and western Colorado . . . . *oreganus concolor*

24a. Body color salmon to red; dorsal body blotches oval with rough edges, blotches fade with age; TBL < 105 cm; restricted to the Grand Canyon of Arizona . . . .
. . . . *oreganus abyssus*

24b. Body color not reddish; TBL > 105 cm 25

25a. Body color dark gray, olive, dark brown, or black; dorsal body blotches poorly defined; TBL to 102 cm; southern Arizona and southwestern New Mexico. . . .
. . . . *oreganus cerberus*

25b. Body color gray, yellowish-brown, or tan; dorsal body blotches dark brown or black, and oval to elliptical in shape; TBL < 110 cm; Great Basin, from southern Idaho and southwestern Oregon south to northwestern Arizona . *oreganus lutosus*

Crotalus adamanteus Palisot de Beauvois, 1799
Eastern Diamond-backed Rattlesnake

RECOGNITION: *C. adamanteus* is the largest (TBL$_{max}$, 259 cm [Ditmars 1931b]; although most adults are 100–150 cm long) and bulkiest (maximum in wild, 7+ kg; in captivity, 12 kg [Campbell and Lamar 2004]; most adults weigh 2.0–2.3 kg) of any venomous snake in the United States. The body is brown with a dorsal pattern of 24–35 (mean, 31) dark, yellow-bordered, diamond-like (rhomboid), broader than long blotches. Abnormal body coloration or patterns sometimes occur, varying from uniform dark coloration (no blotches) or abnormally shaped blotches to various degrees of albinism (Antonio and Barker 1983). The tail has 3–10 (mean, 6) brown and white bands. The appearance of the bands (width, interrupted or not interrupted, etc.) varies, and may aid in identification of individuals in the field (Moon et al. 2004a). The face is adorned with a pair of dark, cream- to yellow-bordered stripes, extending downward and backward from the eyes to the supralabials, and several vertical light stripes on the rostrum. The venter is yellow to cream with some brownish mottling. Dorsal body scales are keeled and pitted; detailed scale topography is shown in Harris (2005). Body scales normally lie in 29–30 (25–32) anterior rows, 27–29 (25–31) midbody rows, and 21 (19–23) rows near the tail. The skin is shed about four times a year (Stabler 1939). The venter has 159–187 ventrals, 20–33 subcaudals, and an undivided anal plate. Most dorsal head scales are small, but the higher than wide rostral, 2 internasals, 10–21 scales in the internasal-prefrontal region, 4 canthals (between the internasals and supraoculars), and 2 supraoculars are enlarged. Prefrontals are absent, but 7–8 (5–11) intersupraoculars are present. Laterally are 2 nasals (the prenasal usually contacts the supralabial, but the postnasal is prevented from touching the upper preocular by loreal scales), 2 (1–3) loreals, 2 preoculars, 2 (3) postoculars, several suboculars and temporals, 14–15 (12–17) supralabials, and 17–18 (15–21) infralabials.

Crotalus adamanteus
(Florida, R. D.
Bartlett)

Crotalus adamanteus
(Florida, R. D.
Bartlett)

The short, bifurcate hemipenis has a forked sulcus spermaticus, many recurved spines along the base, and spines in the crotch between the lobes. The lobes are attenuated (Campbell and Lamar 2004; Klauber 1972).

As in all North American viperids, only the enlarged, hollow fang occurs on the shortened, rotational maxilla; other dentition includes 1–3 (mean, 2.7) palatine teeth, 7–11 (mean, 8) pterygoids, and 9–10 (mean, 9.5) teeth on the dentary (Klauber 1972). Walker (2003) has described the trunk vertebrae.

Adult males have 159–176 (mean, 169) ventrals, 26–33 (mean, 30) subcaudals, and 5–10 (mean, 7) dark tail bands, and reach greater lengths (TBL$_{max}$, 259 cm) than females. Adult females (TBL$_{max}$, 178 cm) have 162–187 (mean, 176) ventrals, 20–28 (mean, 24) subcaudals, and 3–7 (mean, 5) dark tail bands, and are generally shorter than adult males.

GEOGRAPHIC VARIATION: No subspecies are currently recognized. However, individuals from the Carolinas are more contrastingly colored or patterned than those from more southern populations (Campbell and Lamar 2004). Some *C. adamateus* are darker than others, but this seems to occur in all populations. Snakes from the Florida Keys have higher ventral counts (Christman 1980). The content of small basic peptide toxins in the species' venom varies regionally, with that of *C. adamanteus* from South Carolina, Georgia, and northern Florida varying from the venom of individuals from central and southern Florida (Straight et al. 1991).

CONFUSING SPECIES: Only two other rattlesnakes occur within its range. *Crotalus horridus* lacks the diamond body pattern, banded tail (its tail is normally black), and facial stripes. *Sistrurus miliarius* has large plates instead of small scales on its head, and lacks the diamond pattern, banded tail, and facial stripes. Other species of U.S. *Crotalus* can be identified by the key presented above.

KARYOTYPE: Undescribed, but probably similar to that of *C. atrox* (see for details).

Distribution
of *Crotalus
adamanteus.*

FOSSIL RECORD: Most Pleistocene fossils of *C. adamanteus* are from Florida: Irvingtonian (Holman 1991, 1995, 2000a; Hulbert and Morgan 1989; Meylan 1982, 1995), Rancholabrean (Auffenberg 1963; Brattstrom 1953a, 1954a; Christman 1975; Gehlbach 1965; Gilmore 1938; Gut and Ray 1963; Hay 1917; Holman 1958, 1959a, 1959b, 1978, 1995, 1996, 2000a; Holman and Clausen 1984; Hulbert and Morgan 1989; Meylan 1995; Weigel 1962), but have also been found in Augusta County, Virginia, far north of its present range (Guilday 1962). Late Rancholabrean or Early Holocene remains have also been found in Florida (Auffenberg 1963; Holman 1981, 1995, 2000a). The fossil taxa *C. a. pleistofloridensis* Brattstrom, 1953b and *C. giganteus* Brattstrom, 1953b have been synonymized with *C. adamanteus* (Auffenberg 1963; Christman 1975).

DISTRIBUTION: *C. adamanteus* is limited to areas with mild winters and long growing seasons. It ranges from Albermarle Sound, North Carolina, south along the Atlantic coastal plain to the Florida Keys and west along the Gulf Coast to southeastern Mississippi and adjacent eastern Louisiana. The Louisiana population was thought possibly extinct (McCranie 1980a), but Dundee and Rossman (1989) cited recent records from Washington and Tangipahoa parishes.

HABITAT: Its presettlement habitat was probably open canopy, fire-climax pinewoods and savannas, including longleaf pine/wiregrass sandhills, clayhills, and flatwoods (Martin and Means 2000). Today, it is best associated with dry, lowland palmetto or wiregrass (*Astrida beyrichiana, A. stricta; Sabal palmetto, Serenoa repens*) flatwoods, pine (*Pinus* sp.) or pine-turkey oak (*Quercus laevis*) woodlands with gallberry (*Ilex glabra*) and wax myrtle (*Myrica cerifolia*), but it also uses grass-sedge bogs, heaths, temperate hardwood forests (*Acer* sp.; *Asimina* sp.; *Cornus florida; Ilex opaca; Fagus grandifolia; Juglans* sp.; *Liquidamber tulipifera; Magnolia* sp.; *Quercus* sp.; *Smilax* sp.; *Tilia americana, T. floridana; Ulmus* sp.; *Vitis* sp.), tropical hardwood hammocks (*Brassia* sp., *Bursaria simaruba, Coccoloba diversifola*, ferns, *Ficus aurea, Metopium toxiferum, Swietenia mahogani*), pine or scrub oak habitats, and abandoned fields, especially if these are adjacent to pine-dominated habitats. Marshes and swamps are normally avoided, but occasionally it lives along their borders. The snake has been found in mangrove thickets (*Avicennia nitida*, *Rhizophora mangle*), and will swim across narrow stretches of marine waters to reach offshore islands. Animal burrows, such as those of the armadillo (*Dasypus novemcinctus*), pocket gopher (*Geomys pinetis*), or gopher tortoise (*Gopherus polyphemus*) (C. Ernst, personal observation; Funderburg and Lee 1968), hollow logs, old stumps, or windfall mounds from uprooted trees must be present for the snake to use in winter or to avoid surface fires. Habitat specificity to the shrinking longleaf pine (*Pinus palustris*) ecosystem may be a significant contribution to the decline of the species (Waldron et al. 2006a).

Steen et al. (2007) in developing a statistical model to quantify habitat differences between *C. adamanteus* and sympatric *C. horridus* in southern Georgia found that the former species was not associated with any specific habitat type they examined (agricultural land, including food plots; hardwood, natural longleaf pine; pine, including pine plantations and mixed pine-hardwood forest; and wetland). Because riparian zones had previously been reported to be used by both species (Ernst and Ernst 2003), the distance to the nearest riverine system was also determined. Quantification of edge, total length (m) of boundary between habitat types, and road density, as the total length (m/km^2) of roads within each buffer indicated that *C. adamanteus* was associated with roads (also observed by us in Florida).

BEHAVIOR AND ECOLOGY: Southern populations of *C. adamanteus* are active longer than those from the north. It is probably active every month in Florida south of the 18°C isotherm; north of this line short periods of inactivity occur to avoid cold during the 2–4 months of winter (Clamp 1990). Timmerman (1995) reported the snake is surface-active 55% of the time during the winter in Putnam County. In the Florida Panhandle, snakes north of the 12°C January isotherm hibernate up to four months (Martin and Means 2000); the species usually hibernates from November

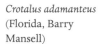

Crotalus adamanteus (Florida, Barry Mansell)

through March at Tall Timbers Research Station (Means *in* Timmerman and Martin 2003). It must retreat underground and hibernate during the winter in the colder northern portions of its range. In North Carolina, *C. adamanteus* has been found in every month except December and January; 68% of observations are from June and August–October, and the latest record of activity is 13 November (Palmer and Braswell 1995).

The same retreats used during the summer to avoid excessive heat are used for overwintering—mammal or gopher tortoise burrows, hollow logs or stumps, or among the roots of wind-felled trees or palmettos (Means 1985, 1986 *in* Timmerman and Martin 2003; Timmerman 1995). It does not live in uplands so does not usually use talus slopes or crevices for hibernacula as do *C. horridus* or *Agkistrodon contortrix*. Florida snakes studied by Timmerman (1995) always used more than 1 retreat during the winter; and 6 of 9 snakes studied by Kain (*in* Timmerman and Martin 2003) used multiple burrows for hibernation, a mean of 3.1 burrows was used during the winter, and one individual used 6 different burrows.

C. adamanteus is mostly diurnal; 60% percent (n = 449) of the 743 observations made by Timmerman (1995) were between 0600 and 1200 hours, confirming our own observations, and 32% more (n = 240) occurred between 1201 and 1800 hours. Only 7% (n = 54) of his observations occurred between 1801 and 0559 hours. Activities recorded for the 689 observations between 0600 and 1800 hours included tightly coiled (49.7%, n = 346); undercover, not visible (33.4%, n = 230); loosely coiled, function unknown, but possibly thermoregulatory (10.7%, n = 74); and crawling (5.7%, n = 39). During 1801–0559 hours, the observations included tightly coiled (50.0%, n = 27); under cover, not visible (42.6%, n = 23); loosely coiled (5.5%, n = 3); and crawling (1.8%, n = 1). The snakes spent from one day to a week in the same tight coil. Means (*in* Timmerman and Martin 2003) thought possibly that some mating and feeding activity may occur at night. Except in winter, when they entered subterranean retreats, Timmerman's (1995) rattlesnakes spent little to no time thermoregulating.

Crotalus adamanteus in the Everglades National Park, Florida, had BTs of 23.5–36.0 (mean, 28.0) °C; many of the BTs were below the ET, indicating attempted thermoregulation (but see below), and none was seen basking (Dalrymple *in* Timmerman and Martin 2003). We have observed them in Collier County, Florida, actively foraging or traveling at ATs of 27–35°C; one recorded by Timmerman (1995) had a BT of 16°C at an AT of 19°C.

Clark and Antonio (2008) reported that spring postcopulatory female basking begins at approximately 0900 hours on sunny days with ATs of 18–20°C and lasts for about 2 hours. While basking, the female presents the back half of her body to the sun. She rarely moves out of the shade at ATs above 32°C. During basking, she raised her BT 3–5°C above the AT; the maximum skin temperature achieved was 36.7°C. During the summer, the male observed began basking about an hour after the female, and only stayed exposed for about 60 minutes. Sometimes he remained in a shaded area 2–3 straight days.

Dorcas et al. (2004) studied the effects of BM and BT on the metabolic rate of *C. adamanteus*. They measured the O_2 consumption of 5 snakes (BM, 0.8–4.98 kg) at 5°C increments from 5 to 35°C. Multiple regressions indicated that the SMR increased with BM and BT. The Q_{10} values, 1.82–4.20, were somewhat high compared to those of other squamates, but similar to values reported for other large rattlesnakes (Beaupre and Duvall 1998a; Beaupre and Zaidan 2001). They predicted that as BT increases, so must prey consumption to meet the energy demands of the SMR; so BT variation probably affects feeding behavior and energy use, and so influences both growth and reproduction. In another study using four *C. adamanteus* under the same ET regimens, Rice et al. (2006) found no difference between heating and cooling rates in the snakes. The rates of BT change also matched those of a biophysical model, further suggesting a lack of thermoregulation.

Water is drunk from that pooled in body coils (Clark and Antonio 2008).

C. adamanteus establishes an initial home range and continues to use it for years. Male and female home ranges usually overlap. Daily underground retreats and hibernacula are the same, and are included within the home range, so no seasonal migration occurs. Also, the snake has favorite places to which it frequently returns and several underground retreats may be used over a period of time. Kain (*in* Timmerman and Martin 2003) reported that of 48 burrows used in southern Mississippi, 37 (77.1%) were mammal burrows, 7 (14.6%) were gopher tortoise burrows, and 4 (8.3%) were unidentifiable.

Males have larger home ranges than females. Several home range estimates have been made at different sites within the snakes' distribution: in northwestern Florida, mean home ranges of males and females were, respectively, 200 ha and 80 ha (Means 1985); in the Florida Everglades, annual convex polygon home ranges of 50 *C. adamanteus* were 120–260 ha (Dalrymple *in* Timmerman and Martin 2003); elsewhere in Florida, mean minimum complex polygon home ranges of males (4) was 84.3 ha and for females (2) 46.5 ha (Timmerman 1995); in southern Mississippi, the mean home range size was 48.1 ha, 59.5 ha for males and 8.2 ha for females (Kain *in* Timmerman and Martin 2003); and in South Carolina, the mean minimum convex polygon home range size was 141.9 ha for 11 males and 87.4 ha for 12 females (Bennett et al. *in* Timmerman and Martin 2003). Waldron et al. (2006a) found no differences in 95% kernel estimates of the home range size of South Carolina *C. adamanteus*: males, 84.82 (range, 16.91–310.48) ha; nonpregnant females, 28.63 (range, 5.42–61.60) ha; and pregnant females, 18.07 (range, 14.92–21.22) ha.

Hoss et al. (2010) found first-year home ranges of 4 males and 6 females in southwestern Georgia to be 19.34–59.94 ha and 7.25–33.81 ha, and composite 2-year minimum convex polygon home ranges of 19.34–76.79 ha and 7.25–68.10 ha, respectively, for the 2 sexes. Although no significant habitat associations were detected, there were trends for a positive association with pine habitat at the landscape scale and a negative association with agriculture within the home range. Individuals in heterogeneous landscapes had small home ranges. They suggest that management programs emphasize preservation of upland pine woods, while maintaining a mosaic of other habitat types and limiting the conversion of forest to agriculture.

The longest movements are usually made in August and September during the mating and birthing season, and males move considerably more often and for longer distances during this period while searching for mates. Means (1985) reported that breeding females make short directed movements to lay down scent trails during this period; several sterol lipids have been recovered from the skin of *C. adamanteus* that possibly serve as sex attractants (Roberts and Lillywhite 1980). At other times, movements are comparable between the sexes. Timmerman's (1995) Florida snakes made movements of 20 (range, 2.9–45.5) m per day in September–November, and their

Crotalus adamanteus (Okeetee, South Carolina; Gladys Porter Zoo, John H. Tashjian)

shortest movements, 10 (range, 0–19.5) m per day, during December–February when they were mostly underground. The mean annual distance moved by Mississippi males was 8,211.0 m versus 3,139.3 m by females (Kain *in* Timmerman and Martin 2003). The snake seems somewhat reluctant to move its tail while crawling, probably so as not to break off its rattle-links (Chiszar et al. 1992).

This snake is a good swimmer that regularly swims across streams and narrow bodies of freshwater, and Clench (1925) reported a case of rafting on a floating water hyacinth mat across saltwater to Sanibel Island, Florida.

This heavy-bodied, cumbersome snake only occasionally climbs into bushes or trees while pursuing prey (Klauber 1972). In tests of subcutaneous interstitial fluid pressure performed by Lillywhite (1993), *C. adamanteus* achieved a fluid pressure of 2.3 μL / mm Hg, much higher than those (range, 0.31–0.51 μL / mm Hg) recorded for arboreal species; presumably, this measurement reflects structural differences related to the requirements of a ground-dweller for counteracting gravitational stresses different from those encountered by climbing snakes.

A male social dominance system, determined during combat dances, is present. This is most evident during the late summer breeding season, and probably involves male territorial possession of females. When two males meet, the identity odor of each is determined by tongue-flicking. The males then raise the anterior 30–40% of their bodies and face each other, as each tries to outstare the other. Bending their heads at a sharp angle to the ground, they entwine their anterior bodies and each tries to push the other off balance and to the ground. Such bouts may last for only a few minutes to almost two hours, and finally end when one male (usually the smaller) is pinned to the ground, unwraps itself from the victor, and crawls away (C. Ernst, personal observation). Wagner (1962) observed a male become involved in a second combat bout almost immediately after winning his first bout. A second male approached him from the rear. The first snake started to crawl away, but the challenger caught up and crawled over him, their bodies being parallel. The anterior parts of their bodies were then raised off the ground, intertwining as they arose. When the two heads were about 30 cm from the ground they started leaning backward. Soon one or both lost balance and fell to the ground. The bout lasted nearly an hour, but at the end the original male was still champion.

REPRODUCTION: Berish (1998) reported the SVL of her smallest mature female to be 109.3 cm; Timmerman and Martin (2003) reported mature females 91–178 cm long, but noted that reproduction usually occurs when a female is about 132 cm long; and Martin (*in* Timmerman and Martin 2003) reported that first-time reproductive females average 132.5 (range, 115.0–142.5) cm. Murphy and Shadduck (1978) described a successful captive mating between a 110 cm female and a 135 cm male. So, the minimum mature TBL of females may be considerably longer than the 76–84 cm range reported by Conant and Collins (1998) and Wright and Wright (1957). Males achieve sexual maturity at about 124.5 cm TBL (Martin *in* Timmerman and Martin 2003).

The sexual cycles of *C. adamanteus* have not been described, but are probably similar to those of other North American pitvipers described by Aldridge and Duvall (2002). It is not known if females reproduce annually or biennially, but Timmerman and Martin (2003) reported that an adult female did not reproduce during an entire year of radio-tracking, and Kain (1995) thought Mississippi females were on a biennial cycle. However, if sufficient prey resources are available, females are capable of producing litters annually (Clark and Antonio 2008). Reproductive frequency is dependent on food availability, annual fat cycles, and the energy allocation dynamics of the particular female involved. It is suspected that females can store viable sperm for relatively long periods after a successful mating (Klauber 1972; Schuett 1992).

Mating occurs in August through September in the northern part of the range (Kain 1995; Means 1985), but in October to December in south Florida (Dalrymple *in* Timmerman and Martin 2003). Captives kept in an outdoor enclosure in Florida were observed mating on 16 and 28 February and 5 March, never in late summer or the fall (Clark and Antonio 2008). Mating activity usually lasts 6–8 weeks at any given locality (Timmerman and Martin 2003). Kauffeld (1939) thought breeding begins in March in South Carolina, and Ashton and Ashton (1981) believed that the snake mates in both fall and spring, but this has not been substantiated. A 30 January copulation has occurred in captivity (Murphy and Shadduck 1978). During the captive mating, the snakes remained joined for about nine hours. Occasional pulsations near the cloaca of the male were seen, but the female remained passive. The 16 February copulation reported by Clark and Antonio (2008) occurred on a 27°C day following a night when the AT dropped to 11°C.

Females seek sheltered places, and usually give birth from mid-July to early October (Klauber 1972), after a GP of possibly 180 to more than 200 days (213 days in captivity; Murphy and Shadduck 1978). Such a long GP seems to indicate a mating period in late winter or early spring; but if females store sperm from fall matings, both the long gestation period and parturition dates are plausible. Most young are born in retreats such as gopher tortoise burrows or hollow logs, and the female may remain with her offspring for a short period until they shed and depart (Butler et al. 1995; Van Hyning 1931; Wright and Wright 1957).

The number of young per litter ranges from 4 to 32 (Timmerman and Martin

Crotalus adamanteus × *C. horridus,* hybrid (R. D. Bartlett)

2003); averages of 12.7 (n = 27; Ernst 1992), 12.8 (n = 4; Clark and Antonio 2008), 13.8 (n = 27; Berish 1992), and 14.8 (n = 19; Klauber 1972) young per litter have been reported. According to Berish (1992), a Florida female could produce 52 offspring by her 10th birthday, but Martin (*in* Timmerman and Martin 2003) thought that females only produce 1.5–2.0 litters during their lifetime. At the time of emergence from the fetal membranes the neonates have TBLs of 30.0–42.4 (mean, 36.8; n = 17) cm and BMs of 32.0–48.5 (mean, 39.5; n = 9) g.

An apparent wild hybridization between *C. adamanteus* and *C. horridus* has been reported (Klauber 1972; see also photograph).

GROWTH AND LONGEVITY: In the wild, growth is dependent on cyclic small mammal populations, optimal habitats, and healthy ecosystems (Clark and Antonio 2008). Martin (*in* Timmerman and Martin 2003) reported the following average TBLs (TLs) by age class for *C. adamanteus*: neonates, 41.4 (3.0) cm; 0.5 year, 57.5 (7.6) cm; 1.5 years, 101.1 (7.9) cm; 2.5 years, 118.6 (10.4) cm; 3 years female, 138.7 (10.4) cm; and 3 years male, 154.7 (15.5) cm. Clark and Antonio (2008) reported the mean neonate TBL and weight to be 41.9 cm and 54.4 g, respectively; one weighed 68.8 g. Ernst and Ernst (2003) reported the neonatal TBL to be 30.0–38.8 cm and the weight to be 35.0–48.5 g. A recaptured South Carolina *C. adamanteus* grew from 60 cm to 109 cm in TBL in about 26 months (Smith 1992). With proper resources growth may be rapid—a juvenile raised by Strimple (1992b) grew from 44.5 cm (64.5 g) to 69.9 cm (284.5 g) in one year. The length of a midbody vertebra is directly proportional to the TBL of *C. adamanteus* ($y = 6.745x - 0.674$) and log BM by log TBL can be calculated with the equation $y = 3.108x + 2.766$ (Christman 1975; Prange and Christman 1976).

During its lifetime a *C. adamanteus* sheds its skin several times, depending on the extent of its feeding, quality of its diet, and the amount it grows betweens ecdysis events. Martin (*in* Timmerman and Martin 2003) listed the following sheds per age: 0.5 year, 1–4; 1.5 years, 5–7; 2.5 years, 8–10; and 3 years, > 11. Stabler (1939) reported that a captive shed four times in one year at an average of three months between sheds.

One *C. adamanteus*, wild-caught when juvenile, survived 22 years, 9 months, and 3 days in captivity (Snider and Bowler 1992). Few wild individuals probably live past 10 years.

DIET AND FEEDING BEHAVIOR: *C. adamanteus* feeds chiefly on small mammals: rabbits (*Sylvilagus floridanus, S. palustris*), squirrels (*Glaucomys volans; Sciurus carolinensis, S. niger*), cotton rats (*Sigmodon hispidus*), woodrats (*Neotoma floridana*), black rats (*Rattus rattus*), brown rats (*Rattus norvegicus*), pine voles (*Microtus pinetorum*), various mice (*Mus musculus; Peromyscus gossypinus, P. maniculatus; Podomys floridanus; Reithrodontomys humulis*), and domestic cats (*Felis catus*). However, birds are also consumed: king rail (*Rallus elegans*), chickens (*Gallus gallus*), young turkey (*Meleagris gallopavo*), common bobwhite (*Colinus virginianus*), brown thrasher (*Toxostoma rufum*), Carolina wren (*Thryothorus ludovicianus*, in captivity) cardinal (*Cardinalis cardinalis*), eastern meadowlark adults and eggs (*Sturnella magna*), and rufous-sided towhee (*Pipilo erythrophthalmus*) (Allen and Neill 1950a; Allen and Slatten 1945; Barbour 1920; Carr 1940; Dalrymple *in* Timmerman and Martin 2003; C. Ernst, personal observation; Funderburg 1968; Howard and Kopf 2003; Klauber 1972; Martin and Means 2000; Mitchell and Ruckdeschel 2008; Stevenson 2003; Timmerman 1995; Timmerman and Martin 2003). The report of one raiding a pileated woodpecker (*Dryocopus pileatus*) nest by Rutledge (1936) is probably erroneous because this snake seldom climbs into trees. Mice and rats are the primary prey of young *C. adamanteus*, while adults seem to prefer rabbits, cotton rats, mice, squirrels, and birds. Dalrymple (*in* Timmerman and Martin 2003) reported that 33% of the adult stomachs he examined contained cotton rat remains, and 19% contained marsh rabbits (*S. palustris*). Carrion may be ingested (C. Ernst, personal observation; Funderburg 1968).

This rattlesnake may actively seek out prey by following scent trails (Chiszar et al. 1986c, 1991b), but it may capture prey from ambush, often lying in wait beside logs or among the roots of wind-felled trees (C. Ernst, personal observation). In addition to their odor, endothermic prey emit infrared heat waves that are detected by the pit on the rattlesnake's face. Abramson and Place (2008) have published an extensive review of olfactory studies involving species of *Crotalus*, and we refer the reader to their paper for details.

Usually only one envenomating bite is delivered, and the wounded animal is released immediately after the strike and allowed to crawl off to die. Biting of the prey sets off a chemosensory search image for the tasted animal, and the snake slowly pursues the struck prey with its tongue frequently flicking until its victim is found. It then examines the prey with tongue-flicks to determine if it is dead, and then swallows it quickly, usually from the head end. Chemical searching may last for 2–62 minutes following a strike, and the snake can distinguish its envenomated prey from that of other individuals (Brock 1981; Chiszar et al. 1991d).

Fecal materials are stored in both the posterior large intestine and cloaca, which are thin-walled and much wider in diameter than is the small intestine (Lillywhite et al. 2002).

VENOM DELIVERY SYSTEM: *C. adamanteus* has a well-developed apparatus for delivering its venom, with fangs as long as 27 mm (Telford 1952); its fangs are the longest of any species in the genus. Five individuals with TBLs of 100–140 cm had FLs of 13.4–15.8 (mean, 14.4) mm, mean TBL/FL of 83%, mean HL/FL of 4.13%, and a mean angle of fang curvature 72° (Klauber 1939). We refer the reader to A. H. Savitsky's (1992) detailed discussion of the embryology of the maxillary and prefrontal bones in relation to the venom apparatus.

VENOM AND BITES: The solid portion of the venom is composed of about 70% protein (McCue 2005). The venom is quite hemolytic, causing strong hemorrhagic, procoagulant, and anticoagulant reactions, but protease activity is low (Tan and Ponnudurai 1991). A neurolytic myotoxin similar to the Mojave toxin β found in *C. durissus* and some *C. viridis* may also be present (Powell et al. 2008; Samejima et al. 1991), although reported not to be by Bober et al. (1988) and Li and Ownby (1994). Straight et al. (1991) reported that the venoms of *C. adamanteus* from different geographical regions (central and northern Florida, southern Georgia, South Carolina) differ in protein composition. The amount of enzyme activity and other aspects of the venom chemistry are discussed by Aird (2005), Al-Joufi and Bailey (1994), Bonilla and Horner (1969), Brown (1973), Gomis-Rüth et al. (1994), Grams et al. (1993), Khole (1991), Kurecki and Kress (1985a, 1985b), Mackessy (1998), Markland and Damus (1971), Markland and Pirkle (1977), Mashiko and Takahashi (1998), McCue (2005), Mebs and Kornalik (1984), Meier and Stocker (1995), Oshima et al. (1969), Pandya and Budzynski (1984), Pérez and Sánchez (1999), Powell et al. (2008), Samejima et al. (1991), Soto et al. (1989), Tsai et al. (2001), Van Mierop (1976a), and Wermelinger et al. (2005). The near ultraviolet absorption spectrum of the venom is illustrated in Horton (1951).

The snake is capable of controlling the amount of venom ejected in a specific bite, and sometimes "dry" bites containing no venom are delivered (lucky is the person bitten in this way). Total dry venom yield may be as high as 848 mg or as low as 20 mg, but averages from about 410 to 700 mg per adult (Brown 1973; Klauber 1972; Russell and Puffer 1970); adult liquid yields may be as high as 1,000 mL. Neonates have an average dry venom yield of 13.5 mg (Glenn and Straight 1982).

The mean intraperitoneal and intravenous LD_{50} values for 16–18 g mice are, respectively, 2.05 (range, 1.80–2.31) mg and 2.26 (range, 1.99–2.57) mg (Arce et al. 2003). The minimum lethal venom dose for a 20 g mouse is 0.04 mg; LD_{50} values in mg/kg BM (19/20 confidence limits) are 1.69 (range, 1.43–1.98) intravenous and 1.89 (range, 1.63–2.19) intraperitoneal; the subcutaneous LD_{50} is about 14.6 µg/g. The minimum lethal dose for a rabbit is 0.25 mg/kg BM. A 350 g pigeon is killed by 0.2–0.3 mg, and the LD_{50} values in mg/kg BM (19/20 confidence limits) for a chick are 1.18 (range, 0.89–1.58) intravenous and 2.38 (range, 2.12–2.67) intraperitoneal (Boquet 1948; Githens and George 1931; Githens and Wolff 1939; Glenn and Straight 1982; Khole 1991; Russell and Emery 1959; Russell and Puffer 1970).

The human lethal dose has been estimated as 100 mg (Dowling 1975). Human deaths from severe untreated bites can occur in 6–30 hours, and the mortality rate may be 40% (Neill 1957). Venom toxicity varies regionally and between individual snakes, even from the same litter.

Symptoms of bites in humans include swelling; pain; weakness; giddiness; respiratory difficulty; hemorrhage; weak pulse or heart failure (or in some cases an increased pulse rate); a drastic lowering of arterial blood pressure and flow; a rise in venous blood pressure, pulmonary artery pressure, and cisternal pressure; changes in both electrocardiogram and electroencephalogram; enlarged glands; soreness; diarrhea (often bloody); convulsion; fainting; shock; and toxemia (Hutchison 1929; Kitchens and Van Mierop 1983; Norris and Bush 2007; Russell and Michaelis 1960; Watt 1985). Necrosis at the site of the bite and occasionally involving much of the injured limb is common, and sensory disturbances may occur, such as a sensation of yellow vision (Minton 1974). Bitten limbs often are at least partially crippled. Case histories of human envenomation by *C. adamanteus* are in Buntain (1983), Dart et al. (1992), Klauber (1972), Norris and Bush (2007), and Parrish and Thompson (1958).

Livestock also suffer from bites by *C. adamanteus*. Campbell and Lamar (2004) reported that a Georgia family lost 10–15 pigs in 1910 to bites by *C. adamanteus*. McCall (*in* Sutcliffe 1952) noted that several bitten mules recovered. Dogs seldom live after a bite.

PREDATORS AND DEFENSE: Recorded predators of the various life stages of *C. adamanteus* include American bullfrogs (*Rana catesbeiana*), river frogs (*Rana heckscheri*), indigo snakes (*Drymarchon corais*), racers (*Coluber constrictor*), common kingsnakes (*Lampropeltis getula*), coachwhips (*Masticophis flagellum*), harlequin coralsnakes (*Micrurus fulvius*), great horned owls (*Bubo virginianus*), red-tailed hawks (*Buteo jamaicensis*), crested caracaras (*Caracara cheriway*), wood storks (*Mycteria americana*), hogs (*Sus scrofa*), raccoons (*Procyon lotor*), black bears (*Ursus americanus*), skunks (*Mephitis mephitis*), northern river otters (*Lontra canadensis*), domestic dogs (*Canis familiaris*), coyotes (*Canis latrans*), domestic cats (*Felis catus*), and bobcats (*Lynx rufus*) (C. Ernst, personal observation; Hinderliter and Lee 2006; Klauber 1972; Neill 1961; Ross 1989; Timmerman 1995; Timmerman and Martin 2003). Adults have little to fear except humans. White-tailed deer (*Odocoileus virginianus*) occasionally stomp *C. adamanteus* to death (Timmerman 1995).

Most will lie quietly coiled when first discovered, displaying a mild temperament for a rattlesnake; it is so calm that Snellings (1986) referred to *C. adamanteus* as "the gentleman of snakes." However, if touched or otherwise provoked, it coils, sometimes hisses while inflating the body (with never more than a 1.5 second timing difference; the hiss is a simple sound with a mean minimum frequency of 0.4 kHz, a mean dominant frequency of 1.2 kHz, and a mean maximum frequency of 4.7 kHz; Kinney et al. 1998; Klauber 1972; Mattison 1996), shakes its tail rattle, and raises the head and neck into a striking position; if further disturbed, it strikes. When handled, it thrashes about and tries to bite. Overall, the snake is extremely dangerous and should be left alone.

It will first try to flee, and then bridge, flip, or inflate its body when encountering snake-eating serpents such as *Drymarchon corais* or species of *Lampropeltis* (Marchisin 1980; Weldon and Burghardt 1979; Weldon et al. 1992).

As far as humans are concerned, the tail rattle is certainly an effective warning device. In addition to the warning role, it is a highly diversionary defense mechanism, calling attention to the tail instead of the head. The SVL-adjusted loudness may be more than 70 dB, frequency of rattling 1.5–18.6 kHz, dominant frequency 6.5–7.1 kHz, and the mean sound band width 12.7 kHz (Cook et al. 1994; Fenton and Licht 1990; Kinney et al. 1998; Young and Brown 1993). Rome et al. (1996) reported that the tail-shaker muscles of *C. atrox* have evolved to permit rattling at about 90 Hz (see their paper for detailed data on the morphology required and the physiological processes involved in tail-rattling).

PARASITES AND PATHOGENS: In addition to predators, *C. adamanteus* is parasitized by the nematodes *Acaris nuda* and *Hexametra boddaërtii* (Ernst and Ernst 2006) and the linguatulid tongue worms (Pentastomida) *Kiricephalus coarctatus* and *Porocephalus crotali* (Forrester et al. 1970; Hill 1935). In addition, the tick *Ambylomma dissimile* occasionally attaches to it (Bishopp and Trembley 1945).

The snake has also suffered from paramyxovirus infections (Jacobson and Gaskin 1992), and coccidiosis caused by the protozoan *Isospora dirumpens* (Bovee 1962). Serotypes of several potentially dangerous *Salmonella* bacteria, including *S. arizonae*, have been isolated from zoo captives (Grupka et al. 2006; Lamberski et al. 2002), and the species has also tested positive for the bacterium *Leptospira ballum* (White 1963). Parrish et al. (1956) found several additional bacteria in its mouth and venom glands.

POPULATIONS: *C. adamanteus* was once common in suitable habitats throughout its range. Large populations formerly occurred in portions of Collier County, Florida (C. Ernst, personal observation), and at Okeetee, South Carolina. A series of 280 snakes caught in 74 days of actual hunting over a period of 6 years at Okeetee included 60 (21%) members of this species (Kauffeld 1957). Timmerman (1995) reported a crude density estimate of one per 5 ha at the Ordway Preserve, Putnam County, Florida. The snake is still considered abundant in parts of Mississippi (Lee 2009). The sex ratio of *C. adamanteus* harvested in Florida is skewed toward males (Berish 1998).

About 1,000 *C. adamanteus* were killed each year on 7 hunting preserves in the Thomasville-Tallahassee area of Florida, where a dollar bounty was offered for each snake (Stoddard 1942). The late Ross Allen reported that over a period of 28 years he had received 1,000–5,000 per year at his Florida snake exhibit, with a grand total of 50,000 (Klauber 1972)! Size records of several thousand *C. adamanteus* purchased by Ross Allen's Reptile Institute between the 1930s and 1960s indicate that the average length decreased by about 30 cm during this period. At a site in the Everglades National Park, Florida, only 1.7% of the 1,782 total snakes seen or collected during 1984–1986 were *C. adamanteus*, indicating a possible decline (Dalrymple et al. 1991). During a 2.8-year road survey of snakes in xeric upland habitats of Hernando County, Florida, Enge and Wood (2002) reported that *C. adamanteus* only comprised 1 (0.004%) of the 228 recorded snakes observed on roads, and only 1 (0.005%) of 213 DOR snakes. Martin (*in* Timmerman and Martin 2003) reported that, during 34 field days in Florida during 1967–1971, 77 individuals were observed at a rate of 2.26 snakes per day; but that in 61 field days during 1992–2001, only 27 individuals were seen at a rate of 0.46 per day.

Today, unfortunately, populations of *C. adamanteus* are declining almost everywhere across its range. Humans are the problem. They often kill the snake on sight, run over them with their motor vehicles, destroy their habitats, and collect them for the leather and meat industries (Berish 1998). Of 51 eastern diamondbacks captured at Camp Shelby Joint Forces Training Camp in Mississippi, 19 were found on roads and 25% of these were DOR (Lee 2009). Rattlesnake roundups in Alabama and Georgia also pose a major hazard for the species (Means 2009).

Collection pressure for the hide, meat, and venom industries and the pet trade is very high. Enge (2005) reported that from 1 July 1990 through 30 June 1994, 42,788 eastern diamond-backed rattlesnakes were reported purchased by hide dealers and taxidermists. Most of the snakes were probably captured during the breeding season when the species is most surface-active. See Fitzgerald and Painter (2000) and Speake and Mount (1973) for discussions of the adverse effects of commercially sponsored rattlesnake roundups on *C. adamanteus*. It can only be hoped that newly implemented reporting requirements in Florida may help to slow this slaughter. Gassing of gopher tortoise burrows to drive out the rattlesnake is now prohibited in most states where the snake occurs.

The species' original range has been shortened and fragmented by agriculture, forestry practices, urbanization, and unfavorable plant succession after fire suppression (Martin and Means 2000). Some caught above ground are killed during prairie fires (C. Ernst, personal observation; Landers 1987). In 1992, when Hurricane Andrew hit extreme southern Florida very hard, it probably did major damage to the snake's Everglade population.

C. adamanteus is now afforded some protection by being listed as endangered in North Carolina (Palmer and Braswell 1995) and as a species of special concern in South Carolina; but in its most populous states, including Alabama, Florida, and Georgia, population decline continues unchecked. It is only common at some sites in Mississippi, and all but extinct in Louisiana (LaClaire and Vandeventer *in* Timmerman and Martin 2003).

REMARKS: A cladistic tree generated from venom Lys-49 phospholipases A_2 comparisons by Tsai et al. (2001) showed the genus *Crotalus* to be closer to North American *Agkistrodon* and Middle or Central American *Bothriechis* and *Porthidium* than to South American *Bothrops*. Minton (1992) also reported *Crotalus* closest to *Agkistrodon* during a study of plasma albumins and immunodiffusion.

The ancestral area of the rattlesnake genera *Crotalus* and *Sistrurus* was probably the Sierra Madre Occidental of Mexico; the most probable vegetation of the ancestral area was pine-oak forest (Place and Abramson 2004). The Miocene mammal fauna of the Mexican region and habitat was rich in actively foraging mustelids and procyonids, prompting Place and Anderson to propose that the tail rattle of these snakes evolved to warn such predators. Gloyd (1940) thought that *C. adamanteus* was a "climax form" derived from *C. atrox*, which he believed most close to the ancestral type for the *atrox*-group of rattlesnakes. Studies by Meylan (1982) support this theory, but Christman (1980) noted there is no reason to believe it arose on the Mexican Plateau and then migrated to Florida. Both Florida and the southwestern United States seem to be refuges where rattlesnakes, like *adamanteus* and *atrox*, evolved more slowly, retaining ancestral characters.

A comprehensive molecular study by Murphy et al. (2002) placed *C. adamanteus* in the "*C. atrox*-group," which includes the species *atrox*, *catalinensis*, *ruber*, and *tortugensis*, but a multigene analysis by Parkinson et al. (2002) showed its closest relative to be *C. tigris* of the "*C. mitchellii*-group" of Murphy et al. Literature and/or systematics were reviewed by Campbell and Lamar (2004), McCranie (1980a), and McDiarmid et al. (1999).

Consumption of rattlesnake meat or capsules containing powder made from ground rattlesnake meat has resulted in human infections with *Salmonella arizonae*. This is particularly true of those products that originate in Mexico, and are marketed under the names "*víbora de cascabel, pulvo de víbora,* and *carne de víbora.*" Cases with drastic symptoms (acquired immunodeficiency syndrome, adenocarcinoma, congestive heart failure, diabetes, erythematosus, hypertension, leukemia, and systemic lupus) have been reported from Arizona, California, and Texas (Riley et al. 1988), so care must be taken with the consumption of rattlesnake products.

Crotalus atrox Baird and Girard, 1853
Western Diamond-backed Rattlesnake
Vibora Serrana

RECOGNITION: *Crotalus atrox* is the second largest venomous snake in northern America, growing to TBL_{max} of 233.7 cm (Curtis 1949; Jones 1997). Dobie (1965) reported it reaches an unsubstantiated TBL of "eight feet or so" (243.8 cm), and Repp and Schuett (2003) and Wilson (2000) reported BMs of 6.5–12.3 kg. Most collected today are about 80–100 cm long and weigh 350–450 g (Repp and Schuett 2003); individuals more than 150 cm long are now infrequently found and one longer than 180 cm is extremely rare (Campbell and Lamar 2004).

    C. atrox is normally grayish-brown, but some individuals are greenish (in the extreme eastern part of the range), reddish, or yellowish; and others have ranged from albino (Cole 2005; Dyrkacz 1981; Gloyd 1958; Hensley 1959; Muir 1990) to melanistic (Pedro Almendariz lava field in Socorro and Sierra counties, New Mexico, Best and James 1984; and the Malpais region of New Mexico near the Mexican border in Dona Ana County, Lewis 1951). The dorsum has 23–45 (mean, 35–36) dark gray-brown to brown, dark brown or black-bordered (and then a second outer light border), diamond-shaped or hexagonal-shaped blotches; but aberrant blotch patterns occasionally occur (Cliff 1954; Nickerson and Mays 1968; Simons 1986; Yancey et al. 1997). Smaller dark blotches lie along the sides. The dorsal body scales are keeled and pitted, and lie in 23–27 anterior rows, 25–27 (23–29) rows at midbody, and 22–23 rows near the tail.

Crotalus atrox (Val Verde County, Texas; University of Texas El Paso, John H. Tashjian)

Crotalus atrox
(Cochise County,
Arizona; Paul
Hampton)

Their microdermoglyphic pattern is one of usually straight, or occasionally with some swirling, ridges (illustrated in Harris 2005 and Stille 1987). Rarely, individuals are born with a reduced number or no scales on their bodies (Bechtel and Bechtel 1991; Mc-Crady et al. 1994; Smith et al. 1996). Normally 4–6 (2–8) alternating wide gray and black bands are present on the tail (individual variation occurs in the banding pattern; Moon et al. 2004a), and the bands are always ventrally incomplete. About 1.4% of the snakes lack a tail rattle (Holycross 2000a; Painter et al. 1999; Rowe et al. 2002). The black and light tail bands are essentially of the same width, and the proximal rattle segment contains at least some black pigment. The venter is white, cream, or pink, with some fine dark mottling along the sides; the underside of the tail is often grayish. The venter is covered with 168–196 ventrals, 16–36 subcaudals, and an undivided anal plate. Two light stripes bordering a darker stripe extend downward and backward from in front and in back of the eye to the supralabials; the posterior one ends well in front of the corner of the mouth. Irregular dark pigmented spots are sometimes present on the top of the head, and some individuals have a light transverse line crossing each supraocular scale. The mental scale and anterior infralabials often are lightly mottled with dark pigment. On the dorsal surface of the head are a higher than wide rostral scale, 2 small internasals, 11–32 additional scales, 2 pairs of canthals (the posterior pair is the largest), 2 large supraoculars, and 4–5 (3–8) intersupraocular scales. No prefrontals are present. Laterally are 2 nasals (the prenasal normally touches the first supralabial; the postnasal usually contacts the upper preocular), 1 (2) loreal, 2 (3) preoculars, 3 (2–6) postoculars, 3–4 (2–4) suboculars, 1–2 interoculabials, 15–16 (12–18) supralabials, and 16–17 (14–21) infralabials. On the chin are a small mental scale, followed by two pairs of chin shields (with the posteriormost largest), which are in contact medially.

As in other *Crotalus*, the hemipenis is bilobed with a divided sulcus spermaticus. Each lobe is adorned with about 64 spines and 57 fringes and has a soft apical projection; there are only a few spines in the region between the lobes (Klauber 1972).

Dentition, other than the fang, consists of 3 palatine, 8 (7–9) pterygoid, and 9–10 (7–11) dentary teeth (Klauber 1972).

Adult males are larger, weigh more, and have longer tails than adult females (Beaupre et al. 1998; Boyer 1957; Duvall and Beaupre 1998; Taylor and De Nardo 2008). Duvall and Beaupre (1998) have studied the strategies of sexual dimorphism in rattlesnakes using several models, and we refer the reader to their paper for details. Males (TBL_{max}, > 200 cm, see above) have an average of 27 anterior dorsal scale rows, 168–193 (mean, 183) ventrals, 19–32 (mean, 25–26) subcaudals, 4–7 (3–8) black tail bands, and a mean TL/SVL of 7.7–8.2% (Fitch and Pisani 1993). Females (TBL_{max}, probably about 150 cm; SVL, 119.3 cm; Fitch 2002) have an average of 25 anterior dorsal scale rows, 173–196 (mean, 186) ventrals, 16–36 (mean, 20) subcaudals, 3–4 (2–6) black tail bands, and a mean TL/SVL of 5.5–6.6% (Fitch and Pisani 1993).

Several alternate explanations exist as to the causes of the sexual size dimorphism in *C. atrox*. Through natural selection, males are larger, giving them an advantage in combat bouts and in securing receptive females. Alternately, females are smaller than males because small females have low metabolic maintenance costs that allow them to quickly recoup energy stores (in the form of lipids) and reproduce frequently. A third hypothesis is that the sex difference is the result of phenotypic plasticity rather than natural selection, with females growing more slowly than males because of their higher energy costs for reproduction, leaving less energy resources to be applied to growth. Male neonates are slightly larger than females, but the difference does not account for the great difference in adult sex size. Testosterone does not seem to promote growth in males. Males and females raised in captivity with controlled diets have essentially the same growth rates, do not diverge in size, and grow to lengths greater than those of wild individuals. Supplementation of the diet to females results in dramatic increases in growth. If the third hypothesis is correct, then the natural selection hypotheses may not be supported (Taylor and De Nardo 2008).

GEOGRAPHIC VARIATION: Although some color (greenish in Arkansas, Perkins and Lentz 1934; pinkish to reddish in western Texas, southwestern New Mexico, and southeastern Arizona, Gloyd 1940; Jameson and Flury 1949; Nickerson and Mays 1969; blackish on lava fields; grayish-brown to tan elsewhere); dorsal blotch shape, color, and border (Cliff 1954; Perkins and Lentz 1934); size; and rattle growth vary between populations, no subspecies are currently recognized.

Campbell and Lamar (2004) reported finding apparent dwarf adults with relatively large heads, bulging eyes, and unusually long rattle-strings. They were part of a confiscation that did not have specific locality data associated with the snakes, and were supposedly from "Charco Cercado, north of San Luis Potosí, Mexico," but Campbell and Lamar were unable to find this locality on any map. Taylor (1936) previously reported a dwarf population from Sonora, and another dwarf population may occur on the island of San Pedro Mátir in the Gulf of Mexico (Cliff 1954). These should be investigated further.

Spencer (2008) examined 37 phenotypic mensural, meristic, and categorical morphological characters in 673 adult *C. atrox* from across the species range. Many traits varied within eastern, western, and middle distributed groups, as well as among them. Mensural traits were highly variable among regions, and there was only a slight trend in body size to be higher in the north. Other characters were found to be clinal; ven-

Crotalus atrox
(R. D. Bartlett)

tral scale counts increased with decreasing longitude, and the number of scales positioned before the supraoculars was higher in the east. Variation in nonmensural data was better explained along a longitudinal cline, possibly due to environmental factors. Morphological traits, types of prey eaten, and reproductive output varied little across the range. Most differences occurred along an east-west continuum.

Genetic variation exists between populations. Castoe et al. (2007) examined the mtDNA sequences of several populations of *C. atrox* to determine their genetic structures and historical demographic patterns across the species range, and to relate these to the broader biogeography of the semiarid and desert regions in the southwestern United States and adjacent Mexico. They found a Late Pliocene divergence between the Baja California peninsular and continental lineages of the species, followed by an Early to mid-Pleistocene divergence across the Continental Divide. The mtDNA data indicated isolation of populations in multiple Pleistocene refugia within the deserts of Sonora and Chihuahua, the southern plains, and the Tamaulipan Plain on both the eastern and western sides of the Continental Divide. The populations east of the Continental Divide particularly showed range extension and population growth. They also found clear evidence of probable recent gene flow, particularly from west to east, across the Continental Divide and secondary contact haplotype lineages, and also suggested that *C. tortugensis* and the supposed undescribed population of diamond-backed rattlesnakes on Isla Santa Cruz be placed in the synonymy of *C. atrox*. In addition, Porter (1994) found two clades of repetitive DNA sequences in Texas *C. atrox*, one dispersed and the other telomeric, with some intergradation. Venom gland DNA of *C. atrox* from McMullen County, Texas, has yielded both P*III*-SVMP and RGD disintegrin (atroxatin) genes (Soto et al. 2007).

CONFUSING SPECIES: Other large rattlesnakes within its range can be identified by the key previously presented. Also see Campbell and Lamar (2004).

KARYOTYPE: The snake's 36 chromosomes consist of 16 macrochromosomes (4 meta-centric, 6 submetacentric, and 4 subtelocentric) and 20 microchromosomes; with females having a pair of ZW macrochromosomes and males a pair of ZZ macro-chromosomes; the Z chromosome is either metacentric or submetacentric and the W chromosome is either submetacentric or subtelocentric (Baker et al. 1972; Stewart and Morafka 1989; Zimmerman and Kilpatrick 1973). Porter (1994) has isolated two families of repetitive DNA sequences from *C. atrox*.

FOSSIL RECORD: Fossils assigned to *C. atrox* are known from the Pliocene (Blancan) of Texas (Rogers 1976); Pleistocene (Irvingtonian) of Texas (Holman and Winkler 1987); and Pleistocene (Rancholabrean) of Arizona (Czaplewski et al. 1999; Mead et al. 1984), California (Holman 1995), Nevada (Tihen 1962), New Mexico (Brattstrom 1954a, 1954b, 1958b, 1964b; Holman 1970; Van Devender et al. 1976; Wiley 1972), Texas (Holman 1966, 1969, 1995; Mecham 1959; Van Devender and Bradley 1994), and Sonora, Mexico (Van Devender et al. 1985).

DISTRIBUTION: *C. atrox* ranges from southeastern and south-central Kansas (Cow-ley, Cherokee, Crawford, Elsworth, Lyon, and Summer counties; Hall and Smith 1947; Matlack and Rehmeier 2002; Riedle 1996) and west-central and southwestern Arkan-sas (Karnes and Tumlison 2003; Trauth et al. 2004) south and westward through east-ern, central, and southern Oklahoma (McAllister et al. 2008; Webb 1970), Texas (except the extreme east; Werler and Dixon 2000), central and southern New Mexico (Degen-hardt et al. 1996), Arizona (Fowlie 1965; Brown 2003a), extreme southern Nevada (Clark County; Anderson and Emmerson 1970; Emmerson 1982), and southeastern Califor-nia (Riverside and Imperial counties; Brown 1997; Klauber 1927, 1929), and southward in Mexico to the states of Querétaro (Dixon et al. 1972), Hildalgo (Campbell and

Distribution of
Crotalus atrox.

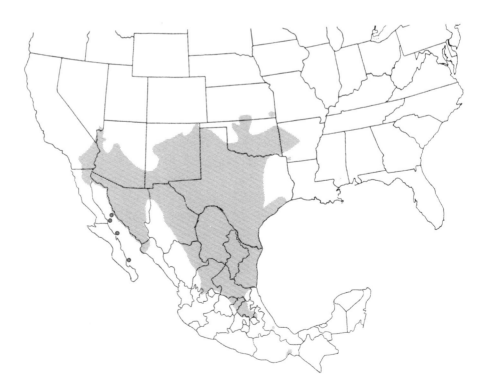

Lamar 1989), Mexico (Matias-Ferrer and Murillo 2004), Veracruz (Pérez-Higareda and Smith 1991), Sonora (Burger and Hensley 1949; Cliff 1954), Sinaloa (Dixon et al. 1962; Hardy and McDiarmid 1969), and Oaxaca (Casas-Andreu et al. 1996; Hartweg and Oliver 1940; Woodbury and Woodbury 1944). It is also found on the islands of San Pedro Mátir, Santa Cruz, Santa Mariá, Tiburón, and Turner in the Gulf of California (Cliff 1954; Grismer 1994b, 1999a).

HABITAT: *C. atrox* lives in a variety of arid or seasonally dry habitats, from near sea level in gravel flats and the tide line of sandy beaches in Sonora (Taylor 1936) to uplands of 2,134 m in New Mexico (Bogert *in* Degenhardt et al. 1996) to about 2,440 m in San Luis Potosí (Klauber 1972; Werler and Dixon 2000), but usually below 1,500 m. It has been found in desert plant–covered sand dunes, grasslands, granite and limestone hills with scattered boulders and rock crevices, mesquite habitats (where it is often confined to flat washes; Ortenburger and Ortenburger 1927 [1926]), shrublands, scrub woods, pine-oak woodlands, and riparian and tropical deciduous forests.

Some reported plants in these various habitats include allthorns (*Koeberlinia spinosa*), ash (*Fraxinus velutina*), blackbush (*Flourensia cernua*), cacti (various, including *Carnegiea gigantea, Ferocactus cylindraceus, Opuntia* sp., and *Stenocereus thurberi*), brittlebush (*Encilia farinosa*), catclaw and tarbush (*Acacia* sp.), cottonwood (*Populus deltoides*), creosote bush (*Larrea tridentata*), crucifixion thorn (*Canotia holacantha*), crucillo (*Condalia spathulata*), featherplume (*Dalea formosa*), grapes (*Vitus arizonica*), grasses and herbs (various, including *Aristida* sp., *Bouteloua* sp., *Euphorbia* sp., *Muhlenbergia emersleyi, Pleuraphis mutica, Setaria* sp., *Spartina* sp., and *Sporobolus* sp.), hackberry (*Celtis laevigata, C. reticulata, C. talagillies*), hopbush (*Dodonaea* sp.), Jerusalem thorn (*Parkinsonia aculeata*), jojoba (*Simmondsia chinensis*), junipers (*Juniperus* sp.), manzanita (composed of *Arctostaphylos* sp. and *Quercus dumosa*), matchbrush (*Xanthocephalum sarothrae*), mesquite (*Prosopis* sp.), mimosa (*Mimosa buncifera*), Mormon tea (*Ephedra* sp.), mulberries (*Morus microphylla*), oaks (*Quercus* sp., especially in riparian zones), palo verde (*Cercidium floridum, C. macrum, C. microphyllum, C. texanum*), pines (*Pinus* sp.), persimmon (*Diospyros texana*), Russian olive (*Elaeagnus angustifolia*), sagebrush (*Artemisia* sp.), saltbush (*Atriplex* sp.), saltcedar (*Tamarix chinensis, T. ramosissma*), sotol (*Dasylirion wheeleri*), sumacs (*Rhus microphilla, R. torreyi*), walnut (*Juglans major, J. microcarpa*), willows (*Chilopsis linearis, Salix* sp.), and various yucca (*Yucca* sp., including the Joshua tree, *Y. brevifolia*). It is often found among scattered rocks and boulders, or rock outcrops with crevices.

In southwestern Nuevo León, *C. atrox* occurs in higher *Juniperus* forest dominated by *Juniperus depeanna* and *J. monosperma*. Other common plants there are cacti (*Cylindropuntia leptocaulis, Opuntia imbricata*), garbancillo (*Perganum mexicanum*), grama grass (*Bouteloua barbata, B. chasei*), mesquite (*P. glandulosa*), saltbush (*Ariplex acanthocarpa*), seepweed (*Suaeda mexicana*), and sunflower muhly (*Muhlenbergia monticolla*) (Lazcano et al. 2007c).

BEHAVIOR AND ECOLOGY: The annual period of surface activity of *C. atrox* is from February–April to October–November, and sometimes December, depending on latitude and elevation. It is principally active from June to September, with a decided peak in August. In the Southwest, rainfall is usually highest from July to September, allowing the snake to reach its greatest activity about a month later, possibly due to increases in rodent populations. Reynolds (1982) found it to be active mainly from July to September with a distinct peak in August in Chihuahua. Ruthven (1907) observed activity more often on cloudy, foggy, or rainy days than on clear ones in Arizona.

In March–April, most hours are spent resting, but as summer continues the periods of activity and alertness become longer than the nonmovement periods, especially during the wet months (15 July–15 October). Fall activity (16 October–30 November) slows until the snake rests as much as it is active or alert. Normally the snake hibernates from December to February, and most time and energy are spent in resting underground (Beck 1995). If the winter weather warms, some *C. atrox* will bask at the mouth of their retreats around noon. Hibernacula are most often rock crevices, small caves, or abandoned mines located on south-facing slopes, but sometimes wood piles, mammal burrows, gopher tortoise (*Gopherus agassizii*) burrows, or deep cracks in the soil are used. Such sites probably maintain, depending on their depth, an ET of 5–10°C during the winter, certainly sufficient for survival of the snake. As many as 100–200 *C. atrox* may regularly congregate at a specific site to overwinter, and a hibernaculum may be used solely by *C. atrox* or be shared with other species of snakes (Mitchell 1903; Ortenburger and Freeman 1930). A period of winter cooling or inactivity is needed for successful breeding in some species of snakes, and this may also be true of *C. atrox* (Tryon 1985).

It appears that *C. atrox* hibernating at higher elevations pick dens with greater insolation and greater angles of mean aspect and slope. Thermal ecology at high-elevation and low-elevation hibernation sites was studied by Hamilton and Nowak (2009) in Arizona. The 28,490 ha high-elevation site in Yauapai County had a mean annual ET of 16°C, and mean January and July ETs of 6°C and 28°C, respectively, and 183 annual frost-free days. The mean elevation of the higher hibernaculum was 1,014 m, that of the lower hibernaculum 866 m. Comparatively, the low-elevation 28,490 ha site in Gila County had a mean annual ET of 20°C, mean January and July ETs of 9°C and 32°C, respectively, and 319 annual frost-free days. The mean insolation of the high hibernaculum was 1,515 (range, 442–2,359) watts/m², but only 1,256 (range, 447–1,844) watts/m² at the lower hibernation den. The mean hibernaculum aspect and slope for the higher and lower dens were 267° and 15° and 66° and 18°, respectively.

Diel activity is determined by air temperature. *C. atrox* is more diurnally active during the spring and fall. It forages in the morning or early evening, and basks some in the morning (Landreth 1973). In the summer, it is more crepuscular (Zweifel and Norris 1955) and nocturnal, resting under some sheltering rock or within shading vegetation during the day. Daytime underground retreats may be shared with other snakes; a retreat in a dirt road bank in Arizona was shared with two *Masticophis bilineatus* (C. Ernst, personal observation).

The snake's heat-sensitive loreal pits can apparently be used to locate cooler spots to rest by sensing nearby surface temperatures (Krochmal and Bakken 2003; Krochmal et al. 2004). Ortenburger and Ortenburger (1927) found Pima County, Arizona, *C. atrox* most active between 1830 and 1930 hours, not after dark, with a few active in the morning during July and August. Those found in the morning quickly retreated into holes, while those seen in the evening were found coming out of their daytime retreats.

Recorded BTs of active individuals range from 18.0 to 36.1°C, usually 27 to 30°C; snakes with BTs of 14°C may be active only because of abnormal disturbance (Beck 1996; Brattstrom 1965; Cowles and Bogert 1944; Landreth 1973). The CT_{max} ranges from 39.0 to 46.5°C (Cowles and Bogert 1944; Mosauer and Lazier 1933). Males and females do not appear to differ in their selected BT when either active or inactive (Malawy 2005).

Beck (1996) fed six radio-tagged Sonoran Desert *C. atrox* rodent meals ranging from 15.9 to 45.0 (mean, 34.0) % of their BM, and monitored their BTs in midautumn. Immediately after feeding, all snakes retreated from open areas into shelters under vegetation or in crevices, but most left these within 24–72 hours and basked nearby on the surface in full or partial sunlight. Others remained in the shelters but exposed the portion of their body containing the food bolus to the sun. While basking, fed snakes had BTs of 26.4–36.1 (mean, 31.9) °C; unfed individuals had BTs of 19.2–28.4 (mean, 25.1) °C. When they were in shelters, the BTs of fed snakes were more variable (average standard deviation, 4.31°C) than those of unfed snakes (2.84 °C). The mean BT selected during basking (31°C) was slightly, but nonsignificantly, higher than the mean activity BT (29°C) recorded earlier in August–September for the species at the site. Most unfed snakes remained in or near retreats during the study period; some were briefly active, during which their BTs approached those of fed snakes basking on the surface. The BTs of fed snakes were significantly higher than those of unfed control snakes during 1200–1600 hours. The postfeeding thermophilic responses of the snakes were restricted by thermal limitations in their environment and by reclusive behaviors following feeding. The elevated BTs selected by fed snakes were similar to those maintained during activity. Malawy (2005) also reported that *C. atrox* does not elevate its BT after a meal.

Taylor et al. (2004b) studied the BTs of 10 female *C. atrox* near Tucson, Arizona, using implanted temperature data-loggers. Mean monthly BTs of the females were May, 26.5°C; June, 30.5°C; July, 30.0°C; August, 30.0°C; September, 29.0°C; and October, 25.5°C. Minimum mean monthly nonrandom BTs for these same months were 18°C, 23°C, 23.3°C, 23.3°C, 21°C, and 18.5°C, respectively; the snake's semicontinuous mean BTs were 23.6°C, 26.2°C, 26.0°C, 26.1°C, 24.0°C, and 22°C, respectively (all BTs interpreted from figures).

The snake's BT affects its SMR. Stinner et al. (1998) kept 5 *C. atrox* (mean BM, 1.256 kg) at constant temperatures of 30°C for 24 hours and 15°C for 48 hours to determine their O_2 consumption rates at these temperatures. At 30°C the snakes had a mean consumption rate of 48.1 mL/kg per hour, and at 15°C the rate was only 6.0 mL/kg per hour. The species' blood contains 99–101 mg% cholesterol and 50–68 mg% reducing sugar (as glucose) (Luck and Keeler 1929).

C. atrox has a single, cooled region centered around its mouth and nasal capsule that extends across the loreal pit membranes at ATs of more than 20°C. Both head and BTs increase linearly with AT, as do also the differences between the head and body temperatures; but the latter differentials decline significantly at higher relative humidities. Rattling snakes have significantly greater head-body temperature differentials than do resting individuals. Respiratory cooling may provide a thermal buffer for the loreal pit organs at high ATs, but neural and behavioral tests must be conducted to prove this (Borrell et al. 2005).

Obtaining enough water in its dry environment is a major problem for the snake. Most water is probably obtained from its prey, and it probably also drinks from puddles during and after rain events. It is also known to lick water from its body coils (Ivanyi et al. 2003; Repp and Schuett 2008), as do other rattlesnakes.

The normal method of movement is by lateral undulations of the body, but *C. atrox* also uses caterpillar or rectilinear movement when it is most appropriate (C. Ernst, personal observation; Mosauer 1933).

Ivanyi et al.(2003), however, have reported a novel method of movement in which an Arizona *C. atrox* formed its body in tight coils and wobbled away from them while maintaining this coil arrangement. It did this by alternately pushing off with the left and right sides of the body coil in contact with the ground. As one side touched the soil, the other was lifted slightly and pulled backward. It appeared that only two zones of contact occurred during the unidirectional motion, which appeared deliberate and was repeated for several steps until the snake had moved a short distance away. The snake continuously faced its disturbers during the event. They thought the snake's motion to be as close as a limbless vertebrate can come to walking.

Oklahoma *C. atrox* studied by Landreth (1973) made annual migrations of 0.7–3.5 km to and from winter hibernacula. During such migrations, the snake does not wander randomly. Instead, it follows a more or less directed pathway, possibly using solar cues (sun compass [?]; Landreth 1973) or following scent trails (Weldon et al. 1990b). In the spring, Landreth's males moved an average of 102.4 m per day, but only averaged 61.2 m and 54.3 m in the summer and fall; females made average daily movements of 82.4 m in the spring, 46.1 m in the summer, and 46.3 m in the fall. However, 3 different females made unexplained midsummer, overnight trips of 72.4–105.6 m. In the winter, the snakes remained close to their hibernacula, only wandering short distances. Average winter movements of males and females were 3.5 m and 2.7 m, respectively.

In southeastern Arizona, Beck (1995) found that *C. atrox* maintained an average home range of 5.42 ha, and traveled a mean distance of 12.9 km in 95 hours of surface activity. During the annual activity period, his snakes moved an average of 50.8 m per day and 94.6 m per activity bout. During his 1996 study, Beck found that fed snakes moved an average of 8.5 (range, 0.6–32.3) m per day during the 9 days after ingestion. During the same period, his unfed snakes averaged 28.5 (range, 5.0–63.3) m per day. In another Arizona study near Tucson, Taylor et al. (2005) found no significant difference between the mean home range of 8 free radio-tagged control females (6.96 ha) and that of 9 supplementally fed female *C. atrox* (7.66 ha). Mosauer (1935b) reported that in the Coachella Valley, California, *C. atrox* has a home retreat to which it regularly returns after foraging.

Beaupre and Duvall (1998a) conducted a laboratory study to determine the variations in O_2 needs between the sexes of 48 *C. atrox* from the Sonoran Desert of Arizona. The BM-elevated SMRs of males and nonreproductive females were similar. O_2 consumption was affected by BM, BT, and time of day, and was about 1.4 times greater in 7 vitellogenic females than in 16 nonreproductive females: vitellogenic females, log −0.4 mL/hour at 15°C to 0.5 log mL/hour at 35°C; nonreproductive females, −0.6 mL/hour and about 2.7 mL/hour at the same temperatures. No differences occurred between males and nonreproductive females. They concluded that differences in the SMR apparently do not contribute to sexual dimorphism in that population. However, estimates of size-dependent maintenance expenditure led them to suggest that adult female body size may represent a compromise between selection or increased litter size (accomplished by increasing body size), and selection for increased reproductive frequency (accomplished by decreasing body size, and, thus, inactive maintenance expenditure). Physical activity probably contributed little variation to the observed patterns of SMR, as their snakes were only physically active less than 4% of the times they were observed while being tested. Stinner et al. (1998) reported O_2 consumption rates (mL/kg per hour) of 6.0 at 10°C and 48.1 at 30°C, and a change in whole body CO_2 stores of -0.04 mmol/kg per °C BT for *C. atrox*. The snake shows high levels of docosahexaenoic acid-containing phospholipids during high-frequency contraction of its ventral and tail-shaker muscles (Infante et al. 2001).

C. atrox climbs and swims well (Cunningham 1955; Klauber 1972; Mitchell 1903).

Males regularly participate in combat bouts, or "dances," during the spring and summer mating seasons. The behavioral sequence involved is similar to that of other rattlesnakes of the genus, and has been well described by Armstrong and Murphy (1979), Lowe (1942, 1948), Repp (1998), and Starrett and Holycross (2003).

REPRODUCTION: Several TBLs have been reported for the size at maturity of both sexes of *C. atrox*: 76.0 cm (Conant and Collins 1998; Stebbins 2003), 76.2 cm (Wright and Wright 1957), and 80 cm (Simons 1986). The shortest mature male examined by Goldberg (2007) had an SVL of 54.1 cm, while the shortest mature female SVLs reported are 64.8 cm (Rosen and Goldberg 2002), 70 cm (both sexes; Taylor and De Nardo 2004), 74.2 cm (Klauber 1972), and 80.0 cm (Tinkle 1962). The heaviest immature female weighed by Tinkle (1962) was 450 g, while the lightest mature female weighed 320 g; most adults weighed more than 500 g. Fat makes up about 10.1% of BM in mature females and fat 6.4% of BM in mature males (Fitch and Pisani 1993). Maturity seems to occur in 30–36 months. See also "Growth and Longevity."

Females are generally postpartum from August on. Immature follicles ($<$ 10 mm) are found from September through the winter. Vitellogenesis and follicle growth begin after emergence from hibernation in the previous nonreproductive year, and the immature follicles grow to a maximum length of 30 mm with considerable yolk deposition prior to the next hibernating period, but little follicular enlargement occurs over the winter. In the spring of the second (reproductive) year of the cycle, follicles continue to grow; by April, most follicles are 30 $\times$ 10 mm. Ovulation and fertilization occur from May to early July. No follicles enlarge when the female is gravid, and postpartum females have follicles no larger than 10 mm. Fat volume slowly increases after parturition (Fitch and Pisani 1993; Rosen and Goldberg 2002; Schuett et al. 2004b; Taylor and De Nardo 2004; Taylor et al. 2004a; Tinkle 1962).

Rosen and Goldberg (2002) examined 43 Arizona females, and reported the following monthly ovarian conditions. Inactive ovaries were found in March (67% of specimens), May (27.3%), June (25%), July–August (100%), September (62.5%), October (50%), and November (25%). Early vitellogenesis was only evident in April and June (14.3% each), and enlarged follicles ($>$ 12 mm) were present in February (100%), March (33%), April (85.8%), May (63.6%), June (50%), September (16.7%), October (50%), and November–December (75%). Oviductal eggs were only present in one May (9.1%) female.

The female serum estradiol level peaks in April; is still high, although less so in May when ovulation occurs; reaches a low level in June; and remains low through the following August. The level of serum progesterone is low in April; starts to rise in May, and continues so during June; peaks in early July; and then drops off during the rest of July through August and September after gestation (some females show an increase in

September). Testosterone levels are practically nonexistent in females during September to mid-July, but rise slightly and are evident in late July through August. Corticosterone is uneven through the year, showing peaks in April and May, in July and August, and at the time of parturition (usually in late August); its lowest levels occur in June and after parturition in late August and September (Schuett et al. 2004b).

From the above data, it is evident that female *C. atrox* probably have a biennial reproductive cycle (Tinkle 1962). Werler and Dixon (2000) reported an annual cycle in the warmer southern range in Texas, but a biennial cycle in northern Texas, where the winters are more severe and the annual activity period is shorter. Correspondingly, Fitch and Pisani (1993) recorded an annual reproductive cycle in Oklahoma females, and Rosen and Goldberg (2002) thought it possible that Sonoran Desert females may sometimes reproduce in successive years. Nevertheless, Price (1988) and Taylor and De Nardo (2004, 2005a) concluded that females give birth at best every three years (a triennial to quadrennial cycle) in Texas and Arizona's Sonoran Desert. Most likely, much of this is controlled by the numbers of available prey when females are trying to recover the lipid and energy stores depleted during the previous gestation period and parturition. Further study is needed to determine the cause of these cyclic variations.

Goldberg (2007) examined the gametic cycles of 67 male Arizona *C. atrox* with 54.1–130.0 (mean, 83.0) cm SVLs. Some males were undergoing complete regression (with spermatogonia and Sertoli cells predominant) in February–April, June–July, and October; early recrudescence (with spermatogonial division and primary spermatocytes predominant) in February, April–May, and July; late recrudescence (with secondary spermatocytes and undifferentiated spermatids predominant) in April–June; early spermiogenesis (with metamorphosing spermatids present) in August; spermiogenesis (with abundant sperm present in the tubule lumina) in June–October; and early regression (with germinal epithelium 1–3 cells thick) in September–October. Sperm were present in 88% of the vas deferens examined between February and October. Arizona males have elevated dihydrotesterone and testosterone levels during February–March and August–September, and baseline levels in April–July. Estradiol is elevated in February–April and September, and baseline in May–August and early February (Schuett et al. 2005; Taylor and De Nardo 2004; Taylor and Schuett 2004; Taylor et al. 2004a). During November–February, the levels of these hormones are not basal when compared to their levels during the active season, but instead are relatively high; however, estradiol decreases progressively (Schuett et al. 2006).

In Mexico, males are probably reproductive during the entire summer. They begin spermiogenesis in July, and by August and September sperm production is in full swing (Jacob et al. 1987). Testes mass does not vary significantly during the summer, but the diameter of the seminiferous tubules is significantly greater in August. The mature sperm remains over the winter in the epididymides in preparation for the following spring/summer mating (Jacob et al. 1987; Klauber 1972; Landreth 1973).

Some matings occur in the spring after hibernation, but most probably take place during the summer months. Spring copulations have occurred in nature as early as 14–29 March (Bogert 1942; Bonine et al. 2004; Mitchell 1903) and as late as 14 May to 20 June (Mitchell 1903; Wright and Wright 1957), and in the summer as late as 29 August (Taylor 1935). Captives have courted and copulated in almost all months.

Typically, a male finds and identifies a female by olfaction through rapid tongue-flicking. Males locate females by following scent trails left by them as they release pheromones containing 1-O-monoalkylglycerols with C_{12}–C_{20} chains from dorsal skin glands (Weldon et al. 1990b). Secretions from cloacal scent glands are also probably involved; these contain alkylglycerol monoethers containing alcohol moieties 12–20 carbons long, as well as cholesterol, which accounts for 30% of total ion content. No difference apparently exists between this secretion in males or females (Weldon et al. 1992; includes photographs of the female cloacal glands).

When a male encounters a reproductive female, he raises his upper body about 50–60 cm and jerks spasmodically, crawls to her, closely examines her back and sides with his tongue, and increases the rate of spasmodic body-jerking. He then rubs his chin sideways on the female's back at irregular intervals, and brings his entire body in contact with hers. At this point, she usually raises her tail and opens her cloacal vent, and the male places his vent next to her open vent and inserts a hemipenis. She may begin rhythmic body spasms at this time, but usually lies still; he may pulsate his anterior tail region during copulation. Copulation may last 15 minutes to 4–5 hours (Armstrong and Murphy [1979] reported one mating lasted 8 hours). Females store sperm from a mating at least over winter until the next ovulation (Jacob et al. 1987; Klauber 1972; Landreth 1973; Repp and Schuett 2003; Schuett 1992; Schuett et al. 2004b).

The GP following fertilization in May normally lasts about 3–5 months. Wiley (1929) reported a captive GP of 167 days. Births have been reported from late June to early October (Armstrong and Murphy 1979; Rosen and Goldberg 2002), but most occur from late July into September in both captivity and nature (Price 1988; Repp and Schuett 2003; Rosen and Goldberg 2002).

Litters average 9.5 young and normally contain 2–25 (n = 104) young. Klauber (1972) reported an unusually large litter containing 46 young—26 alive, 20 dead (most litters contain some stillborn individuals). Litters with the most young probably are born in the wetter portions of the species range, where dietary resources are greater. Litters of only 2–8 (mean, 4.5) young are normal in the drier desert regions (Repp and Schuett 2003), and Rosen and Goldberg (2002) reported litters in 29 Arizona females they examined (based on enlarged follicles, oviductal eggs, and corpora lutea) of only 4–15 (mean, 7.3) young.

Some controversy exists as to whether or not the number of young produced is positively correlated with female size. Rosen and Goldberg (2002, based on counts of enlarged follicles), Fitch and Pisani (1993), and Tinkle (1962) reported that it is correlated. However, Taylor and De Nardo (2004, 2005a) stated that there is no relationship between maternal SVL and litter size, and Rosen and Goldberg (2002) found this to be true if litter size is based on counts of embryos and live-born neonates. Again, the difference may possibly be a reflection of resource availability between the populations from wet areas of the range (presumably having longer females) and those from drier areas (presumably with shorter females). Taylor and De Nardo (2005a) found no significant relationship between female SVL and clutch mass, mean neonate BM, or neonate SVL, and that female postparturient BM is positively correlated with mean neonate mass.

Taylor et al. (2005) reported the mean RCMs of supplementally fed and unfed females from the same Arizona Sonoran Desert site were 30% and 33%, respectively, although the litter sizes were equal. The mean neonate BM and SVL from fed females were 22.9 g and 28.2 cm versus 20.0 g and 28.6 cm for those from unfed mothers.

Neonates average 29.7 (range, 20.5–36.7; n = 43) cm in TBL, and have a BM of

Crotalus atrox ✕
C. viridis viridis,
natural hybrid
(Texas, R. D.
Bartlett)

11.0–24.7 (mean, 12.55, n = 23) g (Lowe et al. 1986; Rosen and Goldberg 2002). The female may remain with her litter up to 6–10 days after parturition, at least until the neonates first shed (Greene et al. 2002; Mitchell 1903; Price 1988; Rosen and Goldberg 2002).

Captive *C. atrox* have mated with *C. adamanteus* (Haast *in* Klauber 1972) and *C. molossus* (Davis 1936), but whether or not these produced offspring is unknown. Jacob (1977) presented evidence that, contrary to apparent morphological overlap, *C. atrox* and *C. scutulatus* do not naturally hybridize in the Big Bend of Texas, southwestern New Mexico, and southeastern Arizona, in contrast to hybridization reported by Campbell and Lamar (2004). A natural hybrid between *C. atrox* and *C. horridus* has been described from Texas (Meik et al. 2008). It has also hybridized with *C. v. viridis* (see photograph).

GROWTH AND LONGEVITY: Growth rates vary between populations according to food supply and the length of the annual activity period. Males show equal or greater growth than females throughout life. Neonates of both sexes are similar in size, but with a slight male advantage, and the juvenile growth rate is essentially uniform between the sexes. Length divergence occurs after maturity, with females hardly growing each year while males show annual detectable growth (Beaupre et al. 1998; Fitch and Pisani 1993; Taylor and De Nardo 2005b).

Taylor and De Nardo (2005b) conducted a two-year experiment to determine the effects of sex and food intake on growth, weight gain, and attainment of sexual maturity; they also measured testosterone levels to determine if this hormone might be involved in male-biased sexual size dimorphism in *C. atrox*. They fed one group of neonates a high-intake diet (a mouse per week) and another group a low-intake diet (a mouse every three weeks). Snakes in the first group grew and gained weight more rapidly than those in the second group, but males did not grow or gain mass faster than females in either group. High-intake snakes matured earlier than low-intake ones, indicating that size, not age, is the critical determinant of reproductive maturity. Males had higher levels of testosterone than females but did not grow faster. This suggests that testosterone does not affect growth in this species, and may not be the proximate determinant of sexual size dimorphism.

Fitch (2002) examined rattle development in a series of 1,011 Oklahoma *C. atrox*. Subadults have tapered rattle-strings with the natal button still evident at the tip. Adults with more than eight rattle segments rarely retain the button, but may have tapered rattle-strings if they still have segments acquired when they were smaller, or they may have parallel-sided strings of uniform-sized rattles if all of the segments acquired during growth have been lost. More than 25% of the rattles of adults are composed of same-sized segments. Some individuals show a reverse taper, with an occasional undersized segment, probably caused by malnutrition. The longest rattles in Fitch's sample had 14 segments. Campbell (1959) reported an approximately 152 cm wild adult had an incomplete string of 15 segments, and Klauber (1972) gave 16 segments as the greatest number for the species. Rowe et al. (2002) reported the average rattle of the species has 5.54 segments and is 13 mm long. The tail muscular morphology involved in the shaking process is described in detail by Savitsky and Moon (2008).

A 50.3 cm (62 g) wild female grew 38.1 mm (17 g) in 43 days between captures, and another 77.4 cm (258 g) wild female grew 50.8 mm (8 g) in 59 days (Laughlin and Wilks 1962).

C. atrox has a potentially long lifespan. Slavens and Slavens (2000) reported a captive longevity of 27 years for one, and a wild-caught juvenile female survived an additional 25 years, 10 months, and 27 days in the personal collection of Charles S. Wallace of Warner, Oklahoma (Snider and Bowler 1992). Minton (1975) reported an approximately 20-year-old individual. Another was still alive after 16 years and 4 months at the San Diego Zoo (Shaw 1969), and a 90 cm adult caught in 1964 was still alive in January 1991 (William W. Lamar, personal communication). Possibly, some wild individuals live more than 15 years (Repp and Schuett 2003).

DIET AND FEEDING BEHAVIOR: The western diamond-backed rattlesnake has a broad vertebrate diet. Being opportunistic, it probably eats what it can find, dead or alive, providing it is a suitable species and of the proper body size.

Once injected, the snake's venom predigests its prey from within, breaking down the circulatory vessels, heart, liver, and kidneys of the bitten prey (Baramova et al. 1989; Thomas and Pough 1979; Willis et al. 1989); the proteolytic activity of the venom also appears to facilitate the entry of digestive enzymes from the snake's stomach into the wound, and inhibits bacterial activity, reducing the possibility of the prey putrefying before it can be digested (Thomas and Pough 1979). Surprisingly, McCue (2007b, 2007d) reported that envenomation had no significant influence on any digestive performance variable he examined.

Known vertebrates taken in the wild include amphibians—frogs and toads; reptiles—lizards (*Cnemidophorus* sp.; *Coleonyx brevis*; *Crotaphytus collaris*; *Eumeces brevilineatus*; *Holbrookia* sp.; *Phrynosoma cornutum*, *P. solare*; *Sceloporus magister*; *Uta palmeri*, *U. stansburiana*) and snakes (*C. atrox*); birds (including eggs and nestlings)—quail (*Colinus virginianus*), gulls (*Larus atricilla*), terns (*Sterna caspia*), boobies (*Sula nebouxii*), black skimmers (*Rynchops niger*), burrowing owls (*Athene cunicularia*), doves (*Columbina* sp.), mockingbirds (*Mimus polyglottos*), towhees (*Pipilo erythrophthalmus*), horned larks

(*Eremophila alpestris*), sparrows (*Amphispiza bilineata, Melospiza melodia*), and western meadowlarks (*Sturnella neglecta*); and mammals—moles (unidentified), shrews (*Cryptotis parva*), murid mice, rats, and voles (*Baiomys taylori; Microtus mexicanus, M. ochrogaster; Mus musculus; Neotoma albigula, N. floridana, N. micropus; Peromyscus eremicus, P. leucopus, P. maniculatus; Onychomys torridus; Rattus norvegicus; Reithrodontomys megalotis; Sigmodon hispidus*), pocket mice (*Perognathus flavescens, P. flavus, P. hispidus, P. intermedius, P. merriami, P. nelsoni* [?], *P. penicillatus*), kangaroo rats (*Dipodomys merriami, D. ordii, D. spectabilis*), pocket gophers (*Geomys bursarius, Pappogeomys castanops*), ground squirrels (*Spermophilus spilosoma, S. variegatus*), prairie dogs (*Cynomys ludovicianus*), fox squirrels (*Sciurus niger*), cottontails (*Sylvilagus audubonii, S. floridanus*), and jackrabbits (*Lepus californicus*) (Beavers 1976; Beck 1996; Best and James 1984; Cottam et al. 1959; Dugan and Melanson 2005; Fouquette and Lindsay 1955; Gates 1957; Grismer 2002; Hermann 1950; King 1975; Klauber 1972; Lewis 1950; Lowe et al. 1986; Marr 1944; McCallion 1945; McKinney and Ballinger 1966; Milstead et al. 1950; Minton 1959; Mitchell 1903; Parker 1974; Pisani and Stephenson 1991; Quinn 1985; Repp and Schuett 2003, 2009; Reynolds and Scott 1982; Sherbrooke 2003; Sherbrooke and May 2008a, 2008b; Smith and Hensley 1958; Tennant 1985; Tinkle 1967; Vorhies 1948; Woodin 1953).

Lubber grasshoppers (*Brachystola* sp.), beetles (Coleoptera), and ants (Hymenoptera) have also been found in the snake's stomach. The beetles and ants were from a snake that also contained mammal hairs and the remains of an iguanid lizard, so they may have been secondarily ingested with other prey (Klauber 1972).

Carrion is readily consumed when available (Gillingham and Baker 1981; Klauber 1972); perhaps this snake at times uses a scavenging feeding strategy.

Small mammals made up the largest diet by weight (94.8%) and frequency of occurrence (86.7%) (birds, 7.6%, 13.3%; lizards, 2.9%, 11.1%) in 205 Texas *C. atrox* (78, 38.1% with food) examined by Beavers (1976). Particular mammals eaten were pocket mice (*Perognathus*) by 39% of the snakes, harvest mice (*Reithrodontomys*) by 9%, and white-footed mice (*Peromyscus*) by almost 7%. Similarly, in 43 *C. atrox* from Chihuahua, Mexico, Reynolds and Scott (1982) found that mammals were the predominant foods (85.7% occurrence—pocket mice [26.5%], white-footed mice [16.3%], and ground squirrels, *Spermophilus* [10.2%]; reptiles [12.2%]; and birds [2.0%]). Prey was selected on the basis of size; animals either too large or too small were not eaten, nor were those that could seriously harm the snake.

C. atrox finds prey by either infrared reception or olfaction. Two hunting strategies are used—active foraging and ambushing. When a fresh prey odor trail is discovered, it is followed with much tongue-flicking (Chiszar et al. 1985; Gillingham and Clark 1981; R. R. O'Connell et al. 1982). When the warm-blooded prey is finally either detected visually or by infrared reception, the rate of tongue-flicking slows. *C. atrox* can detect an artificial infrared stimulus resembling a mouse in both temperature and size as far away as 100 cm, which corresponds to a radiation density of 3.35×10^{-3} mW/cm^2; this is a 3.2 times higher sensitivity to infrared radiation than has been reported for the species in the past (Ebert and Westhoff 2006).

While ambushing, *C. atrox* lies quietly beside a frequently used rodent trail or burrow. When the prey comes within striking range, it is quickly bitten, and then left to wander away to die, to be later followed by its odor. Once the body is discovered, the snake cautiously examines the prey with its tongue to make sure it is incapacitated before it begins to ingest it head first. Envenomated prey seems to be preferred over fresh animals (Chiszar et al. 1978a, 1999b; Duvall et al. 1978). *C. atrox* will accept prey envenomated by a conspecific, but rejects that bitten by another species (*C. oreganus*), indicating that the snake has the ability to recognize the difference in the venom between that of its species and that of other *Crotalus* (Chiszar et al. 2008).

Most of the annual energy budget is needed during May–September (Duvall and Beaupre 1998); yearly maintenance energy requirements can be satisfied in 2–3 large meals with a prey quantity equivalent to 93% of its body mass (Beck 1995).

C. atrox can survive for long periods without food. Apparently, the changes in biochemistry and physiological processes that accompany this during winter hibernation and the annual active season are different. Martin and Bagby (1973) conducted research pertaining to this. After fasting for up to 20 weeks, *C. atrox* exhibits only slight effects on most blood chemical constituents. Cholesterol and NaCl show a tendency to drop, while others (albumin, alkaline phosphatase, blood urea nitrogen, glucose, lactate dehydrogenase, serum glucamate-oxalacetate transaminase, total protein, uric acid) do not essentially change. The time of year apparently has a greater effect on Na^+ and Cl^- than does fasting, as a marked decrease in these two ions occurs during the winter months.

McCue (2007a) found that 16 subadult *C. atrox*, fasted up to 24 weeks under controlled laboratory conditions simulating those of the snake's active season, had a significant reduction in plasma glucose but increased circulating ketone bodies. The snakes also lost BM at a linear rate and increased their relative moisture content during the experiment. Their bodies experienced an increase in the fatty acid unsaturation index and were apparently able to take essential fatty acids effectively from β-oxidation. Endogenous essential and nonessential amino acids were used indiscriminately to provide energy, suggesting that essential amino acids are not preferentially spared during starvation. The ^{15}N signature of excreted nitrogenous waste increased significantly, presumably as a result of shifting amino acid source pools during starvation.

In McCue's (2007c) study, *C. atrox* employed a supply-side / demand-side economic strategy to successfully tolerate starvation. Effective demand-side strategies included the ability to depress the SMR demands by more than 70%. Also, supply-side regulation of food resources was indicated by the ability to spare structurally critical protein stores at the expense of lipid catabolism. These physiological adaptations for minimizing endogenous mass and energy flux during periods of prey limitation might help explain how snakes have survived more than 100 million years, as well as the repeated radiation of snake lineages into relatively low-energy environments.

Feeding the snake supplemental food positively affects growth and reproduction in females (Taylor et al. 2005).

VENOM DELIVERY SYSTEM: *C. atrox* has impressive fangs, probably third in length among the larger North American rattlesnakes after those of its eastern counterpart, *C. adamanteus*, and the Mexican *C. basiliscus* (Klauber 1939, 1972). Klauber (1939) interpreted the fang lengths of 52 *C. atrox* (reported as *C. cinereous*) with 10–14 cm TBLs. The snakes had a mean FL of 11.3 (range, 9.6–12.9) mm, a mean TBL/FL of 106, and a mean HL/FL of 4.64. The mean angle of fang curvature was 82°, the greatest of all the species he measured. The interfang distance between puncture marks after strikes by 20 Arizona *C. atrox* (TBL, 27.9–154.5 cm) was 3.9–24.1 (mean, 11.1) mm (Zamudio et al. 2000). The fangs are replaced every 25–30 days (C. Ernst, personal observation).

High-speed digital videography of both foraging and defensive strikes by *C. atrox* revealed differences in 20 kinematic actions between the 2 strikes (Young et al. 2001). A group of freshly caught snakes was used for foraging trials in which each snake struck at 4 small (14–16 g) and large (24–26 g), live, unrestrained *Mus musculus*. Analysis of the predatory strike revealed little difference between the strikes at the two size classes of mice, and low levels of correlation among kinematic variables. Differences did occur regarding the initial position of the strike (body extended versus body coiled), including the distance from target at launch (means, 6.28 cm and 7.70 cm, respectively), duration of fang penetration (means, 205.4 milliseconds and 167.1 milliseconds), and the angle of fang penetration (means, 92.3° and 83.0°). The strike velocity (means, 337.9 cm/second and 341.3 cm/second) was essentially the same. The second trial involving defensive strikes used a group of freshly caught snakes in which each snake struck at a large stuffed doll, a prekilled adult (mean BM, 27.1 g) *M. musculus*, and a mouse-sized stuffed doll. No significant differences were found between the defensive strikes directed at the mouse or at the mouse-sized doll, but defensive strikes directed at the large doll were significantly different from those directed at either small target. Overall, there was a rather high level of correlation found among the kinematic variables of the defensive strikes. Some important mean differences in the strikes aimed at the three target-types were distance from target at initial launch (mouse, 2.41 cm; mouse-sized doll, 5.52 cm; large doll, 8.90 cm), maximum strike velocity (0.57 cm/second, 225.2 cm/second, and 12.20 cm/second, respectively), duration of fang penetration (1.00 milliseconds, 164.0 milliseconds, and 0.66 milliseconds, respectively), and the angle of fang penetration (0.02°, 87.2°, and 4.21°, respectively). Significant differences occurred between the foraging and defensive strikes directed at similar size mouse classes. These results indicate that *C. atrox* can modulate its strike depending on target size and behavioral mode, but that foraging strikes are more stereotyped than defensive ones. Cundall (2002) and La Duc (2002) discussed other kinematic details of the strike, and we recommend the reader review their papers for additional information on the strike mechanism.

Venom flow per strike is also variable. In most strikes venom expulsion is coincidental with fang penetration, and retrograde flow occurs prior to fang withdrawal. So, the duration of venom flow is always less than the duration of fang penetration. Dry bites make up approximately 35% of strikes by *C. atrox*, and unilateral strikes support a hypothesis for venom pooling in the distal portion of the venom delivery system. No significant difference exists in the temporal or volumetric aspects of venom flow between defensive strikes aimed at small (*Mus musculus*; BM, 20.4–28.8 g) and large (*Rattus norvegicus*; BM, 159.2–185.3 g) rodents. With the species and size of target held constant, the duration of venom flow, maximum rate of venom flow, and total venom volume released are all significantly lower in predatory than in defensive strikes (Young and Zahn 2001). In contrast, Herbert and Hayes (2008) have presented evidence of the release of differential amounts of venom correlated with prey size.

There are two hypotheses that attempt to explain how venomous pitvipers regulate venom expulsion. The venom-metering hypothesis proposes that the snakes have active neural-based regulation, and that this internal control exerts the greatest influence on the amount of venom injected. The pressure-balance hypothesis proposes that the snakes exert little internal regulation over venom expulsion, and that the amount injected is largely determined by the physical interaction between the snake and its target (Young 2008).

To have neural-based regulation, the snake would have to process and integrate by one or more sensory association centers. Such integration is necessary for the differential sensory information to cause differential neural activation of effector organs (muscles, glands) in the venom delivery system. Unfortunately, behavioral evidence of association centers derived from the strike (La Duc 2002) cannot be extrapolated to venom expulsion. Ontogenetic shifts associated with venom delivery occur, and a snake's sensory association center would have to adapt to these. Is there an effector organ capable of controlling venom flow? If so, it is probably the M. compressor glandulae, but to date there is no evidence of a neural connection between it and a sensory association center (Young 2008).

Young et al. (2003) tested the role of peripheral resistance in the strikes of C. atrox in regard to the pressure-balance hypothesis, which proposes that differential venom flow results from the balance of internal forces acting at the venom gland and venom chambers and external forces acting at the exit orifice of the fang. Considerable variation occurs in the trajectory of the fang relative to the target in both foraging and defensive strikes, which would yield wounds with potentially different levels of peripheral resistance. The importance of peripheral resistance is also suggested by the expulsion of venom from the fang after withdrawal from the target (found in 7% of the strikes) and by the forceful ejection of fluid from the target around the embedded fang (in 2.8% of strikes). Young et al. used experimental milking chambers to test the right and left venom delivery systems at different levels of peripheral resistance, and found that significantly less venom was injected into the chamber and significantly more venom was released on the chamber's surface with increased peripheral resistance. Their observations and experimental results are consistent with the pressure-balance hypothesis. The envenomation efficiency suggests that the internal forces of venom injection (fluid pressure within the venom gland and venom chambers) far exceed the external pressure (peripheral resistance from the target tissue).

Additional study by Young and Kardong (2007) has shown that the M. compressor glandulae (the muscle surrounding the venom gland) only explains about 30% of the variation in venom flow, but that lifting (compression) of the fang sheath produces marked increase in venom flow, almost 10 times that recorded from the M. compressor glandulae alone, during a normal strike. Their results suggest that variation in these two aspects of the venom delivery system explains most of the observed difference in venom injection in terms of both magnitude and temporal patterning. The lack of mechanical or functional links between these two components, and the lack of skeletal or smooth muscle within the fang sheath, indicate that it is unlikely that variation in venom flow is under direct neural control. Instead, differential venom ejection results from differences in the pressurization of the compressor glandulae, the gatekeeping effects of the fang sheath and enclosed soft tissue chambers, and differences in the pressure returned by peripheral resistance of the tissue of the target.

Young (2008) has proposed that venom is metered through a synthesis of these two hypotheses, although evidence for this is weak.

Hayes (2008) has pointed out that the two hypotheses represent different forms of analysis, cognitive and physiological, respectively, and are not mutually exclusive. The venom-metering theory has two assumptions: (1) that the snake can accurately assess the target (well-supported; see individual species accounts), and (2) that it possesses cognitive control over its venom delivery system (less proven). Recent videos of venom flow provide evidence of cognitive control of venom release. Other mechanisms probably involved include jaw closure, fang movements, and force and duration of venom gland contraction. Hayes proposed differences in venom release can result from (1) venom-metering (cognitive), (2) differences in strike kinematics (physiological), (3) differences in target features that might affect or constrain the kinematics of venom delivery (physiological), (4) depletion of the amount of venom in subsequent bites (physiological), (5) differences in snake size (ontogenetic), and/or (6) adaptive venom expenditure for different targets or contexts (defense, predatory, a functional consequence). Support for the venom-metering hypothesis is particularly strong; support for the target-features hypothesis (based on the pressure-balance hypothesis) is weak but has some validity. Hayes lists evidence for venom-metering in *Agkistrodon piscivorus*, *Crotalus atrox*, *C. helleri*, *C. oreganus (concolor, oreganus)*, and *C. viridis;* as well as for other pitvipers and vipers.

The metabolic cost of producing venom in *C. atrox* is considerable. Immediately after venom extraction, it experiences an 11% increase in the SMR during the first 72 hours of venom replenishment. This is apparently the result of metabolic costs involved in venom production and is a magnitude greater than that predicted for producing an identical mass of mixed body growth. Extracted liquid venom yield is allometrically correlated with the snake's BM ($4.77W^{0.60}$) and has a mean moisture content of 70.9%. Neither dry nor wet venom yields are correlated with the magnitude of the metabolic increase that occurs during the 72-hour period, suggesting the cost of venom replacement is independent of extracted venom mass. The significant postextraction metabolic increases support existing hypotheses concerning the metabolic costs of venom production, and may help explain why *C. atrox* meters its venom conservatively (McCue 2006).

VENOM AND BITES: The western diamond-backed rattlesnake is extremely dangerous. Consequently, it has been the subject of more research on its venom and on the symptoms and occurrence of human envenomations than most other North American venomous snakes. It ranks second only to the copperhead, *Agkistrodon contortrix*, in the number of annual human envenomations (Parrish 1966), and, within its southwestern range, it is certainly the snake that bites more people annually (Cruz and Alvarez 1994). Ivanyi and Altimari (2004) reported that 4 (9.1%) of 44 nondry bites by 22 venomous snake species that occurred during academic research over a 26-year period were by *C. atrox*, and there are numerous reports in the medical literature of envenomations, many fatal, by captive *C. atrox*.

Abundant studies have been published on the chemistry and the activities of the various components of the venom of *C. atrox*. Some of these are listed below, and the summary presented is based on information in those papers.

The whole venom is composed of approximately 31% solids; 77–80% of these are proteins (Brown 1973); both purines and pyrimidines are involved. Constituents or activities identified in the venom of *C. atrox* include arginine esterase, crotolase-like enzyme, cysteine-rich secretory proteins (catrin 1 and 2, which cause smooth muscle contraction), C-type lectins, deoxyribonuclease, endopeptidase (nonspecific), fibrinogens (catroxobin; α-fibrinogenases [176 amino acids; molecular weight, 20,000]; β-fibrinogenases [molecular weight, 22,900]; atroxase [molecular weight, 23,500], kininase; proteases I–IV), 5′-nucleotidase, hemorrhagic zinc metalloproteinases HT a-g (molecular weights, 24,000–68,000), hyaluronidase, Mojave toxin (not universal), kallikrin-like enzymes (composed of 17 different amino acids), L-amino acid oxidases, NAD-nucleosidase, phosphodiesterase 1 (type 4), phospholipase A_2 (with lysine-49), phosphomonoesterase, proteases (I, IV), ribose 1, trypsin-like activity, and tyrosine esterase (Alfonso et al. 2005; Al-Joufi and Bailey 1994; Bjarnason and Tu 1978; Bjarnason et al. 1983; Du and Clemetson 2002; Fujisawa et al. 2008; Graham et al. 2008; Hite et al. 1992; Holzer and Mackessy 1996; Hung and Chiou 1994; Joseph et al. 2002; Lomonte et al. 2003; Maity and Bhattacharyya 2005; Markland 1991; Markland and Perdon 1986; Mashiko and Takahashi 1998; McCue 2005; Meier and Stocker 1995; Nikai et al. 1983; Ogawa et al. 2005; Ownby et al. 1978; Pandya and Budzynski 1984; Panfoli et al. 2007; Ramírez et al. 1999; Russell 1983; Sant' Ana et al. 2008; Sapru et al. 1983; Serrano et al. 2006; Siigur and Siigur 2006; Soto et al. 1989; Takahashi and Mashiko 1998; Tsai et al. 1983, 2001; Tu 1977; Wagner and Prescott 1966; Willis and Tu 1988; Yamazaki and Morita 2004).

Horton (1951) reported the venom has a near ultraviolet absorption spectrum of 220–320 nm that decreases as the wavelengths became longer; and Johnson et al. (1967) have reported the optical density differences of the raw venom.

The venom is highly hemorrhagic. Almost 53% of its enzymes are concerned with lytic breakdown of the circulatory system (three fibrinogenolytic proteases are present), but another 17% are neurotoxic in action and 30% are digestive proteases (Tennant 1985). Intraspecific differences occur in concentrations of certain proteins; some individuals possess a neurotoxin analogous to the Mojave toxin of *C. scutulatus*, making them more dangerous than most conspecifics (Weinstein et al. 1985). Hemorrhaging from the breakdown of vascular tissues occurs rapidly (Soto et al. 1989), and the venom also contains strong hemaglutinizing enzymes that coagulate red blood cells. The venom of juvenile *C. atrox* is, however, only weakly procoagulant.

The human lethal dose is approximately 60–100 mg (Brown 1973; Minton and Minton 1969), and venom yields are often very high and increase with snake length. Neonate *C. atrox*, like other crotalids, possess fangs and active venom glands that can deliver a serious bite. Their venom may be more potent than that of adults (Theakston and Reid 1978); however, their venom yield per bite is small.

The quantity of venom available naturally increases with the size of the snake, and adults may yield a total of 151–1,150 mg of dry venom (typically about 250–400 mg). The average liquid venom yield is 0.30 mL (do Amaral 1928). These yields are far more than is needed for a fatal human envenomation (Ernst and Zug 1996; Glenn and Straight 1982; Klauber 1972; Minton and Minton 1969; Russell and Puffer 1970). The average number of lethal doses per bite for a 70 kg human is 0.36 (Brown 1973). A 60 cm snake can yield about 0.10 mL of venom per bite (high yield, 0.19 mL), while one 150 cm long could inject 1.27 mL per bite (high yield, 1.88 mL) (Klauber 1972). One 165 cm long *C. atrox* milked by Klauber (1972) yielded a total of 1,145 mg of dried venom (3.9 mL liquid). In contrast, Herbert (1998) reported that the mean venom expenditure of 28 wild *C. atrox* during a defensive bite was only 21.4 (range, 1.1–103.0) mg.

Gregory-Dwyer et al. (1986) found no seasonal variations in the isoelectric properties of the venom proteins from individual *C. atrox* kept under controlled conditions and tested for 20 months; however, intraspecific differences in concentrations of certain proteins were evident.

Geographical differences in the venom occur in lethality, protease activity, hemorrhagic action, and presence of Mojave toxin. Mojave toxin has not been found to be uniformly distributed across the species range (Bober et al. 1988; Minton and Weinstein 1986). Venom from northeastern *C. atrox* has the lowest lethality and highest protease activity, and northern individuals have slightly higher hemorrhagic activity. In contrast, the snakes from the southwestern portion of the range have the opposite lethality and protease activity (Minton and Weinstein 1986).

The venom is quite stable if stored in closed containers in the dark; Russell et al. (1960) reported that venom thus kept lost relatively little of its potency after 16–17 years of storage. A snake's age, however, may play a role in the toxicity of its fresh venom. Minton (1975) reported that as a *C. atrox* in his laboratory aged over an approximate 19-year period its venom became progressively less toxic; the dried venom LD_{50} in mg/kg was at first 11.65 (4.29 lethal doses/mg of venom), but had increased to 28.0 (1.78 lethal doses/mg of venom) after 19 years, with the greatest potency loss occurring during the first 9 years.

Venom from *C. atrox* has been used to determine symptoms, the minimum lethal doses, and LD_{50} values of several animals. The intrasubcutaneous and intramuscular dried venom LD_{50} values of mice (*Mus musculus*) are 7.8–27.8 mg/kg (Minton and Weinstein 1986) and 20 mg/g (Ownby et al. 1975), respectively. Minimum intraperitoneal LD_{50} dried venom doses for mice are 3.71–8.42 mg/kg (Arce et al. 2003; Githens and Wolff 1939; Glenn and Straight 1982; Johnson et al. 1966; Macht 1937; Minton 1956; Russell and Brodie 1974; Russell and Emery 1959; Russell and Puffer 1970). The minimum intraperitoneal lethal dose is 0.12–0.30 mg (Githens and Wolff 1939; Macht 1937), or 6.0 mg/mouse kg (Johnson et al. 1966). The intravenous LD_{50} of mice is 1.0–6.3 mg/kg (Arce et al. 2003; Gingrich and Hohenadel 1956; Russell and Brodie 1974; Russell and Puffer 1970; Theakston and Reid 1978).

The intraperitoneal LD_{50} values for the cotton rat (*Sigmodon hispidus*), wood rat (*Neotoma micropus*), and opossum (*Didelphis virginiana*) are 172, 1,121, and 1,121 mg/kg, respectively (Pérez et al. 1979); this is interesting, as *D. virginiana* is known to possess natural inhibitors and be resistant to crotalid venoms (McKeller and Pérez 2002; Perales et al. 2005; Pérez and Sánchez 1999; Ramírez et al. 1999; Soto et al. 1989). The intravenous LD_{50} of the brown rat (*Rattus norvegicus*) is 0.025 mg/g (Billing 1930).

Sublethal doses of venom (2–3 mg/kg) injected intramuscularly into 2.5–3.3 kg rabbits (*Oryctolagus cunniculus*) resulted in hemorrhage within 24 hours and increased blood platelets. Fibrinogen levels initially rose for 48 hours, and then dropped; the mean levels of fibrinogen-fibrin degradation products essentially did likewise (Grace and Omer 1980; Simon and Grace 1981). Crimmins (1927a) determined that the minimum lethal intramuscular dose for dogs (*Canis familiaris*) was 1 mg/0.5 kg; similarly; Jackson and Githens (1931) reported it to be 1 mg/0.454 kg. Intravenous injection of venom into cats (*Felis catus*) produces a drop in arterial blood pressure and flow; a rise in venous blood pressure, pulmonary artery pressure, and cisternal pressure; and changes in breathing, electrocardiogram, and electroencephalogram (Russell and Michaelis 1960). Intravenous injection of the monkey (*Macaca mulatta*) has caused shock that proved fatal (Fidler et al. 1938).

The minimum lethal dose of dried venom administered intracutaneously to a 350 g pigeon (*Columba livia*) is 0.9 mg, and the usual range is 0.10–0.16 mg (Githens and George 1931).

The venom is not only toxic to warm-blooded animals. A 93 mm SVL (50.5 g) regal horned lizard (*Phrynosoma solare*) was pierced by one fang of a 82 cm SVL *C. atrox* at 0928 hours; a small drop of blood immediately issued from the puncture site. Six minutes later, about 0.5 mL of clear fluid seeped from the wound, but did not spread from the wound onto the observer's hand. The lizard could only move its limbs with difficulty, had further fluid loss, and became lighter in color. At 1004 hours, pinkish fluid (presumably blood and lymph) was exuding from both nares, and the lizard was non-responsive to handling and appeared dead with its head in a strong horns-raised 90° angle (Sherbrooke and May 2008a).

Johnson et al. (1966, 1967) reported that the protozoan *Paramecium multimicro-nucleatum* has a LD_{50} of 0.62–0.73 (mean, 0.67) mg/mL; Philpott (1931) reported that the minimum lethal concentration of *C. atrox* venom needed to kill 14 different species of protozoans ranged from 0.003 to 0.00005 g/mL.

Venomous crotalids harbor many potentially infectious bacteria in their mouths, probably acquired from the feces of the prey they have ingested (Clark et al. 1993). The venom of *C. atrox* has antibacterial properties that keep the incidence of bacterial infection from a bite at a low level in humans. Its greatest antibacterial actions are against aerobic staphylococci and *Citrobacter* sp., *Enterobacter* sp., *Morganella* sp., *Proteus* sp., and *Pseudomonas aeruginosa*. Anaerobic bacteria are less affected (Talan et al. 1991).

C. atrox is not immune to the venom of its own species. Buijs (1988) reported that, when bitten by a conspecific captive, one rolled like a corkscrew, its neck remained twisted for some time, and it finally died four months later. However, death usually occurs within 28 hours of the envenomation (Nichol et al. 1933; Swanson 1946). Nichol et al. (1933) reported a fatal bite that a *C. atrox* inflicted on itself.

Human envenomation by *C. atrox* is not uncommon (Parrish 1964b); within its range in the United States and Mexico, it is responsible for more human envenomations than any other snake. A bite by this species is often very serious. With or without antivenom treatment, death may occur (Crimmins 1927b; Dart et al. 1992; Glass 1969; Hardy 1992, 1997; Russell 1983), and this species is probably responsible for more human deaths than any other snake in the United States (Juckettt and Hancox 2002).

The venom is activated by the pH and BT of the victim. It contains a myocardial depressant factor that negatively affects cardiac muscle contractility (De Mesquita et al. 1991). Symptoms reported in human envenomations include intense burning pain, tingling sensations in the bitten limb, swelling, discoloration of tissues, edema, ecchymosis, hemorrhage, necrosis, hematemesis, hemolytic anemia, lowered blood pressure, cardioarrest (lowered heart rate), increased heart rate, fever, sweating, numbness, weakness, stiffness, giddiness (and occasionally unconsciousness), convulsion, nausea and vomiting, pulmonary arrest (lowered breathing rate), airway obstruction (from a bite on the tongue), shock, and secondary gangrene infection (Burgess and Dart 1991; Crimmins 1927b; Dart et al. 1992; Ehrlich 1928; Gerkin et al. 1987; Glass 1969; Guisto 1995, Hardy 1986, 1997; Holstege et al. 1997; Hutchison 1929; Keyler 2005, Klauber 1972; La Rivers 1976; Loprinzi et al. 1983; Norris and Bush 2001; Rosen et al. 2000; Russell 1960a; Seifert et al. 1997; Werler and Dixon 2000; Young 1940). Like other *Crotalus*, *C. atrox* is capable of delivering several bites consecutively (Russell 1978).

Those working with the venom may develop hypersensitivity to the proteins involved (Zozaya and Stadelman 1930).

PREDATORS AND DEFENSE: Adult *C. atrox* have few natural predators other than humans, but neonates, juveniles, and subadults are probably eaten by a number of other carnivores. However, only predation by the bullfrog (*Rana catesbeiana*), snakes (*Drymarchon corais, Lampropeltis getula, Masticophis flagellum*, and *M. taeniatus*), the roadrunner (*Geococcyx californicus*), wild turkeys (*Meleagris gallopavo*), chickens (*Gallus gallus*), muscovy ducks (*Cairina moschata*), the caracara (*Caracara cheriway*), large hawks (*Buteo albicaudatus* and *B. jamaicensis*), hog (*Sus scrofa*), peccaries (*Pecari tajacu*), and the bobcat (*Lynx rufus*) has been documented (Blair 1954; Clarkson and deVos 1986; Crimmins 1931; Klauber 1972; Mitchell 1903; Ross 1989; Rue 2005; Shaw and Campbell 1974; Sherbrooke and Westphal 2006; Vanderpool et al. 2005; Wilson 1954). In addition, Klauber (1972) reported a case of captive cannibalism of a juvenile by another juvenile *C. atrox*.

Crimmins (1931) observed chickens (*Gallus gallus*) carrying small *C. atrox* in their bills, and after he killed one snake and chopped it up, the chickens and muscovy ducks (*Cairina moschata*) in the pen fought over the pieces. If a snake enters a pen through a chicken-wire fence and then feeds, it may become hopelessly stuck in the fence when it tries to leave (Campbell 1950). Crimmins (1931) also reported that deer (*Odocoileus virginianus*) will attack *C. atrox,* possibly in defense.

Domestic dogs (*Canis familiaris*) and coyotes (*C. latrans*) are not repulsed by anal scent gland secretions from *C. atrox*, and also are probable predators (Weldon and Fagre 1989).

A large, irate *C. atrox* is an awesome adversary. It will coil, raise its neck and head as high as 50 cm above its coils, and continually rattle its tail. One even accidentally discharged venom at its disturber (Madrid-Sotelo and Balderas-Valdivia 2008). Some individuals spray musk from anal scent glands (particularly if handled), and this secretion has been considered a possible deterrent to predators, but this may not be true regarding canids (Weldon and Fagre 1989). *C. atrox* may also hiss (usually associated with trunk inflation) at a frequency range of 400–4,700 Hz and an amplitude of 60 dB (Kinney et al. 1998). If the perceived danger continues, and the disturber get too close, the snake strikes (rarely venom is sprayed; Wheeler 1994), and sometimes even advances toward its would-be enemy, presumably to get within better striking range (Mitchell 1903). No significant differences occur in defensive strikes directed at mouse or mouse-sized models, but strikes directed at both of these are significantly different from those directed at larger models (Young et al. 2001). Collectively, there is a high level of correlation among the kinematic variables of defensive strikes, and significant differences occur between defensive and predatory strikes directed at similar size mouse targets. These suggest that *C. atrox* can modulate its strike depending on target size and behavioral context. Some *C. atrox* may hide their heads when confronted with a roadrunner (*Geococcyx californianus*) (Medica 2009).

C. atrox and other rattlesnakes produce a sustained, high-frequency warning rattle. The length of the rattle and the rate at which it is moving affect the loudness and economy (sound per unit energy input). Rattle loudness increases with rattle length up to 6–8 segments, remains relatively high until a length of 12 segments, and then decreases as segments are added; thus, rattles of 6–8 segments maximize sound output per unit energy cost (Moon and Rabatsky 2008). The SVL-adjusted rattle loudness is over 70 dB, the dominant frequency range is 5.14–6.60 kHz, and the total frequency range is 0.65–20.22 kHz (Cook et al. 1994; Fenton and Licht 1990; Young and Brown 1993). The snake can rattle continuously for hours at frequencies approaching 90 Hz (Rome et al. 1996; Schaeffer et al. 1996). Although of relatively high energy usage, which may affect female reproductive output and other essential behaviors, rattling presumably enhances the snake's survival by deterring potentially dangerous animals; this enhanced survival probably comes at a comparatively low cost (Moon 2001, 2005).

Rattling habituation, a simple form of learning indicated by a decrease in response to a stimulus, may occur. Such habituation is detrimental to the snake in restricting defensive and exploratory behavior to infrequently occurring stimulus patterns, and allowing other species-typical behaviors to occur in possibly dangerous situations in which an antipredator behavior should occur. It does, however, conserve energy. Place and Abramson (2008) tested 10 western diamondbacks in an automated apparatus at 5-minute intervals until they failed to rattle for 10 consecutive trials over 4 consecutive days, and documented both long-term and short-term habituation. Notable variation occurred between individuals and variation within and among individuals was found.

The rattling is achieved by extremely rapid contraction (8–18 milliseconds) of 6 tail-shaker muscles acting as a single motor unit. This is achieved by crossbridge cycling and recycling of elastic strain energy. Tail-shaker muscle contractions produce a mean strain of 3%, which is among the lowest strains ever recorded in vertebrate muscle during movement. The relative shortening velocities (V/V_{max}) are 0.2–0.3 m/second, and are in the optimal range for maximum power generation output. The muscles involved have only limited periods (0.2–0.5 milliseconds and 0.002–0.035%) of active lengthening, indicating little potential for elastic energy storage and recoil. The features show that high-frequency muscles primarily reduce metabolic energy input rather than recycle mechanical energy output (Moon et al. 2003; Schaeffer et al. 1996).

This process minimizes contraction energy costs. The energy expenditure per muscle twitch is 0.015 µmol ATP/muscle g. More than 33% of the ATPs are supplied through glycolysis independent of the O_2 level; the rest come from oxidative phosphorylation. Fatigue is avoided by rapid H^+ and lactate efflux resulting from blood flow rates that are among the highest known for vertebrate muscle. The volume density of the tail-shaker muscle myofiber is 32%, and is consistent with the contraction costs being minimized, in contrast to those of vertebrate locomotory muscles. High rates of rattling are achieved by minimizing contractile use of ATP, which reduces the cost per twitch to among the lowest found in striated muscle. The O_2 cost per muscle contraction is 0.139 ± 0.016 µL per gram of O_2 (Conley and Lindstedt 1996; Kemper et al. 2001; Schaeffer et al. 1996).

Most predator recognition is through vision and thermal cues, but olfaction by way of the vomeronasal system (Miller and Gutzke 1999) is also used. In addition, C. atrox apparently can recognize some airborne sounds. When presented with trial stimuli at a level of 5–10 dB above its perception threshold, the snake ceased body movements, slowed the flicking of its tongue, rapidly jerked its head, and rattled. At least one of these behaviors was observed in 92% of the trials (Young and Aguiar 2002).

If confronted with the presence or scent of ophiophagus snakes (*Drymarchon corais, Lampropeltis getula, Masticophis flagellum, Pseustes sulfureus*), C. atrox will try to flee; if this is not possible, it will either strike and bite, body-bridge, flip, freeze in place, or inflate its trunk (Bogert 1941; Inger *in* Weldon et al. 1992; Marchisin 1980). When tested by Cowles and Phelan (1958) with the odors of human perspiration, *L. getula*, and methyl mercaptan (tested twice) from the skunk *Spilogale putorius*, the snake's heart rate increased 27%, 26%, and 45% and 40%, respectively. Also, the serum corticosterone level becomes elevated when the snake is stressed (Schuett et al. 2004a).

This snake has some learning capability; it can recognize and consequently responds less often (habituates) to a certain signal when the cue is repeated frequently. Place and Abramson (2005) performed a number of experiments to determine if several components of rattling behavior exhibited habituation. Ten *C. atrox* were induced to rattle using an apparatus in which stimuli were automatically presented. The dependent variables were the presence of rattling, latency of rattling, and the duration of the rattling episode. Each individual was induced to rattle every 5 minutes until it failed to respond in 10 consecutive trials or until it reached a maximum of 120 trials. If the 10-trial criterion was met, a dishabituating stimulus was presented to rule out sensory adaptation and effector muscle fatigue. The entire procedure was repeated with each snake on four consecutive days, and individual differences in habituation occurred. In addition to response, rattling latency and duration also decreased over time. However, the rates of decrease differed between individuals and among these behavioral components. Spontaneous recovery occurred in most individuals, but not all. Some also showed retention of response habituation over days.

PARASITES AND PATHOGENS: The western diamond-backed rattlesnake has an impressive array of endoparasitic helminths: trematodes (*Ochetosoma kansense* and unidentified trematode eggs and larval cysts), cestodes (*Mesocestoides* sp. tetrathyridia; *Oochoristica gracewileyae*, *O. osheroffi*; *Proteocephalus perspicua*; and other unidentified cestodes), acanthocephalans (*Pachysentis canicola* and unidentified oligacanthorhynchid cystacanths), and nematodes (*Hexametra boddaërtii*, *Kalicephalus inermis*, *Physocephalus sexalatus*, *Physocephalus* sp., *Physaloptera* sp., unidentified oxyurid nematodes, and unidentified pinworm eggs) (Bolette 1997b; Ernst and Ernst 2006; Flores-Barroeta et al. 1961; Goldberg and Bursey 2002; Goldberg et al. 2002b; Klauber 1972; Loewen 1940; Marr 1944; McAllister et al. 2004; Sprent 1978; Stephenson and Pisani 1991). *C. atrox* also serves as a host of the parasitic liguatulid crustacean tongueworm *Porocephalus crotali* (transmitted through its rodent prey), which attaches in the oral cavity (Buckle et al. 1997; Penn 1942; Soulé and Sloan 1966; Stephenson and Pisani 1991); the lung mite *Entonyssus ewingi* (Hubbard 1939); and we have found a wild individual externally infested with the snake mite *Ophionyssus natricis*, and another with an *Amblyomma* tick attached to its dorsum. The mites *E. halli* and *E. rileyi* also infest it (Reichenbach-Klinke and Elkan 1965). In addition, the parasitic protozoan *Haemogregarina digueti* and other unidentified hemogregarines have been found in its blood (Martin and Bagby 1973; Roudabush and Coatney 1937).

Captives are prone to respiratory infections caused by the bacterium *Aeromonas hydrophila* (Murphy and Armstrong 1978). Wadsworth (1954) described a fibrosarcoma from the neck of a *C. atrox*. Hoff and Trainer (1973) did not, however, isolate any arboviruses from this snake during their survey study. A captive three-year-old male *Crotalus atrox* developed a pliable, firm enlargement of the posterior potion of its abdominal cavity that slowly increased in size until it prevented defecation (Murphy et al. 1975).

POPULATIONS: *C. atrox* is one of the most common snakes in the southwestern United States. At some sites, it is the most frequently encountered species, and is surely the most abundant venomous snake over much of that area.

In southern and western Texas *C. atrox* may be found in dense populations, although not as dense as in the past. About 3,500 individuals were collected at Floresville, Wilson County, Texas, between 1 June and 1 September 1926 (Crimmins 1927b), and a rancher killed 1,200 while clearing about 4,049 ha of cactus and brush in Shackelford County, Texas (0.3/ha) (Wood *in* Klauber 1972). Werler and Dixon (2000) reported that another Texas rancher destroyed 23 *C. atrox* per km^2, and a third killed 28 per km^2. Jones et al. (2008) found three adult females, one adult male, and one juvenile female under a single wood pile in an old farm shed in Caldwell County.

It is extremely abundant in Arizona; of 425 rattlesnakes collected by Klauber (1972), 218 (51.3%) were *C. atrox*, and Rosen and Lowe (1994) reported that 56 of 368 (15.2%) snakes observed on Rt. 85 in 1988–1991 were this snake. In New Mexico, 209 of 454 (46%) snakes collected by Price and LaPointe (1990) while road-cruising between 1975 and 1978 were *C. atrox*. During southern Arizona field trips in 1992, 1995, and 1998, Repp (1999) counted a total of 729 living snakes, of which 370 (50.8%) were *C. atrox*. Mosauer (1932a), however, based on road observations, thought rattlesnakes of all species rare in the area between Carlsbad, New Mexico, and El Paso, Texas, and that a Mr. Smith of Frijole, Texas, only killed 4 in the vicinity of his ranch during 24 years of residence. Of 158 snake captures of 13 species along the Middle Rio Grande, New Mexico, Bateman et al. (2009) recorded only six *C. atrox* (3.8%).

The numbers of *C. atrox* have increased as the semidesert grasslands in Arizona and New Mexico have undergone succession to desert scrub (Mendelson and Jennings 1992). However, Vitt and Ohmart (1978) noted that due to its high trophic position, the species exists in relatively low densities in the lower Colorado River area.

Laughlin and Wilks (1962) recommended the use of sodium pentobarbital to render the species helpless so that they could be examined, measured, and marked without danger of envenomation during population studies, but the amount of damage done to the snake must be assessed. We do not believe this a viable practice.

Probably the sex ratio does not differ significantly from 1:1. Fitch and Pisani (1993) examined 1,011 *C. atrox* collected in Oklahoma roundups and found the sex ratio was 1.4:1 in favor of males, but Boyer (1957) reported a sex ratio for 215 *C. atrox* from southwestern Oklahoma of 1:1.21 in favor of females. The neonate sex ratio is approximately 1:1 (Taylor and De Nardo 2004). Fitch and Pisani (1993) thought that first-year young may constitute up to 40% of the spring population.

Humans are still the major enemy of this snake. Motorized traffic (Campbell 1953b, 1956; Taylor 1936), habitat destruction (including by both amateur and professional snake hunters collecting the species for roundups; Warwick 1990), wanton killing (as noted previously, especially at dens), and the commercial trade in rattlesnake meat and curios have severely decimated populations in some areas. In Mexico, it is sometimes used in folk medicines (Minton and Minton 1991).

One of the worst destructive actions by humans is in the form of the popular commercial rattlesnake roundups in Oklahoma, Texas, and other southwestern states, where *C. atrox* is usually the snake most commonly captured (Campbell et al. 1989b; Fitch and Pisani 1993; Fitzgerald and Painter 2000; Painter and Fitzgerald 1998; Pisani and Fitch 1993; Weir 1992). The snakes involved are usually subjected to harsh treatment, and many are killed. Those that survive are usually sold to commercial dealers for their meat, hides, or rattles. Few are returned to the wild; those that suffer that fate are seldom if ever returned to where they were collected, and so experience a severe reduction in their chance of survival in the new environment.

Encroachment of city suburbs into the habitat of *C. atrox* has brought more of the snakes into contact with humans—a dangerous situation, especially as far as children and pets are concerned. To alleviate this, such snakes are usually killed at once, but a new practice has been tried near Tucson, Arizona, where nuisance rattlesnakes are relocated into proper habitat some distance away (Nowak 1998; Nowak et al. 2002). The success of this operation must still be suitably evaluated, as it is well known that relocated snakes have less chance of survival than resident ones.

It is a shame that because of its venomous nature, *C. atrox* is not protected by the federal governments of Mexico or the United States, or by any state in these countries. Perhaps in the future it will need our protection, as will so many other unlisted reptiles. *C. atrox* is one of the favorite exhibitions in captivity, and all major zoos in the United States, and those that we have visited in Europe, display them. It would be a shame if the species was to slowly disappear from the wild so that only these captive, relatively short-lived, and few individuals were all that remains of the U.S. "King of the Rattlesnakes."

REMARKS: Based on morphology and scalation, *C. atrox* has been thought to belong in a complex with *C. adamanteus*, *C. ruber*, and *C. tortugensis* (Brattstrom 1964a; Gloyd 1940; Klauber 1972). Results of electrophoretic studies of venom proteins by Foote and MacMahon (1977) have agreed with this arrangement, but have also showed a closer linkage to the *C. viridis* complex than shown by morphology. More recent molecular studies using sequences from mtDNA and various RNAs by Murphy et al. (2002) place *C. atrox* in a group containing *C. adamanteus*, *C. catalinensis*, *C. ruber*, and *C. tortugensis*, basically confirming the conclusions of the above studies. Of these species, *C. atrox* is genetically closest to *C. tortugensis* (Murphy et al. 2002). Rodríguez-Acosta et al. (2007) have found a common sequence of amino acids between antigen Uraconina-1 in *C. vergrandis* and a catrocollastatin precursor in *C. atrox*, possibly indicating a mutual venom evolution.

Crotalus basiliscus (Cope, 1864)
Mexican West Coast Rattlesnake
Saye

RECOGNITION: *Crotalus basiliscus* is a large (TBL$_{max}$, 204.5 cm; Klauber 1972; individuals > 150 cm are not uncommon), stout rattlesnake, exceeded in length only by *C. adamanteus* and *C. atrox*. Large individuals possess a prominent anterior vertebral keel formed by elevated vertebral neural spines. Body color varies from brownish-yellow to greenish-brown, greenish-gray, grayish-olive, or olive-green; juveniles are brick-colored, larger adults are more greenish. The dorsal body pattern consists of a series of 19–41, usually 33–35, yellow-, cream-, or white-bordered, diamond-shaped blotches, which are separated from 2–3 rows of lateral blotches; the diamonds of young individuals are lighter in the center. Greenish individuals with reddish, light-bordered diamonds are particularly attractive. Normally no dark paravertebral stripes are present on the neck. Abnormal body patterns sometimes occur, and the snake becomes darker with age, obliterating the blotches. The tail is long, 6.3–9.2% of TBL, with 4–12 (mean, 8.6) dark tail bands and a gray distal peduncle before the rattle. The top of the head is lighter in color. A dark, yellow- to cream-bordered stripe extends obliquely backward from the orbit to the corner of the mouth. The light border may be irregular or discontinuous, and the stripe fades with age. No transverse bar is present in the prefrontal area. The anterior venter is whitish, cream, or yellow with some gray pigment; posteriorly it becomes darker with gray mottling or blotches. The dorsal body scales are keeled and contain apical pits; some tuberculate projections may also be present (see Harris 2005 and Stille 1987 for detailed scale topography). Stille (1987), based on scale microdermatoglyphic characters and character state polarities, placed *C. basiliscus* with 15 species of *Crotalus* and 2 of *Sistrurus* in a cladogram group having linear ridges on the body scales, and had a second cladogram group of 5 species of *Crotalus* with tuberculate scale surfaces most closely related to the first group. The lowest three lateral scale rows are usually unkeeled, although scales in the second row may bear very low keels (Campbell and Lamar 2004). The anterior body scales are in 23–32 rows anteriorly, 24–29 rows at midbody, and 19–23 rows before the anal vent. The venter has 168–206 ventrals, 18–36 subcaudals, and an undivided anal plate. Most dorsal head scales are small, but the higher than wide rostral (in contact with the prenasals), 2 triangular internasals, 2 quadrangular prefrontals, 4 canthals, and 2 supraoculars are enlarged. The approximately 25 scales in the frontal area are smaller, the 2 small postcanthals touch, and 2–3 intersupraoculars are present. Laterally are 2 nasals (the prenasal usually contacts the first supralabial, but the postnasal is prevented from touching the supraocular by loreal scales), 2 (1–5) loreals, 2 preoculars, 2–3 postoculars, several suboculars and temporals, 13–18 supralabials, and 13–19 infralabials.

The hemipenis ends in two elongated, tapering lobes. The tip of each lobe is slightly bulbous. Small spines in adjacent clusters adorn the crotch between the lobes,

Crotalus basiliscus
(Barry Mansell)

Crotalus basiliscus
(R. D. Bartlett)

and 25–50 large spines occur on the proximal third of each lobe. Each lobe averages a total of 162 spines and 36 fringes. The sulcus spermaticus is forked (Klauber 1972).

Dentition consists of a maxillary fang, and 2 (0–1) palatine, 7–8 pterygoid, and 9–11 dentary teeth.

Adult males are smaller (TBL$_{max}$, 170.0 cm) with 168–201 (mean, 190) ventrals, 24–37 (mean, 30) subcaudals, 4–12 (mean, 8–9) dark tail bands, and TLs 9.1% of TBL. Adult females are larger (TBL$_{max}$, 204.5 cm) with 173–206 (mean, 197) ventrals, 18–30 (mean, 24–25) subcaudals, 5–9 (mean, 6–7) dark tail bands, and TLs 6.3–8.5% of TBL.

GEOGRAPHIC VARIATION: No subspecies are currently recognized (Campbell and Lamar 2004). Formerly *C. m. oaxacus* Gloyd, 1948, was considered a subspecies of *C. basiliscus*, but it was transferred to *C. molossus* by Campbell and Lamar (1989) (but see Duellman 1961).

CONFUSING SPECIES: Other species of *Crotalus* occurring in the range covered in this book can be separated from *C. basiliscus* by the above key. The hemipenis of *C. molossus* is 8–10 subcaudals long, that of *C. basiliscus* is 12–14 subcaudals long (Klauber 1972), and that of *C. molossus* is not longer than 120 cm. The smaller rattlesnakes of the genus *Sistrurus* have all of their dorsal head scales enlarged.

KARYOTYPE: Diploid chromosomes total 36; 14 body macrochromosomes (4 metacentric, 4 submetacentric, 4 subtelocentric, and 2 submetacentric-subtelocentric), 2 sex macrochromosomes (ZZ, male; ZW, female), and 20 microchromosomes. The Z chromosome is metacentric-submetacentric, and the W chromosome is submetacentric (Baker et al. 1972; Cole 1990).

FOSSIL RECORD: No fossils have been reported.

DISTRIBUTION: *Crotalus basiliscus* is restricted to western Mexico, where it ranges from western Chihuahua (Lemos-Espinal et al. 2004) and the Río Fuerte watershed in southern Sonora (Bogert and Oliver 1945) southward through Sinaloa (Dixon et al. 1962; Fugler and Dixon 1961; Hardy and McDiarmid 1969; Smith and Van Gelder 1955), Nayarit (Zweifel 1959), Jalisco (Cruz-Sáenz et al. 2009; Davis and Dixon 1957), and Colima (Duellman 1958) to northwestern Michoacán (Duellman 1961, 1965; Schmidt and Shannon 1947) at elevations from near sea level to 2,400 m, but mostly below 600–1,000 m.

HABITAT: The snake normally inhabits arid tropical thorn scrub and forest, but has also been collected or observed in tropical semideciduous and deciduous forest, the transition zone between the deciduous forest and humid pine-oak forest, and, in the Sierra Coalcomán of Michoacán, pine-forest proper (Armstrong and Murphy 1979; Campbell and Lamar 2004; Duellman 1965; Klauber 1972). Rock slides and talus slopes are often utilized in such habitats. For an extensive list of associated plants, see Armstrong and Murphy (1979), Duellman (1965), and Klauber (1972).

BEHAVIOR AND ECOLOGY: Although *C. basiliscus* is common in many parts of its range, few aspects of its life history have been adequately studied; most data are anecdotal. It is active in both the wet and dry seasons, but more so during the wet summer months (Campbell and Lamar 2004; Hardy and McDiarmid 1969). It is usually encountered at night while crossing roads, but both neonates and adults are known to bask during the day, particularly in the morning (Campbell and Lamar 2004; Hardy and McDiarmid 1969), while others remain motionless in the shade or in some other shelter. Mehrtens (1987) reported that it is crepuscular, but that it becomes more nocturnal during the warm, summer months. The snake is not afraid of water, and holds its head and neck as much as 30 cm above the surface as it swims with a lateral-undulatory motion. It can also easily climb a 15–20 cm wall when ascending from the water (Klauber 1972).

Males engage in combat behavior during the mating season. The bout is vigorous

Distribution of
Crotalus basiliscus.

Crotalus basiliscus
(Minas Nuevas,
Sonora; Eric
Dugan)

with one male the aggressor and the other the attacked. Combat behavior may last for hours (Lowe 1948; Perkins *in* Davis 1936; Shaw 1948).

REPRODUCTION: The reproductive biology is not well known. Perkins (1943) observed captives of both sexes engaging in reproductive behavior at an age of 2 years, 1 month, and 20 days (no lengths given); Marcy (1945) reported that a 155 cm female gave birth to a litter in captivity. Mehrtens (1987) thought the average length at maturity is 122–152 cm.

Goldberg et al. (2005) examined the reproductive condition of 44 preserved *C. basiliscus* (6 females, 80.5–128 cm SVL; 6 males, 76–122.5 cm SVL; and 32 juveniles, 28.5–39.5 cm SVL). The seminiferous tubules of a 98 cm male collected on 20 April were regressed with spermatogonia, Sertoli cells, and a few clusters of sperm left over from the previous year; spermatozoa were present in the vas deferens. In agreement, the seminiferous tubules of a 122.5 cm male collected in May were also regressed. One 76 cm long *C. basiliscus* collected 3 July had regressed seminiferous tubules containing spermatogonia and Sertoli cells. The tubules of a 100.5 cm male collected on 1 July were in recrudescence with renewal of germinal epithelium occurring, and primary and secondary spermatocytes and spermatids present, but no mature spermatozoa; mature sperm were present in the vas deferens. Two December males (86.6 and 98.5 cm) were undergoing spermatogenesis, with the lumina of the seminiferous tubules lined with spermatozoa and rows of metamorphosing spermatids, and the vas deferens packed with spermatozoa. These few data suggest that development of the male gametes in this species follows the same cycle as that of U.S. *Crotalus* (Aldridge and Duvall 2002).

Five 80.5–125.8 cm SVL females, collected on 20 April, 10 June, 16 and 18 July, and 22 August, contained inactive ovaries with small follicles not undergoing vitellogenesis. The sixth 128 cm SVL female collected on 27 July contained 12 oviductal eggs (Goldberg et al. 2005). Apparently, not all females reproduce each year; reproduction is most likely controlled by prey availability, as in some U.S. pitvipers. Wiley (1930) described a dicephalic male embryo.

Perkins (1943) observed what he thought was courtship activity in 2-year-old captive *C. basiliscus*. The female had recently shed. At the time the activity was first noticed at 1000 hours, 28 January, the pair was crawling about the cage. At 1150 hours the female was lying quietly with the posterior part of the male's body draped across hers. He was making spastic movements with the anterior half of his body, flicking his tongue along the anterior third of her body, and constantly nudging her forcibly with his chin. His tail entwined hers and they remained in this position throughout the day. About every four minutes the male convulsively jerked with his tail, and once positioned his head and neck as if to strike, remained in this position about a minute, and then resumed nudging the female; but no copulation occurred. Two matings occurred during the next two Augusts; in one of the matings, the male made slight tail movements and was active with the anterior part of his body in one, and in the other he made "breathing" movements with his tail. Perkins (*in* Davis 1936) also reported that a male rubbed the side of his head along the female's side. However, because no copulation was observed during any of these episodes, Lowe (1948) thought they represented male combat behavior rather than heterosexual courtship. Ramírez-Bautista (1994) noted that mating occurred in June in the year prior to parturition, so viable sperm appears to be stored for later fertilization.

The medial-placed caruncle on the upper jaw of the newborn is curved upward, and may be used to slit the embryonic membranes during parturition (Klauber 1972). Birth dates of captives or those in which neonates have been found in the wild were in June (27) and July (7, 10, and 27) during the summer wet season, but newborns have been found through late September (Hardy and McDiarmid 1969; Marcy 1945; Perkins 1943). The data concerning a litter of 20 by Hardy and McDiarmid (1969) found near a burrow probably indicate that young *C. basiliscus* may remain together immediately after parturition, perhaps until completion of their first ecdysis. It is possible that the female was nearby or in the burrow. Marcy (1945) reported ecdysis on the fourth day after parturition.

Seventeen reported litters have contained 12–60 young, and averaged 30.7 (Fitch 1970; Goldberg et al. 2005; Hardy and McDiarmid 1969; Klauber 1972; Marcy 1945; Ramírez-Bautista 1994). The litter of 29 reported by Marcy (1945) also contained 4 atrophied yolked eggs. Neonates are reddish-brown with light-bordered blotches with lighter centers. A transverse band may extend across the posterior half of the canthals and anterior edges of the supraoculars. The mean SVL of 9 neonates reported in the literature was 31.0 (range, 28.5–33.2) cm. Wiley (1930) reported the TBLs of 30 near-term embryos were 17.78–23.02 (mean, 21.30) cm; 17 males: 17.78–22.86 (mean, 21.38) cm, 13 females: 19.53–23.02 (mean, 21.18) cm. TLs of these 30 embryos were 1.90–2.70 (mean, 2.32) cm; males: 2.22–2.70 (mean, 2.49) cm, females: 1.90–2.22 (mean, 2.06) cm.

GROWTH AND LONGEVITY: Growth data are scarce. Twenty-eight young born on 7 July that had been fed pinky mice and hairless baby rats from 14 July were 34.0–39.6 (mean, 36.6) cm long on 1 August of the same year. On 30 August they were measured again and had grown to 37.5–45.0 (mean, 40.4) cm, roughly an average growth of 8.9 cm in 55 days (Marcy 1945). Gloyd (1940) reported that young of the year had TBLs ranging from 30.7 to 36.5 cm—12 males, 31.0–36.5 (mean, 32.4) cm; 14 females, 30.7–32.4 (mean, 31.4) cm.

The 3 survival records reported by Snider and Bowler (1992) and Shaw (1969) were, respectively, 16 years and 14 days for a female of unknown origin at the Fort Worth Zoo; 12 years, 10 months, and 24 days by a female at the Columbus Zoo that had been adult when wild-caught; and 10 years and 8 months at the San Diego Zoo.

DIET AND FEEDING BEHAVIOR: Mammals are probably the primary prey, as seven specimens examined by Klauber (1972) contained mammal, probably rodent, hair. Mehrtens (1987) thought that mammals and birds constituted the diet. In trials, the Mexican beaded lizard, *Heloderma horridum*, recognized and tried to avoid *C. basiliscus* (Balderas-Valdivia and Ramírez-Bautista 2005); perhaps the snake preys on the eggs or young of the lizard. No other data exist. Captives take all stages of house mice (*Mus musculus*) and brown rats (*Rattus norvegicus*), depending on the snake's size.

When exploring with the tongue, it is extended straight out for its full length and held still for several seconds. Then the tips are moved slowly up and down 2–3 times. The tongue is then withdrawn, and again protruded to repeat the behavioral sequence (Wiley 1930).

VENOM DELIVERY SYSTEM: Five *C. basiliscus* with body lengths of 151.5–175.0 cm measured by Klauber (1939) had FLs of 11.7–14.2 (mean, 12.7) mm and mean values for TBL/FL of 126% and HL/FL of 4.97%; the angle of fang curvature was 62°.

VENOM AND BITES: The average total dry venom yield of *C. basiliscus* is 297 mg (Altimari 1998; Brown 1973); Klauber (1972) reported an average of 277 mg for 3 adults.

The venom is highly hemolytic, with possibly the highest anticoagulant activity known for a crotalid snake (Datta et al. 1995). However, it does contain a crotoxin-like presynaptic neurotoxin (Chen et al. 2004). Do Amaral (1929d) commented that the amount of lipid-like constituents is greater than and the yellow pigmentation more accentuated than those of some venom samples from *C. durissus*. The venom contains two hemorrhagic toxins, a bradykinin-releasing kallikrin-like enzyme, three fibrinogenic lysing enzymes (including the unique basilase), and myotoxin α-like proteins (> 2.0 mg/mL). Basilase hydrolyses fibrin directly without activation of plasminogen. The venom exhibits strong protease activity, and can hydrolyze DNA. Papers discussing the biochemical structure of and physiological responses to the venom of *C. basiliscus* include Beasley et al. (1993), Bober et al. (1988), Chen et al. (2004), Datta et al. (1995), de Roodt et al. (2003), Gaffin et al. (1995), Henderson and Bieber (1986), Johnson (1967), Markland and Perdon (1986), Molina et al. (1990), Oshima et al. (1969), Rael et al. (1986), Retzios and Markland (1992, 1994), Shams et al. (1995), Sifford et al. (1996), Stegall et al. (1994), Svoboda et al. (1995), Tu (1977), and Werman (2008).

The LD_{50} for laboratory mice is 2.8 mg/g subcutaneous (Khole 1991; Tu 1977). Githens and Wolff (1939) reported the minimum lethal venom dose for a 20 g laboratory mouse is 0.08 mg and 0.011 mg/g (Criley *in* Klauber 1972). Schoettler (1951) reported a minimum lethal dried venom subcutaneous dose of 0.09 (mean, 2.78) mg/g for mice. Injections of the two venom hemorrhagic toxins into mice caused hemorrhaging and hydrolyzation of hide powder azure, casein, and collagen, and resulted in several types of fibrin chains (Molina et al. 1990). When the kallikrin-like enzyme was injected intravenously into a rabbit (*Orytolagus cuniculus*), it experienced a 7 mm Hg drop in systolic pressure (Gaffin et al. 1995). The Virginia opossum (*Didelphis virginiana*) is immune to the venom of *C. basiliscus* (Pérez and Sánchez 1999). The estimated median LD_{50} for a 350 g pigeon (*Columba livia*) is 0.08 mg (Klauber 1972).

Neither the symptoms nor case histories of human envenomation by *C. basiliscus* have been published, but do Amaral (1929d) reported that the bright yellow portion of the venom causes a slight swelling followed by blood suffusion, ecchymosis, and partial necrosis of the tissues when injected into laboratory animals. Although the human LD_{50} is unknown, an untreated bite by this large snake could probably cause death.

PREDATORS AND DEFENSE: The only record of predation on *C. basiliscus* that we have found is that by an indigo snake, *Drymarchon corais*, in Sonora (Dugan et al. 2006). However, this snake, like all others, probably has other species of large snakes and a variety of large predatory birds and mammals that prey on at least smaller individuals.

C. basiliscus is rather mild mannered and calm for a rattlesnake, and in this regard resembles the eastern diamond-backed rattlesnake, *C. adamanteus*. Wiley (1930) reported that it could be held and even stroked. However, we cannot emphasize enough that this should never be done! The snake is still very dangerous! It is most nervous and irritable when shedding and extra caution should be used when approaching one at such times. If a disturber is detected at a distance, the snake may try to flee or enter a nearby shelter, but many remain motionless with only their tongue flicking. If severely provoked, it will coil, rattle, and strike.

PARASITES AND PATHOGENS: *C. basiliscus* is a host for the microfilariae of the nematode *Macdonaldius oschei*, the nematode *Hexametra boddaertii*, an oligacanthorhynchid acanthocephalan larva, and the linguatulid tongue worms *Porocephalus basiliscus* and *P. crotali* (Pentastomida) (Goldberg et al. 2006; Klauber 1972; Riley and Self 1979; Telford 1965). In addition, the paramyxovirus has been isolated from this rattlesnake, and *C. basiliscus* was the major species affected in a die-off of pitvipers caused by this virus in a collection in the United States (Jacobson and Gaskin 1992). The bacterium *Arizona hinshawii* was found in this snake during a deadly zoo epidemic (Murphy and Armstrong 1978).

POPULATIONS: In the past, the species has been reported to be "common" or "not uncommon" and "endangered" in Jalisco (Cruz-Sáenz et al. 2009, Davis and Dixon 1957), Michoacan (Peters 1954), and Sinaloa (Davis and Dixon 1957; Hardy and McDiarmid 1969), but apparently scarce in Nayarit (Zweifel 1959); Armstrong and Murphy (1979) thought it common in tropical thorn forests, and to a lesser degree in tropical deciduous forests, along the entire Mexican west coast. Because no population surveys have been carried out since, it is not known how the snake is surviving today.

Data concerning population dynamics are few. Hardy and McDiarmid (1969) collected 9 males, 7 females, and 4 juveniles in Sinaloa, a 1.29:1.00 sex ratio and a 0.25:1.00 juvenile to adult ratio, but Fugler and Dixon (1961) collected only 2 males and 3 females in that state.

Some *C. basiliscus* are killed on the roads each year (Zweifel 1959), and habitat alteration is probably occurring within the snake's range. Studies of its status and dynamics are needed before an adequate conservation strategy can be formulated.

REMARKS: *C. basiliscus* has a close relationship with *C. molossus*, particularly where their ranges overlap in Sinaloa (Hardy and McDiarmid 1969). Murphy et al. (2002)

place it in the *C. durissus*-group along with the species *durissus, enyo, molossus, unicolor,* and *vegrandis.* See *C. molossus* for further discussion.

The taxonomy of the species was reviewed by Campbell and Lamar (2004), Gloyd (1940), Klauber (1972), McCranie (1981), and Murphy et al. (2002).

Crotalus catalinensis Cliff, 1954
Santa Catalina Island Rattleless Rattlesnake
Cascabel de la Isla Santa Catalina

RECOGNITION: The Santa Catalina rattleless rattlesnake achieves an SVL_{max} of 75.3 cm (Goldberg and Beaman 2003a), although most individuals are less than 65 cm long. The body is grayish-brown, tan, or brownish with a series of 34–40 brown-, darker brown-, or black-bordered white- to cream-edged blotches that are narrowly separated by 1–2 mid-dorsal scale rows. These are irregularly quadrangular anteriorly, but become more rhomboid or hexagonal toward midbody, and tend to fade toward indistinction posteriorly. Along the sides is a large row with two smaller flanking rows of darker brown blotches. Dorsal body scales are keeled and pitted (linear ridges; Harris 2005; Stille 1987); they occur in rows as follows: 23 (21–23) anterior, 25 (24–26) midbody, and 22 (20–23) near the vent. A series of 5–6 dark brown to black, 4–5 scales wide bands cross the dorsum of the tail; the first rattle segment of the tail is dark brown or black. The venter is whitish, cream, or grayish, with 177–189 ventrals, 18–28 subcaudals, and an undivided anal plate. Dorsally, the head is pale grayish-brown, sometimes with small darker brown markings. A wide, light-bordered (broad but not well-defined), brown stripe extends obliquely backward from the supraocular scales to the corner of the mouth. The supraoculars are crossed by a broad pale line. Dorsal scales consist of a large medial rostral, 2 pairs of canthals (the anterior touch and are smaller than the posterior ones, and the posterior pair prevents the postnasals from

Crotalus catalinensis (Santa Catalina Island, Baja California Sur; Arizona-Sonora Desert Museum, John H. Tashjian)

Crotalus catalinensis
(Santa Catalina
Island, Baja
California Sur;
San Diego Zoo,
John H. Tashjian)

contacting the upper preoculars), 4 kidney-shaped internasals (the first pair are longer than wide), > 8 scales between the internasals and prefrontals, 2 large supraoculars, and 4–5 intersupraoculars. Two nasals (the prenasal is in broad contact with the first supralabial), 2 loreals, 2 preoculars, 2–3 (4) suboculars, 1 interoculabial, 2 (3) post-oculars, several temporals, 14–15 (13–16) supralabials, and 14–15 (13–16) infralabials compose the lateral head scales. The chin has a large mental, and two pairs of chin shields (the second pair is the largest, and both pairs touch at the midline). Neither the hemipenis nor the dentition has been described.

Males (SVL_{max}, 75 cm; Goldberg and Beaman 2003a) are only slightly shorter than females, and have 177–181 ventrals and 24–28 subcaudals. Females (SVL_{max}, 75.3 cm) have 182–189 ventrals and 18–23 subcaudals.

One of the most interesting characteristics of this species is its poorly developed tail rattle, consisting of only one segment (Beaman and Wong 2001; Campbell and Lamar 2004; Klauber 1972; Rabatsky 2006; Radcliffe and Maslin 1975; Rubio 1998; Shaw 1964a). The lobes of the rattle segment are reduced and weak, so that it is lost with each ecdysis and the rattle does not add segments as in other *Crotalus*. See "Predators and Defense" for a discussion of the evolution of rattle loss in this species.

GEOGRAPHIC VARIATION: None.

CONFUSING SPECIES: *C. catalinensis* is the only rattlesnake living on Santa Catalina Island, and the only one that normally has less than two rattle segments. Other North American *Crotalus* can be distinguished by using the above key. Small rattlesnakes of the genus *Sistrurus* have large plate-like scales on the top of their heads. Other nonvenomous and venomous snakes lack even the first rattle segment.

KARYOTYPE: The diploid chromosome total is 36; 16 macrochromosomes (4 metacentric, 8 submetacentric, and 4 telocentric) and 20 microchromosomes. Sex determination is ZZ in males and ZW in females; some variation occurs in the W chromosome

Distribution of
Crotalus catalinensis.

(Stewart and Morafka 1989; Stewart et al. 1990). A photograph of the karyotype of a female is presented in Stewart et al. (1990).

FOSSIL RECORD: No fossils have been found.

DISTRIBUTION: *C. catalinensis* is only found in the Gulf of California on Santa Catalina Island, Baja California Sur.

HABITAT: Santa Catalina Island is xeric and rocky with scattered cacti (*Ferocactus diguetii*, *Pachycereus pringlei*), brush, and trees (*Bursera hindsiana*, *Cercidium microphyllum*, *Ebenopsis confinis*). Cavities beneath rocks, fallen cacti, and plant thickets serve as daily retreats, but the snake can usually be seen from above (Avila-Villegas 2006).

BEHAVIOR AND ECOLOGY: This is one of the poorest studied species of rattlesnakes; few life history data are available.

Grismer (2002) noted an annual activity period lasting from March through October, but this may reflect the timing of his collecting trips as Avila-Villegas (2006) found 28 surface-active individuals during November–February, so it is apparently active during the entire year. Four of the snakes (2 males, 1 female, 1 undetermined sex) were found during the day (1200–1600 hours), and 24 (5 males, 9 females, 10 undetermined) were active at night (1900–2400 hours). When first encountered, 14 were coiled (including the diel observations) and the other 14 were crawling. BTs of 26 snakes recorded with an infrared thermometer were 14.2–30.2 (mean, 18.7) °C; ATs at the point of capture were 18.0–28.0 (mean, 21.3; n = 18) °C. Winter nocturnal capture rates approximated those recorded during the 2004 July–October rainy season.

Three females with SVLs of 63, 60, and 38 cm moved 21.9, 31.8, and 16.3 m, respectively, between captures in one day. A 67 cm female and a 70 cm male only moved around a tree from their original capture rock the first day, and then separated, with the female moving an additional 22 m. A 62 cm male moved only 0.5 m the first day.

After 2 days the 63 cm female had moved at least 88 m, and the 67 cm female had moved 44 m. The movements were generally linear, passing under or over vegetation (Avila-Villegas 2006).

This snake commonly forages off the ground in bushes to a height of 0.6 m, and possibly also in low trees (Avila-Villegas 2006, 2008; Avila-Villegas et al. 2004; Bogert *in* Rubio 1998; Grismer 2002). Seventeen of 45 individuals found by Avila-Villegas (2008) during the dry season and 3 of 159 during the wet season of 2003–2004 were climbing through vegetation. Both sexes were involved (of those sexed, 12 were males and 8 females); mean height above the soil surface was 36 (range, 10–75) cm. The snake may regularly travel through vegetation; Avila-Villegas (2006) observed the snake traveling as much as 3 m in a straight line through plants at heights of 0.6 m.

REPRODUCTION: Reproductive data for this species are scanty. Both sexes are most likely mature at an SVL of 60 cm. Cliff (1954) collected an apparently mature female in March that had not ovulated, but did not report the condition of her follicles. Females have been found gravid from mid-July to early August, with one containing five early embryos, and another collected in early August giving birth to two neonates (Grismer 2002). Goldberg and Beaman (2003a) reported the following observations on the sexual cycles of six *C. catalinensis*. Three females collected 27 April (SVL, 75.3 cm), 24 June (56 cm), and 25 June (54 cm) did not exhibit vitellogenesis, but another 66.2 cm SVL female collected on 19 April contained 4 12–16 mm oviductal eggs. Apparently not all females reproduce annually. The testis of a 69.8 cm SVL male collected on 19 April was undergoing recrudescence; Sertoli cells, spermatogonia, and primary spermatocytes were present, and the vas deferens still contained sperm from a previous spermiogenesis. Another 75 cm SVL male collected on 27 August was undergoing spermiogenesis with clusters of metamorphosing spermatids lining the seminiferous tubules and sperm in the vas deferens. The male cycle is probably like that of U.S. species of *Crotalus* (Aldridge and Duvall 2002).

Crotalus catalinensis (John Ottley, Valley Center, California; John H. Tashjian)

Armstrong and Murphy (1979) described the courtship as involving head-bobbing and tongue-flicking. Copulation has occurred in captivity on 15 January (Armstrong and Murphy 1979). A wild male (SVL, 70 cm) and a wild female (SVL, 67 cm) were found coiled together under a rock at the base of a palo verde tree during 26–27 January, but no copulation was observed. The couple moved around the tree, where they were found still together the following night (Avila-Villegas 2006). Avila-Villegas (2006) thought that winter movements may be largely due to mating activities.

Parturition has occurred in August, September, and October, and the young are surrounded by a fetal membrane (Armstrong and Murphy 1979; Grismer 2002; Shaw 1964b). Neonates are more brightly colored than adults. Shaw (1964b) commented that the prebutton of the 1 October neonate was perfectly normal, and not shrunken and misshapen as in adults.

GROWTH AND LONGEVITY: No growth data are available, but *C. catalinensis* is proportionally shorter than rattlesnakes from the adjacent Mexican mainland (Case 1978). A female, adult when caught, survived an additional 10 years, 3 months, and 22 days at the San Diego Zoo (Snider and Bowler 1992).

DIET AND FEEDING BEHAVIOR: Reported natural prey are lizards (*Dipsosaurus catalinensis, Sceloporus lineatulus, Uta squamata*), birds (*Amphispiza bilineata*), and mice (*Peromyscus slevini*) (Avila-Villegas and Arnaud 2004; Avila-Villegas et al. 2004, 2005, 2007; Bogert *in* Rubio 1998; Grismer 2002). Captives will take house mice (*Mus musculus*) (Chiszar et al. 1978a).

Examination of stomachs and scats of wild *C. catalinensis* caught during the dry (March–June) and wet (July–October) summer seasons from 2003 to 2004 yielded prey data. Only 11 (9.2% of 120 snakes) had prey in their stomachs. Forty-six (37.7%) of 112 palpated individuals showed feces; of these 38 (82.6%) had identifiable remains. Mammals comprised 74% and reptiles 26% of the records (Avila-Villegas and Arnaud

Crotalus catalinensis (Santa Catalina Island; Ali M. Rabatsky and Brad Moon)

2004). Remains consistent with the endemic *Dipsosaurus catalinensis* were found in 13 (15.7%) of 83 scats of *C. catalinensis* examined by Avila-Villegas et al. (2005), but only 2 (11.7%; 25 August, 18 October) of 16 snakes examined by Avila-Villegas et al. (2004) contained *Peromyscus slevini*. One 58 cm SVL snake had tried to swallow a 13 cm SVL, 25 cm TL *Dipsosaurus*, which became lodged in its mouth and esophagus and caused its death (Avila-Villegas et al. 2005).

Avila-Villegas et al. (2007) summarized the feeding data (including those from the above reports) collected during nine visits to Santa Catalina Island from 2002 to 2004. *Peromyscus slevini*, the only mammal eaten, occurred in the stomachs or feces of 70 (70.7%) snakes; 5 and 6 in the stomachs during the dry and wet seasons, respectively; and 18 and 41 in feces during the same respective seasons. Forty-seven (67.1%) of the mice were taken during the wet season when that species was the most important food source. Lizard remains were recovered from 29 (29.3%) snakes; 2 stomach and 4 fecal during the dry season, and 3 stomach and 20 fecal during the wet season. The breakdown for the 3 lizards consumed was *Dipsosaurus catalinensis* (16, 16.2% of all records, 55.2% of lizard records)—dry season, 5; wet season, 11; *Uta squamata* (9, 9.1%, 31.0%)—dry season, 1; wet season, 8; *Sceloporus lineatulis* (4, 4.0%, 13.8%)—wet season, 4. The snakes seem to be more active, or at least more interested in feeding, during the wet season (but see "Reproduction"); perhaps prey is more readily available then. Interestingly, none of these reports mention consumption of birds; perhaps the snake is not as great an avian predator as suspected.

C. catalinensis is an arboreal forager (Bogert *in* Rubio 1998; Grismer 2002); the prey lizards and *Peromyscus slevini* mentioned previously are climbers.

During chemosensory searching experiments conducted by Chiszar et al. (1978a), this species had a latent period of more than 200 seconds when a second mouse was introduced directly after a first was struck, and another latent period of approximately the same length it took to strike the first mouse after it had been swallowed.

VENOM DELIVERY SYSTEM: Specific data on the fangs and venom delivery apparatus of *C. catalinensis* have not been reported, although Rubio (1998) stated that its "fangs are proportionately longer than those of any other rattlesnake, an adaptation for piercing the fluffy feathers of a bird's body."

VENOM AND BITES: As with other aspects of the biology of *C. catalinensis*, its venom has not been adequately studied. The maximum venom yield is 32.6 mg (Altimari 1998; Glenn and Straight 1982). It contains myotoxin α proteins (1 mg/mL of venom), but has relatively small amounts of Mojave toxin (1.3 mg/mL; Bober et al. 1988; Henderson and Bieber 1986). Glenn and Straight (1985b) reported the following enzyme activities for venom from 5 adults and a juvenile: protease—adult, 68–118 (mean, 96); juvenile, 92; esterase—adult, 41–126 (mean, 88); juvenile, 73; and phophodiesterase—adult, 48–164 (mean, 110). Other venom chemistry is discussed in Brown (1973).

The mouse intraperitoneal and intramuscular dry venom LD_{50} values are 2.9–4.1 mg/kg and 10.9 mg/kg, respectively (Glenn and Straight 1982, 1985b). The human LD_{50} is unknown, but the venom must be dangerous.

PREDATORS AND DEFENSE: No observations on predation have been reported for this species.

The rattlesnake rattle is used purely as a warning device against large predators or herbivorous mammals that may trample the snake. There are no such animals on Santa Catalina Island, so selective defensive pressures on the genotype of *C. catalinensis* have not favored the retention of the rattle. Components of the rattling system are already being lost, some more rapidly in some species than in others (Rabatsky 2005). Greene (1997) thought that, as this animal ascends into bushes to hunt, the tail rattle clicking against the vegetation could alert prey; this may have been another pressure for rattle loss to increase foraging efficiency. However, this may not be the reason for rattle loss. There are few records of *C. catalinensis* preying on birds; most prey are mice, which can probably be better caught on the ground. The snake ascends into vegetation less often than supposed (only 7.6% of observations by Martins et al. 2008). Klauber (1972), however, proposed that the rattle loss might have been caused by a defective gene that persisted in the offspring and produced a mutation (genetic drift?). Irregardless, the snake still vibrates its tail when alarmed.

PARASITES AND PATHOGENS: *C. catalinensis* is a host of the parasitic linguatulid tongue-worm *Porocephalus crotali* (Pentastomida) (S. R. Goldberg et al. 2003a). Murphy and Armstrong (1978) reported the successful treatment of a captive that had pneumonia caused by the bacterium *Aeromonas hydrophila*, and Grupka et al. (2006) reported *Salmonella* bacteria in *C. catalinensis*.

POPULATIONS: Avila-Villegas (2006) reported the capture of 38 *C. catalinensis* during the 2004 rainy season (30 person hours, 1.2 snakes/person hour). Of 17 collected for food studies by Avila-Villegas et al. (2004), 10 were adult males, 6 were adult females (1.67:1.00), and 1 was a neonate (neonate to adult ratio, 0.059:1:00).

Because it is endemic to only one island, the species is more vulnerable to extinction than a snake with a broader distribution (Greene and Campbell 1992).

REMARKS: Klauber (1972) and Murphy et al. (2002) using morphology and DNA data placed the species in the *Crotalus atrox*-group, along with *C. atrox*, *C. adamanteus*, *C. ruber*, and *C. tortugensis*. On the basis of biogeography, morphology, and biochemical data, *Crotalus ruber* is the closest living relative of *C. catalinensis*. Allozyme data indicate that it shares its most recent ancestor with *C. ruber*, and that the snake is of more recent origin than many of the other reptiles found on Santa Catalina Island (Murphy and Crabtree 1985a). Stille (1987) placed the species in a clade with 16 other species of *Crotalus*, including *C. ruber*, and 2 of *Sistrurus* that have linear ridge scale microdermatoglyphics.

The taxonomy and literature of *C. catalinensis* were reviewed by Beaman and Wong (2001) and Campbell and Lamar (2004).

Crotalus cerastes, Hallowell, 1854
Sidewinder
Vibora Cornuda

RECOGNITION: *C. cerastes* is a relatively small, thick-bodied rattlesnake; TBL_{max} is 82.4 cm (Campbell and Lamar 2004), but most adults are 50–60 cm long. Its pale body color ranges from pinkish, cream, or tan to gray. Individuals living in Pinacote lava habitats are not noticeably darker than sidewinders from light-colored deserts, as are some other rattlesnake species (Smith and Hensley 1958). Present is a series of 28–47 (mean, 36), broader than long, darker tan, yellowish-brown, orangish, or gray dorsal blotches with irregular borders. Also present alternating with the dorsal body blotches are three longitudinal rows of dark spots (often faded) on each side. The tail has 2–7 dark cross bands, the last of which is usually darkest, and at least some dark pigment occurs on the basal rattle segment. The keeled body scales occur in 21–22 (19–23) anterior rows, 21–23 (19–25) rows at midbody, and 19–21 (18–22) posterior rows. A strongly keeled, tuberculate spinal ridge is present. The dorsal microdermoglyphic pattern of the body scales is one of swirling linear ridges (vermiculate), and is illustrated in both Harris (2005) and Stille (1987). The venter is white to cream, and may have some dark lateral pigmentation. On it are 132–154 ventrals, 13–27 subcaudals, and an undivided anal plate. The head pattern consists of a light-bordered, dark stripe on each side extending diagonally backward from the orbit to just above the corner of the mouth, some dark spots or streaks on the dorsal surface of the head behind the eyes, two broad dark blotches on the occipital region, and a light longitudinal bar crossing the supraocular scale; a small dark supralabial spot is often present below the orbit. Some faded dark pigment may also occur on the infralabials. The most conspicuous feature of the head is its enlarged, pointed, horn-like projections (supraocular scales) dorsal to the orbits. Other dorsal head scales include a broader than long rostral, 2 moderate internasals touching the rostral, and 12–34 scales before the 4–6 intersupraoculars. No prefrontals are present. Laterally are 2 nasals that are in contact dorsal to the naris (the prenasal touches the supralabials, but the postnasal is prevented from contacting the upper preocular by a loreal scale), 1 (2) loreals (the loreal extends to the orbit splitting the preoculars), 2 (3) preoculars, 2 postoculars, 2–3 suboculars, 2 interoculabials, 12–13 (10–15) supralabials, and 12–13 (10–17) infralabials. Beneath is a mental scale followed by two pairs of chin shields; the posterior pair is the largest, and meets at the midline.

Cohen and Myres (1970) suggested that the snake's supraocular horns function as eyelids that protect the eyes while it moves through burrows entangled with such obstructions as creosote roots, rocks, and gravel, which could abrade the eyes' covering scales.

The bilobed hemipenis has a divided sulcus spermaticus, about 54 spines and 21 fringes on each of the short, thick lobes, and numerous spines between the lobes (Klauber 1972).

The dental formula consists of 3 (2–4) palatine, 8 (7–9) pterygoid, and 9–10 (8–11) dentary teeth (Klauber 1972).

Males (TBL_{max}, 66.7; Stephen M. Secor, personal communication) have 132–151 (mean, 142) ventrals, 17–27 (mean, 21) subcaudals, and 3–7 (mean, 5) dark tail bands. Females (TBL_{max}, 82.4 cm) have 135–154 (mean, 144) ventrals, 13–21 (mean, 17) subcaudals, and 2–7 (mean, 4) dark tail bands. According to Klauber (1972), females are about 10% longer than males. However, Stephen Secor (personal communication) has never noticed a distinct sexual size dimorphism in the species; of the 11 largest individuals measuring more than 55 cm that he captured, 5 were males and 6 females.

GEOGRAPHIC VARIATION: Three subspecies are recognized. *Crotalus cerastes cerastes*, Hallowell, 1854, the Mojave Desert sidewinder, is restricted to the Mojave Desert of Mono, Inyo, Kern, Los Angeles, and San Bernardino counties, California, and adjacent

Mojave County, Arizona, southern Nevada (Esmeralde, Ivye, Lincoln, and Clark counties), and Washington County in extreme southwestern Utah. It has a brown proximal rattle-matrix lobe; usually fewer than 141 ventrals in males and fewer than 144 in females; 22 subcaudals in males, 17 or more in females; and 21 midbody dorsal scale rows. *Crotalus c. cercobombus*, Savage and Cliff, 1953, the Sonoran sidewinder or vibora cornuda de Sonora, occurs from south-central Arizona (eastern Yuma, Maricopa, Pinal, and Pima counties) southward to western Sonora and Isla Tiburón in the Gulf of California. It has a black proximal rattle-matrix lobe; usually 141 or fewer ventrals in males, 145 or fewer in females; 20 subcaudals in males, 16 in females; and 21 midbody dorsal scale rows. *Crotalus c. laterorepens*, Klauber 1944, the Colorado Desert sidewinder or vibora cornuda del Desierto de Colorado, ranges in the Colorado Desert from southeastern California and adjacent southwestern Arizona south into northwestern Baja California Norte, and the northwestern panhandle of Sonora. It has a black proximal rattle-matrix lobe; usually 142 or more ventrals in males, and 146 or more in females; 21–22 subcaudals in males, and 17 in females; and 23 midbody dorsal scale rows. Unfortunately, the above scale counts are not adequate characters for differentiating the subspecies (see table insert *in* Klauber 1972), and in some individuals, tail color is also a questionable character. A thorough review is needed of the geographic variation of this snake.

An evaluation of mtDNA sequences from individuals of all three subspecies revealed relatively high levels of molecular diversification. The molecular makeup indicated that the species had a distribution corresponding to the northward displacement of Baja California from mainland Mexico, followed by vicariant separation into five lineages: two in the northern range of *C. c. cerastes*, a northwestern one restricted to Death Valley, California, and another more southeastern one; two eastern south-central clades within the range of *C. c. cercocombus*, one in southwestern Arizona and the other in northwestern Sonora; and a west-central clade consistent with the range of *C. c. laterorepens*, the only well-defined clade. In addition, those clades with close

Crotalus cerastes cercobombus (South of Redrock, Pinal County, Arizona; John H. Tashjian)

Crotalus cerastes cercobombus (R. D. Bartlett)

geographical proximities did not form sister taxon relationships, which reflected the discordance between the molecular clades and the currently recognized subspecies. Deep Pleistocene divisions indicate allopatric segregation of the subclades within refugia and lineage diversification that extended to the Pliocene or Late Miocene (Douglas ct al. 2006).

CONFUSING SPECIES: No other snake in the range covered in this book has such elevated horn-like projections over its eyes.

KARYOTYPE: The total chromosome number is 36; 16 macrochromosomes (4 metacentric, 6 submetacentric, 4 subtelocentric, and the 2 sex chromosomes) and 20 microchromosomes. Males are ZZ and females ZW; the Z sexual chromosome is metacentric, and the W sexual chromosome subtelocentric (Zimmerman and Kilpatrick 1973).

FOSSIL RECORD: Pleistocene vertebrae have been found in Rancholabrean deposits in Yuma County, Arizona (Van Devender and Mead 1978; Van Devender et al. 1991a).

DISTRIBUTION: The sidewinder ranges from extreme southwestern Utah, southern Nevada, and southeastern California southward through southwestern Arizona to northwestern Sonora, eastern Baja California del Norte (Murray 1955), and Isla Tiburon in Mexico.

HABITAT: Prime habitat for this snake is a desert with loose sand dunes and few plants in rather low-lying areas (from below sea level in Death Valley, California, to 1,830 m), but it is sometimes found in hardpan, rocky or gravelly sites, lava flows, stands of Joshua trees within deserts, or even piñon-juniper woodlands at its higher elevations.

Associated plants in its habitat are burr sage (*Ambrosia dumosa*), cacti (*Carnegiea gigantea, Opuntia* sp.) creosote bush (*Larrea divaricata, L. tridentata*), desert croton (*Croton californicus*), grasses (*Hilaria rigida, Oryzopsis hymenoides, Pleuraphis rigida*), Joshua

Distribution of
Crotalus cerastes.

tree (*Yucca brevifolia*) and other yuccas (*Yucca* sp.), junipers (*Juniperus californica, J. os-teosperma*), mesquite (*Prosopis* sp.), palo verde (*Cercidium* sp.), ratany (*Krameria parvi-flora*), Russian thistle (*Salsola kali*), sandpaper plant (*Petalonyx thurberi*), and sumac (*Rhus* sp.).

A detailed description of the habitat of *C. c. cerastes* at Kelso Dunes in the Mojave Desert, California, is provided by Brown and Lillywhite (1992).

BEHAVIOR AND ECOLOGY: Good life history data exist for the sidewinder. In California, *C. c. cerastes* and *C. c. laterorepens* may emerge in the spring as early as late February, but, with increasing daylight and warmth, more and more individuals leave their hibernacula in March and April; fall disappearance occurs in October or November (Brown and Lillywhite 1992; Klauber 1972; Secor 1994). Most spring surface activity takes place in May and June; a fall activity peak occurs in late September or October.

C. c. cerastes begins to enter hibernation in late October and all retreat underground by December (Brown and Lillywhite 1992; Secor 1994), but a few may be surface-active, usually basking, during warm weather. Cowles (1941) speculated that this snake might hibernate at depths below 30 cm, and Brown and Lillywhite (1992) and Secor (1994) recorded hibernation depths of 20–70 cm. It usually hibernates alone within a rodent burrow. Of 15 snakes monitored by Secor (1994), 14 hibernated in kangaroo rat (*Dipodomys*) burrows, and 1 in the burrow of a desert tortoise (*Gopherus agassizii*). In rocky areas, a crevice may be used as a hibernaculum, and during mild winters this snake may merely bury itself in the sand near the surface.

In the spring when the desert surface temperature is no higher than 35°C, *C. c. cerastes* will stay on the surface all day before seeking shelter at sunset. It avoids the summer daytime heat by crawling under rocks, entering animal burrows, or burying beneath the sand so that at most only the outer coil is exposed. It enters a tight coil and then edges or nudges the sand outward from beneath its body or uses its head and neck to pull sand over its coils to form a saucer-shaped, crater-like depression in which it lies with its back flush with the surrounding surface (Cowles 1945; Cowles and Bogert 1944; Secor 1994). Any appreciable drift of sand tends to submerge the snake either by surface movement of rolling grains or by the deposition of wind-blown material; wind often blows away tracks leading to the spot. These hideaways are usually located in the shade of some bush.

Secor and Nagy (1994) reported that the average duration of surface activity of sidewinders in San Bernardino County, California, is 7.2 hours a day. The snake is usually considered nocturnal; almost all summer activity takes place at night, but some (*C. c. cerastes*, *C. c. cercobombus*) make diurnal movements in early spring and in the fall, particularly during the morning or late afternoon (Brattstrom 1952; Stebbins 1943). Suitable daytime ATs for surface activity of *C. c. cerastes* are 32–35°C (Secor 1994). Nocturnal activity occurs during 1930–0500 hours (*C. c. laterorepens*), with peak activity at 1000–0100 hours (Klauber 1944; Moore 1978). Sidewinders are nocturnal despite widely varying ET conditions (Moore 1978). The normal activity BT range for the species is 6.3–40.8°C, and STs are usually higher than 30°C (Brown and Lillywhite 1992; Moore 1978; Secor 1994).

Mean BT varies from month to month (November, 18.4°C; August, 31.4°C), but averages 25.8°C. BTs of active sidewinders have ranged from 8.2 to 40.8°C (Brattstrom 1965; Brown and Lillywhite 1992; Cunningham 1966; Secor and Nagy 1994). Cowles and Bogert (1944) reported a voluntary minimum BT of 17.5°C, an optimum BT of 31.5°C, and a CT_{max} of 41.6°C, and Brattstrom (1965) thought the CT_{min} was −2°C. Mosauer and Lazier (1933) reported that exposure to direct sunlight until the snake's BT is 46°C is fatal.

C. cerastes can often be found coiled near the mouth of a rodent burrow. It usually

Crotalus cerastes cercombus (R. D. Bartlett)

maintains this position until its BT starts to rise in midmorning, and then either retreats into the burrow or crawls to a nearby shaded spot. Due to circulatory adjustment, tightly coiled, inactive individuals may be able to conserve heat more effectively than when uncoiled (Moore 1978). Also in this respect, neonates sometimes aggregate in balls during the daylight hours, which may slightly stabilize their BTs, retard evaporative water loss, act as an antipredator device, and enhance their first ecdysis. The mean core temperature of these aggregations was 31.94 (range, 26–32) °C at the upper limit of the preferred range for the species (Reiserer et al. 2008).

Basking sometimes occurs in the morning or late afternoon, probably following a meal (the region of the body containing the food bolus is usually exposed directly to the sun), and, at night, those individuals near an asphalt road will crawl onto the highway (Klauber 1932) to warm themselves by conduction of the residual heat remaining there.

The mean field metabolic rate (FMR) of *C. cerastes* in San Bernardino County, California, measured (as CO_2 production) by Secor and Nagy (1994) was 0.063 mL/g per hour during the active season (mid-April to mid-October), 0.028 mL/g per hour during the transition seasons (mid-March to mid-April, and mid-October to mid-November), and 0.007 mL/g per hour during hibernation (mid-November to mid-March). The mean rate of water influx during the active period was 7.6 mL/kg per day, and for the transition seasons it was 2.5 mL/g per day. The difference between the FMR and the SMR (the energy allocated to activities and other energy-demanding functions such as digestion) was 65% of the snake's annual energy expenditure. The mean feeding rate, calculated from water influx, was 9.4 kg/day, and the mean annual assimilable energy profit was 177 kJ.

With increasing BT, the O_2 capacity, red cell count, amount of hemoglobin present, and hematocrit are lowered, while the blood pH rises. The hemoglobin system appears well adapted to the thermal variability of the sidewinder's environment, helping to maintain lower O_2 levels, resulting in lower SMR and internal heat production, at higher ETs. Photoperiod does not seem to play a major role in blood adaptations (MacMahon and Hammer 1975a, 1975b).

The opportunity to drink free water often occurs only during the summer monsoon season for *C. cerastes*, and some individuals have adapted a particular behavior for doing this. Rorabaugh (2007) observed two Arizona *C. c. laterorepens* at 1345 and 1355 hours on 16 April (AT and ST, 15°C) during a light rain event. Both snakes maintained a flattened, loose-coiled, posture. The second was apparently drinking water from its skin, as its rostrum was directed into the inside of a coil of its body that was an efficient place to collect rainwater accumulating and running off its skin.

The majority of movements by *C. cerastes* are to find a new ambush site or during the breeding season to locate a mate. Sidewinders may maintain rather large home ranges, probably due to the energy needed to catch widespread desert prey. In the eastern Mojave Desert, male *C. c. cerastes* have home ranges of 8.3–51.2 (mean, 18.2) ha, and females maintain home ranges of 3.7–21.3 (mean, 12.0) ha (Secor 1992, 1994). In the spring, juveniles move an average of 100.2 m between captures, subadults 172.6 m, adult females 76.3 m, and adult males 122.5 m; movements for these same groups in the summer and fall are 163.7, 202.3, 73.0, and 99.1 m, and 140.8, 188.0, 78.5, and 237.5 m, respectively (Secor 1992). One subadult in Secor's (1992) study moved 963 m between captures, and Brown (1970) and Brown and Lillywhite (1992) reported that a sidewinder made a night October trip of 1.27 km. Such long-distance movements probably are not frequent. Most individuals followed by Brown and Lillywhite (1992) moved less than 400 m a night.

Although *C. cerastes* is capable of performing all four typical methods of crawling known for snakes (see Ernst and Zug 1996; Zug and Ernst 2004), its sandy habitat often dictates that it must resort to the specialized method from which it derives its name (Mosauer 1933). Sidewinding is essentially a series of lateral looping movements in which only vertical force is applied and usually no more than two parts of the body touch the sand at any one time. However, when the neck first makes contact, three points (neck, body, and tail) touch the sand. The snake moves diagonally quickly forward, and separate J-shaped tracks, each paralleling the other and angling in the direction of movement, result as it literally skips across loose sand. Its weight seems to shift from head to tail as it moves. Interestingly, this form of locomotion is also used by true vipers of the genus *Cerastes* that also occupy habitats with loose sand (Mosauer 1932b).

Jayne (1986) observed that sidewinding is sometimes alternated with typical lateral undulatory crawling, even on loose sand, and reported maximum mean forward velocities of 0.75–1.7 total body lengths per second. Mosauer (1935b) reported that *C. cerastes* could voluntary sidewind at 0.14 m per second, and that if hassled could reach a speed of 0.91 m per second. Maximal burst speed observed by Secor et al. (1992) was 3.7 km per hour, and the snakes could maintain a speed of 0.5 km per hour for 33–180 minutes. At night, sidewinding is slower, 0.04–0.07 m per second (Secor 1994). When the snake moves through brush or plant debris, lateral undulation is used, while rectilinear crawling is also sometimes used (Bogert 1947), particularly during courtship and the first 1–3 m of nocturnal movement. Sidewinding seems ideally suited for movement in sandy deserts, but also allows some thermal regulation, as only two points of the body are in contact with the hot (or cold at night) sand.

C. cerastes occasionally climbs into desert vegetation to heights of 10–30 cm (Armstrong and Murphy 1979; Stephen Secor, personal communication), and Cunningham (1955) observed a captive *C. c. laterorepens* several times coiled on branches 60–91 cm above its cage floor. However, the only ones found above ground by Stephen Secor (personal communication) had been disturbed, and Baldwin (*in* Klauber 1972) noted that a disturbed *C. cerastes* climbed into a small bush about 30 cm high and later transferred to an even higher one.

When placed in water the sidewinder swims readily with a lateral undulatory stroke, keeping its head and neck well above the surface (Klauber 1972).

Male combat behavior by this species occurs in both spring and summer (Aldridge and Duvall 2002; Lowe and Norris 1950; Wright and Wright 1957). It has been infrequently described, and then mostly from captives. Bouts observed by Lowe and Norris (1950) and Sievert (2002c) were unique in including a sequence of biting not known in other North American crotalids (Thijssen [*in* Bakker 2003] thought it impossible to house several males together without them biting each other). The observation by Lowe and Norris took place in April with both males in breeding condition. While parrying in an upright position, the snakes suddenly bit each other in the neck approximately 50 mm posterior to the head. This behavior was repeated, but it is possible that neither male ejected venom during the biting. Sievert (2002c) reported that a larger male pursued a smaller one relentlessly when the two were placed in the same terrarium until it was exhausted and then bit the smaller snake repeatedly, so much so that the younger snake had to be relocated to another container.

REPRODUCTION: The smallest TBLs of mature individuals of either sex reported by Klauber (1972), Stebbins (2003), and Wright and Wright (1957) were 42.0–43.4 cm. The smallest copulating male found by Stephen Secor (personal communication) had a TBL of 49.5 cm (SVL, 45.3 cm) and weighed 79 g. The smallest male containing mature sperm examined by Goldberg (2004) had a TBL of 36.0 cm (SVL, 33.1 cm), and the smallest reproductive female had a TBL of 40.8 cm (SVL, 38.3 cm). These snakes probably mature at about 3 years of age.

Goldberg (2004) described the gametic cycles of both sexes in a mixed sample of California *C. c. cerastes* and *C. c. laterorepens*. Males had regressed testes in March (63.6%), April (41.3%), May (32.2%), June (5%), and October (50%); none were regressed in July–August. No males were found in recrudescence in September–October, but were undergoing recrudescence in March (36.4%), April (58.6%), May (67.8%), June (50%), July (33.3%), and August (60%). Males were undergoing spermiogenesis in June (25%), July (66.7%), August (40%), September (100%), and October (50%), but not in March–May. The testes are at their maximum size in September (Reiserer 2001).

The ovaries of 43.5% of the females were inactive during February–September. Only one female showed signs of early vitellogenesis in each of the months of January, May, and October, but 40% were undergoing vitellogenesis in April. Enlarged ovarian follicles (> 8 mm) were present in April (30%), May (67%), and June (25%), and eggs were found in the oviducts in June (25%), July (20%), and August (50%). Reiserer (2001) reported a late June ovulatory period. Although Secor (1994) thought it possible that females could reproduce each year, it is more probable that their reproductive cycle is biennial.

Wild *C. cerastes* mate in either spring or fall, and Reiserer (2001) has reported both spring and fall matings in captives. Males are more motile during the breeding seasons than are females (Cardwell et al. 2005). Secor (1994) suspected that the majority of matings are initiated after sunset after the snakes leave the burrows and become surface-active. Spring copulations have occurred on 26 March–4 June; fall matings took place on 20 September and 7–22 October (Bakker 2003; Brown and Lillywhite 1992; Coupe and Dawson 2007; Klauber 1972; Lowe 1942; Secor 1994; Stebbins 1954; Wright and Wright 1957). Sperm from fall matings is stored by the female until ovulation the following spring (Schuett 1992; Secor 1994). All observations of mating were in the morning, so courtship had begun the night before. The male nudges and performs jerky head movements along the female's body. The male wraps his body around that of the female and inserts a hemipenis. After intromission, the male may hold his tail straight up, waving it slowly in a circle, but he does not pulsate or pump. He may drag the female about while they are attached, as his body moves around and over her, while he still nudges her body with his head. This may be followed by a period of little movement before more tail-waving and nudging occur, but not necessarily at the same time. A pronounced pulsating or pumping motion occurs in the female's cloacal region, possibly begun by her, continues at a rate of 1 per second for about 10 seconds, then becomes a little faster or slower until it stops entirely. Eventually, the two snakes break apart and go their separate ways (Lowe 1942; Perkins *in* Klauber 1972). From the onset of copulation, pairs probably stay together on the surface until the warming

ETs of the morning force them to separate and retreat underground (Secor 1994). Copulation may continue for as little as two hours (Lowe 1942) to more than seven hours (Brown and Lillywhite 1992).

The GP takes as long as 150–155 days (Lowe 1942; Perkins *in* Klauber 1972), and apparently parturition occurs underground in rodent burrows (Reiserer et al. 2008). Known birth dates range from 14 May to 28 November, but most occur in September or early October (Bakker 2003; Brown and Lillywhite 1992; Klauber 1972; Reiserer et al. 2008; Secor 1994; Wright and Wright 1957).

Litters average 8.9 offspring (n = 89), and contain as few as 1 (C. Ernst, personal observation) to as many as 20 (Langebartel and Smith 1954), but typically 7–12, young. Fitch (1985a) reported mean litter sizes of 10.8 (range, 7–18) young for *C. c. cerastes* and 9.0 (range, 5–16) young for *C. c. laterorepens*.

Mean neonate TBL is 18.0 (range, 16.1–20.5; n = 82) cm, and their mean BM is about 4.9 (range, 3.0–8.4; n = 33) g. Larger females appear to produce more young per litter; RCM is approximately 40% (Secor 1994). Neonates may remain together for a short time after birth, at least to their first ecdysis, and their mothers may stay close by during this period (Reiserer et al. 2008).

Powell et al. (1990) reported and presented photographs of offspring of a hybridization between a male *C. c. laterorepens* and a female *C. s. scutulatus*. The neonates were patterned more like *C. scutulatus*, but had head scalation typical of *C. cerastes* and moved by sidewinding.

GROWTH AND LONGEVITY: The mean growth rate for 35 California *C. cerastes* studied by Secor (1994) was 0.54 mm per day. He considered his snakes to be in their first full year and possibly part of their second if they had SVLs of 21–40 cm and BMs of 10–50 g; subadults were probably in their second full year or part of the third with SVLs of 41–47 cm and BMs of 45–77 g; his adults (> 3 years old) had SVLs of 49–61 cm and BMs >80 g, and several weighed 200–300 g. The growth rate slows with increas-

Crotalus cerastes laterorepens (R. D. Bartlett)

ing body length. Klauber (1972) noted that the Mojave Desert sidewinder (*C. c. cerastes*) is shorter than the Colorado Desert sidewinder (*C. c. laterorepens*), which may be correlated with a shorter growing season.

A *C. c. cerastes* born in captivity survived more than 28 years (Goodman et al. 1997), and Slavens and Slavens (2000) noted that another *C. cerastes* lived 27 years and 4 months in captivity. Snider and Bowler (1992) reported the following captive survival durations: *C. c. cercobombus* (male, wild-caught as a juvenile), 10 years, 8 months, and 21 days (private collection, Louis Pistoia); *C. c. cerastes* (male, wild-caught when adult), 11 years, 6 months, and 28 days (Columbus Zoo); and *C. c. laterorepens* (female, wild-caught at unknown age), 19 years, 8 months, and 22 days (Los Angeles Zoo). Secor (1994) suspected that the mean life expectancy of wild sidewinders, once they reach maturity, is only 5–7 years.

DIET AND FEEDING BEHAVIOR: *C. cerastes* captures most prey from ambush—lying, partially concealed by sand, outside the entrances of rodent or lizard burrows. One even took an ambush pose under an open-sided cardboard box used to shade a Sherman live trap, probably lured to the site by rodent odor on the trap (Randell and Clark 2007). However, some sidewinders actively forage. Chiszar and Radcliffe (1977) thought that olfaction may only play a minor role compared to visual or infrared clues in prey detection, as they recorded no significant response by naive captive neonate sidewinders to prey odors. However, Secor (1994) observed that olfaction is more important in prey detection than either vision or heat detection, as sidewinders usually tongue-flick before striking and then use olfaction to trail wounded prey. Lizards are usually retained in the mouth when struck, but rodents are released after the strike. Caudal luring has also been reported for *C. cerastes*, but prey selection based on color apparently does not occur (Reiserer 2002).

Adult *C. cerastes* feed mostly on small rodents and lizards. Neonates and small juveniles rely almost exclusively on lizards, although possibly small snakes are also eaten. As the snakes grow, they gradually take larger lizards and rodents (Secor 1994). A moderately feeding captive 40 cm male only ate 6 house mice (*Mus musculus*) between hibernations, but a captive adult female ate voraciously, even during mating (Bakker 2003). Gut passage time is 11–24 days (Secor et al. 1994).

Funk (1965) examined 226 *C. c. laterorepens*; 88 (51.5%) contained mammals (mostly *Dipodomys*, *Perognathus*, and *Peromyscus*), 73 (42.7%) contained lizards (mostly *Cnemidophorus*, *Uma*, and *Uta*), 5 (2.9%) had eaten birds, and 5 (2.9%) contained small snakes. The most important prey species were *Cnemidophorus tigris* (in 14.6% of the snakes), *Dipodomys deserti* (10.5%), *Perognathus penicillatus* (9.4%), and *Uma notata* (8.2%). Klauber (1944) recorded mammal to lizard ratios of 11:5 in wild *C. c. cerastes* and 13:10 in *C. c. laterorepens*.

Reported prey of wild and captive *C. cerastes* are a caterpillar (Lepidoptera); reptiles—lizards (*Callisaurus draconoides*; *Cnemidophorus tigris*; *Coleonyx variegatus*; *Dipsosaurus dorsalis*; *Gambelia wizlizenii*; *Phrynosoma cornutum*, *P. mcalli*, *P. platyrhinos*; *Uma inornata*, *U. notata*, *U. scoparia*; *Urosaurus graciosus*, *U. ornatus*; *Uta stansburiana*), and snakes (*Arizona elegans*, *Chionactis occipitalis*, *Crotalus cerastes*, *Masticophis flagellum*, *Sceloporus magister*, *Sonora semiannulata*); birds—sparrows (*Amphispiza bilineata*, *Chondestes grammacus*, *Passer domesticus*), wrens (*Campylorhynchus brunneicapillus*), and warblers (*Dendroica petechia*); and mammals—shrews (*Notiosorex crawfordi*), kangaroo rats (*Dipodo-*

mys *deserti, D. merriami, D. platyrhinos*), pocket mice (*Perognathus baileyi, P. longimembris, P. penicillatus*), white-footed mice (*Peromyscus eremicus, P. maniculatus*), harvest mice (*Reithrodontomys megalotis*), house mice (*Mus musculus*), ground squirrels (*Ammospermophilus harrisii, A. leucurus; Spermophilus tereticaudus*), woodrats (*Neotoma albigula, N. lepida*), and pocket gophers (*Thomomys bottae*) (Armstrong and Murphy 1979; Bogert 1941, Bouskila 1995, 2001; Brown and Lillywhite 1992; Brattstrom 1996, Coupe and Dawson 2007; Cowles and Bogert 1936; Cunningham 1959, 1966; Funk 1965; Klauber 1932, 1944, 1972; Lyman 2006; Mulcahy et al. 2003; Parker 1974; Revell and Hayes 2009; Secor 1994; Van Denburgh 1922). The caterpillar was found in a juvenile that also contained a lizard (*Uma*), and Klauber (1972) thought that the lizard had either first eaten the caterpillar, or had been captured while eating it. The snake will not reject dead prey; Cunningham (1959) reported that a captive ate a mouse dead for at least two days, and Klauber (1972) found a sidewinder eating a kangaroo rat (*Dipodomys*) that had obviously been killed by traffic. Klauber (1972) found a leaf from a sumac (*Rhus* sp.) in a *C. cerastes*.

Prey capture is not without its dangers. An attempt to ingest too large an animal may result in it becoming lodged in the throat and result in the death of the sidewinder (*Ammospermophilus leucurus*, Coupe and Dawson 2007; *Cnemidophorus tigris*, Mulcahy et al. 2003). In the case of the fatal partially ingested *A. leucurus*, the *C. cerastes* was an adult female with an SVL of 56.7 cm (BM, 133 g). The snake that died while swallowing the *C. tigris* was a neonate with a 22.9 cm SVL; the lizard was 7.35 cm long. Also, catching a meal is not always easy, as kangaroo rats (*Dipodomys*) have evolved avoidance behaviors that allow them to foil predatory attempts by sidewinders (Bouskila 1995, 2001; Fowlie 1965; which see for details).

VENOM DELIVERY SYSTEM: Although small, the sidewinder has relatively long fangs for its size. Five adults with TBLs of 51.8–76.7 cm measured by Klauber (1939) had a mean FL of 6.4 (range, 5.0–8.1) mm, a mean TBL/FL of 98, and a mean HL/FL of 4.83. The mean angle of fang curvature for these snakes was 67°.

VENOM AND BITES: Although not as much as for some North American snakes, data do exist on the properties and action of sidewinder venom.

Total dry venom yield is 18–45 mg per adult snake, with a minimum of 15 mg (Klauber 1928; Russell and Puffer 1970) and a maximum of 67 mg (Githens 1933). The amount injected by an adult snake per bite is low, 0.06 mL (0.018 g dried) (do Amaral 1928). Klauber (1972) milked 176 *C. cerastes* (an equivalent of 119 fresh adults) and obtained a total venom yield of 3,987 mg of dried venom, or an average of 33 mg per adult.

Tests show that sidewinder venom has anticoagulant, arginine ester hydrolase, hemorrhagic, and protease activities (Mackessy 1988; Tan and Ponnudarai 1991), which aid in predigestion of its prey. It apparently does not contain myotoxin enzymes (Bober et al. 1988; Tubbs et al. 2000), but present are 5′-nucleotidase, phosphodiesterase, and phosphomonoesterase (Aird 2005; Tan and Ponnudurai 1991).

Mouse (20–22 g) LD_{50} values are subcutaneous, 5.5 mg/kg (Minton 1956); intraperitoneal, 2.08–4.00 µg/g (Emery and Russell 1963; Russell and Brodie 1974; Russell and Puffer 1970; Tu 1977) or 0.06–0.17 mg/kg (Githens and Wolff 1939; Macht 1937); and intravenous, 1.95–2.60 mg/kg (Emery and Russell 1963; Vick 1971). The minimum lethal dose for 350 g pigeons is 0.9–0.15 mg (Githens and George 1931).

The venom does not lose its lethal potency for mice or cats after 26–27 years of storage, although it does take longer to produce complete neuromuscular blockage than does fresh venom (Russell et al. 1960).

The venom is only moderately toxic compared to that of other rattlesnakes of its genus, but *C. cerastes* is still dangerous. Human envenomation by a sidewinder is usually mild, but fatalities have occurred (Russell 1960a). The human lethal dose is 40–50 mg (Altimari 1998; Dowling 1975); subcutaneous, 385 mg/kg × 70 (Brown 1973). A typical sidewinder contains an average of 0.09 lethal doses based on a 70 kg human (Brown 1973).

Envenomation symptoms include much pain, swelling, an itching sensation, discoloration, weakness, dizziness, slight nausea (after administration of antivenom), paralysis, necrosis of the tissue at the bite site, an increase in body temperature, and tenderness at the site of the bite for several days after being bitten (Hutchison 1929; Klauber 1972; Lowell 1957; Minton 1956; Russell 1960a; Stephen Secor, personal communication).

Most legitimate envenomations take place during the mating periods (Cardwell et al. 2005). Ivanyi and Altimari (2004) noted that bites by the sidewinder made up only 3 of 48 (6.3%) bites during academic research in a 26-year period. Keyler (2005) only reported a single illegitimate bite that occurred among 31 venomous snakebites in Minnesota during 1982–2002; the victim was handling a captive *C. c. cercobombus*, and alcohol was a contributing factor. Probably more illegtimate bites occur annually than legitimate ones.

The sidewinder may be immune to the venom of its own species (Lowe and Norris 1950; Sievert 2002c).

PREDATORS AND DEFENSE: *C. cerastes* has many predators, particularly of the young. The following are considered actual or potential predators: leopard lizards (*Gambelia wizlizeni*), coachwhips (*Masticophis flagellum*), kingsnakes (*Lampropeltis getula*), rosy boas (*Charina trivirgata*), larger sidewinders (*C. cerastes*), falcons and hawks (*Accipiter cooperi*; *Buteo jamaicensis*; *Falco mexicanus*, *F. sparvarius*), owls (*Bubo virginianus*, *Speotyto cunucularia*), roadrunners (*Geococcyx californianus*), loggerhead shrikes (*Lanius ludovicianus*), ravens (*Corvus corax*), grasshopper mice (*Onychomys torridus*), skunks (*Spilogale putorius*), badgers (*Taxidea taxus*), canids (*Canis familiaris*, *C. latrans*; *Urocyon cineroargenteus*; *Vulpes macrotis*), and felids (*Felis catus*, *Lynx rufus*) (Brown and Lillywhite 1992; Funk 1965; Klauber 1972; Ross 1989; Secor 1994).

When confronted with the ophiophagus snakes *Lampropeltis getula californiae* and *Pseutes sulfureus*, *C. cerastes* either tried to flee, strike and bite, body-bridge, hide its head under a coil, or inflate its body (Bogert 1941; Cowles 1938; Weldon et al. 1992). The species is also capable of changing its body melanism to somewhat match that of the substratum (Klauber 1931; Neill 1951); although Neill (1951) thought this is a thermal adjustment, it may also provide some protection from overhead avian predators. It may also conceal its shadow by pressing its body against the substrate (Cloudsley-Thompson 2006).

The sidewinder is a nervous snake, but often remains quietly coiled when first discovered. If prodded it will rattle and usually try to crawl away, but, if further provoked, it will put up a spirited fight while repeatedly coiling and continuously rattling—SVL-adjusted loudness to about 70 dB; frequency range, 3.8–23.8 kHz; and mean dominant frequency, 12.9 kHz (Cook et al. 1994; Fenton and Licht 1990; Young and Brown 1993)—jerking the head, alternately laterally compressing and inflating the body, and striking viciously.

PARASITES AND PATHOGENS: Unidentified oligacanthorhynchid larvae, the tapeworm *Oochoristica osheroffi*, the nematodes *Hexametra boddaërtii* and *Thubunaea cnemidophorus* (possibly accidentally ingested with prey), and hemogregarine protozoans have been found in *C. cerastes* (Alexander and Alexander 1957; Babero and Emmerson 1974; Bursey et al. 1995; Ernst and Ernst 2006; Goldberg and Bursey 2002; Widmer 1966; Widmer and Olsen 1967; Wozniak et al. 1994).

The bacterium *Arizona hinshawii* has been isolated from sick captive *C. cerastes* at the Dallas Zoo, which displayed gaping, inanition and anorexia, and convulsive seizures prior to death (Murphy and Armstrong 1978).

POPULATIONS: *C. cerastes* may be locally common. Armstrong and Murphy (1979) found five sidewinders under the wreckage of a dilapidated building during one search. The snake comprised 4.3% of the snakes observed on Arizona Rt. 85 in 1988–1991 by Rosen and Lowe (1994); Fowlie (1965) found 47 on one moonlit night on a road between Yuma and Gila Bend, Arizona. Repp (1999) recorded 64 (61 alive) *C. c. cercobombus* after a very cool, wet spring in 1998 following an extended dry period from the summer of 1995 to the fall of 1997 in southern Arizona; and Armstrong and Murphy (1979) saw up to 30 per night while driving roads.

Klauber (1926) recorded only 2 (0.1%) individuals in a sample of 1,623 snakes collected in 1923–1925 in San Diego County, California. Also in California, Brown (1970) marked 72 adult *C. c. cerastes*, but recaptured only 8 during an ecological study in the Mojave Desert. Brown and Lillywhite (1992) marked 50 at a study site, and estimated sidewinder density there at 0.29–0.71 snakes/ha. Secor (1994) marked 116 *C. cerastes* at another California locality, and thought the density of his population closer to 1 snake/ha.

Of 116 *C. cerastes* sidewinders captured by Secor (1994), 59 were adult females and 57 adult males, essentially a 1:1 sex ratio, and the ratio of immature to mature snakes was 0.33:1. The 50 individuals captured by Brown and Lillywhite (1992) included 25 males, 20 females (1.25:1), and 5 juveniles (0.11:1). Klauber (1943a) reported a 1.04:1.00 sex ratio (49 males, 47 females) in a sample from San Diego County and one of 1.14:1.00 (24 males, 21 females) in another sample from Los Angeles County.

The inhospitable desert habitat of *C. cerastes* gives it a certain amount of protection as fewer humans will spend much time there. However, many die on the roads, some humans kill them on sight, and the pet trade takes a number of them each year. *Crotalus c. cerastes* is considered a species of special concern and is protected in Utah.

REMARKS: Several scenarios as to the relationships of *C. cerastes* have been proposed. Comparing its morphology to that of other species of *Crotalus*, Brattstrom (1964a) thought the snake most similar to *C. mitchellii*, *C. tigris*, and *C. viridis*, while Klauber (1972) grouped it with *C. durissus*, *C. molossus*, and *C. horridus*. However, Gloyd (1940) cautioned that in comparing such more or less desert-adapted rattlesnakes as *C. cerastes*, *C. enyo*, *C. mitchellii*, and *C. stephensi,* it is difficult to distinguish between fundamentally homologous characters and adaptive parallelisms, and that the sidewinder has gone so far in the direction of such modifications that the various attributes that otherwise might be useful in tracing its relationships are obscured. Dorsal scale micro-dermatoglyphics place it with *C. mitchelli* and *C. pricei* (Stille 1987), and electrophoretic studies of venom proteins show it near *C. pricei*, *C. lepidus*, *C. ravus*, *C. triseriatus*, and *C. willardi* (Foote and MacMahon 1977). More recent studies using DNA and RNA sequences place *C. cerastes* within the *Crotalus polystictus*-group, which also includes *C. intermedius*, *C. pricei*, *C. transversus*, and *C. willardi* (Murphy et al. 2002).

A general summary of the biology of the species was published by Tai-A-Pin (2008).

Crotalus enyo (Cope, 1862)
Baja California Rattlesnake
Cascabel de Baja California

RECOGNITION: *Crotalus enyo* is a common rattlesnake on the Baja California penin-sula. Its TBL$_{max}$ is 89.8 cm (Klauber 1972), and most adults are more than 65 cm long. It has a well-defined vertebral keel. The body is usually tan or brown (pale to dark or grayish); some individuals have a silvery hue. Dorsally are 28–42 (mean, 33) black-bordered, yellowish to reddish-brown, subrectangular (anteriorly) to hexagonal (mid-body) blotches, which become more band-like and faded posteriorly. Below the larger dorsal blotches lies a lateral row of smaller dark brown or black blotches, which may coalesce on the posterior body with the lower edge of the corresponding dorsal blotch. A second lateral row of smaller, often indistinct blotches alternates with the larger lateral blotches. Three to eight indistinct brownish bands are present on the tail, the first rattle segment is dark brown or black, and the rattle is rather long. Body scales are keeled (tuberculate) and pitted (Harris 2005; Stille 1987; Harris displays an electron microscope photograph of the scale surface), and occur in about 25 rows anteriorly and posteriorly, and in 23–27 (mean, 25) rows at midbody. The cream, yellow, or tanish venter is often heavily patterned with gray or brownish mottlings. On it are 157–181 ventrals, 18–31 subcaudals, and an undivided anal plate. The head is relatively narrow with prominent eyes. Its pattern consists of a well-marked, pale, convex, transverse bar across the supraoculars; a dark oblique stripe extending from the orbit posteriorly to the corner of the mouth; a dark, elongated, lateral, parietal blotch on each side, with two prominent smaller dark spots between them in the anterior parietal region; often fusion on each side between the lateral parietal blotch and an elongated neck stripe; and grayish labials that may contain some dark spotting. A broader than long rostral scale, 2 internasals (touching the rostral), 2 large canthals, 13–25 smaller scales in between the internasals and the prefrontal area, 2 large supraoculars (elevated along their lateral borders, causing the frontal area to be depressed), and 4–5 (2–6) in-tersupraoculars comprise the dorsal head scalation. The dorsal scales anterior to the orbits tend to be knobby and rough. Laterally are 2 nasals (the prenasal touches the rostral, and is sometimes also in contact with the first supralabial, but often separated from it by some foveals), 2–3 (1–8) loreals (the lowermost is largest), 2–3 preoculars (the upper does not touch the postnasal), 1 (2) interoculabials, 13–14 (12–15) supra-labials, and 13–14 (11–16) infralabials. Ventrally, is a mental scale and two chin shields, but no interchin shields.

The bilobed hemipenis has a bifurcated sulcus spermaticus with one branch extending to the tip of each lobe. Approximately 134 spines and 33 fringes occur on each lobe, and many small spines are present in the crotch (Klauber 1972).

The tooth count is 1–3 (mean, 2.7) palatines, 8 pterygoids, and 9–11 (mean, 10) dentaries (Klauber 1972).

Males (TBL$_{max}$, 89.5 cm) have 157–177 (mean, 165–166) ventrals, 22–31 (mean, 25) subcaudals, and 3–8 (mean, 6) tail bands. Females (TBL$_{max}$, 80 cm) have 161–181 (mean, 169) ventrals, 18–23 (mean, 19–20) subcaudals, and 4–7 (mean, 4.6) tail bands.

GEOGRAPHIC VARIATION: Three subspecies are recognized. *Crotalus enyo enyo* (Cope, 1862), the Baja California rattlesnake or cascabel de Baja California, is found on the peninsula from El Rosario southward to Cabo San Lucas, and also on several off-shore islands north of Cerralvo Island in the Gulf of California, and on the Pacific coastal islands of Magdalena and Santa Margarita. Its body is tanish, light brown, orangish or pale, or silvery-gray; adults have an SVL/HL of < 27; 157–177 ventrals and 22–28 subcaudals in males, 161–177 ventrals and 18–23 subcaudals in females; and the upper loreal is usually smaller than the lower one. Individuals from Magdalena Island are paler than those on the mainland (Murray 1955). *Crotalus e. cerralvensis* Cliff, 1954, the Cerralvo Island rattlesnake or cascabel de la Isla Cerralvo, is endemic on that island. It has a pale brown body color with distinct dorsal blotches; an adult SVL/HL of > 26; 167–177 ventrals and 27–31 subcaudals in males; 181 ventrals and 23 subcaudals in the only collected female; and the upper loreal is smaller than the lower one. *Crotalus e. furvus* Lowe and Norris, 1954, the dusky Baja California rattlesnake, Rosario rattlesnake or cascabel de Rosario, ranges from the Río San Telmo south to about El Rosario in northwestern portion of the peninsula. Its body color is gray-brown or dark brown; the adult SVL/HL is < 26; 159–162 ventrals and 26–29 subcaudals in males, 165–171 ventrals and 18–26 subcaudals in females; and the upper loreal is usually larger than the lower one. Intergradation occurs between *C. e. enyo* and *C. e. furvus* in the El Rosario region.

Crotalus enyo enyo (South of La Paz, Baja California Sur; Dallas Zoo, John H. Tashjian)

Crotalus enyo enyo (Bahia Los Angeles, Baja California Norte; John Ottley, Valley Center, California, John H. Tashjian)

The major characters used by Cliff (1954) to distinguish *C. e. cerralvensis* from other populations of *C. enyo* were its smaller head and higher number of ventral scales; but, according to Beaman and Grismer (1994), these characters overlap widely with those of mainland *C. enyo*. The snake's pale brown body color does differ from that of the gray-brown mainland snakes, but Grismer (1999a) thought this too was variable, and proposed that no subspecies be recognized in *C. enyo*, and McDiarmid et al. (1999) followed his suggestion pending publication of supporting data. Also, Beaman and Grismer (1994) suggested that because *C. e. enyo* interbreeds with *C. e. furvus* (Grismer 1994a, 1994d) that *furvus* be placed in the synonymy of *enyo*. We retain the subspecies status of both *C. e. cerralvensis* and *C. e. furvus* until more detailed morphological and genetic data are available.

CONFUSING SPECIES: Other rattlesnakes occurring within the range covered in this book can be identified by using the above key. *Crotalus ruber* has boldly marked white and black tail bands, and no dorsal head pattern. *C. mitchellii mitchellii* has a well-marked black-and-white tail band pattern, and may lack the dark diagonal stripe from the orbit to corner of the mouth.

KARYOTYPE: The karyotype consists of 36 chromosomes; 16 macrochromosomes (4 metacentric, 8 submetacentric, and 4 telocentric) and 20 microchromosomes. Sex determination is ZZ in males and ZW in females; some variation occurs in the W chromosome (Stewart and Morafka 1989; Stewart et al. 1990).

FOSSIL RECORD: No fossil *C. enyo* have been reported.

DISTRIBUTION: Most of the distribution of *C. enyo* covers the peninsula of Baja California, Mexico, where it ranges in the northwest from the Río San Telmo near Cabo Colonet on the Pacific Coast, and on the mainland opposite Isla Angel de la Guarda on the Gulf of California, Baja California Norte, southward to Cabo San Lucas, Baja California Sur. It is also found on the following islands in the Gulf of California: Arbajos,

Carman, Cerralvo, Espíritu Santo, Isla Pardo of the Islas Los Candeleros, Los Corona-dos, Partida del Sur, San Francisco, San José, and San Marcos. The snake also resides on Magdalena and Santa Margarita islands off the Pacific coast (Beaman and Grismer 1994; Campbell and Lamar 2004).

HABITAT: *C. enyo* lives mostly under xeric conditions in deserts and adjacent moun-tains in the center of the peninsula; but, on the more mesic San Quintin Plain in the northwest, the coastal fog produces additional moisture and a profusion of ground plants, such as sage (*Artemisia tridentata*) chaparral. Many animal burrows also persist there. In the south, it enters tropical deciduous and pine-oak forests near Cabo San Lucas. Microhabitats include rocky canyons and mesas, cacti, arid thornscrub, poorly vegetated sandy hillsides or mud washes, and even sometimes sand dunes. Near human settlements, it uses trash and other debris for refuges. Distinctive plants in its central distribution include the giant cardon cactus (*Cereus pringeli*), ocotillo (*Fouquieria colum-naris, Franseria* sp.), and the elephant tree (*Bursera microphylla*).

BEHAVIOR AND ECOLOGY: This snake is most active in the early fall during the wet season (Armstrong and Murphy 1979). It is probably surface-active in all months of the year, particularly in the south, but most likely less so during the dry season. No data are available concerning its dry season or winter biology. It is predominantly nocturnal, but some diurnal activity has been observed (Cliff 1954; Lowe and Norris 1954).

Basey (*in* Armstrong and Murphy 1979) collected a young adult female *C. e. furvus* in October at an AT of 21°C, and Lowe and Norris (1954) recorded a BT of 29.7°C. Cliff (1954) found an adult under a piece of tin roofing that was exposed to the sun, and was surprised to find the snake there, as the heat under the tin was extreme. Tryon and Radcliffe (1977) maintained captives either at fluctuating daily ATs of 23–32°C in the summer and 22–28°C in the winter, or at 27–30°C throughout the year.

Distribution of
Crotalus enyo.

C. enyo may adjust its body color in response to the ET, becoming paler at higher temperatures and darker at lower temperatures (Campbell and Lamar 2004; Grismer 2002).

C. e. cerralvensis readily climbs and moves through brush. McGuire (1991) observed one exit a bush (*Pitheceillobium confine*) from 1.5 m above ground, moving easily and rapidly across its surface. He followed it for about 10 m as it ascended into the lower branches of other shrubs and eventually took cover in the crotch of a small bush. He thought the subspecies to be partially arboreal as compared to mainland members of the species, but Huey (*in* Klauber 1972) also observed a small *C. e. enyo* about 76 cm above ground in a bush near Cataviña.

Marmie et al. (1990) found no significant differences in exploratory behavior, as indicated by the mean rate of tongue-flicking, between two litters of captive-born *C. enyo* that were raised in large cages and small plastic boxes, respectively, indicating that the type of container in which they were kept had no serious effect on such behavior.

REPRODUCTION: The smallest reproductive male (with sperm in its vas deferens) examined by Goldberg and Beaman (2003b) had a 47 cm SVL. Klauber (1972) reported that the shortest mature female he examined was 60.8 cm TBL; Goldberg and Beaman (2003b) reported mature females with SVLs of 54.3, 56.3, and 58.2 cm.

Some males have regressed testes with seminiferous tubules containing spermatogonia and Sertoli cells in March–May. Others have testes in recrudescence with a renewal of spermatogenic cells characterized by spermatogonial divisions and occasionally primary and secondary spermatocytes in March–June. Spermiogenesis is in full swing from July through October when metamorphosing spermatids and mature sperm are present. Sperm is present in the vas deferens from March through October (Goldberg and Beaman 2003b). This cycle is similar to that of U.S. pitvipers (Aldridge and Duvall 2002).

Crotalus enyo cerralvensis (Cerralvo Island, Baja California Sur; John Ottley, Valley Center, California, John H. Tashjian)

Data on the ovarian cycle gathered by Goldberg and Beaman (2003b) can be summarized as follows. The ovaries were inactive from March into August, but enlarged follicles were present in females in March (> 15 mm and were ready to be ovulated) and September (> 10 mm). One female contained oviductal eggs in March, and another a full-term embryo in August. Apparently, not all females breed in a given year, but possibly biennially, and this and the ovarian cycle are similar to that of other North American rattlesnakes. Goldberg and Beaman (2003b) proposed that C. enyo could possibly reproduce annually during years of high prey availability, but less frequently when prey was scarce.

Most of what we know of the reproductive behavior of C. enyo was published by Tryon and Radcliffe (1977). Captive courtship activity by C. e. enyo was observed by them on 1 June 1976 during a time when the female was shedding her skin. The male performed alternating tongue-flicks and head-jerks along her body, and copulation occurred. Courtship activity continued into the following day, but no successful mating occurred. Copulation took place again on 3 June for about two hours. Reproductive activity was not observed again until 30 August, when the female shed again; while the male tried, the female remained coiled and unreceptive. The female underwent parturition at 0800 hours on 21 November, 171 days after the first copulation. Another female (27 months old) copulated with a male on 16 November 1974, and gave birth 176 days later on 11 May 1975. This same female was seen mating again on 27 June 1975 and produced a litter on 21 April 1976 after 299 days. Still another female mated on 26 September 1974, 5 July 1975, and 3 August 1975, and gave birth on 1 April 1976 after 242 days. Probably sperm retention and delayed fertilization were responsible for these long periods; Olivier (2008a) reported that captive C. e. furvus that bred mated in the autumn of the previous year with the neonates born in July of the following year.

Armstrong and Murphy (1979) observed courtship by a captive C. e. enyo on 15 January 1975 between 1530 and 1700 hours. The female (71.4 cm, 447 g) had just undergone ecdysis. The male (66.2 cm, 332 g) rubbed his chin on her back with convulsive forward jerks, and vigorously pushed his tail beneath the anal vent of the female. At times, his tail entirely encircled her anal region. He moved forward 1–2 cm on her body by extending his draped coils and bracing his body at her anal region, and performed low-intensity twitches. Copulation, however, did not occur. Olivier (2008a) reported autumn matings in captivity.

Neonates have been found in the wild from late July to mid-October (Grismer 2002), and captive litters have been born in early July (Olivier 2008b). Twelve reported litters contained 1–9 (mean, 5.1) young (Basey in Armstrong and Murphy 1979; Goldberg and Beaman 2003b; Klauber 1972; Olivier 2008b; Tryon and Radcliffe 1977). Emergent young are enclosed in a membrane, and may exhibit a caruncle, an umbilical scar, and only a prebutton rattle segment; 7–15 minutes are often needed for them to escape from the fetal membrane (Klauber 1972; Tryon and Radcliffe 1977). Neonates have 20.6–23.5 (mean, 21.8; n = 8) cm TBLs, and BMs of 9.7–11.1 (mean, 10.2; n = 6) g (Armstrong and Murphy 1979; Klauber 1972; Tryon and Radcliffe 1977). TL/TBL values of two males measured by Tryon and Radcliffe (1977) were 8.5 and 9.8%, those of five females were 5.0–6.8 (mean, 5.9) %; the 21 November 1976 litter reported by Tryon and Radcliffe (1977) had a mean total brood length of 31.2% of the female's TBL.

GROWTH AND LONGEVITY: Data on growth rates are unavailable. A *C. enyo* of unknown age and sex when caught survived an additional 17 years, 1 month, and 13 days at the San Diego Zoo (Snider and Bowler 1992).

DIET AND FEEDING BEHAVIOR: *C. enyo* preys on centipedes (*Scolopendra* sp.); lizards (*Cnemidophorus* sp., *Dipososaurus dorsalis*, *Sceloporus* sp., *Uta stansburiana*); and small mammals—pocket gophers (*Thomomys bottae*), pocket mice (*Chaetodipus spinatus*, *Chaetodipus* sp.), kangaroo rats (*Dipodomys* sp.), and deer mice (*Peromyscus* sp.) (Campbell and Lamar 2004; Stebbins 2003; Taylor 2001). Klauber (1972) found mammal hair, probably from mice, in four specimens he examined. Captives will eat both live and dead house mice (*Mus musculus*) (Olivier 2008a; Tryon and Radcliffe 1977). McGuire (1991) thought the arboreal behavior by *C. e. cerralvensis* that he observed may have been associated with predation on birds that roosted in the low bushes and trees.

Of the 113 *C. enyo* examined by Taylor (2001), 63 (55.8%) contained prey in their stomachs and/or hindguts. Of the 78 prey items identified, 47 (60.3%) were mammals, 26 (33.3%) were lizards, and 5 (6.4%) were centipedes; 50 of 63 (79.4%) snakes contained only one prey item. All ages and sizes took lizards and mammals; but of the 4 that had eaten centipedes, 3 were adults.

Neonates lack prey-chemical preferences. When presented with cotton swabs soaked in water or extracts from potential prey such as lizards (*Sceloporus undulatus*) and mice (*Mus musculus*) or nonprey such as crickets (*Gryllus* sp.), fish (*Lepomis macrochirus*), salamanders (*Ambystoma tigrinum*), and garter snakes (*Thamnophis radix*), the young snakes committed tongue-flicks toward every animal swab, but most were directed toward those with fish, garter snake, and lizard odors, in that order (Chiszar and Radcliffe 1977). Apparently, perception of the proper prey odor is a learned behavior.

C. enyo strike and retain small prey until it stops moving, but strike and release larger prey, and later hunt for it by following its odor trail (Radcliffe et al. 1980; Tryon and Radcliffe 1977). A high rate of tongue-flicking follows the strike and is accompanied by searching for the wounded prey. This increase in tongue-flicks after an envenomation is even greater than during detection of the prey by sight, odor, or its BT (Chiszar et al. 1980b). Tongue-flicking may continue for as long as 150 minutes when a single mouse is struck, and up to 105 minutes if a second one is also struck. A no-strike presentation of a mouse results in tongue-flicking for at least 120 minutes (Chiszar et al. 1982b). Envenomation is apparently essential for the bitten prey to develop the necessary chemical changes sought through tongue-flicking by the rattlesnake (Chiszar et al. 1992). Wild-caught *C. enyo* exhibit higher rates of strike-induced chemical sensory tongue-flicking than do captive-reared individuals (Marmie et al. 1990).

During an eight-week laboratory test designed to reveal how food-deprived *C. enyo* responded with tongue-flicks to prey odors, the snakes were no more responsive in the first four weeks to a cage with mouse (*Mus musculus*) odors than to a clean cage. However, in weeks 5–8 the rate of tongue-flicking in the mouse cage increased significantly compared to the rate in the clean cage. Chiszar et al. (1981a) interpreted this to indicate that increasing hunger resulted in greater sensitivity to and/or responsiveness to prey odors, and that this was the basis for foraging behavior.

VENOM DELIVERY SYSTEM: Five *C. enyo* with 69.6–85.4 cm TBLs had 5-4-5.8 (mean, 5.6) mm FLs; their mean TBL/FL and HL/FL were 135% and 5.59%, respectively. The angle of fang curvature was 63° (Klauber 1939).

VENOM AND BITES: The total amount of dried venom recovered during a first extraction from 18 fresh *C. enyo* and a second milking of one individual was 458 mg; mean dried venom yield from 16 equivalent fresh snakes was 29 mg (Klauber 1972). According to Bober et al. (1988), the venom lacks myotoxin-α proteins. Glenn and Straight (1985b) reported the following relative enzyme activities of adult venom: protease—11 adults, no activity; 3 adults, 11–12 (mean, 12); esterase—11 adults, 66–325 (mean, 204); 3 adults, no activity; and phosphodiesterase—11 adults, 13–139 (mean, 64); 3 adults, 585–669 (mean, 636).

The mean dry venom intraperitoneal and intramuscular LD_{50} values for a standard mouse are 2.8 mg/kg and 4.6 mg/kg, respectively (Glenn and Straight 1985b). The median estimated lethal dose of dry venom for a 350 g pigeon (*Columba livia*) is 0.10 mg (Githens and George 1931).

PREDATORS AND DEFENSE: The only publication of predation on *C. enyo* was that by Grajales-Tam et al. (2003), who reported one eaten by a coyote (*Canis latrans*) in Baja California Sur. However, the snake naturally responds to the odors or sight of ophiophagus kingsnakes (*Lampropeltis getula*) by body-bridging (Bogert 1941; Weldon and Burghardt 1979; Weldon et al. 1992).

If disturbed, and prevented from escaping, *C. enyo* will coil, rattle, and strike if the disturber comes too close. When adjusted for SVL, mean peak rattle frequency is about 7.3 kHz, and mean rattle loudness is just below 70 dB (Cook et al. 1994).

PARASITES AND PATHOGENS: The species is a host for the tetrahyridian larvae of the tapeworm *Mesocestoides* sp., and the cystacanth larvae of an unidentified acanthocephalan (S. R. Goldberg et al. 2003b).

POPULATIONS: The rattlesnake is not uncommon in some areas, but no formal study has been conducted of the size, density, and dynamics of any population. The northwestern subspecies *C. e. furvus* is particularly poorly known because of the hardship of getting into the area where it occurs. The neonate sex ratio in litters reported

Crotalus enyo furvus (Catavinia, Baja California Norte; John H. Tashjian)

by Tryon and Radcliffe (1977) was 6 males to 8 females, or 0.75: :1.00, and by Olivier (2008a) was 3.3:1.0 and 4.5:1.0. Because of their small or fragmented distributions, Greene and Campbell (1992) thought *C. e. cerralvensis* and *C. e. furvus* at least vulnerable to extinction. Murphy and Ottley (1984) mentioned one that had fallen into a well and drowned, so accidents may also take their toll on this species.

REMARKS: The taxonomy and literature of *C. enyo* are reviewed by Beaman and Grismer (1994), Campbell and Lamar (2004), and Klauber (1972). Its systematic relationships are analyzed in Cadle (1992), Minton (1992), Murphy et al. (2002), and Stille (1987). The species is a member of the *durrissus*-group of *Crotalus*, which includes, in addition to those two species, *basiliscus*, *molossus*, *unicolor*, and *vegrandis*; its closest relative is *unicolor* (Murphy et al. 2002). The snakes' origin and evolution in Baja California are discussed by Grismer (1994d, 1999a), Murphy (1983a, 1983b), Murray (1955), and Savage (1960).

Crotalus helleri Meek, 1906
Southern Pacific Rattlesnake
Cascabel del Pacífico Meridional

RECOGNITION: *Crotalus helleri* (TBL$_{max}$, 137.1 cm, Klauber 1972; 162.7 cm, Ashton 2001) is a dark gray, olive-gray, gray-brown, pale brown, or pale yellow snake with a series of 35–36 (27–43) dark, white- or yellow-bordered, angular, and somewhat diamond-shaped dorsal blotches (sometimes centrally pale). These become more wavy and band-like near the tail. Also present are 1–2 rows of conspicuous, alternating, smaller dark blotches. The tail has a series of 2–8 dark bands; the poorly defined last band is about twice as broad as the others. The base of the tail and first rattle segment are yellow to orange in young individuals, but becomes brown or black in adults. The light body markings may fade with age, and older adults may be almost uniformly dark. Much variation in coloration and pattern occurs in all populations. Albinos and other oddly patterned striped individuals have been reported (Klauber 1972). The keeled and pitted body scales have a linear (straight or swirling) microdermoglyphic pattern (Harris 2005; Stille 1987), and occur in 23–25 rows anteriorly, 25 (23–29) at midbody, and 19 (18–22) posteriorly. The venter is cream to grayish with dark mottling. Present on it are 162–189 ventrals, 19–29 subcaudals, and an undivided anal plate. The head is black or dark brown posterior to the supraocular scales (juveniles may have some light dorsal markings that disappear with age). A pale white or yellowish bar normally crosses each supraocular, and a dark stripe angles diagonally downward from the orbit to above the angle of the mouth and is white or yellowish diagonal bordered. Dorsally on the head are a single higher than wide rostral scale, 3–5 internasals that touch the rostral, 2 pairs of canthals (the second pair is largest, prefrontals are absent), numerous small scales before the intersupraocular scales, 2 large supraoculars, and 4–6 smaller intersupraoculars. On each side of the head lie 2 nasal scales (the prenasal contacts the supralabials, the postnasal rarely contacts the upper preocular scale), a single (rarely 2) loreal scale, 2 preoculars, 2 (3) postoculars, 1–3 suboculars, 14–16 (12–18) supralabials, and 15–16 (13–20) infralabials. On the chin are a single small mental scale and two pairs of chin shields.

The bilobed hemipenis has a branch of the sulcus spermaticus extending up each lobe. Each lobe has about 56 spines and 28 fringes; no spines occur in the crotch between the lobes (Klauber 1972).

Dentition, other than the two maxillary fangs, consists of 3 (2–4) palatines, 7–8 (6–10) pterygoids, and 9–10 (6–12) dentaries (C. Ernst, personal observation).

Males (TBL$_{max}$, 162.6 cm) have 170–174 (162–184) ventrals, 24–25 (19–29) subcaudals, and a mean TL/TBL of 7.4. Females (TBL$_{max}$, 145.0 cm) have 175–178 (166–189) ventrals, 18–19 (15–25) subcaudals, and a mean TL/TBL of 5.6.

GEOGRAPHIC VARIATION: *Crotalus helleri helleri* Meek, 1906, the southern Pacific rattlesnake or cascabel del Pacifico Meridional, ranges from southwestern San Luis Obispo and Kern counties, California, south to northern Baja California Norte, and also occurs on Santa Catalina (Bryant 1915) and Coronado del Sur islands (Zweifel 1952). It is a large snake (TBL, > 100 cm), and is described above. *Crotalus h. caliginis* Klauber, 1949b, the Coronado Island rattlesnake or cascabel de Isla Coronado, is restricted to South Coronado Island off northwestern Baja California Norte. It is a short (TBL$_{max}$, 71 cm; Stebbins 2003), grayish to tan snake with a dark rostral mark and prominent brown dorsal blotches that are darker on the sides and become more band-like to the rear. The posteriormost tail bands are not very dark, and the body blotches may fade with age until older adults may be essentially patternless. Some dark spotting is present anteriorly on the supraoculars that sometimes forms a cross band.

CONFUSING SPECIES: Other species of *Crotalus* within or near its range can be distinguished from *C. helleri* by the key presented above. The smaller rattlesnakes of the genus *Sistrurus* have large plate-like scales on the dorsal surface of their heads. *Pituophis catenifer* lacks both loreal pits and a tail rattle.

KARYOTYPE: The karyotype is probably the same or similar to that described by Baker et al. (1972) and Zimmerman and Kilpatrick (1973) under the name *C. viridis*; see that species for details.

Crotalus helleri helleri (San Marcos, San Diego County, California; John H. Tashjian)

Crotalus helleri helleri, juvenile (Riverside County, California; Jeffrey E. Lovich)

FOSSIL RECORD: *C. helleri* has a rich fossil history, all from Pleistocene (Rancholabrean) sites in Southern California (Brattstrom 1953b, 1954a, 1958a; Gilmore 1938; Holman 1995; Hudson and Brattstrom 1977; Klauber 1972; La Duke 1991; Lundelius et al. 1983; Miller 1942; Miller 1971).

DISTRIBUTION: The species ranges south from southwestern San Luis Obispo and Kern counties, California, to at least northern Baja California Sur (Smith et al. 1971), and possibly the midlatitudes. It also occurs on the islands of Santa Catalina and South Coronado. Its total elevation range is from sea level to 3,296 m (Ewan 1932).

HABITAT: *C. helleri* is generally a resident of arid habitats: flat deserts with hard-packed soil (it is absent from the southeastern California deserts); dry coastal sagebrush; chaparral; non-native grassy areas; orchards and vineyards; upland mixed wooded habitats, such as the pine-oak forests of Baja California Norte; rocky canyons; and riparian zones. At these sites, it shelters in logs and crevices in rock outcrops, or beneath flat rocks or brush. Neonates prefer microhabitats that are densely vegetated and continuous (Figueroa et al. 2008).

Some typical vegetation found in its habitats include agave (*Agave* sp.), Baja elephant-tree (*Pachycormus discolor*), bursage (*Ambrosia* sp.), cacti (*Cereus* sp., *Opuntia* sp., *Stenocereus thurberi*), cardon (*Pachycereus pringlei*), cirio and ocotillo (*Fouquieria* sp.), copal (*Ailanthus altissima*), cottonwood (*Populus fremontii*), creosote bush (*Larrea tridentata*), ferns, junipers (*Juniperus* sp.), limberbush (*Jatropha cinerea*), manzanita (*Arctostaphylos* sp.), mesquite (*Prosopis* sp.), mountain lilac (*Ceanthus cordulatus*), oaks (*Quercus* sp.), pines (*Pinus* sp.), sagebrush (*Artemisia* sp.), and yucca (*Yucca* sp.).

BEHAVIOR AND ECOLOGY: The biology of wild *C. helleri* has been the least researched of the three current species in the *C. viridis*-complex. Consequently, much of our information is from scattered, diverse reports, some of which is anecdotal. Intensive long-term studies are needed, and some are currently underway.

In Southern California, some surface-active individuals can be found in every month, providing the weather is clement (Armstrong and Murphy 1979; Klauber 1926, 1972), but when the fall and winter weather turns cool and rainy, most seek shelter. Those from the vicinity of the San Gabriel River, Los Angeles County, do not den up or hibernate as such, but do congregate in south- and east-facing rock slides or outcroppings

Distribution of *Crotalus helleri*.

Crotalus helleri caliginis (Coronado Island, Pacific Ocean, Baja California Norte; Houston Zoo; John H. Tashjian)

during the rainy winter months (Armstrong and Murphy 1979). At Chino Hills State Park in Los Angeles County, the snake has an annual activity period of March–November; but is most active in the spring and fall, possibly due to a bimodal reproductive strategy, and least active from December to February (Dugan and Hayes 2005). The earliest fall disappearance on record is 15 September (Grinnell and Grinnell 1907), but November and December retreats are probably more normal. Neonates studied by Figueroa et al. (2008) remained active into December and had a longer active season than adults in the same area. Most individuals become active again in March (Dugan and Hayes 2005).

Nine radio-equipped male California *C. h. helleri* tracked through 2 annual active periods (2003–2004) and the winter of 2003–2004 had minimum convex polygon (MCP) and fixed-kernel (FK) home ranges as follows: active 2003—MCP, 4.69 ha; FK, 6.07 ha; winter 2003–2004—MCP, 0.08 ha; FK, 0.12 ha; and active 2004—MCP, 6.95 ha; FK, 5.00 ha. The snakes exhibited mean daily movements of 16.8 m per day, 2.7 m per day, and 23.9 m per day, respectively, for the 3 seasons (Dugan et al. 2008).

Winter retreats are normally only temporary, and, unlike those of *C. oreganus* or *C. viridus*, often involve only 1–2 snakes. Mammal burrows, shallow rock crevices, large flat rocks, or piles of leaves are used. At higher elevations these are more likely found in the northern and eastern parts of the range (Klauber 1972).

In the warmer months, Southern California *C. helleri* are nocturnally active from sunset to at least midnight (Klauber 1972), but are more diurnally active during the cooler spring and fall months. This same pattern has been reported for Baja California Norte populations; highland snakes tend to be diurnal, while those living in the lowlands are more nocturnal (Armstrong and Murphy 1979). Mexican *C. helleri* often emerge to bask immediately after thunderstorms (Armstrong and Murphy 1979).

BTs recorded from active Southern California *C. helleri* were 9.3–37.8 (mean, 25.4) °C and 21.0–33.4 (mean, 28.9) °C, and 27.6–30.8 (mean, 29.4) °C, and for those from Los Angeles County tested in a thermal gradient (Brattstrom 1965; Cunningham 1966). The snake is a confirmed basker, particularly in the mornings (C. Ernst, personal observation).

Figueroa et al. (2008) followed neonate *C. helleri* fitted with radio-transmitters. The little snakes generally moved in a linear position with movements being greater in September, indicating dispersal from the natal area. One individual moved 61 m in one day. All of the neonates that retained their radios survived their first winter, but it is not known whether or not they hibernated.

Both neonate and adult *C. helleri* will climb into bushes or trees, sometimes as high as 7.6 m, while foraging or possibly basking (Cunningham 1955; Figueroa et al. 2008; Klauber 1972; Shaw 1966). Klauber (1972) placed individuals in a freshwater pond; they did not seem frightened by the water, swam readily with lateral undulations of the body using their tails for propulsion (rattles held upright to avoid getting them wet), and could climb a wall 15–20 cm high to get out of the pool.

C. Ernst has witnessed both spring and fall male-male combat in captivity, and Klauber (1972) reported natural observations made in March and July. The behavior was similar to that reported for other species of the *C. viridis*-complex.

REPRODUCTION: The shortest gravid female *C. h. caliginis* examined by Klauber (1972) had a TBL of 52.8 cm, and that of the shortest reproductive *C. h. helleri* was

59.6 cm. The shortest male undergoing spermatogenesis found by Aldridge (2002) had a 62.5 SVL.

The female reproductive cycle is probably similar to that reported for *C. oreganus* and *C. viridis,* with vitellogenesis beginning in the summer, follicles overwintering at an intermediate stage, final vitellogenesis occurring in the spring prior to ovulation, and fertilization taking place in the spring (Aldridge 1979b). A female captured on 25–26 June expelled a partially formed embryo on 14 August (Brown *in* Armstrong and Murphy 1979). Females are capable of storing viable sperm for later fertilizations (Klauber 1972; Schuett 1992). They can store it for either a short period after spring matings or for longer periods over winter after late summer–fall matings (Schuett 1992).

Aldridge (2002) described and illustrated the male reproductive cycle, in which the various stages are also similar to those of *C. oreganus* and *C. viridis.* During March to mid-April, the seminiferous tubules contain only spermatogonia and Sertoli cells. From April to June most males contain both primary and secondary spermatocytes. Spermiation begins in late June and lasts through September, with sperm first appearing in seminiferous tubules during the second half of June. All males contain sperm in the seminiferous tubules in July–September, but spermatogenesis slows and ends in September. The vas deferens at the level of the kidney contains sperm throughout the active season. Morphological changes also occur in the seminiferous tubules during the various gametic stages. In the early season (spring) the seminiferous tubules are at their smallest diameter (mean, 193 µm), by midseason they have increased in diameter (mean, 203 µm), and they reach their maximum size late in the season (mean, 288 µm). The mean diameter of the sexual segment of the kidney also changes during the reproductive cycle in a bimodel pattern: early season, 189 µm; midseason, 150 µm; and late season, 183 µm. Its hypertrophied period corresponds to the mating season. Male-male combat only occurs during this kidney-segment stage.

C. helleri has two mating seasons, spring and late summer–fall (Dugan et al. 2008), each coordinated with female vitellogenesis. Klauber (1972) reported the following mating dates: *C. h. caliginis*, 5 March; *C. h. helleri*, 7 May (captivity) and 17 September. Females are located through olfaction, especially after they have recently shed. The species' courtship and mating behaviors are similar to those described for *C. oreganus* and *C. viridis* (C. Ernst, personal observation).

A litter of *C. h. caliginis* was born in captivity on 26 August; parturition in *C. h. helleri* normally occurs in late September into late October (12 September–25 October; Klauber 1972), but a female captured the previous August gave birth in captivity on 20 December (Cunningham 1959). Litters of *C. h. caliginis* contain 1–4 (mean, 2.57; n = 7) young (Klauber 1972), while females of the larger *C. h. helleri* may produce litters of 1–16 (mean, 7.47; n = 17) young (Cunningham 1959; Klauber 1972). Neonate *C. h. helleri* have 25.0–30.5 (mean, 26.5; n = 18) cm TBLs. The largest neonate *C. h. caliginis* measured by Klauber (1972) had a 22 cm TBL, and the average TBL at birth was only 19 cm. Neonates of both subspecies resemble adults, but are of a brighter hue with more pronounced markings and a bright yellow to orange tail base. Embryos have a caruncle (egg tooth) attached to their snout.

Natural hybridization with *C. ruber* has been reported by Armstrong and Murphy (1979), and also in captivity by Klauber (1972) and Perkins (1951); a color photograph is in Klauber (1972). Stebbins (2003) reported natural hybridization with *C. scutulatus* in the Antelope Valley of California, and Glenn and Straight (1990) reported venom hybridization between the two species. A male *Pituophis catenifer* copulated with a female *C. helleri* at the San Diego Zoo, resulting in her death (Klauber 1972).

GROWTH AND LONGEVITY: Klauber (1972) presented a growth curve based on 829 *C. helleri* from San Diego County, California. At birth the young have 27.5–28.0 cm TBLs. Neonates found in the wild at the end of September and October are 30–34 cm long, and Klauber thought that some of this initial growth results from the absorption of the residual yolk supply, as neonates are "puffed up" with yolk at birth. Most enter their first hibernation at 35 cm with only a rattle button. By the following mid-April, the average length is 38 cm, but most individuals are 31–46 cm; a few have 3 rattle segments, but 1–2 are more prevalent. The average TBLs in May–September are, respectively: 40 (range, 31–50) cm with mostly 2 (range, 1–4) rattle segments, 43 (range, 31–58) cm with 2 (range, 2–4) segments, 45–50 cm with 3 segments, 50 cm with 3–4 segments, and 54 cm with 4 segments. Total increase in length during the first year is about 94%, and year-old snakes enter hibernation at about 60 cm TBL with 4–5 rattle segments. During their second year, the snakes grow to 70 cm or slightly more by June; in September males somewhat exceed 80 cm, but females only average about 72 cm. They enter hibernation that year at about these same lengths with 8–9 segments on complete rattles. Based on his generated growth curve, Klauber thought that males would average 90 cm with 9–10 rattle segments by the following June.

The species is potentially long-lived. An adult female captured in the wild survived for another 24 years, 1 month, and 25 days at Zoo Atlanta (Snider and Bowler 1992). Another lived 19 years and 5 months in captivity at the San Diego Zoo (Shaw 1969).

DIET AND FEEDING BEHAVIOR: Prey are detected by three senses: olfaction to find prey trails and potential ambush sites, and to locate envenomated prey; vision to locate prey, especially while foraging during the day; and infrared detection for finding

Crotalus helleri caliginis (Coronado Island, Pacific Ocean, Baja California; Reptilienhaus, Uhldingen, Germany; John H. Tashjian)

prey, especially at night and at ambush sites. The visual and infrared data are combined in bimodal neurons of the optic tectum that receive input from both the retina and the pit organ (Newman and Hartline 1981, 1982). This is very important during crepuscular or nocturnal foraging. Ambush sites include the sides of rodent trails, fallen logs, and under brush and bushes. Neonates seek out sites with lizard chemosensory cues.

The snake has consumed Jerusalem crickets (*Stenopelmatus fuscus*, but the snake's stomach also contained a toad that may have first ingested the insects), toads (*Bufo californicus*), salamanders (*Aneides lugubris*), lizards, snakes (*C. helleri*, captivity), birds (*Picoides pubescens*), and mammals have been reported as prey (Atsatt 1913; Cunningham 1959; Grinnell 1908; Grinnell and Grinnell 1907; Grismer 2002; Klauber 1972; Mahrdt and Banta 1997; Mitchell 1986; Powers 1972; Ruthling 1916; Storer and Wilson 1932; von Bloeker 1942; Wright and Wright 1957). At Chino Hills State Park prey consists mostly of small (20–150 g) rodents (*Microtus californicus*, *Perognathus* sp., *Thomonys bottae*) (Dugan and Hayes 2005). Neonates and juveniles < 50 cm TBL of both subspecies prey predominantly on lizards (*Cnemidophorus tigris*; *Elgaria multicarinata*; *Eumeces skiltonianus*; *Sceloporus graciosus*, *S. occidentalis*, *S. orcutti*; *Uta stansburiana*), changing over to mostly mammals as they grow (Campbell and Lamar 2004; Klauber 1949b, 1972; La Bonte 2008; Mackessy 1988), with those snakes > 50 cm almost entirely mammal eaters (*Dipodomys agilis*; *Microtus californicus*; *Neotoma albigula*; *Perognathus longimembris*; *Peromyscus maniculatus*, *P. pseudocrinitus*; *Reithrodontomys megalotis*; *Spermophilus beecheyi*; *Sylvilagus auduboni*, *S. bachmani*; *Tamias merriami*, *T. speciosus*; *Thomonys bottae*) (Ashton 2000; Campbell and Lamar 2004; Cunningham 1959; Dugan and Hayes 2005; Grismer 2002; Klauber 1949b, 1972; La Bonte 2008). Klauber (1972) also reported the consumption of birds (*Callipepla californica*, *Gallus gallus*, *Geococcyx californicus*, *Meleagris gallopavo*, *Oreortyx pictus*, *Zenaida macroura*) by adults. Cannibalism has been observed on several occasions (Mitchell 1986). Powers (1972) observed cannibalism of a carrion neonate by another captive neonate, and Cunningham (1959) also reported the consumption of carrion. A captive *C. helleri* ingested a *C. mitchellii pyrrhus*, and another a *C. helleri* (Klauber 1972; Powers 1972). Secondarily ingested (?) gravel and small stones have been found in both *C. h. caliginis* and *C. h. helleri* (Klauber 1972).

VENOM DELIVERY SYSTEM: The fang dimensions of 5 *C. h. helleri*, (reported as *C. v. oreganus* from San Diego County) measured by Klauber (1939) were FL, 7.8–9.4 (mean, 9.0) mm; mean TBL/FL, 113; and mean HL/FL, 5.04. The same measurements for 5 *C. h. caliginis* (reported as *C. v. oreganus* from Cornados Island) were FL, 4.9–6.1 (mean, 5.5) mm; mean TBL/FL, 110; and mean HL/FL, 5.22. Klauber did not give the mean angle of fang curvature for either subspecies.

VENOM AND BITES: *C. helleri* is involved in more than 80% of the human envenomations in Southern California (Wingert and Chan 1988), and as such presents a major public health problem.

The venom yield varies between the two subspecies. A 61 cm *C. h. caliginis* yielded 19 mg of dry venom that was composed of 60% protein; normal yields of dry venom for *C. h. helleri* have ranged from 75 to 390 (mean, 112) mg (Brown 1973; Glenn and Straight 1982; Klauber 1972 [combined sample with *C. oreganus*]; Minton and Minton 1969; Russell 1960a; Russell and Brodie 1974; Russell and Puffer 1970). Herbert (1998) and Rehling (2002) have reported expenditures per bite of 0–235 mg. Individual *C. helleri* contain venom volumes allowing for the envenomation of several consecutive prey, but the amount ejected is reduced with each additional bite (Rehling 2002).

The venom is neurotoxic (Bush and Siedenburg 1999; Klauber 1972; Grenard 2000), and quite potent (McCue 2005). The neurotoxic reactions are caused by the presynaptic Mojave toxin (French et al. 2004). The venom has two color variations, white and yellow (Galán et al. 2004; Johnson et al. 1987). The white venom contains fewer low molecular weight components and is considerably less toxic, producing no effects in mice when injected intravenously at concentrations up to 10 mg/kg (Johnson et al. 1987). However, Galán et al. (2004) reported a LD_{50} of 2.95 mg/kg, whereas the LD_{50} of yellow venom is 1.60–1.84 mg/kg. Both venom types are hemorrhagic, but the white causes less intradermal hemorrhage in mice. White venom protease and phospholipase A_2 activity is much less than that of the yellow venom. Both venom types produce myonecrosis at 1, 3, and 24 hours after intramuscular injection, and some Mojave toxin activity is associated with each type.

Recent articles in the popular press have reported that the venoms of several southwestern species of *Crotalus*, especially *C. helleri* and *C. scutulatus*, may be becoming more virulent. There is no scientific basis for this speculation (Hayes and Mackessy 2010).

Seventy-one percent of the solids in the venom are composed of protein (Brown 1973; McCue 2005). The pH is approximately 5.5–5.6 (Mackessy and Baxter 2006).

The following important enzymatic activities and those of other venom components have been identified in the venom: arginine esterase, catrin, catrocollastatin, catroxase I, chynotrypsin-like activity, defibrizyme, dipeptidyl peptidase IV, deoxyribonuclease, esterase activity, hemorrhagic toxin II (HT-2), kinin-releasing activity, L-amino acid esterase, Mojave toxin (which has 97% of the DNA sequence of that produced by *C. scutulatus*; some geographic variation occurs, see below), NAD-nucleosidase, 5′nucleotidase activity, peptide C, phosphodiesterase activity, phospholipase A_2, phosphomonoesterase activity, protease activity, ribonuclease 1, and trypsin-like activity (Aird 2005, 2008; Al-Joufi and Bailey 1994; Bober et al. 1988; French et al. 2004; Galán et al. 2008; Glenn and Straight 1985b; Gregory-Dwyer et al. 1986; Henderson and Bieber 1986; Jurado et al. 2007; Maeda et al. 1978; McCue 2005; Meier and Stocker 1995; Rael et al. 1986; Sanchez et al. 2007; Schaeffer et al. 1972a, 1972b; Wingert and Chan 1988). The southern-most populations lack Mojave toxin (Glenn and Straight 1985b; Rael et al. 1986), but do contain a low molecular weight myotoxin (peptide C) homologous to crotamine (Aird et al. 1991; Rael et al. 1986; Straight et al. 1991), and possibly a component that can activate factor X in the coagulation process (Ouyang et al. 1992; Russell 1983).

The content of dipeptidyl peptidase IV varies geographically (Aird 2008). Interestingly, the venom of *C. h. caliginis* tested by Glenn and Straight (1985b) showed no phosphodiesterase or protease activity, while that of *C. h. helleri* showed esterase, phosphodiesterase, and protease activity; they found no Mojave toxin activity in either subspecies. The populations at the southern end of the distribution are essentially hemotoxic, while the northern populations in northern Baja California Norte and Southern California have a more neurotoxic venom.

Ontogenetic variation occurs in the development of the venom. Juveniles have highly toxic venom with low protease activity, which efficiently incapacitates lizards and young rodents. Protease increases significantly, and L-amino oxidase and exonuclease activities also increase somewhat with growth. Phospholipase A_2 activity, however, decreases significantly, as does venom toxicity (Mackessy 1988). Studies by Gregory-Dwyer et al. (1986) indicated that seasonal changes in venom toxicity do not exist, but that individual variation in toxicity does occur.

Laboratory mice (*Mus musculus*) have been the favorite test animal for the determination of the venom's lethality. The intramuscular LD_{50} is 3.56 mg/kg (Minton 1956). Kocholaty et al. (1971) reported an intraperitoneal LD_{50} of 2,440 µg/kg for *C. h. helleri*; other reported intraperitoneal mouse LD_{50} values by this subspecies have been 1.40–2.44 mg/kg (Githens and Wolff 1939; Glenn and Straight 1982, 1985b; Macht 1937; Minton 1956; Russell and Brodie 1974; Russell and Puffer 1970; Weinstein et al. 1992). Two *C. h. caliginis* had LD_{50} values of 2.0–3.0 (mean, 2.6) mg/kg (Glenn and Straight 1985b). Kocholaty et al. (1971) gave the intravenous LD_{50} as 844 µg/kg; other reported intravenous LD_{50} values are 1.00–2.13 mg/kg (Glenn and Straight 1982; Russell and Brodie 1974; Russell and Puffer 1970; Schaeffer et al. 1973).

All laboratory rats (*Rattus norvegicus*) injected with 4 mg/kg of *C. h. helleri* venom died within 15 hours; others injected with a lower dosage survived. Venom injection produced a drop in the numbers of blood platelets, but did not seem to damage the platelet precursors in the bone marrow (Wingert et al. 1981).

Venom injected into the diaphragm and bladder of guinea pigs (*Cavia porcellus*) resulted in a decrease in contractions of the striated muscle of the diaphragm and some decrease, but more immediate, contraction of these muscles in the bladder (Metsch et al. 1984).

The intravenous LD_{50} for cats (*Felis catus*) and dogs (*Canis familiaris*) is 1.00 and 0.05 mg/kg, respectively (Schaeffer et al. 1973). Venom injected in a 1.0 mL volume into blood vessels of dogs, at a dose level of 50 µg/50 mL, caused smooth muscle contraction in all vessels tested (Pattabhiraman et al. 1976). Domestic livestock fatalities occur in no more than 20% of envenomation by *C. h. helleri*; cattle, horses, and burros usually recover (Klauber 1972).

The LD_{50} for a 350 g pigeon (*Columba livia*) is normally 0.10–0.12 (0.06–0.14) mg (Githens and Butz 1929; Githens and George 1931; Klauber 1972).

Most human envenomations in Southern California are by *C. helleri,* and most of those bites occur during the mating season when males are more aggressive, have high testosterone levels, and are actively searching for females (Cardwell et al. 2005). Fang contact during bites of humans averages 0.22–0.23 second (Herbert 1998; Rehling 2002).

Envenomations have resulted in death (Norris 2004), and some of the fatal envenomations attributed to *C. oreganus* in 1928–1929 reported by Hutchison (1930) probably involved *C. helleri* instead. The estimated dry venom human LD_{50} is 1.0 mg/kg (Russell 1960a, 1960b), or about 50–70 mg (Minton and Minton 1969). Brown (1973) reported a subcutaneous LD_{50} of 252 mg/kg × 70, and an estimated number of LDs per bite of 0.49.

Interestingly, venom from Southern California *C. helleri* seems more neurotoxic than in the past. However, this observation may be due to more snakebite cases as humans expand into its habitat. Also, there is the possibility that the snake has received some genes for more potent venom through hybridization with the Mojave rattlesnake, *C. scutulatus* (Glenn and Straight 1990; Stebbins 2003).

Reported symptoms of human envenomation by *C. h. helleri* are blebbing and discoloration at the bite site, changes in heart rate and potential cardiovascular failure, chills and fever (possibly from serum sickness), difficulty in speaking, dizziness and giddiness, drawn feeling in scalp, drooling, hematuria, hypotension, instant pain (or in some cases, no pain), nausea (or not), paleness, numbness, pain and partial paralysis, profuse sweating, respiratory difficulty and potential failure, shock, swelling, tingling sensations (mouth, scalp, digits, bite site), and weakness (Bush 2006; Bush and Siedenburg 1999; Davidson 1988; Happ 1951; Holstege et al. 1997; Klauber 1972; Norris and Bush 2007; Russell 1960a, 1960b; Russell and Michaelis 1960; Sanchez et al. 2007; Schaeffer et al. 1973, 1979; Wingert and Chan 1988).

In addition, the bacteria (*Alcaligenes faecilis*; *Bacterioides thetaiotaomicron, B. fragilis*; *Clostridium carnis, C. perfringens, C. sordellii*; *Enterobacter cloacae*; *Eubacterium* sp.; *Lactobacillus jinsenii*; *Pseudomonas aeruginosa*; *Staphyloccous epidermidis*), several of which potentially could cause serious infections in humans, have been isolated from the fangs and oral cavity of *C. helleri* (Talan et al. 1991). The snake's venom shows antibacterial activity toward most of these.

The venom may be neutralized with the commercial antivenoms Antivipmyn (Fab$_2$H) and CroFab (FabO[ovine]), although CroFab is more effective (Galán et al. 2004). Interestingly, sera from the gray woodrat (*Neotoma micropus*) and cotton rat (*Sigmodon hispidus*) also neutralize the venom, but sera from the South American opossum (*Didelphis marsupialis*) and the Mexican ground squirrel (*Spermophilus mexicanus*) do not. Serum from the kingsnake, *Lampropeltis getula*, also neutralizes the venom (Weinstein et al. 1992).

PREDATORS AND DEFENSE: This species apparently experiences a relatively high predation rate; 55% of the radio-telemetered *C. h. helleri* followed by Dugan and Hayes (2005) were predated. Many potential predators occur within its range: snakes (other *Crotalus* sp.; *Lampropeltis getula, L. zonata*; *Masticophis flagellum, M. lateralis*; *Pituophis catenifer*), roadrunners (*Geococcyx californicus*), predatory hawks (*Buteo* sp.), owls (*Bubo virginianus*), raccoons (*Procyon lotor*), mustelids (*Mustella frenata, Spilogale gracilis, Taxidea taxus*), canids (*Canis latrans, Urocyon cineroargenteus, Vulpes macrotis*), and felids (*Felis catus, Lynx rufus, Puma concolor*). However, reports of actual predation in nature are very rare, although Anthony (1893) witnessed a predatory attack on a *C. helleri* by a Swainson's hawk (*Buteo swainsoni*). Both *C. helleri* and *C. ruber* have ingested this snake in captivity (Klauber 1972; Powers 1972), and *Masticophis flagellum* in the wild (Tabor and Germano 1997).

The body color and pattern tend to be cryptic (Sweet 1985), especially hiding the snake when it is lying in shaded vegetation. It also forms a mimicry complex with the similarly patterned nonvenomous gopher snake, *Pituophis catenifer*, in which that species mimics the rattlesnake (Sweet 1985). Fang contact during a defensive bite lasts an average of 0.22 (range, 0.97–0.53) second (Herbert 1998).

When confronted with the potential predator, *Lampropeltis getula californiae*, *C. helleri* will strike and bite the other snake or body-bridge to avoid being grasped and constricted (Klauber 1972; Weldon et al. 1992).

C. helleri is a very dangerous snake. It tends to be nervous and quick tempered, rattling furiously and striking, especially if startled or attacked; but variations in temperament occur between individuals ranging from extremely nervous to very calm (C. Ernst, personal observation). During the 0.07–0.53 (mean, 0.22) second of fang contact during a defensive strike it can expend up to 3.8–235.2 (mean, 71.8) mg of venom (Herbert 1998), which is much more than needed for a lethal human envenomation.

PARASITES AND PATHOGENS: Several parasites have been recorded from *C. helleri*. The cestodes *Mesocestoides* sp. and *Oochoristica crotalicola* have been found in it (Alexander and Alexander 1957; Bursey et al. 1995; Mankau and Widmer 1977; Widmer and Hanson 1983; Widmer and Specht 1992). Wood and Wood (1936) reported an unidentified hemogregarine protozoan in its blood, and Klauber (1972) found fly maggots in a wound on one.

Klauber (1972) reported that an individual at the San Diego Zoo was diagnosed with the paracolon bacterium (#10). A *C. helleri* had a cystic hemangioma in its cloaca (Wadsworth 1956).

POPULATIONS: The species is not uncommon over most of its northern mainland distribution, but less so in Baja California (Armstrong and Murphy 1979). Klauber (1926) collected 171 individuals in San Diego County from January 1923 through December 1925; the species represented 10.5% of the grand total of 1,623 snakes collected. By 1956, Klauber (1972) had collected 2,751 *Crotalus* sp., of which *C. h. helleri* made up 45.7% (1,258 individuals). Armstrong and Murphy (1979) thought the populations on the lower slopes of the San Gabriel Mountains in Los Angeles County to be as dense as any of the genus they knew.

A sample of 638 *C. h. helleri* from San Diego County yielded an almost perfect 1:1 male (318) to female (320) ratio (Klauber 1943b), and Klauber (1972) reported an essentially 1:1 male (102) to female (107) ratio among 209 neonates from San Diego County.

Today, the snake experiences ever increasing habitat destruction, especially around urban centers in Los Angeles and San Diego counties; if found near human inhabitations it is usually killed. Also, as the human population increases in Southern California, more and more *C. helleri* are killed on its roads (C. Ernst, personal observation; Klauber 1932). Another hazard the snake now encounters more frequently in Southern California is brush fire, and some individuals are probably killed and others injured during such episodes. Some are collected for the pet trade each year, as the species is not protected in California.

C. h. helleri is less common on Santa Catalina Island since the introduction of pigs (*Sus scrofa*), but its existence there may be assured because most of the island is now protected and introduced animals are being eradicated (Ashton 2000).

On South Coronado Island, *C. h. caliginis* is probably limited to about 1.5 km² of habitat because of the steep sides of the island; once common, the snake now faces an uncertain future there (Ashton 2000). Although the island has few human inhabitants, goats and mules have been introduced, which can potentially destroy the snake's habitat (Rubio 1998).

REMARKS: The evolutionary and systematic relationships within the *Crotalus viridis*-complex, of which *C. helleri* is a member, are discussed under that species. Beaman and Hayes (2008), in the most recent checklist of rattlesnakes, include both subspecies of *C. helleri* among those of *C. oreganus*.

Crotalus horridus Linnaeus, 1758
Timber Rattlesnake

RECOGNITION: Adult *C. horridus* have reached a TBL$_{max}$ of 189.2 cm (McIlhenney *in* Klauber 1972). Body color varies from pale to bright yellow to gray, dark brown, or black (mostly in northeastern populations), and the body tends to be darker posteriorly. The light morph makes up 40–93% of eastern populations; dark morphs tend to predominate in montane, densely forested sites, particularly in the Northeast (Martin 1988). Individuals from southern populations have a red to reddish-orange vertebral stripe, and such a marking may be found in the Midwest from Texas to Minnesota. Present are 15–34 (mean, 24) chevron-like or V-shaped, usually light-bordered, dorsal body bands that are more blotch-like anteriorly, but become more chevron-like posteriorly. Lateral body stripes are rarely present (Dundee 1994; Gloyd 1935b; Nickerson and Mays 1968). The tail is dark gray, brown, or black; neonates and juveniles may have 3–6 caudal cross bands, and some faded cross bands may be still present in adults. The basal tail segment is dark gray, dark brown, or, usually, black. About 6.6% of the snakes lack a rattle (Rowe et al. 2002), and rattles with 19–22 segments have been reported (Brimley 1923; Fitch 1985b), but more normally there are only 6–12 segments present (Fitch 2002). Abnormal colorations, including albinos, have been reported (Dyrkacz 1981; Grube 1963; Harris 2006b; Hensley 1959; Hudson and Carl 1985; Klauber 1972; Neill 1963; Ortenburger 1922; Petersen 1970). At one site in the hills overlooking Lake George, New York, during one day, William S. Brown and C. Ernst captured individuals with body colors varying from pale xanthic to black. Odum (1979) reported the following color phases in New Jersey's pine barrens: yellow, 56%; brown, 28%; and black 16%. The various color hues are not sex-related (Gloyd 1940; Schaeffer 1969). Dorsal body scales are keeled and pitted, and form 25–27 (23–29) rows anteriorly, 23 or 25 (21–26) midbody rows, and 19 (17–21) posterior rows. Their microdermoglyphic pattern is made up of curved and twisted tubercles, and is illustrated in both Harris (2005) and Stille (1987). The venter is pink, white, cream, or yellow with small, dark stipples present; juveniles have a more distinct pattern of dark marks. Present are 154–183 ventrals, 13–31 subcaudals, and an undivided anal plate. On each side of the head is a dark reddish-brown, dark brown, or black stripe that extends from the eye backward to beyond the corner of the mouth, and some individuals may have round occipital spots. Dorsally, the head is unpatterned, and covered with numerous small scales. Dorsal head scales include a higher-than-broad rostral, 2 internasals, 4 canthals (1 pair of which may be in contact medially; only 1 pair is positioned between the internasals and supraoculars), and 2 supraoculars. Laterally are 2 nasals (the prenasal touches the first supralabial; the postnasal is separated from the preoculars by the loreals), 2 (1–3) loreals, 2 preoculars, 4 (2–6) postoculars, several suboculars, 2 (1–3) interoculabials, 13–15 (10–17) supralabials (separated from the anterior suboculars by interoculabials), and 14–16 (11–19) infralabials. On the chin are a mental scale and two pairs of chin shields, with the larger second pair in contact medially with no small scales separating them.

The hemipenis is illustrated in Dowling (1975). It is bifurcated with a forked sulcus spermaticus. Present along the length of each lobe are more than 70 recurved spines and 30 fringes, but no spines are present in the crotch between the lobes (Klauber 1972).

In addition to the maxillary fang, 2–3 palatine, 8–9 (10–11) pterygoidal, and 10–11 (12–13) dentary teeth are present on each side (Klauber 1972).

Adult males are larger than females (TBL$_{max}$, 189.2 cm, sex not originally given; but average 90–150 cm), and have 154–179 (mean, 171) ventrals, 18–31 (mean, 24–26) subcaudals, and TLs 5–14 (mean, 8) % of TBL. Females (TBL$_{max}$, 135 cm; but average 80–100 cm) have 154–183 (mean, 170) ventrals, 13–28 (mean, 19–21) subcaudals, and TLs 4–15 (mean, 6) % of TBL. Fitch (1999) reported that 4-year-old Kansas males have SVLs of 87.5–100.3 (mean, 92.5) cm, while females are 76.0–90.0 (mean, 85.6) cm; Galligan and Dunson (1979) reported that in Pennsylvania 42% of males surpass 90 cm SVL, but only 10% of females grow to this size. Female *C. h. horridus* retain the dark postorbital stripe throughout life, although its intensity may decrease; the stripe becomes obscured in mature males (Storment 1990).

GEOGRAPHIC VARIATION: Formerly, two subspecies were recognized based on body color patterns and scale patterns (see Brown and Ernst 1986 and Pisani et al. 1973 for discussions of their validity). One is *Crotalus horridus horridus* Linnaeus, 1758, the timber rattlesnake, formerly in southern Ontario and southern Maine (Hunter et al. 1999) but currently ranging from New Hampshire, Vermont, and northeastern New York west to southeastern Minnesota (Breckenridge 1944; Le Clere 1996), southern Iowa, and southeastern Nebraska south to northern Georgia, northwestern Arkansas, and northeastern Texas. The body color varies from yellow to gray or black, the distinct reddish vertebral stripe is absent, there are 23 (21–26) midbody scale rows, and 15–34 dorsal body bands. The second subspecies is *Crotalus h. atricaudatus* Latreille, *in* Sonnini and Latreille, 1801, the canebrake rattlesnake, ranging from southeastern Virginia along the Atlantic Coastal Plain to northern Florida, westward to eastern Texas, and northward in the Mississippi Valley to southern Illinois. The body color is pinkish-brown or gray, and has a distinct reddish-orange vertebral stripe, 25 (21–25) midbody scale rows, and 21–29 dorsal body bands.

Crotalus horridus; yellow phase, northern *horridus* morph (R. D. Bartlett)

Crotalus horridus;
yellow phase,
northern *horridus*
morph (Pennsylva-
nia; Ft. Worth Zoo;
John H. Tashjian)

Conant and Collins (1998) use the term *color morphs* to describe these populations instead of the term *subspecies*: a "yellow variation" and a "black variation" in the Northeast (the *horridus* morph), a "southern variation" in the lowlands of the South (the *atricaudatus* morph), and a "western variation" found west of the Mississippi River from the Ozarks northward (a *horridus* × *atricaudatus* morph?).

We have seen photographs of about 20 live and DOR individual *C. horridus* of varying lengths and patterns from Prince William Forest Park, Prince William County, Virginia (Martin et al. 1992; Mitchell 1994b). A small population is present there, but whether it is natural or introduced has not been determined (Ernst et al. 1997; Mitchell 1994b). These snakes seem intermediate in coloration and pattern between the northern *horridus* and southern *atricaudatus* morphs. Oldfield and Keyler (1997) have reported that Minnesota timber rattlesnakes, thought to be the *horridus* color phase, more closely resemble the *atricaudatus* morph.

Villarreal et al. (1996) isolated and characterized 6 microsatellite loci from 32 populations in eastern Pennsylvania, southern New Jersey, North Carolina, South Carolina, and Alabama. The allelic frequencies varied among the geographically separated populations. The isolated population from Ocean County, New Jersey, showed reduced heterozygosities at three of the four loci examined, in comparison with a population from Berks County, Pennsylvania. For two of the loci, all six snakes from New Jersey were homozygous for the same allele in comparison to heterozygosities of 0.4 and 0.2, respectively, for the Pennsylvania snakes. Such high levels of homozygosity within the New Jersey population may indicate it is experiencing a genetic bottleneck, a conclusion also reached by Bushar et al. (2005).

The subspecies question was finally settled by A. M. Clark et al. (2003), who examined 310 base-pair fragments of mtDNA cytochrome b in 123 individuals of *C. horridus* from across the species range. Neighbor-joining and parsimony analysis revealed only a shallow gene genealogy (d_{max} = 0.024) and the sharing of haplotypes among the purported subspecies. Analysis of molecular variance demonstrated that traditional subspecific divisions explain only 3.5% of variation within the species, while the alternative geographic classification, northern (the *horridus* morph), southern (the *atricaudatus* morph), and western (the *horridus* × *atricaudatus* morph?) regions explain 18.6% of the genetic variation. The better performance of the regional approach to intraspecific variation can be attributed to an east-west phylogeographic partitioning at the Appalachian and Allegheny mountain ranges, which were probably uninhabitable at higher elevations during glacial intervals. Distribution of haplotypes and climatic data suggest that a radiation into the more northern portions of the range occurred after the Wisconsinan glaciation. So, the mtDNA data indicate distinct population segments across the range of the species, but do not show evolutionary separations that support subspecific designations. Allsteadt et al. (2006) also found that *C. horridus* has high levels of polymorphism in morphological traits, especially in coloration and pattern, but that these variances are strongly clinal, and too extensive and complex to support the recognition of subspecies.

CONFUSING SPECIES: Other *Crotalus* can be identified by the key previously presented. Pygmy rattlesnakes and massasaugas (*Sistrurus*) have nine enlarged plates on top of the head.

KARYOTYPE: Each body cell has 36 chromosomes, consisting of 16 macrochromosomes, including the Z and W sex chromosomes (4 metacentric, 6 submetacentric, 4 subtelocentric), and 20 microchromosomes. Sex determination is ZZ in males and ZW in females (Zimmerman and Kilpatrick 1973).

FOSSIL RECORD: Fossil *C. horridus* are known from the Pliocene (Blancan) of Nebraska (Rogers 1984); the Pleistocene (Irvingtonian) of Arkansas (Dowling 1958), Maryland (Holman 1977a, 1980), Nebraska (Holman 1995), Pennsylvania (Holman 1995), Texas (Holman and Winkler 1987), and West Virginia (Holman 1982; Holman and

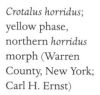

Crotalus horridus; yellow phase, northern *horridus* morph (Warren County, New York; Carl H. Ernst)

Grady 1989); and the Pleistocene (Rancholabrean) of Alabama (Holman et al. 1990), Georgia (Holman 1967), Indiana (Holman and Richards 1981; Richards 1990), Massachusetts (Van Frank and Hecht 1954), Missouri (Holman 1965, 1974; Parmalee et al. 1969), New York (Steadman and Craig 1993; Steadman et al. 1993), Pennsylvania (Guilday et al. 1964, 1966; Holman 2000a), Tennessee (Brattstrom 1954a; Holman 1995; Van Dam 1978), Virginia (Guilday 1962; Holman 1995; Martin 1996a), and West Virginia (Holman 1995; Holman and Grady 1987).

In addition, a Nebraska Miocene (Hemphillian) fossil has been identified as close to *C. horridus* (Parmley and Holman 1995), and a Rancholabrean vertebra has been recovered at a Kansas site that may be from either this species or *Agkistrodon contortrix* (Johnson 1975).

DISTRIBUTION: *C. horridus* once lived in southern Maine (Hunter et al. 1999; Palmer 1946) and still occurs in New Hampshire and Vermont (Baarslag 1950; Conant and Collins 1998), but its main distribution today extends from northeastern and southern New York west to southeastern Minnesota, southern Iowa, and southeastern Nebraska; and south (except on the Delmarva Peninsula; Harris 2007) to northern Florida and eastern Texas. Although much development has occurred in the western Piedmont of North Carolina, some *C. horridus* still exist there (Platt et al. 1999; Stroupe and Dorcas 2001), and scattered colonies exist in eastern Massachusetts and Ohio. The species formerly occurred in southwestern Ontario, Canada, but has been extirpated there; the last one was taken at Niagara Glen in 1941 (Cook 1984). The species' distributional elevation ranges from sea level to about 1,200 m (Barbour 1950).

HABITAT: The northern highland populations of the *horridus* morph inhabit upland woodlands, usually south-facing with nearby rock ledges (with crevices) or talus slopes. In Pennsylvania, the snake uses a gradient of habitats, from mature forests with numerous fallen logs to young forests with predominantly leaf litter cover (Reinert 1984b). Dark snakes prefer the former, but yellow ones are more frequently found

Distribution of
Crotalus horridus.

in the latter. Such habitat separation of the color variations probably is primarily concerned with background matching for camouflage.

Reinert (1984a, 1984b) found that Pennsylvania gravid females use more open, generally warmer microhabitats than do males or nongravid females; but in North Carolina, both sexes were more often located in close canopy areas, and only one male was found more often at open sites (Davis 2010). In the pine barrens of southern New Jersey, males and nongravid females utilize forested habitats with greater than 50% canopy closure, thick surface vegetation (approximately 75%), and few fallen logs, while gravid females use less densely wooded sites with approximately 25% canopy closure, an equal mixture of vegetation and leaf litter covering the ground, frequent fallen logs, and generally warmer conditions. The habitat preferred by gravid females is of lesser extent, being mostly restricted to road borders (Reinert and Zappalorti 1988a). Similar sexual microhabitat differences occur in South Carolina and Indiana (Gibson et al. 2008; Waldron et al. 2005). Gravid females in Indiana often utilize the insides of large hollow logs for retreats (Gibson et al. 2008). *C. horridus*, although primarily a forest-dweller, apparently requires some open microhabitats for successful breeding.

In North Carolina, males and nongravid females use deciduous woods with nearly closed canopies, but gravid females prefer rockier, less forested, and more open canopied sites (Sealy 2002). *C. horridus* on southern Georgia's Gulf coastal plain are closely associated with hardwood habitat and riverine edges, but do not usually use roads or the ecotonal edges between habitats (Steen et al. 2007). Kentucky *C. horridus* are most often found in dry, brushy habitats, but also in mixed woodlands. In the eastern mountains, caves are sometimes used for summer retreats and the snakes are usually found near the entrance where the floor is moist (Briggler and Prather 2002).

Typical plants in its wooded eastern habitats include ash (*Fraxinus americana*); basswood (*Tilia americana*); beech (*Fagus gradifolia*); birches (*Betula* sp.); bittersweet (*Celastrus scandens*); blackberry (*Rubus* sp.); blackgum (*Nyssa sylvatica*); blueberry (*Vaccinium* sp.); butternut (*Juglans cinerea*); cedar (*Chamaecyparis thyoides*); cherry (*Prunus serotina*, *P. virginiana*); cypress (*Taxodium distichum*); dogwood (*Cornus* sp.); grape (*Vitis aestivalis*); greenbrier (*Smilax* sp.); hackberry (*Celtis occidentalis*); hemlock (*Tsuga canadensis*); hickory (*Carya ovata*); hornbeam (*Ostrya virginiana*); ironwood (*Carpinus caroliniana*); laurel (*Kalmia angustifolia*, *K. latifolia*); maples (*Acer* sp.); oaks (*Quercus alba*, *Q. ilicifolia*, *Q. falcata*, *Q. marilandica*, *Q. phellos*, *Q. prinus*, *Q. pungens*, *Q. rigida*, *Q. rubra*, *Q. stellata*, and *Q. velutina*); pawpaw (*Asimina triloba*); pines (including *Pinus echinota*, *P. palustris*, *P. rigida*, and *P. virginiana*); poison ivy, oak, and sumac (*Rhus radicans*, *R. thyphina*, *R. toxicodendron*, *R. vernix*); rhododendron (*Rhododendron* sp.); rose (*Rosa* sp.); slippery elm (*Ulmus rubra*); sphagnum moss (*Sphagnum* sp.); spicebush (*Lindera benzoin*); spruce (*Picea rubens*); sweetgum (*Liquidambar styraciflua*); sycamore (*Planus occidentalis*); walnut (*Juglans nigra*); Virginia creeper (*Parthenogenesis quinquefolia*); and various species of ferns.

In the South, males have no seasonal preference for a specific habitat, but nongravid females seem to utilize pine savannas (*Quercus falcata*, *Q. lyrata*, *Q. marilandica*, *Q. stellata*; *Pinus australis*, *P. elliottii*, *P. palustris*, *P. taeda*) (Waldron et al. 2006a). Coastal plain stands of cane (*Arundinaria gigantea*) are also heavily used (Dundee and Rossman 1989; Platt et al. 2001a), and have given the *atricaudatus* morph the common name canebrake rattlesnake. In Florida, *C. horridus* inhabits wiregrass (*Poa compressa*) flatwoods (Carr 1940).

In the Upper Midwest, cultivated row crop and old fields are often used for foraging, but riparian woodlands or secondary wooded areas with hilly outcrops are the usual habitat (Fogell et al. 2002a). Grasslands may also be occasionally utilized. At a site in western Wisconsin, gravid females were observed in prairie (28%), secondary woodland (19%), sumac/dogwood/blackberry shrub (13%), and marsh woodland (3.1%) habitats, but not in agricultural habitats (Sajdak et al. 2005). Males were found most often (39%) in mature oak woods, secondary woods (24%), and swamp woodland (24%), but less frequently in sumac/dogwood/blackberry shrub (5%), marsh (4%), prairie (2%), and agricultural habitats (1%). Juveniles occupied secondary woodland (39%), prairie (15%), and sumac/dogwood/blackberry shrub (11%) habitats, but not marsh woodland or agricultural habitats.

Texas *C. horridus* occur mostly in moist lowland woods and wooded hills near waterways where outcrops or ledges with crevices are present, but in Texas the snake may also be found in such rock-free habitats as palmetto lowlands, cane thickets, old fields, and woodland clearings littered with decaying logs and stumps (Werler and Dixon 2000). In Kansas (Fitch and Pisani 2006), Nebraska (Fogell et al. 2002a), and Wisconsin (Oldfield and Keyler 1989), some plants found in its mixed wooded habitat are ash (*Fraxinus* sp.), bittersweet (*Celastrus scandens*), blackberry (*Rubus ostryiafolius*), cedar (*Juniperus virginiana*), cherry (*Prunus serotina*), coralberry (*Symphoricarpos orbicularis*), cottonwoods and poplars (*Populus* sp.), dogwood (*Cornus drummondii*), elm (*Ulmus americana*), hackberry (*Celtis* sp.), locust (*Gleditsia triacanthos*), mulberry (*Morus rubra*), Osage orange (*Maclura pomifera*), sumac (*Rhus glabra*), and walnut (*Juglans nigra*). The Kansas woods studied by Fitch (1999), Fitch and Pisani (2006), and Fitch et al. (2004) is surrounded by flats dominated by goldenrod (*Solidago* sp.) and grasses (*Andropogon gerardi, Sorghastrum nutans*), where the snakes forage. Other flatland plants there are aster (*Aster* sp.), dogbane (*Apocynum cannabinum*), eupatorium (*Eupatorium altissimum*), ironweed (*Veronia interior*), oxeye daisy (*Chrysanthemum leucanthemum*), ragweed (*Ambrosia artemisifolia*), spurge (*Euphorbia maculata*), thistle (*Cirsium altissimum*), and verbena (*Verbena stricta*).

BEHAVIOR AND ECOLOGY: The life history of the timber rattlesnake is one of the most well known and researched of the North American species of *Crotalus*. This is in large part due to the efforts of three researchers: William S. Brown, Howard K. Reinert, and William H. Martin.

The recent use of radio-tracking and passive integrated transponders (PIT tags) has opened many secret doors to the snake's natural history, and most modern studies now include the technique. We refer the reader to the papers of MacGregor and Reinert (2001) and Reinert (1992) for excellent summaries of the technique.

The duration of the annual activity period varies throughout the snake's range. *C. horridus* is usually surface-active from March to mid-April (occasionally early May) when ATs climb to at least 14°C and the ST to 18°C (Fox and Hamilton 2007), to October–November (possibly occasionally early December) when ATs dip to 10°C or below. The farther south, the longer the annual activity period; the farther north, the shorter the time spent above ground. In New York, it is annually active for 4.6–5.2 months, and hibernates for 7.4 months (Brown 1992), while it may be active for 9–10 months in the South (Martin et al. 2008). Published early and late national records for this snake are 5 March (Wright and Wright 1957) and 5 November (Martin 1988).

Crotalus horridus, subadult male; dark phase, northern *horridus* morph (Warren County, New York; Carl H. Ernst)

The duration of dormancy usually depends on how warm the early spring or fall days are. A few individuals may change den sites during the winter (Sealy 2002). Snakes leave the hibernacula over an extended 18-day to 2-month period. In North Carolina, 88% of its records are from June to October (Palmer and Braswell 1995).

Formerly, northeastern *C. horridus* congregated in large groups of as many as 50–200 snakes at suitable hibernacula, usually south-facing rock crevices or talus slopes. Some overlook waterways, and may be shared with *Pantherophis obsoletus* and *Agkistrodon contortrix.* Unfortunately, snakes at many of these hibernation dens were killed off or have abandoned them due to harassment.

In Virginia, the mean date of hibernacula ingress is 12 October (range, 1–18 October), but stragglers continue to arrive until 5 November after other *C. horridus* have gone underground (Martin 1988). Some temporary and sporadic emergence usually occurs in the spring before general emergence. Commonly the first snakes are seen about 14 April, but the annual range of emergence is from 8 March to 2 May, depending on the warmth of the early spring. During a limited study in Wisconsin, Oldfield and Keyler (1989) observed the first timber rattlesnakes on 1 May and the last on 11 September. In another Wisconsin study, Berg and Bartz (2005) found that most of the Upper Midwest snakes enter hibernacula during late September and early October and exit in May; the earliest and latest dates for ingress were 10 September and 19 October.

Berg and Bartz (2005) reported that 11 *C. horridus* in Wisconsin averaged 10.2 (range, 3.7–19.4) % weight loss during hibernation, but the percent loss was not correlated to differences in weight classes. Females lost an average of 9.9%, while males lost 10.5% of their fall body weight. Such weight loss is probably due to a combination of evaporative water loss and that from the depletion of lipid energy stores.

Snakes from a radius of several kilometers migrate over set pathways (Neill 1948) to reach such hibernation dens. Some heavily populated dens still exist in remote or protected areas of the Northeast, but many have been reduced or extirpated. Hibernaculum fidelity is high (Bushar et al. 1998). Homing to previous hibernacula is limited; only 1 of 11 Pennsylvania snakes translocated to a distant hibernaculum returned to that site the next fall (Reinert and Rupert 1999). In the South, the *atricaudatus* morph hibernates individually or in small groups in mammal burrows, old logs and stumps, or shallow rock crevices.

Due to a lack of rocks in the New Jersey pine barrens, rattlesnakes there are forced to burrow as deep as 60 cm into the sandy soil or to use underground rodent burrows and the natural spaces under the root systems of cedar, blackgum, and red maple trees to hibernate. These overwintering sites have no directional orientation, and as many as 30, but usually only 10–15, snakes share them (Pittman *in* Burger 1934; Zappalorti and Reinert 2005). One postpartum female moved about 0.8 km southwest to another stream system, the first observation of a pine barrens *C. horridus* shifting stream corridors to hibernate (Zappalorti and Reinert 2005). Browning et al. (2005) have developed a complicated GIS-based model to characterize timber rattlesnake hibernacula, and we refer the reader to their paper for details.

Across the range, neonates and adults share the same hibernacula. The individuals that congregate at a specific hibernaculum are probably more closely related to each other than to conspecifics from distant hibernacula. This probably explains some of the ability of neonates to follow conspecific odor trails to their maternal hibernaculum.

BTs of hibernating New York *C. horridus* were 4.3–15.7 (mean, 10.5) °C from September to May, and mean rate of BT decline was 0.5°C per week through February; in March the BT stabilized at 4.3°C, and then rose by 0.6°C per week in April–May (Brown 1982). In Wisconsin, the mean BT of hibernating individuals was 10.3 (range, 5.0–18.9) °C, the average BT at 6 dens was 10.7°C, and the average BT recorded at the main study hibernaculum between October and May was 11.1°C (Berg and Bartz 2005). The lowest BTs were recorded in March.

C. horridus sometimes conceals itself in leaf litter, possibly to ambush prey, but more likely to thermoregulate or to retard evaporative water loss. Water loss from the body through evaporation from the skin and mouth is a problem facing terrestrial animals, and *C. horridus* is no exception. Agugliaro and Reinert (2005a, 2005b) studied permeability of the skin and its lipid content in neonates and adults from two localities in New Jersey and Pennsylvania. Skin permeability did not differ according to locality, but rates were significantly different between the age groups. Permeability of the adult skin was greater than that of neonates. Lipid content did not differ by locality, but showed the same age difference. Neonate shed skin experienced a greater percentage (2.2 times) total water content lost per hour than that of adult skin. Resistance to cutaneous water loss may be advantageous to neonates in relation to their greater surface to volume ratio. Evaporative water loss may be made up by drinking free water in the environment, through ingestion of prey, or by lapping water from the skin during rain events (Anderson and Drda 2005).

In the spring and fall *C. horridus* is primarily diurnal, but when the days become hot in summer and the AT rarely falls below 20°C the snake is often active at any time of the day (Beaupre 2008; Sealy 2002) but may be more crepuscular or nocturnal. Wright (1987) reported that most *C. horridus* near the Peaks of Otter, Virginia, are seen crossing the roads in the afternoon, and that only one was seen at night.

Cloudy, windless days with ATs of 20–25°C and a BT of about 25°C are preferred by *C. horridus*. Recorded BTs of active snakes are 5.5–33.3 (average, 27–30) °C (Brown et al. 1982; Oldfield and Keyler 1989; Reinert and Zappalorti 1988a; Wills and Beaupre 2000). Maximum BTs of basking Pennsylvania *C. horridus* recorded by Rach and Delis (2005) were 30.6–32.2°C. Differences between BT and ET in Arkansas *C. horridus* during August were 1.0–4.5°C at 1900–0800 hours and −2.0 to −8.0°C at 0900–1600 hours, but no significant difference occurred between 1700 and 1800 hours (Wills and Beaupre 2000). Foraging Arkansas snakes in August and September had BTs of 15°C to more than 28°C (Beaupre 2008).

The BT of gravid Arkansas females was significantly higher in the early morning and again for most of the late afternoon and first half of the dark segment of the light/dark cycle than that of nongravid females. The mean overall BT for both groups of females during the 24-hour period was 26.4°C. Mean hourly BTs of gravid snakes was 25.2–31.7°C, spanning 6.5°C; those of nongravid females were 19.6–28.8°C, spanning 9.2°C. Gravid females had a higher overall mean BT, 28.6°C, than nongravid females, 24.2°C, during the 24-hour period (Gardner-Santana and Beaupre 2009).

The annual climatic variables for active *C. horridus* in New Jersey were recorded by Reinert and Zappalorti (1988a). Mean data for 119 males, 50 nongravid females, and 88 gravid females, respectively, were as follows: AT—24.8, 25.8, and 28.3°C; surface temperature—25.6, 25.9, and 29.1°C; soil temperature—18.2, 18.1, and 21.0°C; air relative humidity—59.9, 59.4, and 59.0%; surface relative humidity—67.3, 65.4, and 65.5%; and illumination at snake—5,382, 4,198, and 9,041 lux. Obviously, gravid females seek warmer, drier, more illuminated (open) microhabitats.

Apparently neonates born in the fall initially find a hibernaculum by following adult odor trails or by traveling with other *C. horridus*, including their mothers (Brown and MacLean 1983; Cobb et al. 2005; Reinert and Zappalorti 1988b). In a study in which neonates were introduced to the scents of their own species, *Agkistrodon contortrix*, *Pantherophis obsoletus*, and *Nerodia sipedon*, and a "no scent" control, the young snakes recognized the odor of sympatric heterospecifics, but trailed only conspecifics with no statistical preference for their mothers (Rowe and Sealy 2005). Neonates can also distinguish between siblings and nonsiblings; captive-raised young females associated more frequently with female siblings than with nonsiblings (Clark 2004b).

Mean migration distances in Virginia recorded by Martin (1990) were 2.45 km for adult males, 2.16 km for nongravid females, 0.5 km for gravid females, and 1.73 km for juveniles. Radio-equipped New York adults moved an average of 504 m (females, 280 m; males, 1,400 m), and gained a mean 102 m in elevation from hibernaculum to summer feeding range (Brown 1987; Brown et al. 1982). The maximum single migratory movements were 7.2 km by a male and 3.7 km by a nongravid female (W. S. Brown, personal communication). Gravid females remain close to the hibernaculum; one only moved 39 m in 46 days (Brown et al. 1982). Two females tracked by Brown et al. (1982) used the same migratory routes when returning to the den in autumn as they had in leaving it the previous spring. Pennsylvania snakes may remain at one spot for a period of time, then move a considerable distance, and finally settle down again for some time (Galligan and Dunson 1979). Perhaps these erratic movements are stimulated by prey availability.

Within a week of birth, neonates begin to venture away from their mothers. In Tennessee, 4 neonates ventured from 3–154 (mean, 56.8) m in the first 10 postbirth days compared to only 22 m by their mother. The neonates moved sporadically throughout the rest of the active season, and then returned to their maternal hibernaculum in the fall (Cobb et al. 2005).

Males have 40–207 ha foraging home ranges and nongravid females 17–42 ha, but gravid females only occupy 4–22 ha home ranges (Brown 1993; Reinert 1991; Reinert and Zappalorti 1988a; Sealy 2002).

In New Jersey, radio-telemetry revealed that males had the largest activity ranges (mean length, 1,463 m), but the sizes of their ranges were positively correlated with the number of days they were monitored. This was not true of either nongravid females (mean home range length, 995 m) or gravid ones (mean length, 665 m). Time series analyses indicated that movement patterns of males and nongravid females consisted of constantly shifting, nonoverlapping ranges. In most cases, these snakes moved during the active season in a looping pattern that returned them in the fall to the same hibernaculum from which they departed in the spring. Gravid females had more static overlapping activity ranges and shorter dispersal distances from their hibernaculum (Reinert and Zappalorti 1988a).

At another New Jersey site, a military reservation, 3 males traveled a mean distance of 6.2 km (mean distance moved per day, 43.7 m) while radio-tracked. Their mean home range length was 2.4 km, and the convex polygon area averaged 61.5 ha; the same mean parameters for 3 radio-tracked females were 3.2 km, 23.7 m, 1.3 km, and 40.1 ha, respectively (R. M. Smith et al. 2008).

The average home range radii by sex in Virginia based on recaptures were adult males, 2.47 km; nongravid females, 2.19 km; gravid females, 0.48 km; and < 3-year-old juveniles, 1.74 km (Martin 1988). In northwestern Arkansas, the mean area used by 6 gravid females, 0.4 ha, was significantly less than that of 5 nongravid females, 2.4 ha. The mean distance moved per day by gravid females, 8.8 m, and overall distance traveled, 384 m, were also significantly less than those of nongravid females, 18.2 m and 827.4 m, respectively (Gardner-Santana and Beaupre 2009).

In North Carolina, males moved 13.7–42.2 m per day; overall, females moved 6.2–19.2 m per day; and 2 gravid ones moved 6.2 m and 8.9 m, respectively (Sealy 2002).

Eastern Texas yearling *C. horridus* had a minimum convex polygon activity range of 12.10 ha and a maximum distance between capture points of 0.97 km during the first year of tracking after release. In the second full year of tracking, the young snakes had a mean activity range of 22.01 ha and a maximum distance between captures of 0.06 km. For those still alive after 2 years, the mean activity range was 7.27 ha, and the maximum distance between captures was 0.16 km (Conner et al. 2003).

C. horridus, although it will cross roads within its habitat, usually does not, and when it does the snake crawls at a perpendicular angle, which minimizes crossing time. It moves across roads more slowly (logspeed about 0.5–1.4 cm/second) than do nonvenomous snakes. If a vehicle approaches, the snake will freeze in place 50% of the time before it passes (Andrews and Gibbons 2005). Clark et al. (2009) studied four New York colonies of *C. horridus* associated with specific hibernation dens. Habitat modification via roads created a barrier for snake movement throughout the total range of the hibernaculum colony. Rattlesnakes in hibernacula isolated by roads had significantly lower genetic diversity and higher genetic differentiation than those in hibernacula in contiguous habitat. They determined that interruption to seasonal migration by avoidance of crossing the roads was the underlying cause of these genetic patterns, and stressed the need for mitigating the effects of roads.

Gravid females occupy small, warmer home ranges with relatively open canopy and much leaf litter near the hibernaculum. Their home ranges are more static and overlapping than those of males and nongravid females (Reinert and Zappalorti 1988a). Males and nonreproductive females have larger, distant home ranges; canopy cover and vegetation are more extensive, and logs are often present (Peterson 1990; Reinert 1984a, 1984b; Reinert and Zappalorti 1988a).

The bioenergetics of free-ranging *C. horridus* have been studied by Beaupre and Zaidan (2001), who measured the CO_2 production rate of 83 snakes in response to BM, BT, time of day, sex, and geographic locality (northwest Arkansas and coastal Virginia). The effects of BM, BT, time of day, and the BT-by-time interaction were extremely similar to those reported for other rattlesnakes. The overall BM scaling of CO_2 production (mL/hour) less neonates was volume $CO_2 = 0.03597BM$. The snakes had a relatively high Q_{10} (3.71–4.78), but the adaptive significance of this, if any, remains to be determined. Once the effect of BM was statistically adjusted, no sex-specific effects were apparent, but there was a significant locality-by-time effect of equivocal biological significance. Neonates had SMRs 200–400% greater than expectations from the BM scaling of yearlings and older individuals. Beaupre and Zaidan interpreted this as evidence for a cost of synthesis in growing neonates.

C. horridus often climbs into bushes or small trees usually less than 5 m high, possibly while pursuing prey (Collins 2003; Fogell et al. 2002b; Linsdale 1927; Muir 1982; Saenz et al. 1996; Sealy 2002; see photograph in Sajdak and Bartz 2004). Rudolph et al. (2004) have observed them as high as 14.5 m in trees in eastern Texas, and documented frequent arboreal activity. Most arboreal activity was by < 90 cm subadults (also noted by Muir 1982). Even neonates may ascend as high as 1 m (H. K. Reinert, personal communication).

C. horridus is a good swimmer (Dunn 1915), and often swims from the mainland to islands in Lake George, New York (W. S. Brown, personal communication).

Adult males, like other *Crotalus*, engage in combat dances during both the spring,

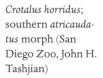

Crotalus horridus; southern *atricaudatus* morph (San Diego Zoo, John H. Tashjian)

and, predominantly, in the late summer to early fall breeding periods (Aldridge and Brown 1995; Aldridge and Duvall 2002; Anderson 1965; Collins 1974; C. Ernst, personal observation; Martin and Wood 1955; McIlhenny *in* Klauber 1972; Merrow and Aubertin 2005; Sutherland 1958).

REPRODUCTION: Males normally require 4–6 years to mature, but in Kansas they may mature in 2 years (Fitch 2002). Female maturity takes longer, 4–13 years. Females of the northern *horridus* morph are mature at TBLs of 90–93 cm, SVLs of 67.7–90.0 cm, and BMs of 500–820 g (Brown 1991; Galligan and Dunson 1979; Keenlyne 1978; Martin 1988; Sealy 2002). SVLs of recently matured females of the southern *atricaudatus* morph are usually > 100 cm, but males are mature at SVLs of 90–100 cm (Gibbons 1972).

Females normally have either biennial or triennial reproductive cycles, with triennial most common, but some may take as long as six years before they next produce a litter (Berg et al. 2005; Brown 1991, 1993; Fitch 1985b; Galligan and Dunson 1979; Gibbons 1972; Keenlyne 1978). These cycles are probably determined by prey availability.

The numbers of gravid females in a population vary from year to year, probably determined by prey availability. If fat reserves are low, female reproduction may be delayed one or more years (Gibbons 1972). Proportions of gravid females each year vary throughout the range: 27–75% in New York (Brown 1991), 18–51% in Virginia (Brown and Martin 1990), and up to 84% in Pennsylvania (Reinert 1990).

Follicular development and vitellogenesis occur from late July to October (Brown 1991, 1995a; Gibbons 1972; Keenlyne 1978; Martin 1993; Sealy 2002). The eggs are ovulated the following May–June in the north, but South Carolina females may contain embryos in the spring (Gibbons 1972). Parturition occurs in August–October.

Lutterschmidt et al. (2009) investigated seasonal and reproductive variation in basal and stress-induced hormone profiles in free-ranging *C. horridus* in north-central Pennsylvania. Baseline corticosterone concentrations varied seasonally; and were significantly lower during summer sampling (July) than in either spring (May) or early fall (September), and increased significantly in both sexes after stress. A significant negative relationship was noted between baseline corticosterone and testosterone in males, while baseline corticosterone and estradiol tended to be positively correlated in females. The correlations between the increased corticosterone response and stress suggest that adrenocortical function is seasonally modulated. Although having similar baseline corticosterone levels, nonreproductive females and postparturient females responded differently to capture stress, with significant increases in corticosterone in females that had given birth. Collectively, Lutterschmidt et al.'s data suggest that the sensitivity of the hypothalamus-pituitary-adrenal axis varies both seasonally and with changing reproductive status (see Lutterschmidt et al. 2009 for specifics).

Male spermiogenesis begins in June and July, with maximum production in late July–September, when the seminiferous tubules fill to their greatest diameters. Sperm then passes into the epididymides and vas deferentia for overwinter storage; tubular diameters are small in the spring (Aldridge and Brown 1995).

Some minor breeding activity may take place in the spring (Aldridge and Duvall 2002; Anderson 1965; Keenlyne 1978), but most mating occurs from mid-July to October with a peak in August. Sperm is stored until ovulation the next spring (Brown 1987, 1991; Martin and Wood 1955; W. H. Martin 1988, 1990, 1992, 1996b; McGowan

2004; Merrow and Aubertin 2005; Mitchell and Webb 2000; Puskar 1999; Schuett 1992; Sealy 1996, 2002; Waldron et al. 2006b). In northeastern New York, the mating season is concentrated in 72 days from 15 July to 23 August, and in southeastern New York most mating occurs in a 5-week period from late July to late August (McGowan and Madison 2008). Heterosexual grouping coincides with female vitellogenesis (Brown 1995a). Males probably depend on scent trails left by the females to find mates (Brown and MacLean 1983), and the snake may court above ground in bushes (Bartz and Sajdak 2004; Coupe 2001).

Release of pheromones from the female's skin is the main cue that stimulates courtship activity in males, and the males return to sites where they have found reproductive females in the past (Coupe 2002). Merrow and Aubertin (2005) observed a sequence involving female ecdysis, male-male combat, and copulation that occurred within a few hours. Larger males are probably more successful in securing mates than are shorter ones (McGowan 2004), and large to medium-sized males successfully defend females at outcrop ecdysis sites (McGowan and Madison 2008). Smaller males paired with females at more distant forest sites.

Missouri males mate at longer distances from hibernacula (172–2,240; mean, 1,001 m) than females (195–1,148; mean, 511 m), and random gene exchange between the sexes from different hibernacula may occur. Some gravid females move considerable distances between locations, but most returned to the vicinity of their hibernaculum before parturition. Distance of the birth site from the hibernaculum averaged 246 (range, 17–661) m, and parturition sites tended to be shorter than the distance between adjacent hibernacula. Genetic testing revealed only a small significant proportion of the genetic differentiation in the total population among hibernacula, but Bayesian clustering algorithms could not sort individuals or hibernacula groups into discrete demes. This suggests that patterns of genetic differentiation among hibernacula colonies of *C. horridus* are better explained by behavioral and demographic factors that result from natal philopatry, sex-biased dispersal, and/or a limited number of breeding adults, than by balance of gene flow and genetic drift (Anderson 2010).

Once a female is found, the male first rubs her neck with his chin. He positions himself alongside her and begins a series of quick, rapid jerks of his head and body. Next, he curls his tail beneath hers until the vents touch, and finally inserts his hemipenis. Some pumping motions of the male's tail near the vent may occur during copulation, which may last several hours (in captivity; C. Ernst, personal observation).

C. horridus is ovoviviparous, and parturition usually takes place between 4 August and 15 October (Martin 1996b; Puskar 1999). The southern *atricaudatus* morph possibly gives birth earlier; a captive produced a litter on 20 July (Kauffeld *in* Klauber 1972). Birthing rookeries are sometimes located at the hibernaculum, but may be as far away as 0.5–1.0 km (Martin 1988; Reinert and Zappalorti 1988b).

An old folk tale relates that rattlesnakes lay eggs. This is untrue, as all *Crotalus* and *Sistrurus* are live-bearers. An interesting paper telling of the debunking of this myth in Minnesota *C. horridus* was recently published by Cochran (2010).

Trapido (1939) described the birth of 10 neonates in captivity by a freshly caught 122 cm New York female on 8 September. The entire parturition took about 220 minutes; the first 3 young were born between 1000 and 1100 hours, and the remainder between 1225 and 1405 hours. When they first emerged, most neonates were coiled within a transparent membrane, but some were already entirely or partially free of the membrane. Those fully or partially surrounded by the membrane took 1–39 minutes to totally separate themselves from it. On the first day after parturition, the umbilical stalk dried and broke off.

Most broods consist of 6–10 young (mean, 10.6; n = 105 literature records), but litters of 1–20 are known (Brimley 1923; Edgren 1948; Fitch 2002; Palmer and Braswell 1995; Puskar 1999; Trapido 1939). The largest litters are produced by southern females. Fitch (2000) estimated the total annual production by 26 adult Kansas females was 104 neonates, with an average litter size of 7.2 young.

The female remains with her young for 7 to 10 days before all disperse (Clark 2005a; Cobb et al. 2005; Martin 1988, 1990, 1996b). Litter size is directly proportional to female SVL. Seigel and Fitch (1984) reported an RCM of 33.5%. Neonates are believed to trail adults, probably through odor detection, away from the birthing area, and, in the fall, back to the hibernaculum (Reinert and Zappalorti 1988b).

Neonates are patterned like adults, but gray in hue, and have 19.5–40.9 (mean, 32.8; n = 94) cm TBLs, and have 11.2–39.0 (mean, 25.5; n = 48) g BMs. Fitch and Pisani (2006) reported an SVL range of 28.3–40.3 cm for 23 captive-born Kansas neonates. Newborn snakes possess a button-like terminal scale on the tail that is exposed after the first shedding of the skin (normally in 7–10 days). They are dangerous even at this small size, having fangs 2.6–3.8 mm long (Stewart et al. 1960) and a ready supply of venom (Ditmars 1931b).

Chiszar (*in* Schuett 1998) and Schuett et al. (1998) reported an apparent case of automictic parthenogenesis by a female *C. horridus*. She had been in captivity since birth and never with a male, but produced a litter of three triploid males (one live, two stillborn).

Cases of bicephaly and axial bifurcation resulting from abnormal embryological development have been reported in Alabama and Arkansas *C. horridus* (Lasher 1980; Wallach 2007).

A *C. horridus* repeatedly copulated with an *Agkistrodon contortrix* in captivity (Smith and Page 1972). The species has hybridized with *C. adamanteus* (Klauber 1972).

GROWTH AND LONGEVITY: Fitch (1985b, 1999) reported the following SVL/age correlates for Kansas *C. horridus*: fall of birth, 31.0–34.5 cm; 1st spring, 32.4–59.1 cm; 1st year, 54.8–71.2 cm; 2nd year, 50.4–83.4 cm; 3rd year, 64.4–99.9 cm; 4th year, 76.0–100.3 cm; 5th year, 88.5–108.0 cm; 6th year, 96.5–114.7 cm; 7th year, 100.0–119.6 cm; 8th year, 123.0–124.8 cm; 9th year, 103.8–104.4 cm; and 15th year, 127.0 cm. Minton (1972) estimated that a 81 cm Indiana *C. horridus* was almost 2 years old, and 4 others were 70 and 95 cm in their 2nd year and 100 and 110 cm in their 3rd year, but W. S. Brown (personal communication) believed these ages to be vastly underestimated. In Virginia, juveniles average 43.2 cm at 1 year, 58.4 cm at 2 years, and 70.6 cm at 3 years (Martin 1988). Shortly after birth in South Carolina, juveniles have 35–43 cm SVLs, and grow to 50–60 cm by the following June; at 2 years they are 65–75 cm, and at 3 years 80–90 cm (Gibbons 1972).

The growth rate depends on the frequency of feeding and the quality of the diet, and can be accelerated with increased feeding; a captive grew from a TBL of 34.3 cm to 75.6 cm in only 8 months (Schwab 1988). An adult Kansas male, first captured at an estimated age of 4 years, recaptured 55 months later had increased its SVL from 89.1 cm to 100.7 cm and its BM from 745 g to 1,125 g (Fitch et al. 2004).

Rate of growth of eastern Texas neonates (mean TBL, 33.2 cm; mean SVL, 30.9 cm; mean BM, 28.7 g) raised for as long as 327 days in captivity and later released were recorded by Conner et al. (2003). The snakes were fed young live *Mus musculus* weekly while in captivity in an attempt to maximize their growth. They consumed a cumulative mouse mass of more than 1 kg at an average of 873.5 g per snake. After the feeding regime, the snakes had the following mean measurements: TBL, 88.3 cm; SVL, 81.7 cm; and BM, 461.8 g. Individual increase in BM was directly proportional to the prey mass consumed ($Y = 2.0416X - 66.516$, $R^2 = 0.98$), as was also their growth in TBL ($Y = 0.0021X^2 - 1.1265X + 142.04$, $R^2 = 0.97$). BM increased proportionally with increased TBL ($Y = 0.0957X^2 - 4.5229X + 75.063$, $R^2 = 0.97$). On average, for every 1 g of prey mass consumed the snakes increased their BM by 0.48 g and their TBL by 0.63 cm.

Beaupre and Zaidan (2001) examined the role of metabolism in the growth of neonate *C. horridus* (see discussion in "Diet and Feeding Behavior").

Growth is accompanied by shedding (ecdysis) of a snake's skin, and both growth and shedding are resource dependent. Neonate Virginia *C. horridus* average 1–8 molts per year for the first 5 years; afterward, adult males average 1.3–1.5 molts per year and adult females 1.2–1.4 molts per year (Brown 1988; Martin 1988, 1990; Sealy 2002; Schwab 1988). Wisconsin individuals shed an average of 1.74 (range, 1.0–2.7) times a year: juveniles (> 65 cm SVL), 1.85; males, 1.80; and females, 1.54 (Berg et al. 2005). In North Carolina, ecdysis peaks in June, when 57% of the snakes shed their skin (Sealy 2002). Rattlesnakes add a segment to the rattle at each ecdysis; female ecdysis during the breeding period may enhance mating activity in males (Merrow and Aubertin 2005). Fitch (1985b) reported the following average relationships between age and the number of rattle segments present for Kansas *C. horridus*: < 1 year, button only; year 1, 2 segments; year 2, 3–4 segments; year 3, 5–6 segments; year 4, 7–8 segments; year 5, 9–10 segments; and year 6, 11–12 segments.

C. horridus is capable of a long life, both in captivity and in the wild. The longevity record for captives is 36 years, 7 months, and 27 days (Cavanaugh 1994); other captives have survived 30 years and 2 months (Slavens and Slavens 2000) and 28 years and 1 month (Shaw 1969).

Pertinent survival data on wild individuals are also available. A Kansas male survived 24 years after its first capture and was probably 28 years old when last captured (Fitch and Pisani 2002, 2006); in July 2007 W. S. Brown (personal communication) recaptured 2 New York females that were first marked in 1981 and 1982, 26 and 25 years earlier. When first marked they were immature and around 5–7 years old, so, pending a more careful estimate, the snakes were in their early thirties.

DIET AND FEEDING BEHAVIOR: Much data have accumulated on the prey species and predatory behavior of the timber rattlesnake.

C. horridus is a "sit and wait" ambusher at a place where small mammals will probably pass (Brown and Greenberg 1992; Reinert et al. 1984, Sealy 2002). Ambush sites are selected because of the odors both of conspecifics and of the prey species associated with them (Clark 2004c, 2007a, 2007b). The snake remains at the ambush site an average of 17 hours (Clark 2006b). Prey capture in South Carolina occurs from April to November (Waldron et al. 2006b), but the foraging season is shorter in more northern populations and in those at high elevations (C. Ernst, personal observation).

Prey is detected through thermal cues (especially at night), odor, and probably vision (especially during the day). Whitt (1970) studied the mechanisms by which the species responds to stimuli. He compared the responses of control snakes with no sense organ blocked to those having the eyes covered, eyes and both loreal pits covered, and both pits occluded. The snakes in each group were subjected to water at room temperature, 3–5°C, and 37–40°C; 40% formalin; acetic acid; and various substrate vibrations. Control snakes responded positively to all of the stimuli. Those with the eyes covered responded positively to all stimuli except water at room temperature, with the exception that if the 3–5°C and 37–40°C water was on their right side they did not respond, but if to the left of them they did. Those with the eyes and both pits covered responded positively to all stimuli except water at room temperature. Snakes with the eyes uncovered but the pits occluded responded positively to only the formalin, acetic acid, and substrate vibrations. Whitt concluded that the species uses its eyes to good but restricted advantage. Although the hearing apparatus is much modified, the snake can sense substrate vibrations through body contact with the ground, and the senses of taste and smell are well developed via its tongue and vomeronasal organ. He thought the most important of the senses is that of temperature detection. As long as the objects with differential temperatures are in front of the snake, and within 50 or so degrees lateral to the midsaggital line of the body, the snakes could detect the various temperature regimes.

Clark (2004c) exposed 24 naive, captive-born New York neonates to aqueous extracts from the skins of the green frog (*Rana clamitans*), great plains skink (*Eumeces obsoletus*), brown rat (*Rattus norvegicus*), cotton rat (*Sigmodon hispidus*), dwarf hamster (*Phodopus sungorus*), chipmunk (*Tamias striatus*), white-footed mouse (*Peromyscus leucopus*), and domestic dog (*Canis familiaris*) with tap water as a control, to determine if their odors would stimulate tongue-flicking search behavior followed by the assumption of an ambush posture. All of the extracts elicited tongue-flicks, and all but the tap water and skink extracts stimulated ambush behavior. The ambush response only occurred toward the dog once and the green frog twice; the other extracts elicited the following numbers of ambush postures: white-footed mouse, 11; chipmunk, 8; cotton rat and dwarf hamster, 5 each; and the brown rat, 4. The species is a well known rodent eater, and the white-footed mouse and chipmunk are natural prey in New York. These responses indicate *C. horridus* has the ability to recognize and distinguish among the odors of potential prey species at birth.

Ackerman et al. (2007) found that in laboratory tests *C. horridus* actually located and grasped the less odoriferous of three strains of *Mus musculus* quicker, the opposite of what was expected. However, novel chemical cues on prey carcasses do not alter the snake's predatory behavior. Melcer et al. (1988) reported that *C. horridus* exposed to water-diluted perfume and plain water misted mouse (*Mus musculus*) carcasses struck, grasped, and ingested the bodies of both types of treated mice, but took somewhat longer to strike the perfumed ones although the differences were nonsignificant.

Favored ambush positions are coiled beside a fallen log with the head positioned perpendicular to the log's long axis and the chin rested on the side of the log, or lying beside a tree with the head elevated and held vertically against the trunk. Some prey may be actively sought, as evidenced by the abundant records of *C. horridus* climbing trees (previously discussed), presumably seeking birds or squirrels.

Cundall (2002) and Cundall and Beaupre (2001) reported the kinematic details of predatory strikes by *C. horridus*, and we refer the reader to their paper for details. Post-strike behavior only occurs after the prey is struck; missed strikes do not elicit further behavior. Typical stereotyped poststrike behavior includes the release of the prey immediately after the strike, poststrike immobility, chemical search with much tongue-flicking, location of the chemosensory trail, trail-following, and prey-swallowing (Chiszar et al. 1985, 1992; Clark 2006a; C. Ernst, personal observation). Only about 13% of strikes are successful (Clark 2006a).

Minton (1969) reported that four *C. horridus* struck toward the thoracic region of mice 29%, the lumbar region 20%, the neck 17%, the head 9%, the inguinal region 5%, the hindlimb 3%, and the forelimb 2% of the time. Incapacitation of the prey took a few seconds to more than an hour, but in 50% of the cases was less than 2 minutes; the time interval between the strike and death was variable, but occurred in less than 7 minutes in about half of the envenomations.

As white-footed mice and chipmunks are the preferred prey in the East (Brown 1987; Martin 1988), feeding must occur both day and night. Neill (1960) suggested the young may use their tails as a lure for small prey, but this has not been ascertained. According to Brown (1987) and Fitch (1982), New York and Kansas *C. horridus* only eat 6–20 meals per year, and annually consume at least 2.5 times their body weight. In a more recent study, Clark (2006a, 2006b) reported the New York snakes feed 12–15 times and typically consume 1.25–1.55 kg of prey annually.

Three potential prey species were observed by Clark (2005b) to exhibit conspicuous visual displays to timber rattlesnakes that were either basking or hunting—the wood thrush (*Hylocichla mustelina*), chipmunk (*Tamias striatus*), and gray squirrel (*Sciurus carolinensis*). The displays lasted from 1 (thrush) to 28 (chipmunk) minutes before ending when the snake moved away.

Zaidan and Beaupre (2003) investigated the effects of BM, meal size, fasting time, and temperature on the specific dynamic action (SDA, the total energy expenditure associated with digestion and assimilation of ingested food and related biosynthesis) of 26 *C. horridus*. This requires data on the cost of digestion, which can represent up to 30% of the proportion of ingested energy of infrequent feeders such as snakes. They examined the effects of BM, ET (20°C and 30°C), fasting time (1 and 5 months), and prey size (10–50% of snake BM) on SDA by measuring the hourly CO_2 production rate for 1–17 days after feeding. The tested snakes showed large and ecologically relevant increases in SMR correlated with feeding. Depending on treatment and the individual snake, the CO_2 production increased 2.8–11.8 times the SMR within 12–45 hours after ingestion, and decreased to baseline within 4.3–15.4 days. Snake BM, meal mass, and fasting time all caused significant effects. A rise in ET decreased the time required to complete digestion, but had little effect on total energy expended on SDA. The energy expended on SDA increased with increasing fasting time, snake BM, and prey mass. Simple allometric relationships explained almost 97% of the variation in total CO_2 production via SDA. The energy devoted to SDA may approach 20% of the annual energy budget of wild *C. horridus*. The SDA coefficient (SDA expressed as a percentage of ingested energy for a 50 g, 400 kJ meal for *C. horridus* of differing BM) shows a strong correlation between SDA and snake BM (Beaupre 2005). Mean passage time for a food bolus is 12.8 days (Zaidan and Beaupre *in* Lillywhite et al. 2002). Hill et al. (2008) found fermentation bacteria (*Lactobacillus fermentum*, *L. hammesii*, *L. lactis*) in the intestines of *C. horridus*, which may play a role in digestion.

Wild Arkansas timber rattlesnakes studied at times of low food availability spent more time foraging; had reduced growth, lower field metabolic rates (FMRs), and poorer body condition; and did not reproduce. In contrast, those with high food intake foraged less, grew faster, had higher FMRs and better body condition, and engaged in extensive mate search and courtship (Beaupre 2008).

The feeding behavior of females is strongly related to their reproductive condition (Keenlyne 1972; Reinert et al. 1984). Gravid females feed very little, if at all, while those with maturing follicles eat more often. Keenlyne (1972) found the following mean fat to BM indexes in Wisconsin: males, 5.83%; immature females (containing no follicles), 5.36%; nonreproductive adult females (with unyolked follicles), 8.77% in the form of visceral fat; and gravid females, 5.07%. Nongravid females contained an average of 73% more fat than did gravid females.

Warm-blooded prey is preferred. Most prey is taken alive, but carrion is eaten (Nicoletto 1985). Mice composed 38%, squirrels and chipmunks 25%, rabbits 18%, shrews 5%, and birds (mostly songbirds) 13% of the prey taken by 141 Virginia *C. horridus* (Uhler et al. 1939); and in central Georgia, 90% of the prey is made up of mammals, 4% birds, 4% amphibians, and 2% reptiles (Parmley and Parmley 2001). In Pennsylvania, mammals composed 94% of the diet (Surface 1906). Prey by volume found in the three Kentucky *C. horridus* examined by Bush (1959) consisted of 80% *Peromyscus leucopus*, 10% *Sciurus carolinensis*, and 10% songbirds.

A wide variety of animals are eaten. Those recorded include insects; amphibians—anurans (*Bufo* sp., *Rana* sp.) and salamanders (*Eurycea cirrigera*); reptiles—lizards (*Cnemidophorus sexlineatus*; *Eumeces fasciatus*, *E. laticeps*; *Scincella lateralis*) and snakes (*Coluber constrictor*, *Thamnophis sirtalis*); birds (eggs, young, and adults)—galliforms (*Bonasa umbellus*, *Colinus virginianus*, *Gallus gallus*, *Melaegris gallopavo*), rails (*Rallus longirostris*), woodpeckers (*Melanerpes carolinus*, *Sphyrapicus varius*), and passeriforms (*Ammodramus savannarum*, *Bombycilla cedrorum*, *Coccyzus americanus*, *Dendroica caerulescens*, *Hylocichla mustelina*, *Melospiza melodia*, *Minus polyglottos*[?], *Passer domesticus*, *Pipilo erythrophthalmus*, *Seiurus aurocapillus*, *Setophaga rutilla*, *Spizella pusilla*, *Thryothorus ludovicianus*, *Toxostoma rufum*, *Turdus migratorius*, *Zonotrichia albicollis*); and mammals—opossums (*Didelphis virginiana*), bats (*Eptesicus fuscus*, *Myotis* sp.), shrews (*Blarina brevicauda*, *B. carolinensis*, *B. hylophaga*; *Cryptotis parva*; *Sorex cinereus*, *S. fumeus*, *S. longirostris*), moles (*Scalopus aquaticus*), murid rodents (*Clethrionomys gapperi*; *Microtus chrotorrhinus*, *M. ochrogaster*, *M. pennsylvanicus*, *M. pinetorum*; *Mus musculus*; *Neotoma floridana*, *N. magister*; *Ochrotomys nuttalli*; *Oryzomys palustris*; *Peromyscus gossypinus*, *P. leucopus*, *P. maniculatus*; *Reithrodontomys humulis*; *Rattus norvegicus*, *R. rattus*; *Sigmodon hispidus*; *Synaptomys cooperi*), jumping mice (*Napaeozapus insignis*, *Zapus hudsonicus*), pocket gophers (*Geomys bursarius*), squirrels (*Glaucomys volans*; *Marmota monax* [young]; *Sciurus carolinensis*, *S. niger*; *Spermophilus tridecemlineatus*; *Tamiasciurus hudsonius*; *Tamias striatus*), rabbits (*Sylvilagus aquaticus*, *S. floridanus*), raccoon (young) (*Procyon lotor*), and mustelids (*Mephitis mephitis*; *Mustela frenata*, *M. vison*). Barbour (1950) took a snail shell from a stomach, but thought it probably had been secondarily consumed in a chipmunk pouch.

The references for the above-mentioned prey are Anderson (1965), Ashton and Ashton (1981), Babcock (1929), Bailey (1946), Barbour (1950, 1971), Breckenridge (1944), Brown (1979a), Brown and Greenberg (1992), Bush (1959), Campbell and Lamar (2004), R. F. Clark (1949); R. W. Clark (2002, 2005b), Collins (1993), Conant (1951), Dundee and Rossman (1989), C. Ernst (personal observation), Ernst et al. (1997), Fitch

(1982, 1999), Fitch and Pisani (2006), Fitch et al. (2004), Gibbons and Dorcas (2005), Grant (1970), Green and Pauley (1987), Hamilton and Pollack (1955), Harding (1997), Hibbard (1936), Huheey and Stupka (1967), Hulse et al. (2001), Hurter (1911), Jackson (1983), T. R. Johnson (1987), Keenlyne (1972), Kennedy (1964), Keyler and Oldfield (1992), King (1939), Klauber (1972), Le Clere (1996), Linsdale (1927), Linzey and Clifford (1981), Linzey and Linzey (1968), MacGregor and Reinert (2001), Martin and Wood (1955), Martin et al. (2008), McCauley (1945), Minton (1972, 2001), Mitchell (1994a), Myers (1956), Nicoletto (1985), Odum (1979), Palmer and Braswell (1995), Parmley and Parmley (2001), Platt et al. (2001b), Price (1998), Reinert et al. (1984), Sajdak and Bartz (2004), Savage (1967), Schorger (1968), Sealy (2002), Smyth (1949), Surface (1906), Swanson (1952), Tennant (1998), Trauth et al. (2004), Uhler et al. (1939), Vogt (1981), Wright and Wright (1957), and Zappalorti and Reinert (2005).

The bacteria *Lactobacillus fermentum* and *L. lactis* have been identified from the gastrointestinal tract of *C. horridus*, and probably play a role in digestion (Hill et al. 2008).

VENOM DELIVERY SYSTEM: Six adult *C. horridus* with TBLs of 91.2–1,162.0 cm measured by Klauber (1939) had 8.7–10.4 (mean, 9.3) mm FLs, a mean TBL/FL of 107, and a mean HL/FL of 4.61. The mean angles of fang curvature were 70° for the *horridus* morph and 60° for the *atricaudatus* morph. Replacement fangs for those lost or broken are already present at birth; neonate functional fangs measured by Barton (1950) were 3.2–3.3 mm long, the first reserve fangs were 3.0–3.1 mm, the second reserve set 2.4–3.0 mm, and the third reserve fangs 2.0–2.1 mm. The embryonic development of the motile maxillary bone in relation to the fangs is described by A. H. Savitsky (1992).

VENOM AND BITES: Envenomation by *C. horridus* is among the most commonly reported of bites by rattlesnakes in the United States. At least in the past, the snake occurred in populous areas of the East, and frequently the paths of human and snake crossed. Presently, many populations of the snake have been extirpated near human settlements, so the incidence of envenomations by it has decreased in most areas over the years; however, it still occurs and has been reported from all states in which the timber rattlesnake resides.

The species' venom has been well researched. A typical *C. horridus* contains 75–210 mg dry weight of venom (Russell 1983; Russell and Brodie 1974; Russell and Puffer 1970); the minimum and maximum known dry yields are 42 mg and 300 mg, respectively (Brown 1973; Glenn and Straight 1982). Minton (1953) reported wet venom volumes of 0.23–0.71 mL per snake. McCue (2006) reported liquid venom amounts of 100–1,160 mg per snake, and Johnson et al. (1968a) noted that adult raw venom concentrations varied from 199.2 mg/mL to 302.4 mg/mL. Although Johnson et al. (1968a) found no apparent correlation to snake BM, McCue (2006) reported an allometrical correlation of $4.77W^{0.60}$ between liquid volume and snake BM. Klauber (1972) extracted a total of 1,392 mg of dried venom from 13 fresh snakes (10 adults); the average yield per fresh adult was 139 mg with a maximum yield of 229 mg. Do Amaral (1928) reported the following average amounts of venom secreted at one time from various age groups of *C. horridus*: young snakes, 0.21 mL or 0.060 g; adult snakes, 0.32 mL or 0.090 g; and old specimens, 0.63 mL or 0.180 g. Githens and Butz (1929) noted an average venom discharge of 45 mg.

The species' crude venom contains many protein components and several metallic ions, but possesses low protease activity. Juvenile venom has a much lower L-amino acid oxidase component and exhibits lower protease activity than does that of adults (Bonilla et al. 1973). It is composed of 20–30% solids that contain approximately 80% protein (Brown 1973), 0.61% nitrogen, and 0.091% amino acids (Hansen 1931). The mean moisture content is 70–71% (McCue 2006). Venom from the *atricaudatus* morph has a slightly acidic pH of 6.06 (Johnson et al. 1968a), although Bonilla and Fiero (1971) isolated highly basic proteins from the species' venom.

A detailed spectographic analysis of crude venom from the *atricaudatus* morph by Johnson et al. (1968a) indicated that it is composed (ash percentages) of the following elements for adults (251–693 g) and neonates (19–27 g), respectively: aluminum, 0.035 and 0.13; boron, < 0.01 and 0.011; calcium, 0.077 and 0.048; chromium, 0.027 and 0.020; cobalt, < 0.001 and < 0.002; copper, 0.0077 and 0.027; iron, 0.54 and 0.084; lead, 0.0081 and 0.045; lithium, < 0.01 and trace 0.03; magnesium, 0.81 and 0.047; manganese, 0.0021 and 0.0038; molybdenum, < 0.01 and < 0.003; nickel, 0.0085 and trace < 0.003; phosphorus, 0.037 and < 0.10; potassium, < 0.40 and < 1.0; silicon, 0.005 and 0.14; silver, < 0.0001 and < 0.0003; sodium, 2.7 and 2.7; strontium, < 0.0001 and < 0.0003; tin, 0.010 and 0.030; tungsten, < 0.04 and < 0.10; zinc, 0.037 and < 0.03; and other elements, nil and nil.

Additional reported chemical constituents of timber rattlesnake venom include a small basic peptide (functioning as a myotoxin separate from phospholipase A_2), a bradykinin-releasing enzyme, canebrake toxin, cholinesterase, crotalocytin (a platelet-activating enzyme), crototoxin, C-type lectins (CHH-A, CHH-B, GP-Ib), defibrizyme, deoxyribonuclease, hemorrhagic metalloproteinase (HP-IV), hyaluronidase, L-amino acid oxidase, Mojave toxins α and β, monoesterase, myotoxin-α, NAD-nucleosidase, 5'-nucleotidae, phosphodiesterase, phosphomonoesterase, phospholipase A_2, procoagulant esterase (a fibrinogen-clotting enzyme that produces defibrination), and ribonuclease-1 (RNase) (Bober et al. 1988; Bond and Burkhart 1997; Bonilla et al. 1973; Deutsch and Diniz 1955; Glenn and Straight 1985a; Glenn et al. 1994; Hawgood 1982; Johnson et al. 1968a; Lee and Zhang 2003; McCue 2005; Meier and Stocker 1995; Moran and Geren 1979; Norris 2004; Ogawa et al. 2005; Powell et al. 2008; Rael et al. 1986; Shu et al. 1988; Straight et al. 1991; Soto et al. 1989; Tubbs et al. 2000; Van Mierop and Kitchens 1980; Werman 2008; White 2005)

Straight et al. (1991) subjected the venom of individual *C. horridus* from throughout its range to electrophoresis, and found that the small basic peptide (SBP) venom toxins vary geographically in this snake. Venoms containing SBP were present from New York to Texas, but the toxin content of the Mississippi, northern Florida, Kansas, Louisiana, and Wisconsin snakes was only weakly positive. The highest toxin content (5–15%) was in venoms from New York, Maryland, North Carolina, and eastern Tennessee. Considerable individual SBP toxin variation was detected in venoms from snakes in South Carolina, Georgia, and northern Florida, varying from 0.0 to 0.5% in 7 Florida *C. horridus* while venom from other states contained 5–10% SBP toxins. Such variations may be related to north to south regional Pleistocene glacial division. Further testing by Glenn et al. (1994) demonstrated that four venom patterns occur within the range of the species: (1) type A is largely neurotoxic (canebrake toxin) and found variably in southern populations (South Carolina, Georgia, northern Florida, Louisiana, southern Arkansas, and Oklahoma); (2) type B is hemorrhagic and proteolytic and found throughout the northern portion of the range and variably in the Southeast; (3) type C is a relatively weak venom having none of the components of types A and B found in the southeastern portion of the range; and (4) type D has properties of both A and B venoms (a mixture of hemorrhagic and neurotoxic) and is found in areas of intergradation between the *horridus* and *atricaudatus* morphs.

The metabolic cost of venom production is high. McCue (2006) found that individual snakes have a 11% increase in SMR (range, 1.731–11.595 kJ), during the first 72 hours of venom replenishment, the apparent result of the metabolic costs involved in venom production, and an order of magnitude greater than that predicted for producing an identical mass of body growth. Neither wet nor dry venom yields are correlated with the magnitude of metabolic increase during the first 72 hours, suggesting that the cost of venom replenishment is independent of extracted mass. The metabolic costs of production may be why rattlesnakes meter their venom conservatively.

Minimum lethal venom doses for several test animals have been calculated, and the venom is particularly virulent to mammals, the primary prey of *C. horridus*. Mean LD_{50} intraperitoneal and intravenous values for 16–18 g mice reported by Arce et al. (2003) were 0.26 (range, 0.19–0.36) mg and 0.80 (range, 0.64–1.01) mg, respectively. Glenn and Straight (1982) reported the following LD_{50} values (mg/kg) for mice by treatment with raw venom: intravenous, 1.64–3.09; intraperitoneal, 2.84–7.25; and subcutaneous, 3.00–9.15. Friederich and Tu (1971) noted the intravenous LD_{50} of laboratory mice (*Mus musculus*) was 2.6–3.1 µg/g. Githens and Wolff (1939) reported the minimum intraperitoneal lethal dose for a 20 g mouse is 0.08 mg, and Macht (1937) found it to be 0.11 mg for a 22 g mouse, but Russell and Brodie (1974) and Russell and Puffer (1970) reported mean intraperitoneal and intravenous LD_{50} values of 2.91 and 2.63 mg/kg, respectively, and Russell and Emery (1959) listed these same toxicities as 2.95 (range, 2.72–3.18) and 2.63 (range, 2.36–2.91) mg/kg, respectively. Johnson et al. (1968a) recorded a LD_{50} of 0.4 mg/100 g with a 95% confidence interval of 0.21–0.78/100 g. Schoettler (1951) reported the minimum lethal subcutaneous dose as 15.0 mg/kg, with a mean lethal dose of 24.9 mg/kg. Sexual and size differences, as well as differences between gravid and nongravid females, occur in venom volume, weight (mg), and toxicity between individual snakes (Minton 1953). The bite of a timber rattlesnake may have killed a young black bear (*Ursus americanus*) in the George Washington and Jefferson National Forest in western Virginia (Klenzendorf et al. 2004), and Hurter (1911) reported that a young cat (*Felis catus*) bitten by this snake died in about 15 minutes. The venom is also fatal to dogs, with an intravenous toxicity of 0.5 mg (Brown 1973; Essex and Markowitz 1930).

Some mammals, however, apparently have an innate immunity to the venom of *C. horridus*: *Didelphis virginiana* (adults?), *Neotoma micropus*, and *Sigmodon hispidus* (Huang and Pérez 1980; Kimon 1976; Pérez and Sánchez 1999).

Githens and Butz (1929) and Githens and George (1931) reported minimum intravenous lethal doses of 0.30 and 0.35 (range, 0.02–0.40) mg for a 320–360 g pigeon (*Columba livia*), and Russell and Emery (1959) reported the intraperitoneal and intravenous LD_{50} values of chicks (*Gallus gallus*) as 3.71 (range, 3.40–4.04) and 0.59 (range, 0.50–0.69) mg/kg, respectively.

Timber rattlesnake (*horridus*) venom (mg/g of body weight) injected into *Charina bottae* (0.515), *Coluber coluber* (0.925), *Crotalus horridus* (*atricaudatus* morph) (0.515), *Diadophis punctatus* (1.22), and *Nerodia sipedon* (0.122 and 0.961) by Keegan and Andrews (1942) killed the snakes, but *Heterodon platirhinos* (0.224), *Pituophis catenifer*

Crotalus horridus; southern *atricaudatus* morph (La Cygne, Linn County, Kansas; Staten Island Zoo; John H. Tashjian)

(0.0913), and *Thamnophis sirtalis* (0.229–0.449) survived venom injections. The two *C. horridus* (*atricaudatus* morph) died 1.3 hours (0.764 mg/g) and 24 hours (0.515 mg/g) after being injected. Some timber rattlesnakes, however, show immunity to the venom of their species; 4 of 5 individuals injected with venom from *C. horridus* by Swanson (1946) survived, while *Agkistrodon contortrix* and *Sistrurus catenatus* died from the venom. A *C. horridus* survived a bite by a *Sistrurus miliaris* with few effects (Munro 1947), but the venom of *A. contortrix* was lethal to other *C. horridus* (Swanson 1946).

Ranid frogs (*Rana palustris*, *R. pipiens*) are not greatly affected by the venom of *C. horridus* (Essex and Markowitz 1930; Mitchell 1860). Johnson et al. (1968a) reported an LD_{50} range of 3.70–4.66 mg for *Paramecium multimicronucleatum*.

The venom over most of the range is strongly hemolytic, but that of the southern *atricadatus* morph is more neurotoxic and contains phospholipase A_2 in the form of canebrake toxin. Human envenomation has resulted in swelling, pain at the site, blebs (blood blisters, ecchymosis), general weakness, breathing difficulty, myonecrosis, blood coagulation, hemorrhage (thrombocytopenia, a decrease in blood platelets), bleeding gums, weak pulse and lowered blood pressure, heart pain, increased heart rate, heart failure, renal dysfunction or failure, rhabdomyolysis, nausea, paralysis, giddiness, unconsciousness (stupor), shock, foaming at the mouth, paresthesia (tingling of limbs, face, scalp), gastric disturbance, and diarrhea (Bond and Burkhart 1997; Carroll et al. 1997; Furman 2007; Hutchison 1929; Keyler 2008; Kitchens et al. 1987; Norris 2004; Norris and Bush 2007; Rao et al. 1998; Sheldon 1929). Also, several types of bacteria occur in the mouths of timber rattlesnakes, which could cause secondary infections following a bite (Parrish et al. 1956).

Ivanyi and Altimari (2004) reported that only 1 of the 36 (2.8%) venomous reptile bites that occurred during academic research in the United States during a 26-year period involved *C. horridus*. Researching this snake can be dangerous, however; both W. S. Brown and H. K. Reinert have suffered, but survived, severe envenomations.

The snake's venom is capable of killing a human (Barbour 1950; Brown 1987; Carroll et al. 1997; Cochran 2008; Ditmars 1931b; Furman 2007; Hutchison 1929, 1930; Guidry 1953; Kitchens et al. 1987; Klauber 1972; Norris and Bush 2007; Parrish and Thompson 1958; Russell 1983). The estimated lethal venom dose for a human is 75–100 mg (Minton and Minton 1969; Russell 1983), well within the range of venom contained in the glands of an adult *C. horridus*. Brown (1973) reported the subcutaneous LD_{50} for a 70 kg human is 616 mg/kg, and that a typical *C. horridus* releases 0.09 lethal dose per bite.

Other cases of human envenomation have been reported by Bond and Burkhart (1997), Carroll et al. (1997), Dart et al. (1992), Furlow and Brennan (1985), Hinze et al. (2001), Hutchison (1930), Keyler (2005, 2008), McPartland and Foster (1988), Norris and Bush (2007), Oldfield and Keyler (1997), Otten and McKimm (1983), Parrish and Thompson (1958), Rao et al. (1998), Sheldon (1929), and Watt (1985). A strange envenomation in Minnesota involved a snake spraying venom into the human's eye, resulting in blurred vision for several days after the eye had been irrigated (Keyler 2005). The practice of handling timber rattlesnakes during services at some Appalachian churches has resulted in envenomations and several deaths.

PREDATORS AND DEFENSE: As with other snakes, the smaller size classes probably experience the most predation.

Known or suspected natural predators, particularly of young *C. horridus*, are snakes (*Agkistrodon piscivorus*; *Coluber constrictor*; *Drymarchon corais*; *Lampropeltis getula*, *L. triangulum*; *Pantherophis obsoletus*), hawks (*Buteo jamaicensis*), owls (*Bubo virginianus*, *Strix varia*), chickens and turkeys (*Gallus gallus*, *Meleagris gallopavo*), opossums (*Diadelphis virginiana*), hogs (*Sus scrofa*), bobcats (*Lynx rufus*), dogs and foxes (*Canis familiaris*, *C. latrans*; *Vulpes vulpes*), weasels (*Mustela frenata*), and skunks (*Mephitis mephitis*) (Brimley 1923; Keegan 1944; Klauber 1972; Klemens 1993; McGowan and Madison 2008; Mitchell 1994a; Odum 1979; J. B. Sealy, personal communication; Vogt 1981).

C. horridus is a mild tempered rattlesnake that will retreat if disturbed. However, the blood level of corticosterone increases during stress (Lutterschmidt et al., 2009); and, if prevented from escaping, it will form a loose coil, rattle, raise its head, and strike (sometimes with its mouth closed, but do not count on it). Its predatory strike lasts on average only 123 (100–200) milliseconds, with the time between initiation of the strike and prey contact possibly only as short as 50–100 milliseconds (Cundall and Beaupre 2001; Kardong and Bels 1998). It is possible that the snake's defensive strike is even faster.

The SVL-adjusted loudness of the rattle is 70 dB, and its frequency is 1.47–20.63 kHz (Cook et al. 1994; Fenton and Licht 1990; Young and Brown 1993). Forbes (1967) reported the following mean respiratory activities for tail and midbody epaxial muscles, respectively, of 10 *C. horridus* during defensive tail movements: the rate of O_2 consumption—0.7 and 7.5 VO_2 (µL/hour/mg), succinic dehydrogenase activity—2.9 and 25.7 (ΔOD [?] $\times 10^{-4}$/minute/mg), and cytochrome oxidase activity—21.4 and 627 (ΔOD [?] $\times 10^{-4}$/minute/mg) all these measures are significantly higher than those of the colubrid snakes he also tested. The rattling can be sustained at a rapid rate for at least 30 minutes (Perry 1920).

C. horridus instinctively recognizes the odor and sight of potential ophidian predators, such as the indigo snake (*Drymarchon corais*) and some kingsnakes (*Lampropeltis calligaster, L. getula*), and even the southern watersnake (*Nerodia fasciata*). In such cases, it either tries to flee or responds by freezing in position, bridging or flipping its body, inflating it, or striking (Gutzke et al. 1993; Marchisin 1980; Meade 1940; Weldon et al. 1992).

PARASITES AND PATHOGENS: Endoparasites identified from *C. horridus* include the protozoans *Caryospora bigenetica*; *Cryptosporidium serpentis*; *Hepatozoon horridus*, *H. sauritus*; and *Isopora naiae*; the trematode *Ochetosoma kansense*; the nematodes *Capillaria hepatica, Capillaria* sp., *Hexametra boddaërtii*, and *Kalicephalus costanus*; and the pentastomid crustacean tongue worm *Porocephalus crotali* (Bowman 1984; Ernst and Ernst 2006; Fantham and Porter 1954; Levine 1980; Penn 1942; Riley and Self 1979; Soloman 1974; Telford et al. 2008; Wacha and Christiansen 1982a, 1982b). In addition, we have found a tick, *Dermacentor* sp., attached to one in Kentucky, and Wolfenbarger (1952) has reported an infestation by larvae of the mite *Trombicula alfredugesi*.

Like other venomous snakes, *C. horridus* is not immune to disease. The bacterium *Arizona hinshawii* has been implicated in the death of a canebrake rattlesnake at the Dallas Zoo (Murphy and Armstrong 1978), and *Pseudomonas* sp. has been found in the liver (Jacobson and Gaskin 1992). *C. horridus* has also developed adenocarcinomas on its intestines and pyloric tumors (Kauffeld 1955; Wadsworth 1956). Orr et al. (1972) reported a fibroma in the species.

Parrish et al. (1956) found the following bacteria in the mouth of *C. horridus*: *Aerobacter aerogenes*; *Clostridium* sp.; *Corynebacterium* sp.; *Escherichia coli*; *Micrococcus pyogenes* (var. *aureus*); *Paracolon bacterium*; *Proteus aeruginosa, P. vulgris*; and *Streptococcus* sp.; some of these may be harmful to the snake, as well as cause infections in human bite victims. McAllister et al. (1993) reported a case of fungal dermatitis in an Arkansas *C. horridus*.

POPULATIONS: Timber rattlesnakes vary in numbers according to the natural suitability of the habitat, and whether or not the habitat has been greatly disturbed by humans.

At a Pennsylvania site, Harwig (1966) recorded 1,628 *C. horridus* in 10 years, finding as many as 17 at one time at a 6 m rock, and as many as 40–80 neonates on a single September day. A survey of Pennsylvania dens by Martin et al. (1990) showed that 78 (25.6%) dens contained 0–15 snakes; 117 (37.5%), 15–30; 72 (23%), 30–60; 37 (11.9%), 60–120; and only 8 (2.6%) of the dens had more than 120 rattlesnakes.

From 1973 to 1987, Martin (1988, 1990) examined 509 sites in the Shenandoah National Park, Virginia, and collected 5,195 *C. horridus*, including 1,271 neonates, and estimated the total population to be 5,400–6,700 individuals. Individual dens there have populations of 10–205 snakes, and the population increases by an average of 63% each fall with the birth of the young (Martin 1988).

In contrast, samples of 2,083 snakes from northern Louisiana and 7,062 snakes from 3 counties in northeastern Kansas, *C. horridus* comprised only 30 (1.4%) and 13 (0.18%), respectively (Clark 1949; Fitch 1992). At the Fitch Natural History Reservation of the University of Kansas, several density and population biomass calculations were as follows: 1949–1963, 0.306 snakes/ha and 0.16 kg/ha (Fitch 1982); only 0.2 snakes/ha and 0.143 kg/ha for 1960–1963 (Fitch and Echelle 2006); and 0.119 kg/ha for the total 50-year study in which 113 different *C. horridus* were captured (Fitch 2000). From 1990 to 2006, Fitch and Pisani (2006) captured 49 adult males, 35 adult females (an

Crotalus horridus, overwintering den (Warren County, New York; Carl H. Ernst)

adult sex ratio of 1.40:1.00), 9 neonates, and 34 immatures (a juvenile and neonate to adult ratio of 0.51:1.00).

The adult sex ratio favored males: 1.79:1.00 and 1.84:1.00 in 2 Pennsylvania snake hunts (Galligan and Dunson 1979; Reinert 1990), 1.69:1.00 in Kansas (Fitch 1982), and 1:1 in Wisconsin (Oldfield and Keyler 1989). The operational sex ratio of the 2 timber rattlesnake populations in southeastern New York studied by McGowan and Madison (2008) were highly male-biased: 49:17 and 12:14. The ratio of males to females in newborn litters is essentially 1:1 (Odum 1979). However, a Minnesota litter had a 2.5:1.0 ratio (Edgren 1948) and 2 Pennsylvania combined litters had a 1.8:1.0 ratio (Galligan and Dunson 1979). The juvenile to adult ratio at Fitch's Kansas site was 1.00:1.50 (Fitch 1982).

Survivorship of young-age classes is probably low, although Reinert (2005) reported only a maximum 6.25% mortality rate from birth to first hibernation for 16 radio-telemetered New Jersey neonates, and thought that use of unknown hibernacula may be confused with mortality. Adult females also seem to be at higher risk than adult males (Fitch 2000), possibly due to their poorer condition as they enter hibernation after parturition. Also, the practice of implantating radio transmitters into a wild individual close to hibernation may increase the snake's chance of winter mortality and should be avoided (Rudolph et al. 1998).

From 1978 to 2002, Brown et al. (2007) conducted a mark-recapture study of *C. horridus* in New York that included 588 neonates (407 wild-born, 181 laboratory-born). Recapture rates declined from about 10 to 20% over time while increasing from young to older age classes. Estimated survival rates in the first year were significantly higher among wild-born individuals (dark phase, 0.773; yellow phase, 0.531) than among laboratory-born snakes (dark phase, 0.441; yellow phase, 0.301). Brown et al. thought that lower birth BMs coupled with a lack of field experience until release contributed to the lesser survival rates in the laboratory-born neonates. Survival of 1- to 4-year-old snakes declined over the years in both color phases, while it remained almost constant or slightly increased for 5-year-old or older adults. Survival rates after the first year were 2- to 4-year-old snakes, 0.845 (dark phase, 0.084; yellow phase, 0.999); and for 5-year-old or older snakes (dark phase, 0.958; yellow phase, 0.822).

Martin (1988, 1990) estimated overwinter mortality rates in Virginia to be 50% for all age classes, and 61% for young of the year. The death rate for 2-year-olds was estimated at 40%, for the third year 25%, and for the fourth year 17.5%. Five- to 14-year-old females and 5- to 17-year-old males experienced 10% annual mortality. Only about 17% of Martin's total population was composed of snakes 15 years of age or older.

The minimum survival rates of 9 neonates (7 males, 2 females) raised in captivity and later released and radio-tracked for 6 years in eastern Texas by Conner et al. (2003) were 1 year, 8 (6 males, 2 females); 2–3 years, 4 (3 males, 1 female); 4–5 years, 3 (2 males, 1 female); and 6 years, 2 (1 male, 1 female). The maximum possible survival rates during the same period were 1 year, 8 (6 males, 2 females); 2–4 years, 7 (5 males, 1 female); 5 years, 6 (5 males, 1 female); and 6 years, 5 (4 males, 1 female). Fitch (1985b) estimated that only about 17% of his Kansas population survived through the fifth year.

Although a few may be trampled by deer (*Odocoileus virginianus*) each year (Minton 1972), adults have few enemies except humans (Bringsøe 1998; Brown 1995b; Cochran 2010; Stechert 1981). By far, more die on our highways (Wright 1987) or are blown apart by guns or otherwise killed (Schorger 1968; Sealy 2002; Tyning 1987); such human-caused mortality may be greater in the northern dark morphs of the snake (Martin 1988).

Some short-distance translocations of *C. horridus* to remove them from human danger have been successfully attempted in North Carolina (Sealy 1997), but mortality of translocated individuals is usually more than 50%, mostly occurring when the snakes cannot find suitable hibernation sites (Reinert and Rupert 1999). Translocated snakes also wander over greater distances than resident individuals, thus opening them to greater danger from human contact, motorized traffic, and natural predators.

Most timber rattlesnakes, however, perish from destruction of their habitat (Ballard 2004; Brown 1995b; Cochran 2006; Foster et al. 2006; Freda 1977; Martin et al. 2008; Platt et al. 1999; Stroupe and Dorcas 2001; Tyning 1987). In New Jersey, less than 2% of the original long-leafed pine habitat remains (Waldron et al. 2006a). Repeated defoliation of eastern oak-hickory forests by gypsy moths (*Lymantria dispar*) leads to replacement by other trees and reduces the numbers of small mammal prey at the site (Peterson 1990), and treatment with insecticides reduces the insect foods of the amphibians and lizards occurring there (C. Ernst, personal observation), further reducing the prey supply of young rattlesnakes.

Collecting for the pet trade, bounty-hunting (largely in the past; Brown 1995b; Furman 2007; Oldfield and Keyler 1997), organized rattlesnake hunts (such events may seriously damage the reproductive capacity of a population by removing a large proportion of the gravid females; Berish 1998; Reinert 1990)—all at levels that are nonsustainable—are major threats to *C. horridus* throughout its range.

C. horridus is a classical K-selected species with a low reproductive replacement rate and a long generation time. Adult mortality is likely naturally low, with a relatively low turnover, making it susceptible to small amounts of exploitation. Given its demographic characteristics, there is a good case to be made for giving the species total protection throughout its range. Some states have recognized its problems of fragmented distribution and small numbers of often isolated denning colonies by listing it as endangered (Connecticut, Massachusetts, New Hampshire, New Jersey, Ohio, Vermont, Virginia [*atricaudatus* morph]), threatened (Illinois, Indiana, Minnesota, New York), or a species of special concern (Pennsylvania, West Virginia, Wisconsin).

Modern studies are using genetic markers to determine the population structure of *C. horridus*, especially at specific hibernacula, and this tool should yield much valuable data from which adequate conservation plans may be formulated (Anderson 2006; Clark et al. 2008; Shine 2008; Villarreal et al. 1996).

REMARKS: Using electrophoretic and morphological data, Foote and MacMahon (1977), Gloyd (1940), and Stille (1987) assigned *C. horridus* to the *C. durissus* subgroup of large rattlesnakes, and thought it closest to *C. molossus*. Plasma electrolyte concentrations (Punzo 1976) and cross-species microsatellite amplification using skin DNA (Bricker et al. 1996; Bushar et al. 2001) also indicate that it is close to the black-tailed rattlesnake. More recent mtDNA and RNA evidence, however, has led Murphy et al. (2002) to include it in the *C. viridis* group along with *C. helleri*, *C. oreganus*, and *C. scutulatus*.

The literature and taxonomy of the species were reviewed by Collins and Knight (1980), and Ahrenfeldt (1955) reviewed the anatomical study of the species based on dissections by Edward Tyson in 1682. Summaries of the snake's life history requirements and conservation were published by Brown (1993), Ernst (1992), Ernst and Ernst (2003), and Furman (2007).

Crotalus lepidus (Kennicott, 1861)
Rock Rattlesnake
Cascabel Verde

RECOGNITION: *Crotalus lepidus* is the first of three accounts covering the small, mostly montane, cool temperature–tolerant rattlesnakes in the portion of Mexico and the United States presented in this book; the other two species are *C. pricei* and *C. willardi*.

This attractive rattlesnake (TBL$_{max}$, 82.8 cm; Klauber 1972) has a ground color of gray, grayish-green, light green, olive, pink, tan, or reddish-brown. The body is patterned with 13–38 narrow, dark brown or black, irregularly spaced dorsal cross bands, which may or may not reach the venter. The anterior body bands are usually less distinct, but become more vivid posteriorly. They are often light-bordered (particularly anteriorly), and small spots or irregular gray mottling may occur between the bands. Dorsal body scales are pitted and keeled with a microdermoglyphic dorsal pattern of mostly straight linear ridges (Harris 2005; Stille 1987; illustrated in Harris 2005). They lie in 23–25 (21–30) anterior rows, 23 (20–25) midbody rows, and 17–18 (14–20) rows near the tail. The tail contains 2–3 (1–6) dark bands and 9–11 (8–13) rattle-fringe scales; the end of the tail is reddish or yellow-orange, and the proximal rattle segment is dark gray, brown, or black (some individuals are rattleless; Christman et al. 2004; Rowe et al. 2002). The venter is pink to light tan anteriorly, but more grayish toward the tail; a variable amount of dark mottling is present. The snake has 147–173 ventrals, 16–33 subcaudals, and an undivided anal plate. A pinkish-brown stripe runs backward from the orbit to beyond the corner of the mouth, light pigment is often present on both the supralabials and infralabials, and a pair of dark occipital blotches (often coalesced) are present. The dorsal surface of the head may be heavily dark-mottled in some Mexican populations. Dorsal head scalation consists of a rostral (usually broader than long), 2 large internasals that touch medially, 5–15 scales in the internasal-prefrontal area, a pair (occasionally two pairs) of canthals, no prefrontals, 2 supraoculars, and 2 (1–4) intersupraoculars. Laterally on the face are 2 nasals that are in contact dorsal to the nostril (the prenasal touches a supralabial, and usually curves beneath the postnasal; the postnasal does not contact the upper preocular), 1 (2) loreal, a series of 3 (0–7) small prefoveal scales anterior to the preoculars, 2 (3–4) preoculars (the upper is usually vertically divided, with the posterior portion curving in front of the supraocular), 2 (1–5) postoculars, 3 (2–5) suboculars (variably in contact with the supralabials), 1 (0–2) interoculabials, 12–13 (10–15) supralabials, and 11–12 (9–13) infralabials. The chin has a small, broader than long, mental scale; 2 pairs of chin shields with the posterior pair the largest; and no scales between the chin shields.

The short hemipenis is bifurcate with cylindrical to bulbous lobes, a divided sulcus spermaticus, about 30–54 (mean, 36–44) sharp medium-length basal spines, and 19–26 (mean, 22–25) fringes per lobe, but no spines in the crotch between the lobes (Dorcas 1992; Klauber 1972). It is illustrated in Dorcas (1992).

Males are larger than females (TBL$_{max}$, 82.8 cm) and have 147–172 (mean, 158–161) ventrals, 20–33 (mean, 25) subcaudals, and TLs 6.9–10.4 (mean, 8–9) % of TBL; females (TBL$_{max}$, probably about 65 cm) have 149–173 (mean, 159–164) ventrals, 16–25 (mean, 19–20) subcaudals, and TLs 5.4–8.5 (mean, 7.0)% of TBL.

Sexual dichromatism is present from birth in *C. l. klauberi*, with the males more pale greenish or olive-colored, and the females grayish, bluish-gray, or pinkish-gray (Armstrong and Murphy 1979; Jacob and Altenbach 1977; Prival 2008; Van Devender and Lowe 1977). In addition, the females also have much mottling between the body bands, whereas this space is almost plain-colored in males (mottling is usually less extensive in the males of all the subspecies).

Crotalus lepidus lepidus (R. D. Bartlett)

Crotalus lepidus klauberi (Chiricahua Mountains, Arizona; Dallas Zoo; John H. Tashjian)

GEOGRAPHIC VARIATION: Four subspecies have been described; the following physical characters and scale patterns have been adapted from Campbell and Lamar (2004), Klauber (1972), and Tanner et al. (1972).

Crotalus lepidus lepidus (Kennicott, 1861), the mottled rock rattlesnake or cascabel de las rocas, is distributed from western Texas (except the El Paso area) and southeastern New Mexico southward on the eastern Mexican Plateau to San Luis Potosí. It has 13–24 (mean, 18–19) dark body cross bands, 150–168 (mean, 162) ventrals in males and 149–164 (mean, 161) in females, 21–29 (mean, 24–25) subcaudals in males and 17–23 (mean, 19–20) in females, and an average TL/TBL of 8.6% in males and 7.1% in females. It is patterned with a dark gray or brown stripe extending backward from the orbit to the corner of the mouth, a pair of dark blotches on the nape of the neck that are usually joined, a dorsal body pattern of faded and only slightly distinct narrow cross bands that extend to the ventrals, much dark mottling between the cross bands, and a relatively dark venter. Variations in body coloration and pattern between populations of *C. l. lepidus* may be the most extensive of any North American pitviper. Its ground color is set at birth, and seems to be an adaptation to the dominant substrate color of its habitat (Axtel 1959; Campbell and Lamar 1989; Dixon 1956; Quinn 1981; Vincent 1982a, 1982b). In addition, most individuals from the Sierra Madre Oriental populations do not have a divided upper preocular scale.

Crotalus l. klauberi Gloyd, 1936b, the banded rock rattlesnake or cascabel rayada de piedra, resides in the area from extreme southwestern Texas, southwestern New Mexico, and southeastern Arizona southwestward along most of the western Mexican Plateau to northern Aguascalientes (Anderson and Lidicker 1963) and Jalisco, extending from southeastern Arizona. It has 13–21 (mean, 17) dark body cross bands that reach the ventrals, 153–172 (mean, 161–163) ventrals in males and 155–170 (mean, 160–162) in females, 20–29 (mean 25) subcaudals in males and 16–24 (mean, 19–21) in females, and an average TL/TBL of 8.1% in males and 6.6% in females. This subspecies often lacks a dark stripe extending backward from the orbit, and has a pair of dark,

Crotalus lepidus klauberi (Arizona-Sonora Desert Museum; Terry R. Creque)

often joined blotches on the nape of the neck, a darker ground color crossed by well-defined dark bands with little mottling between, and a light-colored venter. This subspecies displays sexual dichromatism; males are greenish and females grayish (Jacob and Altenbach 1977; Prival 2008). This difference may have evolved for background matching through natural selection influenced by predation pressure, and is a result of a sex-linked gene that either balances the percent of color morphs in the population or functions in sex recognition. However, Prival (2008) found most Arizona's Chiricahua Mountain gravid females on south- and west-facing slopes, while nongravid females and males occupied more northern slopes. The microhabitat of gravid females had warmer ATs, more grass cover, and less canopy cover than those of nongravid females and males. He thought this microhabitat diversity could possibly explain the color difference in the sexes, as the greenish color of males could camouflage them more on the cooler, more moist slopes inhabited by them and nongravid females, but could increase the chance of predation on these snakes on the warmer, drier southern slopes inhabited by gravid females. In addition, we expand on Prival's hypothesis by suggesting that on the northerly slopes males would come into more contact with nongravid, ready to breed females, while the increased warmth of the slopes where gravid females are found could enhance the development of their embryos.

Crotalus l. maculosus Tanner et al., 1972, the Durangan rock rattlesnake or cascabel de piedras de Durango, is found in Durango and Sinaloa on the western highlands of Sierra Occidental; the report of the snake occurring on Volcán de Tequila in northern Jalisco by Ponce-Campos et al. (2000) has been questioned by Campbell and Lamar (2004) on biogeographic and ecological grounds. Cruz-Sáenz et al. (2009) thought it endangered in Jalisco. This subspecies has 23–38 (mean, 30–33) dark body cross bands that do not reach the ventrals, 159–169 (mean, 164–165) ventrals in males and 157–173 (mean, 165) in females, 26–33 (mean, 29–30) subcaudals in males and 20–25 (mean, 22–23) in females, and an average TL/TBL of 9.4% in males and 7.4% in females. It has a dark brown or black, light-bordered stripe extending backward from the orbit to the level of the corner of the mouth, a pair of separate dark blotches on the nape of the neck, body cross bands consisting of small spots or elongated blotches, and a pinkish venter. Harris and Simmons (1972) reported an intragrade *C. l. maculosus* × *C. l. klauberi* from Durango, Mexico.

Crotalus l. morulus Klauber, 1952, the Tamaulipan rock rattlesnake or cascabel café de las rocas, occurs in the humid pine-oak forest and the upper cloud forest (it may not occur in the lower cloud forest; Martin 1958) of the Sierra Madre Oriental in southwestern Coahuila, Nuevo León, and southwestern Tamaulipas. It has 24–34 (mean, 29) dark body cross bands, 156–167 (mean, 159–160) ventrals in males and 160–171 (mean, 164) in females, 25–30 (mean, 28) subcaudals in males and 20–25 (mean, 22) in females, and an average TL/TBL of 9.0% in males and 7.2% in females. It has a well-marked white- to cream-bordered dark brown or black stripe extending backward from the orbit to the level of the corner of the mouth; a pair of usually separated dark blotches on the nape of the neck; anterior body cross bands that are large, blotch-like, and normally well-defined anteriorly and posterior cross bands that may reach the venter; and a dark venter. Some mottling is present between the cross bands. Males have yellow to orange mid-dorsal pigment.

CONFUSING SPECIES: Other rattlesnakes within the distribution covered by this book are distinguished by the lack of a vertically divided upper preocular scale.

KARYOTYPE: The diploid karyotype is made up of 36 chromosomes, 16 macrochromosomes and 20 microchromosomes. Sex determination is ZW in females and ZZ in males. The ZW genetic heteromorphism occurs on the female's fourth pair of macrochromosomes, where the Z chromosome is metacentric-submetacentric and the W chromosome subtelocentric (Baker et al. 1972; Cole 1990).

FOSSIL RECORD: A Pleistocene (Irvingtonian) vertebra from Arizona has been assigned to *C. lepidus* (Brattstrom 1955a; Holman 2000a), and another vertebra from the

Crotalus lepidus klauberi (Barry Mansell)

Distribution of
Crotalus lepidus.

Pliocene (Blancan) of Arizona may be from *C. lepidus* (Brattstrom 1955a). Van Devender and Bradley (1994) reported a Late Holocene fossil from Maravillas Canyon Cave in Trans-Pecos, Texas.

DISTRIBUTION: The rock rattlesnake ranges from southwestern Texas, southern New Mexico, and southeastern Arizona south in Mexico through Coahuila (Gloyd and Smith 1942), western Nuevo León, Chihuahua, and northeastern Sonora to southwestern Tamaulipas, western San Luis Potosí, Aguascalientes (Anderson and Lidicker 1963), northern Jalisco (Ponce-Campos et al. 2000), Zacatecas, southeastern Sinaloa, and possibly eastern Nayarit (but see "Reproduction" and comments by Campbell and Lamar 2004). Its total distribution occurs between the elevations of about 600 and 2,592 m.

HABITAT: This snake lives at medium elevations in arid and semiarid habitats. It is most often associated with steep rocky areas, particularly talus slopes (but not *C. l. maculosus*; Armstrong and Murphy 1979), rock-strewn hillsides, rock outcrops with crevices, basalt rubble, dry arroyos, and rocky canyons and streambeds in brushy habitats (agave-creosote-cactus to piñon pine-oak-juniper), open areas in pine forests, boreal-tropical forests, and tropical deciduous forests, but is also found in mesquite grasslands, and even the Chihuahuan desert (only *C. l. lepidus*). Reported plants in its various habitats include agave (*Agave lechuguilla, A. palmeri*), cacti (*Opuntia* sp.), guajillo and catclaw (*Acacia berlandieri, A. greggii*) and other acacias (*Acacia wrightii, Acacia* sp.), creosote (*Larrea tridentata*), grasses (*Bouteloua* sp., *Hilaria mutica, Stipa speciosa*), hackberry (*Celtis pallida*), junipers (*Juniperus* sp.), manzanita (*Arctostaphylos* sp.), mascalbean (*Sophora secundiflora*), mesquite (*Prosopis glandulosa*), mimosa (*Mimosa biuncifera*), ocotillo (*Fouquieria splendens*), various oaks (*Quercus* sp.), palo verde (*Cercidium* sp.), persimmon (*Diospyros texana*), quaking aspen (*Populus tremuloides*) and other cotton-woods (*Populus* sp.), various pines (including *Pinus cembrioides*, *P. discolor*, and *P. edulis*), silk tassel (*Garrya wrightii*), sotol (*Dasylirion scariosa*), tarbush (*Flourensia cernua*), wal-nuts (*Juglans nigra*), and yuccas (*Yucca elata*).

BEHAVIOR AND ECOLOGY: Almost all of the reported life history parameters have been from studies in the United States on *C. l. lepidus* and *C. l. klauberi*. Such data are almost totally unknown for the Mexican subspecies *C. l. maculosus* and *C. l. morulus*.

C. l. lepidus and *C. l. klauberi* are active from March to October in New Mexico, where Degenhardt et al. (1996) made 29% of their collections during the height of the mating season in August. Wright and Wright (1957) gave the annual activity period of *C. l. lepidus* as 1 May to 11 November. In the United States the two subspecies are forced to hibernate during the winter. In the northern part of the species' range, win-ter is spent under rocks, in rock crevices, within old logs or stumps, in animal burrows, and possibly also in caves. Mexican *C. l. maculosus* may be active throughout the year if the AT is at least 24°C, but most are seen during the rainy season; 17 were found on a single day after the first major rainfall of the season (Armstrong and Murphy 1979).

During the spring and fall, most activity is diurnal, but shifts to crepuscular during the hotter periods of summer. In Texas, *C. l. lepidus* is usually found in the open during the cool morning hours, and in the shade until 0930–1000 hours; the hottest hours are spent under cover (Telotte *in* Conant 1955). Beaupre (1995a) found this subspecies surface-active from 0700 to 1400 hours and at 1800–2200 hours in western Texas. Prob-ably the snake is surface-active for more daylight hours in the spring and fall.

Activity periods may differ between populations, depending on environmental conditions: Beaupre (1995a) found *C. l. lepidus* living in a hot, dry habitat moved only 20–35 (mean, 33) % of the time, while those in a cooler, wetter habitat were found active 40–95 (mean, 62) % of the time. Beaupre's snakes moved an average of 20.4 m per day. Night captures of *C. l. lepidus* are less frequent in Texas, and most nocturnal observations have occurred in August. In the arid areas west of Texas and in south-eastern Arizona *C. l. klauberi* is more often found on roads at night (Campbell and Lamar 2004; C. Ernst, personal observation). High-elevation populations in Mexico are almost exclusively diurnal (Campbell and Lamar 2004).

Some data regarding the thermal ecology of the species have been reported. Living at relatively high elevations, this snake may be adapted to cooler ETs and its BT is 87% related to ET (Beaupre 1995b). Basking or otherwise active individuals have been found at ATs of 24–35°C (Klauber 1972); the maximum voluntary BT recorded by Beaupre (1995a) was 38°C. BTs of *C. l. lepidus* at Beaupre's (1995a) 560 m, 15.1% veg-etation cover, hot site averaged 29.9 (range, 21.5–35.5) °C; compared with 28.8 (range, 17.4–38.0) °C at his 1063 m, 23.2% vegetation cover, cool site. At the hot site, BTs were 2.0–4.5°C higher during July–August than at the cool site (Beaupre 1995b). The aver-age, minimum, and maximum BTs peaked between 1500 and 1800 hours; decreased after sunset at about 2100 hours; and slowly decreased until sunrise at about 0700 hours (Beaupre 1995b). Nocturnal surface activity was practically nonexistent at the hotter site, but was rarely below 10% at the cooler site due to more cloud cover (Beau-pre 1995b). Snakes from the hot site were 64% less surface-active in 1995.

Bryson et al. (2008) recorded the following mean temperatures for Mexican *C. l.*

maculosus: BT, 26.0 (range, 19.4–32.6) °C; AT, 22.5 (range, 20.2–26.1) °C; and ST, 24.0 (range, 22.5–25.5) °C; and *C. l. morulus*: BT, 24.7 (range, 18.2–31.8) °C; AT, 21.9 (range, 16.9–25.2) °C; and ST, 23.6 (range, 17.8–28.0) °C.

Borrell et al. (2005) conducted a study of respiratory cooling in two *C. lepidus lepidus* from Texas, and the partitioning of the effects of AT, relative humidity, and activity levels on head-body temperature differences in the snakes. They found a single, cooled region centered around the mouth and nasal capsule that extended across the pit membrane at ATs above 20°C. Both the temperatures of the head and body increased linearly with AT, as did head-body differences, but these declined significantly at higher relative humidities. Snakes rattling had significantly greater head-body temperature differentials than did those not rattling. Borrell et al. concluded that respiratory cooling may form a thermal buffer for the pit organs at high ATs, but cautioned that this adaptive hypothesis should be tested with direct neural or behavioral studies.

Average daily field respiration calculated by Beaupre (1996) was significantly lower (only about 50%) at the hotter habitat than at the cooler habitat—2,019 J/day compared with 4,872 J/day. The annual average energy budget for a 100 g adult was 113.3% of BM/year (16.4 g food/month) at the hot site, and 193% of BM/year (32.7 g food/month) at the cooler site. The snakes' O_2 consumption increased with rising ET (Beaupre 1993), and was lowest at both sites at 20°C (0700–1300 hours) and highest at 35°C in the afternoon. However, nongravid females had lower SMRs at 25°C than either gravid females or males. Snakes at the hotter site had a lower growth rate and achieved smaller adult body lengths (Beaupre 1995a), even though the SMR of snakes from both populations was similar (Beaupre 1995b). Five *C. lepidus* (subspecies not given) acclimated to 24°C had heart and breathing rates averaging 13.4 (range, 5.7–50.0) beats/minute and 1.76 (range, 0.48–8.60) breaths/minute (Jacob 1980).

Beaupre's (1995a, 1995b, 1996) two study sites differed in available water, resulting in different water balance physiologies between the two populations. Usually, snakes from both sites experienced negative water balance—the average water influx rate was 13.2 mL/kg per day at the hot, dry site and 9.3 mL/kg per day at the cooler, more moist site. Frey (1996) found a New Mexico *C. l. klauberi* completely submerged in a 1 m deep pool of water (WT, 16°C; AT, 35°C), and thought it was either thermoregulating or possibly foraging for small fish.

The rock rattlesnake occasionally climbs above ground level. Telotte (*in* Conant 1955) found a *C. l. lepidus* resting about 1.4 m high on a tree stump in western Texas, and another lying on the edge of a concrete stock watering trough at a height of about 61 cm. Rossi and Feldner (1993) found a juvenile Arizona *C. l. klauberi* on 9 September at 0930 hours wedged behind the loose bark of a pine tree in complete shade 61 cm above the ground. If put into water, *C. lepidus* is a good swimmer (Frey 1996; Klauber 1972).

Like other pitvipers, males participate in ritualized combat bouts (Carpenter et al. 1976), particularly during the mating season (*C. l. klauberi*; C. Ernst, personal observation).

REPRODUCTION: *C. lepidus* matures at approximately 3 years of age and a TBL of 38 cm (Stebbins 2003). The smallest male producing mature sperm and the smallest female undergoing secondary vitellogenesis examined by Goldberg (2000a, combined subspecies) had SVLs of 35.2 cm and 34.0 cm, respectively. The smallest gravid female *C. l. lepidus* found by Beaupre (1995a) had SVLs of 37.4 cm and 41.8 cm.

Crotalus lepidus maculosus (West of Durango, Durango; Dallas Zoo; John H. Tashjian)

The male testes are regressed with spermatogonia and Sertoli cells in May, June, and October, and are in spermiogenesis with maturing spermatids and mature sperm from June to October. Mature sperm is stored over winter in the vas deferens. The sexual segment of the kidney is enlarged and contains secondary granules from May to October. Females have inactive ovaries from July through October, and undergo vitellogenesis from July through August and in October; enlarged follicles (6 mm) are present from April through October (Goldberg 2000a).

Beaupre (1995a) found gravid west Texas female *C. l. lepidus* on 17–18 May, and another Texas female examined by Stebbins (1954) contained 5 large ova of approximately 34 mm × 17 mm on 1 July. A female *C. l. klauberi* from Santa Cruz County, Arizona, examined by Goldberg (2000a) had 6 oviductal eggs in April, and 2 females with 34.3 cm and 38.2 cm SVL from Cochize County, Arizona, each had 2 enlarged follicles on 10 and 21 October, respectively. Prival (2008) captured gravid female *C. l. klauberi* between 16 July and 16 August in the Chiricahua Mountains of southeastern Arizona. A 53.8 cm SVL Durango *C. l. maculosus* collected on 25 August had 5 enlarged follicles (Goldberg 2000a). Normally, females are biennial reproducers, but Beaupre (1995a) reported that a *C. l. lepidus* produced litters in 3 consecutive years. Probably the nutritional condition of the female dictates whether or not she can reproduce annually.

In nature, courtship and mating activity occurs from late July to early October. It has occurred as early as 21 February (Armstrong and Murphy 1979) and as late as 11 October (Barker *in* Armstrong and Murphy 1979) in captive *C. l. klauberi*. While courting, the male directs head-bobs (3–5 every 5 seconds) on the female's back and tongue-flicks at the same rate (Armstrong and Murphy 1979). Females store sperm at least over winter (Dancik *in* Schuett 1992).

The species experiences a long GP, 240–363 days (Swinford 1989; Tennant 1985, 1998), possibly correlated with its cooler habitat, and the young are usually born in late June–August, but April and October births are known. A Mexican *C. l. maculosus* × *C. l. klauberi* has given birth on 14 April (Harris and Simmons 1972), and Armstrong and Murphy (1979) observed a Sinaloan female of this subspecies with 11 young on 10 July. A female *C. l. morulus* underwent parturition in late August, and neonates of the same subspecies have been collected in late August near the Nuevo León-Coahuila border (Armstrong and Murphy 1979).

C. l. lepidus and *C. l. klauberi* produce litters of 2 (Goldberg 2000a) to 9 (Liner and Chaney 1986) young, with an average of 4 young (n = 49); Seigel and Fitch (1984) reported a mean RCM of 41.4%. Litters of *C. l. maculosus* contain 1–11 young (Armstrong and Murphy 1979; Harris and Simmons 1972). *C. l. morulus* produces litters of 4–10 young (Porras *in* Campbell and Lamar 2004; Sánchez et al. 1999); one litter had a RCM of 35.3% (Vitt and Price 1982). The report by Armstrong and Murphy (1979) of a female having 11 young with her indicates that females may remain with their litter for at least a short time after parturition.

Neonates have 12.5–24.3 (mean, 17.0; n = 44) cm TBLs and 3.1–8.0 (mean, 6.0; n = 29) g BMs; those of *C. l. morulus* are the shortest and weigh the least—TBL, 12.5–19.0 cm; BM, 3.1–5.7 g (Campbell and Lamar 2004). Neonates are colored and patterned like the adults of their subspecies, but have yellowish tails (see "Diet and Feeding Behavior"). Females likely attend their neonates (Prival 2008).

A possible hybrid between *C. lepidus* and *C. triseriatus* (USNM 46333) from Santa Teresa, Nayarit, Mexico, exists. Campbell and Lamar (2004) assigned this specimen to *C. lepidus* based on information in Smith (1946), but its identity is still questionable until more specimens from the area can be examined for comparison.

GROWTH AND LONGEVITY: Growth data are scarce and are mostly based on captives. A *C. l. lepidus* kept by Falck (1940) grew from 40.7 cm to 47.6 cm in 409 days, and Strimple (1993a) had captive-born *C. l. lepidus* that grew an average of 17.8 cm (1.5 cm/month) and 37.2 g (4.2 g/month) during their first year, and 9.2 cm (0.8 cm/month) and 50.7 g (3.1 g/month) the second year. One grew from 23.5 cm (13 g) to 52.1 cm (106.8 g) in 2 years, and to 63.9 cm (171.2 g) in 3 years. Woodin (1953) reported that 2 young *C. l. klauberi*, 20.5 and 21.0 cm when collected, grew to 21.4 and 21.8 cm, respectively, in 23 days. Kauffeld (1943b) noted that a male of this subspecies, 20 cm at birth, grew to 31.2 cm in 118 days. It fed readily, as opposed to a female that had to be force-fed and grew only 3.2 cm during the same period.

Growth slows once wild females mature, possibly due to their feeding less often when gravid. Beaupre (2002) has developed a model predicting the effects of food availability and thermal relationships to growth based on *C. l. lepidus* (which see for details).

Vandeventer (1977) reported that a recently obtained adult male *C. l. lepidus* experienced a double pre-ecdysal period. Its eyes became opaque, remained so for five days, and then cleared. A newly formed basal rattle segment was clearly visible. However, the snake did not shed its skin. Soaking in water for 24 hours and manual manipulation failed to loosen the skin, and it appeared that no new underlying skin layer had developed even though the snake had formed a new rattle segment. Eleven days after its eyes had initially cleared, they became opaque again and the condition lasted for two days before the eyes again cleared. Ecdysis occurred two days later, but only one skin layer was shed. The number of molts per year is normally tied to the amount of prey taken and growth achieved during that period.

Survival records for captives include 33 years and 7 months for the species (Slavens

Crotalus lepidus morulus (North of Cerro Potosi, Nuevo León; Dallas Zoo; John H. Tashjian)

and Slavens 2000); 24.5 years, and 23 years, 3 months, and 24 days for *C. l. klauberi*; 17 years, 11 months, and 10 days for *C. l. lepidus*; 11 years and 1 month for *C. l. maculosus*; and 9 years, 2 months, and 21 days for *C. l. morulus* (Russell 1983; Snider and Bowler 1992).

DIET AND FEEDING BEHAVIOR: *C. lepidus* is an opportunistic predator that seems to prefer cold-blooded prey, particularly lizards, to mammals. Holycross et al. (2002a) reported that the diet of *C. l. klauberi*, mostly (91% of specimens examined) from the northern Sierra Madrean Archipelago, consisted of 55.4% lizards, 28.3% scolopendromorph centipedes, 13.8% mammals, 1.9% birds, and 0.6% snakes. Species of *Sceloporus* comprised 92.4% of the lizards eaten, with *S. jarrovii* 82.3% of these. Diet did not vary significantly geographically, or by sex or age class. However, snake SVL differed significantly among prey taken: those that ate birds were the longest, followed in order by those that ate mammals, lizards, and centipedes. Collection date also differed significantly among the prey groups; the mean date for centipede consumption was later than those for lizards, birds, or mammals. No significant difference was noted among the various prey in the elevation of collection sites.

Few species of mice of suitable prey size live at the elevations inhabited by *C. lepidus*, while lizards of various types abound, and a preference for lizard prey may have resulted from natural selection favoring the more plentiful animal.

Reported wild prey are centipedes (*Scolenpendra* sp.), insects (caterpillars, grasshoppers), anurans (*Syrrhophus marnocki*); reptiles—lizards *(Anolis nebulosus; Barisia imbricate; Cnemidophorus gularis, C. sacki* [?]; *Cophosaurus texanus; Eumeces brevirostris; Phrynosoma cornutum, P. hernandesi; Sceloporus clarki, S. grammicus, S. jarrovii, S. merriami, S. microlepidotus, S. minor, S. poinscttii, S. torquatus, S. virgatus; Urosaurus ornatus),* and snakes (*Gyalopion canum, Hypsiglena jani,* and an additional unidentified specimen); birds (unidentified); and rodents (*Chaetodipus* sp., *Dipodomys* sp., *Geomys bursarius, Perognathus* sp., *Peromyscus pectoralis, Peromyscus* sp., *Sigmodon* sp.). (Beaupre 1995a; Bryson et al. 2002a; Campbell 1934; Campbell and Lamar 2004; Dickerman and Painter 2001; Forstner et al. 1997; Geluso 2007; Gloyd and Smith 1942; Holycross et al. 2002a; Kauffeld 1943a; Klauber 1972; Marr 1944; Milstead et al. 1950; Stebbins 1954; Telotte *in* Conant 1955; Woodin 1953). In addition, captives have consumed salamanders (*Ambystoma tigrinum*), frogs (*Acris crepitans; Pseudacris triseriata; Rana pipiens, R. sylvatica*), lizards (*Anolis carolinensis, Eumeces laticeps, Sceloporus undulatus*), snakes (*Crotalus lepidus, Virginia striatula*), and rodents (*Microtus* sp., *Mus musculus, Peromyscus* sp.) (Armstrong and Murphy 1979; Axtell 1959; Barker 1991; Bryson et al. 2002a; Campbell 1934; Falck 1940; Harris and Simmons 1977b; Holycross et al. 2002a; Kauffeld 1943b; Lazcano et al. 2004; Mata-Silva et al. 2010; Milstead et al. 1950; Strimple 1993a; Vandeventer 1977; Williamson 1971; Woodin 1953). As the species has consumed smaller individuals in captivity, the possibility exists that it is also sometimes cannibalistic in the wild.

Juvenile rock rattlesnakes have yellow tails, which they wave as a lure to attract lizards (Kauffeld 1943b; Neill 1960; Starrett and Holycross 2000). Adults may actively forage, but often ambush prey. They lie in an *S*-shaped coil with the head pointed upward along the side of a rock, or across the open surface of a boulder or a gap in vegetation (Beaupre 1995a). Lizards are struck on the body (multiple times; Chiszar et al. 1986c), held until comatose, and then quickly swallowed. Mice are struck, released, and later trailed. In tests, chemoreception (definitely the vomeronasal system, and possibly the nasal system as well) was used to find the dead rodents, but the species scored significantly lower than *C. v. viridis*, which in nature depends more heavily on rodent prey (Chiszar et al. 1986d).

VENOM DELIVERY SYSTEM: Five *C. l. klauberi* with 52.0–59.5 cm TBLs had 3.2–3.6 (mean, 3.4) mm FLs, a mean TBL/FL of 162%, and a mean HL/FL of 7.54%; the angle of fang curvature was 58° (Klauber 1939).

VENOM AND BITES: The venom of *C. lepidus* has been one of the least studied of *Crotalus* rattlesnakes, but some data are available. Fresh adults may yield as little as 0.1 mL (0.03 g) (do Amaral 1928) or as much as 129 mg of dry venom (Glenn and Straight 1982). Thirty-one fresh *C. l. klauberi* yielded a total of 177 mg of dried venom with an average yield of 10 mg per fresh adult (18) when milked by Klauber (1972). Minton (1977) reported yields of 24, 25, and 34 mg in 3 extractions from a *C. l. klauberi*.

Chemically, the venom contains 22 distinct protein bands (Forstner et al. 1997), including esterase (BAEE), phosphodiesterase, and phospholipase A_2, and shows caseinolytic and hemorrhagic activities (Glenn and Straight 1987). In addition, some populations contain Mojave toxin proteins (both α and β) that make them more dangerous (Glenn and Straight 1985a, 1987; Powell et al. 2008; Rael et al. 1992); see below.

Texas populations of *C. l. lepidus* have different venom chemical profiles. Those from the Rio Grande Valley, though geographically separated, are chemically homogeneous (supporting Vincent's 1982a, 1982b hypothesis of dispersal of the species along river canyons). These populations are also intermediate between the mountains and eastern plateau populations. Venom of *C. l. lepidus* from populations in the Edwards and Stockton plateaus differs from that of populations from the Rio Grande and western mountains of Texas. Venom toxicity across the Texas range of this species varies from 0.72 to 2.20 mg/mouse kg, and the LD_{50} varies from 0.15 to 0.64 mg/kg (Forstner et al. 1997).

Similarly, *C. l. klauberi* from different populations differ in their venom chemistry and toxicity (Glenn and Straight 1987). The venom LD_{50} values of populations (Arizona and New Mexico, and Chihuahua and Durango, Mexico) containing the Mojave-like toxin are intraperitoneal, 0.9–6.3 (mean, 0.63) mg/mouse kg, and intramuscular, 2.8–13.5 (mean, 8.8) mg/kg. The venom of *C. l. klauberi* from the Magdelena Mountains of New Mexico lacks the Mojave-like toxin or only contains a small amount (0.06 mg/mL) (Bober et al. 1988). Snakes from Aguascalientes, Nuevo León, and Zacatecas, Mexico, also lack the Mojave-like toxin, and have 0.9–6.3 (mean, 2.6) mg/kg intraperitoneal and 8.5–23.4 (mean, 15.4) mg/kg intramuscular LD_{50} values. Repeated venom extractions from Zacatecas *C. l. klauberi* exhibited the highest LD_{50} values: intraperitoneal, 4.0–6.3 mg/kg and intramuscular, 19.0–23.4 mg/kg. Minton (1977) reported intravenous and subcutaneous LD_{50} values of 9.0 and 23.95 mg/kg, respectively, and a minimum hemolytic dose of 1.56 μg for a *C. l. klauberi* from Zacatecas.

Such variance in the venom of populations of *C. l. lepidus* and *C. l. klauberi* is probably related to the vulnerability of the prime prey species within their specific ranges.

The venom of this snake may be more specific for vertebrates than invertebrates; it took more than an hour to kill a scolopendromorph centipede (Rodríguez-Robles 1994). The lethal intravenous venom dose for a typical mouse is 0.02–5.0 mg (Githens and Wolff 1939; Tu 1982b), and the intraperitoneal dose is 0.90 mg (Githens and Wolff 1939). Minton (1977) reported mean mouse intravenous and subcutaneous LD_{50} values of 9.00 mg/kg and 23.95 mg/kg, respectively, for venom from *C. l. klauberi* (also see the mouse LD_{50} values of the various subspecies and populations previously discussed). The LD_{50} for a typical pigeon is 0.01 mg (Githens and George 1931).

The venom is extremely hemorrhagic (Rael et al. 1992; Soto et al. 1989), and local necrosis may occur, so *C. lepidus* should be treated with caution. Fortunately, envenomation by this snake is rare. Only a few case histories have been published.

Norris (2005) supplied data in the newest and most detailed symptomatic report involving a full two-fang *C. l. lepidus* bite to the top of the left thumb of a male apparently handling the snake. Within minutes, the thumb became painful and swollen. A rubber band tourniquet was applied within five minutes to the wrist above the thumb, and the victim drove himself to the hospital, arriving an hour after the envenomation.

On arriving at the emergency room, he was experiencing moderate pain and swelling in his left hand, but had no systemic symptoms. When examined, his blood pressure (BP) was 141/89 mm Hg, heart rate 103 beats/minute, and breathing rate 18/minute; he was afibrile (had no spontaneous contraction of muscles); and he had no appreciable swelling (adenopathy) of the axial lymph nodes.

The swelling advanced to the wrist, and he could not touch his thumb and index finger due to the swelling. The rubber band was removed and the arm elevated. During the next hour the swelling advanced into the upper arm, and the man developed a sensation of "cold water dropping" on his lips and chin. An electrocardiogram and laboratory blood tests revealed no abnormalities.

The patient refused antivenom (Wyeth-Ayerst Laboratories) treatment. At 135 minutes after the bite, he received 1 g of cefazolin intravenously to prevent infection. He was reluctant to take medication for pain, but eventually agreed to a small dose (2 mg) of morphine sulfate intravenously at 180 minutes after being bitten. Fifteen minutes later, he became pale, began to sweat, and vomited. His vital signs at this point were BP, 105/61 mm Hg; heart rate, 91 beats per minute; and breathing rate, 20 breaths per minute. These symptoms were thought to have been brought on by the morphine, not the venom.

The man was admitted to a telemetry unit bed for observation, and his arm was kept elevated. Overnight, the swelling progressed all the way up his arm and onto the trunk of his body and left side of his neck. Three more episodes of nausea and vomiting were experienced during the first 24 hours, but no other systemic signs or symptoms. Over the next two days hemorrhagic areas developed on the posterior-medial aspect of the bitten appendage, but no blood blisters appeared. His blood coagulation remained normal, although a mild, gradual, decrease in blood platelets and hematocrit occurred. The patient was discharged on the fifth day without having received antivenom treatment. An examination more than a year later revealed no necrosis and no permanent damage to the limb.

Hardy (1992) reported a legitimate envenomation by a wild Sonoran *C. l. klauberi*. The bite was on the right distal index finger of a 16-year-old male student biologist. An Extractor™ pump was immediately applied over the single puncture wound and continued for 4 minutes, during which time a clear plastic cup was filled 2.5 times with serosanguineous fluid. Swelling extended to the victim's wrist and after 48 hours a blood blister formed over the puncture wound, but no necrosis occurred. He consulted a physician 96 hours after the envenomation, and as the symptoms were resolving, no specific therapy was given. Recovery was complete, including normal range of motion of the finger, with no tissue necrosis.

Other less detailed case histories have also been published. The late W. W. Wright was bitten on the thumb by both fangs of a *C. l. lepidus* while holding the snake. His arm swelled considerably and axial lymph glands were affected, but he fully recovered (Wright and Wright 1957). Klauber (1972) described the symptoms from an envenomation of a herpetologist by *C. l. klauberi*. One fang pierced the middle finger. By the next day swelling had advanced to the forearm, and on the second day, to the shoulder. An intense and continual burning pain developed at the site of the puncture; it began on the day following the bite and became almost unbearable by the second day, but then lessened. Within a day or so after the bite, a large (12.5 mm × 19.0 mm) blood blister formed at the site of the puncture wound. The swelling remained for five weeks, with numbness and tingling sensations.

These case histories indicate the severity of this snake's venom and the foolishness of handling them or any other venomous snake.

PREDATORS AND DEFENSE: Predatory data are scarce. Tennant (1985) reported that one was eaten by a copperhead (*Agkistrodon contortrix*), and another was seized by a collared lizard (*Crotaphytus collaris*) (Klauber 1972). Captives have cannibalized other members of their own species (Harris and Simmons 1977b; Williamson 1971), so large individuals could also eat smaller ones in nature. Ophiophagous snakes (*Drymarchon corais, Lampropeltis* sp., *Masticophis* sp.), birds of prey (*Bubo virginianus, Buteo* sp.), and carnivorous mammals (*Canis latrans, Lynx rufus, Mephitis sp., Nasua narica, Procyon lotor, Puma concolor, Vulpes macrotis*) probably eat juveniles on occasion.

C. lepidus has a rather calm temperament. It will often remain quiet when first discovered, partially camouflaged by its body color matching that of the substrate (Beaupre 1995a; Conant 1955; Vincent 1982a). If given the chance, however, it will escape into a rock crevice or under some object. When it cannot escape, it will coil and strike, and attempt to bite if handled (see above), rattling all the time. The frequency of rattling varies from 2.4 to 23.1 Hz (Young and Brown 1993), and the SVL-adjusted loudness is over 70 dB (Cook et al. 1994).

In laboratory tests conducted by Miller and Gutzke (1999), *C. lepidus* with unimpaired vomeronasal organs detected and reacted negatively to the odors of several subspecies of the potential predator *Lampropeltis getula* by body-bridging and hiding their heads, but those with the organs sutured shut failed to react.

PARASITES AND PATHOGENS: *Crotalus lepidus* is a host for parasitic acanthocephalans—unidentified oligacanthorhynchid larval cystacanths, and nematodes—*Abbreviata terrapenis* (possibly an accidental parasite) and *Physocephalus sexalatus* (Campbell and Lamar 2004; Ernst and Ernst 2006; Goldberg and Bursey 1999a; Goldberg et al. 2002a; McAllister et al. 2004). A *C. l. klauberi* was found with symptoms of visceral gout, characterized by large deposits of uric acid crystals in the kidney tubules, and on the surface of the heart, liver, and peritoneum (Wallach and Hoessle 1967). A paramyxo-like virus has been isolated from a female *C. lepidus* with progressive central nervous system disease (Jacobson et al. 1980).

POPULATIONS: *C. lepidus* is common at some sites, with large numbers living among the rocks of talus slopes, but most populations seem small. Cruz-Sáenz et al. reported it endangered in Jalisco. A maximum of 106 *C. l. klauberi* were recorded at any of more than 40 Arizona localities sampled by Johnson and Mills (1982), and of these only 6 sites had more than 7 individuals. Ehret et al. (2005) captured and marked 132 *C. l. klauberi*, 49 of which were recaptured a total of 73 times, in 1999 and 2002–2004, in the Chiracahua Mountains of southeastern Arizona. In 3 years, Reynolds (1982) collected only 3 *C. lepidus* (0.72%) in a sample of 418 snakes captured along a Chihuahuan highway.

Some data on sex ratios are available. Beaupre (1995a) recorded ratios of 1.33:1.00 and 1.92:1.00 in 2 Texas populations of *C. l. lepidus*. Lazcano et al. (2005) reported captive

Crotalus lepidus morulus, juvenile (Nuevo León; John H. Tashjian)

adult male to female and juvenile to adult ratios of 1.40:1.00 and 0.58:1.00, respectively, for *C. l. klauberi*, and 1.24:1.00 and 1.15:1.00, respectively, for *C. l. morulus*.

During a study of the effects of prescribed burning on rattlesnake populations in the Coronado National Forest in the Peloncillo Mountains of extreme southern Arizona and New Mexico, Smith et al. (2001) compared the movements of five radio-equipped *C. l. klauberi* before the fire; immediately after the fire, but before the summer rainy season; and during the summer rainy season. One, which had moved 60 m from its prefire position, was found dead within a crevice immediately after the burn, apparently a victim of the fire, but at least three others survived. Two exhibited smaller daily activity areas than before the burn (two others were not found during this intermediate period, and one of these was never located during the rest of the study). Three of the remaining males were found during the rainy season, and all had increased their daily movement area. These data show that fires, even prescribed ones, are very dangerous to rattlesnake populations.

Humans cause the most problems for this species, with their mining, grazing, road-building, traffic, legalized scientific and illegal pet trade collecting, recreational and urban development, and fires (Ehret et al. 2005; Johnson and Mills 1982; Smith et al. 2001). Most of these activities involve habitat destruction, but with the exception of widespread grazing, road-building, and wildfires, are of local impact.

The species is protected by law in Arizona and New Mexico.

REMARKS: Electrophoretic studies of blood and venom proteins by Foote and MacMahon (1977) and Harris (2006a) and of morphology by Dorcas (1992) and Gloyd (1940) indicate that *C. lepidus* is closely related to the Mexican rattlesnake, *C. triseriatus*. However, recent studies of mtDNA and RNA by Murphy et al. (2002) show that *C. lepidus* is not as close to *C. triseriatus* as formerly suspected, but they still place it in the *C. triseriatus* group of rattlesnakes, along with *C. aquilus* and *C. pusillus*. Harris (2006a) found that *C. aquilus* is situated between *C. lepidus* and *C. triseriatus*.

Crotalus mitchellii (Cope, 1861)
Speckled Rattlesnake
Vibora Blanca

RECOGNITION: *Crotalus mitchellii* grows to a maximum TBL of at least 137 cm (Campbell and Lamar 2004; Case 1983 reported a TBL_{max} of 141 cm), but most individuals are shorter than 10 cm. Ground color is quite variable, ranging from gray, bluish-green, cream, yellowish-pink, salmon, tan, or pale brown to black (*C. m. pyrrhus*; lava flows at Pisgah, San Bernardino County, California, and Pinacate, Sonora, Mexico). Usually a series of 23–46 (mean, 35) dorsal blotches is present that range in color from orangish to reddish, dark gray, brown, or black, and contain liberal amounts of dark pigment. Anteriorly the dorsal blotches are hexagonal or diamond-shaped, but posteriorly they fuse with 1–2 series of smaller lateral blotches to form dark cross bands. The scales between the various blotches are marked with dark speckles. The tail has 2–9 (mean, 4.5) dark bands, and the TL is 5–9% of the TBL. The pitted and keeled dorsal body scales are microdermoglypically adorned with swirling ridges (Harris 2005) or are foveate with raised cell borders/vermiculate (Stille 1987); Stille (1987) presents an electronmicrograph of the scale surface. The body scales lie in 23 (25) anterior rows, 23 or 25 (21–27) midbody rows, and 23 (25) posterior rows. The venter is white, pink, cream, yellowish-brown, orange, or reddish with some grayish-black mottling. On it are 162–190 ventrals, 16–28 subcaudals, and an undivided anal plate. The head is indistinctly patterned, but is marked dorsally with dark specks or spots, and the anterior infralabials and mental may contain black pigment. If a dark postocular stripe is present, it extends diagonally backward from the orbit to the corner of the mouth. Dorsal head scales include a wider than high rostral that contacts 2–4 internasals but is separated from the prenasals by small scales, up to 51 scales in the internasal and prefrontal area (no prefrontal scales are present), 2 supraoculars, and 1–8 intersupraoculars. Lateral scales are 2 nasals (the prenasal may, particularly in the north, or may not, in the south, contact the first supralabial; the postnasal does not meet the upper preocular), 1–2 (0–5) loreals, 2 (3) preoculars, 2 (3) postoculars, several suboculars, numerous temporals, 2 interoculabials (separating the anterior suboculars and supralabials), 15–16 (13–19) supralabials, and 15–16 (13–19) infralabials. On the chin are a triangular mental and 4 chin shields (the posterior pair are largest), but no interchin scales.

The bifurcate hemipenis has a divided sulcus spermaticus and 0–3 spines in the crotch; each lobe has 30–47 spines and 31 fringes (Klauber 1972).

The dentition includes 2–3 palatines, 8–9 (7–10) pterygoids, 8–9 (7–10) dentaries, and the 2 maxillary fangs (Klauber 1972).

Males (TBL_{max} to at least 137 cm) have 177–178 (range, 162–187) ventrals, 23–24 (range, 20–28) subcaudals, 5 (range, 3–9) dark tail bands, and mean TL 7.0 (range, 6.7–8.0) % of TBL. Females (TBL_{max} 100 cm) have 178–179 (163–190) ventrals, 18–20 (16–23) subcaudals, 4 (2–6) dark tail rings, and mean TL 6.0 (range, 5.0–6.8) % of TBL.

GEOGRAPHIC VARIATION: Four subspecies are recognized; only one occurs north of Mexico. *Crotalus mitchellii mitchellii* (Cope, 1861a), the San Lucan speckled rattlesnake or vibora blanca, is found in Baja California Sur and on adjacent islands. Its TBL_{max} is about 94 cm, and it has a pale gray or tan body, 26–41 variably shaped body blotches that are usually paler in the center, a darkly mottled buff venter, 3–5 tail bands, and the rostral and prenasals separated by at least 2 scales. *Crotalus m. angelensis* Klauber, 1963, the Angel Island rattlesnake or vibora blanca de Isla Angel, is restricted to Angel de la Guarda Island in the Gulf of California. This large, ponderous, taxon (TBL_{max} to possibly 141 cm) is characterized by having a subadult body color of pinkish-cinnamon with 36–46 brown blotches on the lateral sides of the ventral scutes, 4–8 tail bands, and the rostral and prenasals usually separated by a single scale. Its rattle may be somewhat reduced (Rabatsky 2006). *Crotalus m. muertensis* Klauber, 1949b, the El Muerto Island rattlesnake or vibora blanca de Isla Muerto, is restricted to that island in the Gulf of California. It is the smallest subspecies, having a TBL_{max} of about 64 cm, a grayish body, 32–39 poorly bordered brown body blotches that are hexagonal anteriorly but become cross bands posteriorly, a cream to buff venter, 2–6 tail bands, and the rostral separated from the prenasals by a row of small scales. *Crotalus m. pyrrhus* (Cope, 1867), the southwestern speckled rattlesnake or vibora blanca de suroeste, occurs in extreme southwestern Utah, adjacent southeastern Nevada, western Arizona, adjacent southeastern California, and northern Baja California Norte. It has a TBL_{max} of 132 cm (Lowe et al. 1986), and is extremely variable in body color, being

Crotalus mitchellii mitchellii (South of La Paz, Baja California Sur; Dallas Zoo; JohnH. Tashjian)

Crotalus mitchellii mitchellii (Monseratte Island, Baja California Sur; John Ottley, Valley Center, California; John H. Tashjian)

white, gray, pink, or tan to orange-red. It has dark, usually bordered or spotted, and often divided by light pigment, brown blotches that resemble cross bands, and 4–9 tail bands. The rostral is usually separated from the prenasals by small scales but sometimes in contact on one side, and no pitting or furrowing occurs on its smooth-edged supraocular scales. Douglas et al. (2006, 2007) reported that two distinct mtDNA clades exist within *C. m. pyrrhus*, and two more in Mexican *C. m. mitchellii* mainland populations. Grismer (1999a, 2002) considered both *C. m. angelensis* and *C. m. muertensis* separate species (*C. angelensis*, *C. muertensis*).

Crotalus stephensi was formerly considered the subspecies *C. m. stephensi* of *C. mitch-*

ellii. A determination of the mtDNA sequence divergence of mainland populations of *C. mitchellii* and *C. stephensi* by Douglas et al. (2007) revealed that *C. m. mitchellii* is separated from *C. m. pyrrhus* by 5.0–6.4%, and that the two subspecies are identical in all interon base pairs. *C. m. mitchellii* has diverged from *C. stephensi* by 6.4–7.3%, while *C. m. pyrrhus* is 5.2–6.7% divergent from *C. stephensi. C. stephensi* differs from both subspecies by a single nucleotide polymorphism. This molecular evidence, coupled with morphological differences, caused Douglas et al. (2007) to recommend the elevation of *C. m. stephensi* to a full species, *C. stephensi.* It apparently diverged from *C. mitchellii* about 3.26–3.69 million years ago (Douglas et al. 2007). See its account for more details.

CONFUSING SPECIES: Other *Crotalus* rattlesnakes within the range of *C. mitchellii* can be distinguished from it by the key presented above. The smaller rattlesnakes of the genus *Sistrurus* all have enlarged dorsal head scales.

KARYOTYPE: The karyotype's 36 chromosomes are made up of 16 macrochromosomes (5 metacentric, 6 submetacentric, and 4 subtelocenteric; the metacentric W chromosome has not been described) and 20 microchromosomes; sex determination is ZW for females and ZZ for males (Zimmerman and Kilpatrick 1973).

FOSSIL RECORD: Pleistocene (Rancholabrean) fossil vertebrae from the Lower Grand Canyon, Arizona, may be from either *C. mitchellii* or *C. oreganus* (Van Devender et al. 1977), and another Holocene fossil is from the Grand Canyon (Mead and Van Devender 1981).

DISTRIBUTION: *C. mitchellii* ranges from extreme southwestern Utah and adjacent southeastern Nevada, western Arizona, adjacent Southern California, western and northwestern Sonora south through Baja California, Mexico. It also occurs on the islands of Angel de la Guarda, Carmen, Cerralvo, El Muerto, Espíritu Santo, Monserrate, Piojo, Salispuedes, San José, and Smith in the Gulf of California, and Santa Margarita

Crotalus mitchellii mitchellii (Monserratte Island, Baja California Sur; Ali M. Rabatsky)

Distribution of
Crotalus mitchellii.

off the Pacific Coast of Baja California. It extends as far north in the Grand Canyon to Coconino County, Arizona (Brown 2003c). The total range lies between elevations near sea level or a few hundred meters to 2,440 m.

HABITAT: *C. mitchellii* is primarily a desert-dweller (sometimes in areas with loose sand), where it normally occupies the hottest, driest, rocky microhabitats, such as canyons, foothills, buttes, and erosion gullies vegetated with various cacti, thickets of chaparral, creosote bush (*Larrea tridentata*), mesquite (*Prosopis glandulosa*), sagebrush (*Artemisia tridentata*), thornscrub, saltbush (*Atriplex* sp.), Joshua tree (*Yucca brevifolia*), and piñion-juniper woodlands, but has also been found in some lowland brushy habitats. Klauber (1972) thought that those *C. mitchellii* occasionally found along the coast were strays presumably carried down from the uplands by river floods. It is found at elevations of 305–1,646 m in the Joshua Tree National Monument, California (Loomis and Stephens 1967).

Competition or some other factor of habitat exclusion may occur between *C. mitchellii*, and *C. atrox* and *C. scutulatus*, as the other two rattlesnakes will only occupy the microhabitat of *C. mitchellii* when it is absent.

BEHAVIOR AND ECOLOGY: Most reported life history data have been of *C. m. pyrrhus*; the Mexican subspecies have been practically ignored.

Active *C. mitchellii* have been found from late February to December over its range (Wright and Wright 1957). Klauber (1926) found them active from March to September in San Diego County, California. The species is most surface-active from April into early October; in the southwestern portion of the range, the late summer rainy season is the primary time of activity (Armstrong and Murphy 1979). In Riverside County,

California, *C. m. pyrrhus* initially becomes surface-active in April, and the number of hours of daily activity increases in each successive month through September (Moore 1978). Activity is then reduced, and between December and March most northern populations hibernate below the frost line in crevices or animal burrows, but caves or abandoned mines are also used. Although mostly solitary during the activity period, *C. m. pyrrhus* often congregate (20 to more than 100 individuals; Klauber 1972) at suitable hibernacula.

Surprisingly, *C. m. pyrrhus* choose winter hibernacula that receive less solar radiation. Possibly in regions with only mild winters, where low ETs never become lethal, it is best for the snake to be as cold as possible while hibernating to keep its metabolic rate low during the several months of winter fasting. This could occur in a hibernaculum that experiences low irradiance (Greenberg 2005).

From June to September, foraging is usually nocturnal, particularly between dusk and shortly after midnight (Klauber 1972; Moore 1978), but some diurnal activity occurs in the spring and fall (Armstrong and Murphy 1979; Campbell and Lamar 2004). Since desert surface temperatures sometimes fall rapidly after sunset, the snake may be limited to only a few suitable hours of activity each night.

Daytime ATs in its microhabitats are usually at least 32°C, and BTs of active individuals have ranged from 18.8 to 39.3°C (Moore 1978). Brattstrom (1965) reported BTs of 26.3–31.8 (mean, 30.3) °C for active *C. mitchellii*. Courtship and mating have occurred at ATs of 26, 29, and 31.8°C; STs of 31.0 and about 40°C (nearby in full sunlight); and a BT of 31.8°C (Gartner and Reiserer 2003; Warren *in* Brattstrom 1965). However, the overall preferred BT (April–December) seems to be about 31°C (Moore 1978). The CT_{min} is -2°C (Brattstrom 1965). Tightly coiled, inactive *C. mitchellii* may be able to conserve heat more effectively through circulatory adjustments than uncoiled snakes (Moore 1978).

C. mitchellii may bask in the morning, particularly after feeding, but the hottest summer hours are spent undercover beneath rocks or bushes, or in rock crevices,

Crotalus mitchellii mitchellii
(R. D. Bartlett)

caves, or animal burrows. Greenberg (2005) reported that *C. m. pyrrhus* occupy summer shelters that receive high solar radiation, but thought these retreats were sufficiently deep that the ST has little effect on its BT. Such sites may be warmer at night when the snake becomes active. *C. m. pyrrhus* collect rain water in their coils as do other species of *Crotalus*, and sometimes drink it from structural features in their habitat, such as rock surfaces. It does this when thermal conditions are not optimal, such as during cold ETs, and presumably may emerge from hibernation to drink rain water (Glaudas 2009).

Both planimetric (two-dimensional) and topographic (three-dimensional) home ranges were calculated for 14 *C. mitchellii* on the south side of the Coachella Valley near Palm Desert, Riverside County, California, by Greenberg and McClintock (2008). Topographic measurements were greater than planimetric ones by an average of 14% (8% for females, 19% for males). Males used more variable terrain than females, so planimetric calculations (16.1 ha) underestimated the average home range sizes as compared to the topographic estimate (19.7 ha) by 22%.

During field studies by Gartner and Reiserer (2003) on *C. m. pyrrhus* in San Bernardino County, three males were found crossing roads in June many kilometers from rocky retreats, apparently having crawled long distances through creosote flats. Despite road cruising from May to October, no other individuals were encountered away from rocky areas; the males were most likely searching for females. Young *C. m. pyrrhus* have been observed using a sidewinding-type locomotion similar to that of *C. cerastes*, possibly to avoid the extreme heat of the desert surface (Cowles 1941), and adults sometimes practice rectilinear locomotion (Gartner and Reiserer 2003). The subspecies occasionally climbs as high as 90 cm into vegetation (Gander *in* Klauber 1972).

Male *C. m. pyrrhus* have been seen engaged in combat dances in both the wild and captivity (Klauber 1972).

REPRODUCTION: Reproductive data for *C. mitchellii* are few. Both sexes are apparently mature at a TBL of 58 cm (Stebbins 2003); the shortest gravid female *C. m. mitch-*

Crotalus mitchellii angelensis (Angel de la Guarda Island, Gulf of California; California Academy of Sciences; John H. Tashjian)

Crotalus mitchellii muertensis (El Muerto Island, Gulf of California; John H. Tashjian)

ellii, C. m. muertensis, and *C. m. pyrrhus* examined by (Klauber 1972) were, 79.7 cm, 43.1 cm, and 57.3 cm, respectively. Smith and Hensley (1958) reported that a 65 cm TBL Sonoran female *C. m. pyrrhus* contained 3 11 mm follicles. Goldberg (2000c) reported a mature female with an SVL of 55.2 cm, and a spermiogenic male with an SVL of 51.2 cm. Probably 2–3 years of growth are needed to reach this size.

The gametic cycles of both sexes were studied by Goldberg (2000c). Some males had regressed testes in March–June and were in recrudescence in March–August. Spermiogenesis occurred in individuals from April into September. Goldberg's observation of 44% of the May males undergoing spermiogenesis suggests that spermiogenesis occurs primarily in the spring in this snake, in contrast to the cycles of several other U.S. species of *Crotalus*. The kidney sexual segments were enlarged and contained secretory granules in 94% of the males undergoing spermiogenesis, 100% of those in regression, and 70% of males in recrudescence; mating in snakes coincides with hypertrophy of the kidney sexual segment (Saint Girons 1982). Secretory granules were found in the following proportions of monthly males: March, 100%; April, 89%; May, 83%; June, 86%; July, 67%; August, 75%; and September, 100%. Mature sperm were present in the vas deferens of 100% of the males examined in March through May, in 88% in June, 67% in July, and 100% again in August–September, suggesting that mating could occur from March to September.

Goldberg (2000c) found reproductively inactive females from March to November, but others (44% of females examined) undergoing early vitellogenesis in April and May, and some with > 10 mm follicles in April–June. A female *C. m. pyrrhus* in San Bernardino County contained small ova on 10 June (Cunningham 1959), possibly in the early stages of vitellogenesis (Goldberg 2000c). Another female collected on 3 August was postovulatory, gravid, and contained active sperm, suggesting a recent mating (Gartner and Reiserer 2003), and Van Denburgh and Slevin (1921a) found a female containing three young in September. Female *C. mitchellii* apparently have a biennial reproductive cycle.

The natural breeding period of *C. m. pyrrhus* is mid-April (18 April; Brattstrom 1965; Lowe et al. 1986) to mid-June (11 and 13 June; Gartner and Reiserer 2003; Goldberg 2000c); Grismer (2002) reported mating in *C. m. mitchellii* occurs from late April to late May. Potentially breeding could extend into early August, near the time of parturition (see above). Captives have mated as early as January and as late as October (Armstrong and Murphy 1979; Peterson 1983; v.d. Velde 1995a). One captive pair mated four times between 8 January and 13 May in six years (Peterson 1983).

On 13 June 1995 at 2248 hours, a copulating pair of *C. m. pyrrhus* was found in the shade of a small bush in San Bernardino County. The larger male rattled and assumed a defensive posture, but remained attached to the smaller female. On 23 May 2000 at 2300 hours, a female was found crossing a road with a male 2 m away crawling toward her. The snakes were captured, and, upon release, the male immediately pursued, courted, and copulated with the female. Mating continued for more than 30 minutes, at which time the observers left. On the same evening, another presumably courting pair was also found on a road. Armstrong and Murphy (1979) observed a 7-hour (0800–1500), 13 October, captive copulation by *C. m. mitchellii*, during which the right hemipenis was inserted, and a prominent bulge extending 20 scale rows developed anterior of the female's vent.

Klauber (1972) watched a courting male *C. m. pyrrhus* that frequently tongue-flicked (1+/second) and sometimes touched the female's body with his tongue, but did not elevate or depress the tongue. While this was being done, he jerkily advanced his head along the female's back. Klauber also reported a 7.75-hour copulation.

Gestation lasts approximately 140 days, and parturition usually occurs from July into October in the wild (Gates 1957; Goldberg 2000c; Lowe et al. 1986; Strimple 1992a), but a rare litter may be produced in late June (Goldberg 2000c). Litters contain 1 (Armstrong and Murphy 1979) to 12 (Wright and Wright 1957) young, and average 5.2 (n = 34). However, the number of neonates per litter varies with the size of the subspecies; normal counts are *muertensis* (the smallest subspecies), 1–3; *mitchellii*, 3–5; and *pyrrhus* (the largest mainland subspecies), 3–12 (Klauber 1972; Wright and Wright 1957). Neonates (all subspecies combined) are 14.3–30.5 (mean, 22.8; n = 20) cm in TBL, and weigh 3.3–26.6 (mean, 14.7; n = 16) g; the smallest subspecies, *muertensis*, only produces 3.3–7.0 g young. Females may remain with their young for several days, and only leave the site after the young have dispersed (Greenberg *in* Greene et al. 2002).

A captive male *C. m. pyrrhus* at the San Diego Zoo repeatedly tried to mate with a male *C. helleri*, bringing on combat behavior in the latter snake (Shaw 1951).

GROWTH AND LONGEVITY: No growth rates have been reported. Snider and Bowler (1992) noted that a male *C. m. pyrrhus*, wild-caught as an adult, lived an additional 22 years, 16 days at the Los Angeles Zoo; and that a male *C. m. mitchellii* of unknown reproductive stage when wild-caught survived 19 years, 8 months, and 15 days at the Houston Zoo.

DIET AND FEEDING BEHAVIOR: Natural prey include ground squirrels (*Ammospermophilus leucurus, Spermophilus* sp.), woodrats (*Neotoma* sp.), kangaroo rats (*Dipodomys agilis*), pocket mice (*Perognathus* sp.), white-footed mice (*Peromyscus crinitus, P. guardia* [?], *P. maniculatus, P. truei*), cotton rats (*Sigmodon* sp.), rabbits (*Sylvilagus auduboni*), lizards (*Callisaurus draconoides; Cnemidophorus maximus, C. tigris; Eumeces skiltonianus;*

Crotalus mitchellii muertensis (El Muerto Island, Gulf of California; John H. Tashjian)

Crotalus mitchellii pyrrhus (Barry Mansell)

Petrosaurus mearnsi; *Sauromalus obesus*; *Sceloporus* sp.; *Streptosaurus mearnsi*; *Uta stansburiana*), goldfinches (*Carduelis tristis*?), and insects(?) (Armstrong and Murphy 1979; Camp 1916; Campbell and Lamar 2004; De Vault and Krochmal 2002; C. Ernst, personal observation; Klauber 1972; Lowe et al. 1986; Stebbins 2003).

Juveniles feed primarily on lizards. Adults seem to prefer small mammals, either fresh or as carrion (De Vault and Krochmal 2002), but will also eat lizards. Klauber (1972) reported that of the specimens containing food that he examined, 18 had eaten mammals, 9 lizards, one a bird, and another snake had ingested both a mouse and a lizard. Most mammals taken are small, but Shaw (*in* Klauber 1972) found a large

C. m. pyrrhus that had taken a nearly grown cottontail rabbit (*S. auduboni*). Although birds are often listed as food, the only one yet identified as prey has been the goldfinch mentioned above (Batchelder *in* Klauber 1972). Captive adults will eat *Mus musculus* and young *Rattus norvegicus* (C. Ernst, personal observation; Rossi and Rossi 1995), and Porter (1983) has related three instances of cannibalism by captive juveniles.

The speckled rattlesnake captures its food by either actively foraging or by lying in ambush. If a lizard is struck, the snake usually holds on to it until it dies and then swallows it immediately, but when a small mammal is bitten the snake normally releases it and later trails the wounded mouse by olfaction until it is found dead or incapacitated, and then swallows it (C. Ernst, personal observation).

VENOM DELIVERY SYSTEM: Klauber (1939) reported that 7 76.0–89.6 cm TBL *C. m. mitchellii* had 5.2–5.7 (mean, 5.4) mm FLs, a mean TBL/FL of 110%, a mean HL/FL of 5.22%, and a mean fang curvature angle of 70°. He also reported that 5 82.6–111.4 cm TBL *C. m. pyrrhus* had 7.8–10.8 (mean, 8.7) mm FLs, a mean TBL/FL of 107%, a mean HL/FL of 5.01, and a mean 70° angle of fang curvature.

VENOM AND BITES: An adult *C. mitchellii* may contain an average of 227 mg of dried venom (Klauber 1972). Klauber (1972) extracted a total of 1,775 mg of dried venom from 64 adult *C. m. mitchellii*, an average of 32 mg per fresh snake; and 39,524 mg from 298 *C. m. pyrrhus*, an average of 227 mg per fresh snake. Other reported maximum venom yields include 60 (juvenile ?)-265 mg (do Amaral 1928), 245 (mean, 135) mg (Glenn and Straight 1982), and 350 mg (Altimari 1998; Campbell and Lamar 2004). On average young individuals yield only about 0.06 mg (0.18 mL) of venom per bite, while adults may inject 0.1 mg (0.3 mL), and older, larger snakes 0.16 mg (0.48 mL) (do Amaral 1928).

The venom is 50–79% protein (Glenn and Straight 1985b). Venom races exist within the species. Some populations of *C. m. pyrrhus* (Baja de los Angeles, Baja California Norte, Arizona?) have tested positive for Mojave toxin, but others (near San Felipe, Baja California Norte; Arizona; and southwestern Utah) have not (Bober et al. 1988; Henderson and Bieber 1986; Li and Ownby 1994). Venom from *C. m. mitchellii* (Baja California Sur) does not contain the toxin (Bober et al. 1988), but four other individuals of *C. m. mitchellii* (populations not stated) have tested positive for the toxin (Henderson and Bieber 1986). Venoms of the two subspecies that contain myotoxin act on the nervous system, those lacking it are hemorrhagic. Chen et al. (2004) have isolated a crotoxin-like presynaptic neurotoxin different from Mojave toxin from the venom of *C. m. mitchellii*, which was later identified as Mojave toxin β (Powell et al. 2008).

Other enzymes (activity ranges) found in the venom are esterase (*C. m. mitchellii*, 303–435; *C. m. pyrrhus*, 14–251), phosphodiesterase (*C. m. mitchellii*, 171–409; *C. m. pyrrhus*, 33–156,) phospholipase A_2 (*C. m. pyrrhus*, 19.6), and protease (*C. m. mitchellii*, 0–1.0; *C. m. pyrrhus*, 66–157) (Glenn and Straight 1985a, 1985b; Holzer and Mackessy 1996; Mackessy 1998). In addition, the venom of *C. m. pyrrhus* contains several peptide inhibitors that help stabilize the venom stored in the lumen of the venom gland (Munekiyo and Mackessy 2005). The pH of freshly extracted venom from adult *C. m. pyrrhus* is 5.52 (Mackessy and Baxter 2006), and that of its phosphodiesterase, 8.5 (Mackessy 1998).

The venom is potent, particularly for birds; the minimum lethal doses of intravenous dried venom from *C. m. mitchellii* and *C. m. pyrrhus*, respectively, for a 350 g pigeon are only 0.002–0.04 mg and 0.04–0.06 mg (Githens and Butz 1929; Githens and George 1931; Klauber 1972). Various recorded dried venom minimum lethal doses for mice are 22 g mice (intraperitoneal—*C. m. mitchellii*, 0.045 g; *C. m. pyrrhus*, 0.075 g; Macht 1937); and 20 g mice (*C. m. mitchellii*, 0.0035 g; *C. m. pyrrhus*, 0.05 g; Githens and Wolff 1939). Reported LD_{50} values for mice in mg/kg are for *C. m. mitchellii*: intraperitoneal—0.13–0.24 (mean, 0.18), 2.05, 0.18, 0.40; and intramuscular—0.20–0.47 (mean, 0.30); and for *C. m. pyrrhus*: intraperitoneal—1.35–4.00 (mean, 2.7), 3.41, 2.5; intramuscular—4.1–23.5 (mean, 9.6) (Githens and Wolff 1939; Glenn and Straight 1982, 1985b; Khole 1991, Macht 1937; Russell et al. 1960). Lethality of dried venom from *C. m. mitchellii* is not even diminished with time; Russell et al. (1960) found it had not weakened after 26–27 years in storage.

A human envenomation by *C. m. pyrrhus* is very painful, and results in considerable swelling (edema) and discoloration that move up the extremity bitten (Hartnett 1931; Klauber 1972). When the victim is treated with antivenom, recovery is usually complete and uneventful.

PREDATORS AND DEFENSE: Only four reports of predation exist, apparently all on *C. m. pyrrhus*: a kingsnake (*Lampropeltis getula*) attacked one (Klauber 1972), an individual was eaten by a Late Holocene ringtail (*Bassariscus astutus*) (Mead and Van Devender 1981), a gray fox (*Urocyon cinereoargenteus*) ate another (Zweifel *in* Stebbins 1954), and notes accompanying a *C. m. pyrrhus* in the UCLA collection describe a bobcat (*Lynx rufus*) attack on it (Cunningham 1959).

Temperament varies between individual *C. mitchellii*; a few lie quietly with only an occasional rattle when first discovered, but most seem very nervous, and, if they can not escape or are challenged, will coil, raise a loop of the body above the ground, inflate the trunk, and strike, while continuously rattling. One even ejected venom into a handler's eyes (Madrid-Sotelo and Balderas-Valdivia 2008). The rattle frequency is 1.9–20.0 kHz (Young and Brown 1993), and the SVL-adjusted loudness about 75 dB (Cook et al. 1994). *C. mitchellii* reacts defensively to the odor of the California kingsnake (*Lampropeltis getula californiae*) by body-bridging (Bogert 1941; Weldon et al. 1992).

PARASITES AND PATHOGENS: *Crotalus mitchellii* is the host for parasitic cestodes *Mesocestoides* sp. (tetrathyridia larvae) and acanthocephalans (unidentified oligoacanthorhynchid cystocanth larvae) (Ernst and Ernst 2006; Goldberg and Bursey 2000).

C. m. pyrrhus has suffered from severe suppurative dermatitis in captivity. Cultures from the necrotic areas included the bacteria *Citrobacter freundii*, *Corynebacterium xerosis*, *Pseudomonas aeruginosa*, coagulase positive staphylococci, and rare *Streptococcus*. *Enterobacter aerogenes* and *Escherichia coli* were also isolated from the skin area (Murphy and Armstrong 1978). In addition, serotypes of potentially dangerous *Salmonella* bacteria have been isolated from captive *C. m. mitchelii* (Grupka et al. 2006). One suffered from an adenosarcoma in its pancreatic duct (Effron et al. 1977).

POPULATIONS: Few data concerning populations of *C. mitchellii* have been published. Klauber (1926) collected only 28 (1.7%) *C. m. pyrrhus* in a total of 1,623 snakes and 352 rattlesnakes (8.0%) of 4 species in San Diego County in 1923–1925. *C. m. pyrrhus* is protected in Utah.

REMARKS: Morphologically, *C. mitchellii* seems most closely related to *C. tigris* and *C. viridis* (Brattstrom 1964a; Gloyd 1940; Klauber 1972). In contrast, electrophoretic

Crotalus mitchellii
pyrrhus (Robert E.
Lovich)

studies of rattlesnake venoms by Foote and MacMahon (1977) and a microdermato-glyphic study of the dorsal scales by Stille (1987) indicate that it is closest to *C. pricei* and *C. cerastes*. However, more recent mtDNA analyses place the species in the *C. mitchellii*-group of rattlesnakes that also includes its closest relative, *C. tigris* (Murphy et al. 2002).

The species has been reviewed by McCrystal and McCord (1986).

Crotalus molossus Baird and Girard, 1853
Black-tailed Rattlesnake
Cascabel Serrana

RECOGNITION: *Crotalus molossus* (TBL$_{max}$, 152.4 cm; Platt and Rainwater 2008, but most adults are shorter than 100 cm) is olive-gray, greenish-yellow, cream, yellow, reddish-brown, or even blackish (dark individuals are usually from New Mexico lava areas with dark substrates; Lewis 1949) with 20–43 chestnut-brown, dark brown, or black, pale-bordered, rhomboid or sometimes diamond-shaped blotches, a dark brown to black tail beyond the vent, 3–8 tail bands and often black pigment on the snout from the rostral to near the supraoculars. Anteriorly, the body blotches contain some internal paler scales bordering the midline. Posteriorly, they are broader and may join with lateral blotches to form irregular cross bands. Dorsal body scales are keeled and contain linear ridges that may swirl (see Harris 2005 for a microdermograph), and have apical pits. The scales occur in 25–27 (23–29) anterior rows, 25–27 (23–31) midbody rows, and 21–23(19–27) rows near the tail. The venter is white, cream, or pale gray anteriorly, but more greenish toward the tail, and may be mottled with gray. On it are 164–201 ventrals, 16–30 subcaudals, and an undivided anal plate. Dorsally, a few pale longitudinal lines may be present on the head. A dark band lies between the eyes, and a light-bordered, dark stripe extends diagonally backward from the eye to the corner of the mouth. Dorsal head scales include a slightly higher than wide rostral scute, 2 large internasals touching the rostral, 5–7 (4–18) scales in the internasal-prefrontal area, 4 canthals (the anterior pair is largest, the second pair lies between the supraoculars and is sometimes missing), 2 large supraoculars, and 2–5 intersupraoculars. Laterally on the face are 2 nasals (the prenasal rarely touches the first supralabial, the postnasal rarely touches an upper preocular), 2–3 (1–9) loreals, 2–3 preoculars, 2 (1–3) interoculabials (separating the anterior suboculars and supralabials), several suboculars, 3–5 (6–7) postoculars, 16–18 (13–20) supralabials, and 16–18 (14–21) infralabials. Beneath are a mental scale and two elongated chin shields that usually meet at the midline; only rarely do any small scales lie between the chin shields.

The short, blunt, bilobed hemipenis is illustrated in Klauber (1972). It has a divided sulcus spermaticus, about 68 spines and 21 fringes per lobe, and no spines in the crotch between the lobes.

The typical dentition, other than the 2 fangs, is 2 (1–3) palatine teeth, 8–9 (6–10) pterygoid teeth, and 10 (9–11) teeth on the dentary (Klauber 1972).

Males (TBL$_{max}$, 133 cm) have 164–199 (mean, 187) ventrals, 21–30 (mean, 26) subcaudals, TLs 5.8–8.6 (mean, 7) % of TBL, and 5–6 (3–8) dark tail bands. Females (TBL$_{max}$, 100 [?] cm) have 168–201 (mean, 190–193) ventrals, 16–26 (mean, 22) subcaudals, TL 4.6–6.7 (mean, 5–6) % of TBL, and 5–6 (3–6) dark tail bands.

Crotalus molossus molossus, yellow morph (R. D. Bartlett)

GEOGRAPHIC VARIATION: Four suspecies are recognized (Campbell and Lamar 2004). *Crotalus molossus molossus* Baird and Girard, 1853, the northern black-tailed rattlesnake, or cascabel de colinegra norteña, is found in Arizona, New Mexico, and Texas; in the Mexican states of Coahuila, Chihuahua, Sonora, Nuevo León, and northern Durango; and on Isla Tiburón in the northern Gulf of California. It is characterized by its large size (TBL$_{max}$, 133 cm); dark brown or black snout; yellow, olive-green, greenish-gray, or blackish body color; 23–41 (mean, 30–31) dark reddish-brown to black body blotches, which may form irregular cross bands posteriorly; 5–6 (4–8) dark tail bands; the completely black base of the tail; 27 mean midbody scale rows; 188–189 mean ventrals in males and 193 in females; 24–25 mean subcaudals in males and 21 in females; 17–18 mean supralabials; and 17–18 mean infralabials. *C. m. molossus* from the uplands of western Texas are generally darker than those from New Mexico and Arizona, which are more yellowish in color; *C. m. molossus* living on dark lava flows are melanistic, apparently adapted to matching the substrate color (Lewis 1949; Prieto and Jacobson 1968). *Crotalus m. estebanensis* Klauber, 1949b, the San Esteban Island rattlesnake or cascabel de la Isla San Esteban, is restricted to that island. It is short, with a TBL$_{max}$ of 89 cm; lacks a dark snout; and has a paler brownish body color; 31–40 smaller body blotches, which are lighter and may fade or disappear entirely toward the rear; a rattle reduced only to 1–2 segments (Rabatsky 2006), which is often compressed longitudinally and transversely (Klauber 1949b); 27 mean midbody scale rows; 188 mean ventrals in males and 192 in females; 25 mean subcaudals in males and 22 in females; 18 mean supralabials; and 16–17 mean infralabials. *Crotalus m. nigrescens* Gloyd, 1936a, the Mexican black-tailed rattlesnake or palanca, occurs in the Mexican highlands from northern Durango, southern Chihuahua, and northern Nuevo León south to western Veracruz and northern Oaxaca. It has a TBL$_{max}$ of 109.2 cm; a black snout; dark brown, reddish-brown, or brownish-olive body color; 24–34 (mean, 28) body blotches, and some irregular posterior cross bands; 5–6 (4–9) dark tail cross bands; black tail base; 25 mean midbody scale rows; 174 mean ventrals in males and 177–178

Crotalus molossus molossus, yellow morph (Cochise County, Arizona; Steve W. Gotte)

Crotalus molossus molossus, dark morph (West Texas, Carl H. Ernst)

in females; 24 mean subcaudals in males and 19–20 in females; 16 mean supralabials; and 16–17 mean infralabials. *Crotalus m. oaxacus*, Gloyd 1948, the Oaxacan black-tailed rattlesnake or tleua, is found in the highlands of central Oaxaca and Puebla. It has a TBL_{max} of 74 cm; a yellow-brown to olive-brown body color; 29–30 dark body blotches; 5 (3–5) dark tail bands; a gray-black tail base; mean 25 midbody scale rows; mean 174–175 ventrals; 20 mean subcaudals; 13–15 mean supralabials; and 14 mean infralabials.

Campbell and Lamar (1989, 2004) reported clinal variation in hemipenial morphology—individuals from the United States and adjacent Mexico have straight, thick, stubby lobes with more than 50 small spines at the base of each lobe; those from farther south in Mexico have slender lobes with fewer spines.

Possible intergradation between *C. m. molossus* and *C. m. nigrescens* has been reported from Chihuahua, Coahuila, Durango, Nuevo León, and San Louis Potosí (Dixon et al. 1962; Fouquette and Rossman 1963; Fugler and Webb 1956; Gloyd 1940; Tanner 1985; Van Devender and Lowe 1977).

Because of its different rattle morphology, Grismer (1999a) elevated *C. m. estabanensis* to a full species, but we concur with Campbell and Lamar (2004) and have kept this taxon as a subspecies of *C. molossus*.

CONFUSING SPECIES: No other sympatric rattlesnake has the base of its tail totally dark gray or black. Other rattlesnakes within the geographic area covered by this book can be differentiated by the key presented above.

KARYOTYPE: *C. molossus* has 18 pairs of diploid chromosomes—16 macrochromsomes (4 metacentric, 6 submetacentric, 4 subtclocentric), and 20 microchromosomes; sex determination is ZZ/ZW (Baker et al. 1971, 1972; Zimmerman and Kilpatrick 1973). Baker et al. (1971, 1972) reported that the Z macrochromosome is submetacentric and the W macrochromosome is subtelocentric, but Zimmerman and Kilpatrick (1973) thought these chromosomes to be metacentric and submetacentric, respectively.

Distribution of
Crotalus molossus.

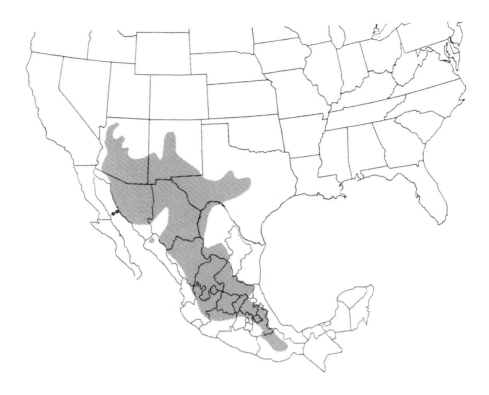

FOSSIL RECORD: No fossil *C. molossus* have been reported.

DISTRIBUTION: *C. molossus* ranges from northern and western Arizona (Holycross 2001; Koenig and La Grone 2000) south and east through southern Arizona, central and southern New Mexico (Degenhardt et al. 1996; Gehlbach 1965; Lewis 1949, 1950; Mosauer 1932a) and western Texas to the Edwards Plateau of central Texas (Werler and Dixon 2000), and into Mexico through Sonora (Alt 2002; Bogert and Oliver 1945; González-Romero and Alvarez-Cardenas 1989; Taylor 1936), eastward into most of Chihuahua (Tanner 1985; Walker *in* Armstrong and Murphy 1979) and Coahuila (Fugler and Webb 1956; Gloyd and Smith 1942; Klauber 1938a), south to Oaxaca (Gloyd 1948) and Nuevo León (Lazcano et al. 2007c). It is also found on the islands of San Esteban (Lowe and Norris 1955) and Tiburon (Campbell and Lamar 2004) in the Gulf of California. The total range encompasses elevations from near sea level to about 3,150 m.

HABITAT: The black-tailed rattlesnake is mostly a highland species. There its major habitats are upland pine-oak or boreal forests, where the snake is found in rocky sites, such as talus slopes, the sides of canyons, crevices in outcrops, caves (Taylor 1936), and rocky riparian zones. At lower elevations its habitat consists of mesquite-grassland, chaparral, or even desert. It has been found in woodrat (*Neotoma* sp.) stick nests (Wood 1944).

Some representative plants occurring in the snake's habitat over its distribution include gramma grass (*Bouteloua* sp.), agave (*Agave lechguilla*, *A. parryi*), barberry (*Berberis trifoliolata*), blackbrush (*Coleogyne ramosissima*), *Brickellia* sp., cactus (*Echinocactus wislizenii*, *Fouquieria splendens*, *Mammillaria* sp., *Opuntia* sp.), catclaw (*Acacia greggii*), creosote (*Larrea tridentata*), crucillo (*Condalia spathulata*), *Fendlera rupicola*[?], golden weed (*Haplopappus laricifolius*), *Lippia wrighti*[?], manzanita (*Arctostaphylos pringlei*), mesquite (*Propis glandulosa, P. juliflora, P. velutina*), mountain mahogany (*Cercocarpus betuloides*), sotol (*Dasylirion wheeleri*), tarbush (*Flourensia cernua*), balsam fir (*Abies concolor*), cedar (*Juniperus* sp.), Douglas fir (*Pseudotsuga menziesii*), mimosa (*Mimosa bunciflora*), oaks (*Quercus arizonica, Q. dumosa, Q. dunnii, Q. emoryi, Q. gambelii, Q. glaucoides, Q. gravesii, Q. grisea, Q. havardii, Q. hypoleucoides, Q. mohriana, Q. oblongifolia, Q. pungens, Q. rugosa, Q. toumeyi, Q. turbinella*), persimmon (*Diospyros texana*), pines (*Pinus cembrioides, P. edulis, P. engelmannii, P. leiophylla, P. ponderosa*), quaking aspen (*Populus tremuloides*), sweetgum (*Liquidamber styraciflua*), *Ungnadia speciosa*[?], and willows (*Salix* sp.). Near its southeastern limit in Nuevo León, the snake inhabits microphyll desert scrub dominated by creosote bush (*L. tridentata*), crown of thorn (*Koeberlinia spinosa*), goatbush (*Castela texana*), and tree yucca (*Yucca filifera*) (Lazcano et al. 2007c).

BEHAVIOR AND ECOLOGY: Annually *C. molossus* is surface-active from April to October, and occasionally as late as November or December, but it is most typically seen in the late summer or fall (Greene 1990; Reynolds 1982). The earliest recorded dates of appearance are 4 April (Wright and Wright 1957) and 21 April (Gates 1957). The latest observations reported are October (Greene 1990; Wright and Wright 1957), late November when there was a thin covering of snow on the ground (Kauffeld 1943a), and December (Lowe et al. 1986). Lowe et al. (1986) reported surface activity from March to November and in February and March in Arizona, but gave no specific dates. In Chihuahua, the snake is active from July through September (Reynolds 1982).

Hibernation takes place from November through March in desert scrub, creosote flats, chaparral, woods, and upland rocky areas. The snakes retreat below the frost line into animal burrows, the base of rock outcrops, ledges, rock crevices, caves, and possibly abandoned mines. Most such sites face south or southwest. At some Arizona hibernacula, the rattlesnake overwinters alone (Greene 1997). At others it is a communal hibernator, sharing its den with *Crotalus oreganus cerberus* and some colubrid

Crotalus molossus molossus, dark morph (Texas; R. D. Bartlett)

snake species (Spille and Hamilton 2006). Hamilton and Nowak (2009) studied hibernating *C. molossus* at 1,432–1,897 (mean, 1,458) m hibernacula in Javapai County, Arizona. The mean annual ET of the area was 12°C, the mean January and July ETs were 2°C and 23°C, respectively, and the site averaged 148 frost-free days. Insolation at the dens averaged 2,252 (range, 1,425–2,645) watts/m^2, the mean aspect was 177 (range, 102–237) °, and the mean slope of the hibernacula was 29 (range, 12–48) °. Not all individuals remain entirely torpid during the winter, as some bask on warm days in January–March (Greene 1990). Spille, Schuett, and Nowak (*in* Spille and Hamilton 2006) have noted high site fidelity at communal hibernacula.

Spring and autumn activity is mostly diurnal, but *C. molossus* may shift to a more crepuscular or nocturnal pattern during the hotter summer months. According to Beck (1995), most summer activity by *C. molossus* in Arizona's Sonoran Desert occurs during the hours 1600–2200 and 0800–1000. The desert nights may be quite cold at higher elevations, restricting spring and fall activity to only the warmest days, but the snake is more crepuscular or nocturnal during the hotter summer months. It usually remains concealed during overcast or cold days. We have observed crepuscular activity in May in the Chihuahuan Desert of southeastern Arizona during a very dry spring.

Active Sonoran Desert *C. m. molossus* average 4,796 surface hours/year—19.7 daylight hours during the summer (reduced to 12 hours in dry periods), 12.8 hours in the fall, but only 5–8 hours in the winter. Surface activity often occurs after summer rains; probably the snake drinks water from puddles at such times. Greene (1990) observed one drinking water from the film seeping over a rock face. In the arid highlands of southern Puebla, Mexico, the snake is mostly nocturnal (Campbell and Lamar 2004).

Some data now exist on the snake's thermal ecology. During the annual foraging period in the Sonoran Desert, the mean active BT is near 29.5 (range, 21.8–29.5) °C, but in the winter (December–February), the mean BT of resting individuals is only 20.3°C; in the spring (March–April) it is 22.8°C, dry summer (May–15 July) 29.3°C, wet summer (16 July–15 October) 25.1°C, and in the fall (16 October–30 November) it is 22.4°C (Beck 1995). The loreal pit of *C. molossus* enables it to thermoregulate by detecting thermal cues, allowing the snake to select the proper BT range, but its supranasal sac lacks this ability (Krochmal et al. 2004).

A study of respiratory cooling in three Texas *C. m. molossus* revealed a single, cooled region around the mouth and nasal capsule that extended across the pit membrane at ATs of 20°C or higher. This possibly indicates that respiratory cooling provides a thermal buffer for the thermoreceptive pit organs at higher ATs. Both head and body temperatures of the three snakes studied increased linearly with AT. The difference between the temperatures of the head and body also increased with AT, but declined significantly at higher humidities. Those rattling had significantly greater differences between the head and body temperatures than did resting individuals (Borrell et al. 2005).

Beck (1996) monitored the BT of three Arizona *C. m. molossus* before and after consuming rodents constituting about 45% of their BM in midfall. The snakes showed a preference for an elevated BT after feeding, and had significantly higher BTs than unfed controls.

One individual survived at 4°C for about 10 days (Bogert *in* Klauber 1972), so the CT_{min} of this snake is lower than that temperature. The CT_{max} is unknown, but one exposed for 10 minutes to bright summer sunlight died (Wright and Wright 1957).

Those found in the morning are usually basking, apparently trying to warm after the chilling effects of night ATs, and nighttime basking on warm paved roads is a common occurrence.

At 24°C, *C. molossus* had apneic and ventilatory heart rates of 11.3 (range, 5.8–23.1) and 12.5 (range, 5.8–24.0) beats per minute, respectively, and a breathing rate of 1.36 (range, 0.06–8.82) breaths per minute (Jacob 1980).

Beck (1995) calculated the annual maintenance energy budget for a 300 g Sonoran Desert *C. molossus*. When active in the winter, the snake used 240 mL O_2/km × 0.736 km (0.20 kJ) over 5.62 hours at a BT of 30°C, but an alert one used only 5.06 mL O_2 (3.2 kJ) over 651 hours in December–February, and, when resting for 1,503 hours during 90 days, only 4.6 mL O_2 (4.3 kJ). In contrast, in the spring, dry summer, wet summer, and fall, the snake uses 240 mL O_2/km × 1.49 km (0.4 kJ), 240 mL O_2/km × 1.51 km (0.41 kJ), 240 mL O_2/km × 8.16 km (2.2 kJ), and 240 mL O_2/km – 3.09 km (0.83 kJ), respectively. Yearly, an active snake (114.4 hours, 1.31% of time) would need 4.04 kJ of energy, an alert one (4,457.8 hours, 50.9%), 45.35 kJ, and one resting (4,187.8 hours, 47.8%), 18.42 kJ. Beaupre (1993) also researched O_2 consumption in *C. molossus*, and found the metabolic rate to be BM specific (we refer the reader to his paper for details).

The electrolyte components and pH of the blood plasma may play roles in water retention in reptiles. Punzo (1976) examined these factors in 13 *C. molossus* from Presidio County, Texas, maintained in the laboratory at 30°C. The mean pH was 7.51, and the following electrolyte means (mEq/L of serum) were recorded: sodium, 152.9; chloride, 129.7; bicarbonate, 26.8; potassium, 3.4; calcium, 3.4; and phosphorus, 1.6. Sodium, chloride, and bicarbonate constituted more *than* 80% of the osmotically significant components in the serum, which is typical for reptiles (Dessauer 1961). The plasma electrolyte concentrations were similar to those previously reported for the rattlesnakes *C. atrox*, *C. horridus*, *C. oreganus*, and *C. viridis* (Carmichael and Petcher 1945; Dessauer 1970; Luck and Keeler 1929).

Crotalus molossus nigrescens (Charles Hanson, Tucson, Arizona; John H. Tashjian)

Beck (1995) monitored movements by Sonoran Desert *C. m. molossus*. Their mean home range is 3.5 ha, and the snakes move an average of 88.7 m per day, with an average maximum movement of 103.6 m per activity bout. Most movement occurs during the wet summer and fall months. During the dry summer, *C. m. molossus* travels an average of 19.9 m per day; during the wet summer, 88.7 m per day; in the spring, 24.4 m per day; in the fall, 67.2 m per day; and in the winter, only 8.2 m per day. In late summer, the snake migrates from its summer foraging range to a hibernaculum located in a rocky slope, arroyo, or creosote flat. Total distance traveled during the year averages about 15 km. In another movement study in the Pelocillo Mountains of southern Arizona and New Mexico, a radio-equipped male had a shorter daily activity period after a fire than before it, and an even briefer one once the summer rains began, but the snake had recently fed when found during the rainy season (L. J. Smith et al. 2001).

C. molossus is an accomplished climber that sleeps and forages in bushes and trees to a height of about 3 m (Allen 1933; Campbell and Lamar 1989; Klauber 1972; Lowe et al. 1986). Hastings (*in* Klauber 1972) witnessed one climb a stone wall. It also swims well when placed in water (Klauber 1972).

Male dominance combat may occur in late summer, sometimes when a female is present. During such bouts each male attempts to pin his opponent's head to the ground, and the loser crawls away undamaged (Greene 1990). Although combat events normally occur on land, one observed bout occurred in the water of a running creek (Hamilton and Wrieden 2004).

REPRODUCTION: Few data are available concerning the reproductive biology of this fairly common snake. The smallest reproductive male and female *C. m. molossus* examined by Goldberg (1999c) had 57.6 cm and 65.3 cm SVLs, respectively; the smallest mature female known to Klauber (1972) had a TBL of 70.3 cm.

The gametic cycles of both sexes were studied by Goldberg (1999c). Males have regressed testes with spermatogonia and Sertoli cells in March–June; spermatogonial division, with primary and secondary spermatocytes and a few spermatids, takes place during recrudescence in April–August; and spermiogenesis, with maturing spermatids and mature sperm, occurs from May to September, with peak activity (83%) in July–September. Sperm is stored over winter for use in the next summer's matings (Dancik *in* Schuett 1992).

The seasonal patterns of the plasma sex steroids testosterone, 5α-dihydrotestosterone, and 17ß-estradiol were studied in male Arizona *C. m. molossus* by Schuett et al. (2005). The levels of all three hormones were at baseline in March–April and September; but were elevated in May–August, corresponding to the summer mating season.

Females with enlarged follicles (> 12 mm), oviductal eggs, or embryos have been found in April, June–July, and September–October. Some may contain oviductal eggs from late April to early June, and contain full-term young in mid-July. Hulse (1973) collected two gravid females on 20–21 June that each contained five developing embryos, and a female collected on 5 July by Gates (1957) also had well-developed embryos. Because some females have enlarged follicles in July and September–October but no oviductal eggs, this suggests at least a biennial cycle, as some females have inactive ovaries in any given year.

C. molossus has a single annual breeding season in the summer, which lasts from mid-July to early September (Aldridge and Duvall 2002; Schuett et al. 2005). Armstrong and Murphy (1979) observed copulation by captive *C. m. molossus* on 2 March and 28 May, but did not notice any pulsations or other movements during the act. A possible mating attempt performed on a dead female by a male *C. m. molossus* occurred on 7 August (Wright and Wright 1957).

Armstrong and Murphy (1979) found a pair of *C. m. nigrescens* mating in Mexico on 1 February at 0920 hours. The diameter of the female's cloaca was 15 cm and was distended due to the male's deep purple hemipenis. The shoulder spines of the hemipenis were visible. The female also had a noticeable bulge in her body extending 35 mm anterior to her cloaca. The copulation occupied 105 minutes, and release of the hemipenis lasted 100 seconds. During the mating, both snakes tongue-flicked and head-bobbed.

After mating, females apparently are capable of storing the sperm for a relatively long time (Greene *in* Schuett 1992; Schuett et al. 2005).

The male sometimes remains with the female after mating, perhaps guarding her from courtship by another male. A heterosexual pair of *C. m. molossus* radio-tagged after being found in a woodrat (*Neotoma*) nest traveled and basked together for several weeks, but by early fall they had moved apart and entered separate hibernacula in the same rocky bluff (Greene 1990). Males may also occur with females when they are with their litters (Dunkle and Smith 1937).

Parturition in wild *C. m. molossus* occurs from late July to September. Thirty litters of this subspecies reported in the literature averaged 5.7 (3–16) young, but most contained less than 10 young. Neonate *C. m. molossus* are 22.9–31.8 (mean, 27.32; n = 25) cm in TBL, have BMs of 11–28 g, and have dark tail bands.

Five young *C. m. nigrescens* were born in captivity on 9 June. Their mean measurements were TBL, 30.4 (range, 29.1–31.6) cm; SVL, 28.4 (range, 26.7–29.0) cm; and BM, 26.6 (range, 25.4–27.9) g (Armstrong and Murphy 1979). Another female *C. m. nigrescens* went through parturition on 24 June, producing 9 young, 6 of which were alive (Sánchez-Herrera 1980). This subspecies may also produce litters of 16 young (Dunkle and Smith 1937). The neonate tail contains dark bands.

Savary (1999) reported an apparent case of a female *C. m. molossus* defending five young in an opening at the base of a rock outcrop. She advanced rattling from the opening, then retreated and lay on top of the neonates. Female *C. m. nigrescens* have also been found with their neonates (Dunkle and Smith 1937; Smith 2001).

C. molossus apparently often hybridizes with *C. basiliscus* in the foothills along the northwestern coast of Mexico (Campbell and Lamar 2004; McMillan and Lieb 2005), and a hybrid mating between a male *C. molossus* and a female *C. atrox* occurred in captivity (Davis 1936).

GROWTH AND LONGEVITY: Data on growth rates have not been reported. A wild-caught adult *C. m. molossus* of unrecorded sex survived an additional 20 years, 8 months, and 24 days at the Columbus Zoo in Ohio (Snider and Bowler 1992), and another wild-caught adult male lived for 12 years at the University of Kentucky. A wild-caught adult male *C. m. nigrescens* survived 15 years, 2 months, and 25 days at the Columbus Zoo (Snider and Bowler 1992).

DIET AND FEEDING BEHAVIOR: The preferred prey of this snake are small mammals, but birds and lizards are occasionally taken, and suitable carrion is probably not

Crotalus molossus nigrescens (North of Uruapan, Michoacan; Terry Basey, Arcadia, California; John H. Tashjian)

rejected. Prey recorded from *C. molossus* include mammals—bats (*Antrozous pallidus*); rodents (*Dipodomys merriami*; *Neotoma albigula*; *Perognathus flavus*, *P. intermedius*; *Peromyscus eremicus*, *P. maniculatus*, *P. pectoralis*; *Rattus rattus*; *Sciurus* sp.; *Spermophilus* sp.; *Tamias* sp.), and rabbits (*Sylvilagus* sp.); birds (*Sialia mexicanus*); reptiles—lizards (*Heloderma suspectum*, and unidentified scales), and snakes; anurans; and insects (Campbell and Lamar 2004; Degenhardt et al. 1996; Forks and Hughes 2007; Funk 1964; Gehlbach 1956; Greene 1990; Klauber 1972; Lemos-Espinal et al. 2007; Lowe et al. 1986; Miller *in* Gloyd and Smith 1942; Milstead et al. 1950; Minton 1959; Powell et al. 2004b; Reynolds and Scott 1982; Sánchez-Herrera 1980; Vermersch and Kuntz 1986; Werler and Dixon 2000; Woodin 1953; Yarrow 1875).

In Arizona, typical prey are woodrats (*Neotoma*) and cottontails (*Sylvilagus*) (Greene 1990), but Chihuahuan *C. molossus* consume 33.3% *Perognathus*, 25.0% *Peromyscus*, 16.7% *Dipodomys*, 16.7% birds, and only 8.3% *Neotoma* (Reynolds and Scott 1982). Captives have readily fed on *Mus musculus*, *Neotoma albigula*, and *Perognathus intermedius* (Beck 1996).

C. molossus probably captures its prey by either ambushing it or by active hunting. We have seen them actively foraging at dusk in Arizona. In the spring and fall most foraging is done in the morning or late afternoon, but crepuscular or nocturnal hunting is the summer norm. Prey are struck and later scent-trailed, found, and swallowed when dead (C. Ernst, personal observation). To fulfill its annual energy requirements, *C. molossus* needs to ingest prey quantity equal to 93% of its BM; this can be met in 2–3 large meals (Beck 1995).

VENOM DELIVERY SYSTEM: Klauber (1939) measured the fangs of 44, 80–120 cm TBL, *C. m. molossus* and interpreted the following characteristics from regression lines: mean FL, 11.7 (range, 9.6–13.5) mm; mean TBL/FL, 86%; and mean HL/FL, 4.07%. The average angle of fang curvature for both *C. m. molossus* and *C. m. nigrescens* was 55°.

VENOM AND BITES: Klauber (1972) made 64 venom extractions from *C. m. molossus* for a total of 13,734 mg of dried venom; the average dried venom yield per fresh adult was 286 mg, and the maximum dry yield was 540 mg. After 32 days, the liquid venom yield is only 47% of the first milking (Klauber 1972). A single extraction from a large adult *C. m. estabanensis* yielded 32 mg of dried venom (Klauber 1972). The average liquid venom yield from the species is 0.60 mL (= 0.18 g dried, n = 5) (do Amaral 1928).

Some data are available on the chemistry of the venom. The average protein content of venom from *C. m. molossus* is 225 mg (Hardy et al. 1982), and the isoelectric patterns of the venom proteins do not vary over a 20-month period (Gregory et al. 1984; Gregory-Dwyer et al. 1986). Myotoxin α-like proteins have been detected in the venom of both *C. m. molossus* (western Texas; > 2.0 mg/mL, 0.1 μg/mL) and *C. m. nigrescens* (Durango, Mexico; > 2.0 mg/mL, 0.5 μg/mL) (Bober et al. 1988), although a *C. molossus* from Texas tested by Li and Ownby (1994) was not positive for Mojave toxin. Phosphodiesterases and phospholipase A_2 are present, as are two proteases that may cause hemorrhagic systems, N-benzoyl-L-arginine ethyl esterase and p-tosyl-L-arginine methyl esterase (Beasley et al. 1993; Stegall et al. 1994; Tsai et al. 2001). The fibrinolytic/hemorrhagic protein CMM4 has been isolated; its specific hemorrhagic activity decreases as the specific fibrinolytic activity increases (Sánchez et al. 2001). Venom from smaller (younger) *C. m. molossus* has less fibrinolytic activity than that of larger (older) snakes (Rael et al. 1997). A comparative enzymatic study of *Crotalus* venoms, including that of *C. m. molossus*, was conducted by Soto et al. (1989), and we refer the reader to their detailed paper.

Minton (1959) reported the venom is only of moderate toxicity, with a mouse LD_{50} of 17.4 mg/kg. Minimum lethal doses for 22–28 g mice are 0.06–14.0 mg/kg of dried venom (Githens and Wolff 1939; Macht 1937). Russell found the intravenous LD_{50} of mice to be 2.63 (range, 1.98–3.20) mg/kg, and Glenn found the intraperitoneal and intramuscular LD_{50} values to be 2.44 mg/kg and 8.36 mg/kg, respectively (*in* Hardy et al. 1982). Ramírez et al. (1990) reported a mouse intraperitoneal LD_{50} of 2.35 mg/kg. The minimum lethal dose for a 350 g pigeon is 0.40 mg of dried venom (Githens and George 1931). The human lethal dose has not been determined.

The venom of *C. m. molossus* is considered basic and relatively mild by rattlesnake standards. Envenomations by it are rather uncommon, and have mostly occurred while the snake was handled. However, the snake has a proportionally large head (Klauber 1938a) and can potentially deliver a significant amount of venom per bite, so envenomation can be serious. The venom is strongly hemorrhagic (7.5 mm/μg [diameter of the hemorrhagic spot divided by the amount of protein injected], one of the most hemorrhagic of Crotalinae venoms (Sánchez et al. 2001), and includes fibrinolytic and platelet-aggregating properties. A coagulant effect, although present, is much less marked.

Human envenomation by *C. m. molossus* produces local pain, marked swelling and ecchymosis (often with blebs) at the area bitten, elevated blood pressure, thrombocytopenia, and, in one case, hypofibrinogenemia (Altimari 1998; Corrigan and Jeter 1990; Hardy et al. 1982; Norris 2004; Russell 1969). Victims normally recover after treatment with antivenom, but the potentiality of death is present.

The venom can be fatal to a conspecific (Nichol et al. 1933).

PREDATORS AND DEFENSE: Predation data concerning *C. molossus* are scanty. The Sonoran whipsnake (*Masticophis bilineatus*) is a known predator (Enderson 1999; Klauber 1972). Other probable natural enemies of *C. molossus*, particularly of the young, are ophiophagous snakes (*Lampropeltis*, *Pituophis*), large hawks (*Buteo* sp.), owls (*Asio* sp., *Bubo virginanus*), roadrunners (*Geococcyx californianus*), and carnivorous mammals (*Canis latrans*, *Conepatus mesoleucus*, *Didelphis virginiana*, *Lynx rufus*, *Mephitis mephitus*, *Pecari tajacu*, *Spilogale gracilis*, *Taxidea taxus*). A *C. molossus* was eaten (possibly as carrion) by a captive *C. atrox* (Klauber 1972).

The black-tailed rattlesnake is usually mild-mannered, and depends greatly on its cryptic coloration and pattern to conceal it. If detected, it often remains calmly coiled, either quietly or with occasional rattling, or tries to retreat into some shelter. However, some will begin rattling when a human is still some distance away, and a few aggressively strike when first disturbed. Females guarding their young may be aggressive (see above). Armstrong and Murphy (1979) reported that a provoked Mexican *C. m. nigrescens* gave an open-mouth threat in which it displayed its folded fangs for more than five minutes, and it may even spit venom into the eyes of its disturber (Madrid-Sotelo and Balderas-Valdivia 2008).

If confronted with either the sight or odor of snakes of the genera *Lampropeltis* or *Coluber*, both *C. m. molossus* and *C. m. nigrescens* will bridge a body coil (Weldon and Burghardt 1979; Weldon et al. 1992). When it does rattle, the sound is rather loud (about 70 dB; Cook et al. 1994) with a dominant frequency range of 6.9–9.8 kHz, and total frequency range encompassing 0.6–19.9 kHz (Young and Brown 1993).

PARASITES AND PATHOGENS: The species is parasitized by the tetrathyridia of the cestode *Mesocestoides* sp., and the digestive tract nematodes *Kallicephalus inermis* and *Ophidascaris labiatopapillosa* (Ernst and Ernst 2006; Goldberg and Bursey 1999a; Klauber 1972; Prado Vera 1971). Captive *C. molossus* at the Dallas Zoo have suffered from swellings on the rostrum that sometimes extended posteriorly to the frontal area of the head. Bacterial cultures from the swollen site revealed *Escherichia coli* and *Pseudomonas aeruginosa* (Murphy and Armstrong 1978).

POPULATIONS: No study of the population structure or dynamics of *C. molossus* has been published, and most data are anecdotal. Of 425 rattlesnakes collected in central Arizona, only 27 (6.4%) were of this species (Klauber 1972). In 3 seasons Crimmins (1927b) collected 85 rattlesnakes in the mountains around El Paso, Texas, of which 24 (28.2%) were *C. molossus*. Engelhardt (1932) noted that the species was the prevailing rattlesnake in the Davis Mountains of western Texas. Sánchez-Herrera (1980) reported an equal sex ratio for the eight *C. m. nigrescens* he collected in Federal District of Mexico.

Humans are this snake's worst enemy, often killing it on sight, but many more are destroyed through habitat destruction, on the roads, or by overcollecting for the commercial and pet trades. In Mexico, *C. molossus* and other rattlesnakes are captured, killed, and dried to be used as folk medicines (Minton and Minton 1991).

In an attempt to alleviate some of the wanton killings, a relocation program has been set up at Montezuma Castle National Monument in the Verde Valley of Arizona (Nowak 1998). *C. m. molossus* and other species of rattlesnakes that are found near areas of heavy human activity in the Monument or in nearby towns are captured and removed to a distant place within the species' natural habitat and range. However, the

Crotalus molossus
nigrescens
(R. D. Bartlett)

effectiveness of the program is not known, as rattlesnakes removed from their natural foraging range and the area of their hibernaculum probably suffer reduced survivorship.

In spite of these problems, the species is still fairly common in most parts of its range, and is not protected in any state of its occurrence in the United States. It is considered endangered in Jalisco (Cruz-Sáenz et al. 2009).

REMARKS: Gloyd (1940) thought *C. molossus* part of a complex with *C. horridus* and *C. durissus*, but actually closely related to the Mexican *C. basiliscus*. Electrophoretic studies of rattlesnake venom by Foote and MacMahon (1977) show it closest to *C. scutulatus*, *C. tigris*, and *C. horridus* (*C. basiliscus* was not tested). A more recent mtDNA and RNA study by Murphy et al. (2002) places *C. molossus* in the *C. durissus*-group of large rattlesnakes, which also includes *C. basiliscus*, *C. enyo*, *C. unicolor*, and *C. vegrandis*. Considering both morphological and molecular characters, its closest relative is *C. basiliscus*.

The taxonomy and literature of the species were reviewed by Price (1980).

Crotalus oreganus Holbrook, 1840
Northern Pacific Rattlesnake

RECOGNITION: This is the most variable of North American rattlesnakes, making a composite description difficult; so, after reading the following description, also examine those of its five subspecies presented in "Geographic Variation."

C. oreganus (TBL$_{max}$, 165.0 cm; Stebbins 2003) is quite variable in ground color: pinkish, pinkish-gray, pinkish-tan, salmon, yellow, orange, tan, light or dark gray, grayish-brown, reddish-brown, olive-brown, olive-green, or black. Albinism occurs occasionally (Dyrkacz 1981; Hensley 1959; Klauber 1972), as does melanism (Klauber

1972). The dorsal pattern consists of 33–42 (20–49) large white-, gray-, or yellowish-bordered blotches separated by narrow interspaces. The blotches vary in shape from diamonds, ovals, and rectangles, to squares. A small pale area may be present in the center of each blotch. Two rows of lateral blotches are present along each side; the upper may be poorly developed. Two to 12 cross bands are present on the tail; the two most distal and the base of the rattle are usually black. The body scales are pitted and keeled; their microdermoglyphic pattern consists of linear (both straight and swirling) ridges (Harris 2005 [see illustration]; Stille 1987). These occur in 23–25 anterior rows, 25–26 (23–29) rows at midbody, and 19–20 (18–22) posterior rows. The cream to yellow venter is mottled with small brown flecks; present are 161–196 ventrals, 14–31 subcaudals, and an undivided anal plate. The rostral area usually contains a large dark brown blotch that ends posteriorly in a pale border forming the transverse bars on the supraocular scales. On each side of the head a conspicuous dark, white-bordered, postocular stripe extends diagonally backward from the orbit to above the corner of the mouth (which may be faded and merge laterally with the ground color in southern populations). Dorsal head scales consist of a higher than wide rostral, 3–5 (1–8) small to moderate internasals, no prefrontals, 2 pairs of canthals, 15–30 (7–45) scales between the canthals, 4–6 (1–9) intersupraoculars, and 2 supraoculars. Laterally are 2 nasals (the prenasal may or may not contact the supralabials; the postnasal only rarely contacts the upper preocular), 1 (1–2) loreals, 2 (3) preoculars, 14–15 (11–19) supralabials, 4–6 suboculars, and 15–16 (13–19) infralabials. On the underside of the jaws are a mental scale and two pairs of chin shields.

The bilobed hemipenis has a fork of the sulcus spermaticus extending up each lobe, and approximately 43–63 spines and 23–29 fringes per lobe (Klauber 1972).

The dental formula is 2–3 (4–5) palatines, 7–8 (6–10) pterygoids, 9–10 (6–11) dentaries, and a maxillary fang on each side (Klauber 1972).

Males (TBL$_{max}$, 165.0 cm) have 171–179 (161–190) ventrals, 23–25 (15–29) subcaudals, 5–8 (3–12) tail cross bands, and a mean TL/TBL of 7.3 (6.9–7.8) %. Females

Crotalus oreganus oreganus (Klickitat County, Washington; Brad Moon)

Crotalus oreganus
oreganus
(R. D. Bartlett)

(TBL$_{max}$, 132.0 cm; Klauber 1972) have 175–184 (164–196) ventrals, 18–20 (15–25), subcaudals, 3–7 (2–8) tail cross bands, and a mean TL/TBL of 5.7 (5.4–5.9) %.

GEOGRAPHIC VARIATION: Five subspecies are recognized (except where noted, TBL$_{max}$ and scalation adapted from Klauber 1972). Much variation occurs in the scalation and number and size of the tail bands within the individual populations of any subspecies (Klauber 1943b). *Crotalus oreganus oreganus* Holbrook, 1840, the northern Pacific rattlesnake, ranges west of the Rockies from British Columbia south to San Luis Obispo and Kern counties, California. It has a TBL$_{max}$ of 162.6 cm, and is dark gray, olive, yellowish-brown, brown, or black. It has 33 (20–41) hexagonal, oval, or almost circular dark dorsal blotches with well-defined light borders (the upper row of lateral blotches is absent); usually a single loreal scale; 161–190 ventrals; 15–29 subcaudals; 15 (11–18) supralabials; and 16 (13–20) infralabials.

Crotalus o. abyssus Klauber, 1930b, the Grand Canyon rattlesnake, is found only on the floor and sides of the Grand Canyon north to at least Antelope Island (Brennan and Holycross 2004) in Arizona. This pale snake has a TBL$_{max}$ of 102.5 cm (Durham 1956), and is flesh-colored, pink, salmon, red, or sandy brown. It has 41–42 (36–48) oval, rough-edged blotches that fade with age (the upper row of lateral blotches is faded or indiscernible); 2 loreals; 173–191 ventrals; 18–29 subcaudals; 15–16 (13–19) supralabials; and 16 (14–18) infralabials. Analysis of venom proteins indicates that *C. o. abyssus* is most closely related to the subspecies *C. o. lutosus* (Young et al. 1980).

Crotalus o. cerberus (Coues, 1875), the Arizona black rattlesnake, ranges from Mohave and Yavapai counties southeast to Apache, Graham, and Pima counties, in Arizona, and in adjacent Grant County, New Mexico (Christman et al. 2000; Klauber 1934; Mello 1978). It has a TBL$_{max}$ of 105.0 cm, and is dark gray, olive, dark brown, or black. It has 35 (25–46) large, poorly defined dark blotches (the lateral blotches in the lower row are almost indiscernible); 2 loreals; 161–184 ventrals; 16–26 subcaudals; 15–16 (13–18) supralabials; and 16 (13–19) infralabials.

Crotalus oreganus abyssus (Grand Canyon, Arizona; Arizona-Sonora Desert Museum; John H. Tashjian)

Crotalus oreganus abyssus (R. D. Bartlett)

Crotalus o. concolor Woodbury, 1929, the midget faded rattlesnake, occurs in the Colorado and Green river watersheds in southwestern Wyoming, eastern Utah, western Colorado, and northwestern Arizona (at least south to Halls Crossing; Brennan and Holycross 2004). This small subspecies (TBL$_{max}$, 75.6 cm; Parker and Anderson 2002), is cream, yellowish-brown, or tan. It has 42–43 (34–52) rectangular to oval only slightly darker (often faint or absent) body blotches; 2 loreals; 163–183 ventrals; 16–29 subcaudals; 14–15 (11–17) supralabials; and 15 (13–18) infralabials. Oyler-McCance et al. (2005) isolated 5 polymorphic microsatellites from the DNA of this subspecies, 4 of which had relatively high levels of diversity consisting of 8–9 alleles; no 2 loci were linked. The validity of the name *C. o. concolor* Woodbury, 1929, was debated by Gloyd (1940), who placed it in the synonymy of *C. viridis decolor* Klauber, 1930b. Woodbury (1942) defended the validity of *concolor*, and this name was officially validated by the International Commission on Zoological Nomenclature in 1955 (Woodbury 1958).

Crotalus o. lutosus Klauber, 1930b, the Great Basin rattlesnake, ranges between the Rocky and Sierra Nevada mountains from southern Idaho, southwestern Oregon, and northwestern California, south through western Utah and Nevada to northwestern Arizona. It has a 150.0 cm TBL$_{max}$ (Nussbaum et al. 1983), and is gray, yellowish-brown, tan, or greenish-brown (Smart 1951 has given a detailed analysis of the color variations). It has 40 (32–49) elliptical to oval to hexagonal, brown or black, blotches widely separated by lighter pigment (the lower lateral row of blotches is borderless and almost indiscernible); 2 loreals; 171–196 ventrals; 16–29 subcaudals; 15–16 (12–19) supralabials; and 15–16 (13–19) infralabials.

Intergrade zones occur where the subspecific ranges meet, especially in Arizona around the rim of the Grand Canyon and upper reaches of the Colorado River, and in Colorado and Utah (Eaton 1935; Gloyd 1940; Lowe et al. 1986; McKee and Bogert 1934; Tanner 1930, 1958).

CONFUSING SPECIES: Other species of *Crotalus* within its range may be distinguished from *C. oreganus* by using the previously presented key. The smaller rattlesnakes of the genus *Sistrurus* have enlarged scales on the dorsal surface of their heads. Nonvenomous snakes of the genera *Heterodon* and *Pituophis* have round pupils, and lack both loreal pits and rattles.

KARYOTYPE: Monroe (1962) described the 36 chromosomes in the karyotype of *C. o. lutosus* and *C. o. oreganus*. Ten are V-shaped, including a medium pair and two pairs each of large and small chromosomes, respectively. The medium pair is submetacenteric, and the smallest pair is generally larger than the longest pair of 16 rod-shaped chromosomes. The shortest three pairs of rod chromosomes are acrocenteric with

Crotalus organus abyssus, neonate (R. D. Bartlett)

only two arms in anaphase I of mitosis. The rest of the karyotype is made up of micro-chromosomes. The karyotype is probably the same or similar to that reported under the name *C. viridis* by Baker et al. (1972) and Zimmerman and Kilpatrick (1973).

FOSSIL RECORD: Pleistocene (Rancholabrean) remains of *C. oreganus* abound in California (Brattstrom 1953b, 1954a; Holman 1995; Lundelius et al. 1983; Rage 1984), Idaho (Lundelius et al. 1983; McDonald and Anderson 1975), and Nevada (Brattstrom 1954a, 1954b, 1958b; Holman 1981; Lundelius et al. 1983; Mead and Bell 1994; Mead et al. 1982, 1989).

DISTRIBUTION: *C. oreganus* is found from south-central British Columbia, Washington, southwestern Idaho, and west-central and southwestern Utah south to northern San Luis Obispo and Kern counties in California, all but the southern portion of Nevada, Utah, and western Colorado to southeastern Arizona and Grant County, New Mexico (see "Geographic Variation" for individual ranges of the five subspecies). The total range encompasses elevations of sea level to 3,962 m (Arment *in* Klauber 1972; Basey 1988; Bogert *in* Klauber 1972; Hamilton and Richard 2010).

HABITAT: Generally, *C. oreganus* prefers habitats with much cover and abundant prey odors (Theodoratus and Chiszar 2000). Because of the extensive north-south distribution of the species, it ranges through several climatic and vegetational zones, making it difficult to describe a general habitat for the species. Therefore, we have taken a subspecific approach in describing the various habitats.

In British Columbia *C. o. oreganus* resides throughout the interior dry belt, where it is found in sagebrush habitats containing boulders and talus slopes. In southwestern Idaho it is found in semiarid mixed stands of sagebrush (*Artemisia tridentata*) and winterfat (*Cerratoides lunata*), but grasses (primarily *Bromus tectorum* and *Poa sandbergii*), shadscale (*Atriplex confertifolia*), and greasewood (*Sarcobatus vermiculatus*) are also present (Diller and Johnson 1988). *C. o. oreganus* from southwestern Utah have a Horn's index habitat niche breadth of $R_0 = 0.21$ (Diller and Wallace 1996).

Distribution of
Crotalus oreganus.

The habitat of *C. o. abyssus* consists of desert scrub (characterized by *Acacia* sp., *Atriplex* sp., *Lycium* sp., *Prosopis* sp., and *Yucca* sp.) on the floor of the Grand Canyon grading into pinion-juniper at higher elevations along the canyon's sides.

C. o. cerberus occurs in chaparral in southern Arizona, but also in upland situations, where it is more of a woodland snake found in pine (*Pinus* sp.)-juniper (*Juniperus* sp.) habitats, in deciduous riparian zones, and around rock outcroppings.

C. o. concolor is mostly found in areas where rocky ledges and outcrops occur and sagebrush is the chief vegetation. Hibernacula are usually nearby, not requiring a lengthy migration.

The habitat of *C. o. lutosus* has been the best documented. Its distribution encompasses the Great Basin Desert, where the snake lives in arid to semiarid sagebrush steppe, in rocky canyons, around lava flows, and on talus slopes, but it ranges upward into forested zones. Its habitats often have a south, southwest, or west exposure (Millard *in* Klauber 1972). Predominant plants reported in this subspecies' habitats include cacti (*Ecinocereus* sp., *Opuntia polycantha*), cliffrose (*Cowania mexicana*), curl-leaf cercocarpus (*Cercocarpus ledifolius*), desert mallow (*Sphaeralcea ambigua*), desert paintbrush (*Castilleja chromosa*), desert primroses (*Oenothera* sp.), desert tea (*Ephedra nevadensis*, *E. viridis*), grasses (*Agropyron cristatum*, *A. spicatum*; *Bromus tectorum*), greasebush (*Forsellesia nevadensis*), greasewood (*Sarcobatus vermiculatus*), hop sage (*Grayia spinosa*), horsebrushes (*Tetradymia spinosa*), junipers (uplands, *Juniperus* sp.), locoweeds (*Astragalus* sp.), lupines (*Lupinus* sp.), penstemons (*Penstemon* sp.), phacelias (*Phacelia* sp.), pinyon pines (uplands, *Pinus* sp.), prince's plume (*Stanleya pinnata*), rabbitbrush (*Chrysothamnus nauseosus*), sagebrush (*Artemisia* sp.), shadescale (*Atriplex confertifolia*), summer cypress (*Kochia scoparia*), thistle (*Salsola kali*), and winterfat (*Ceratoides lanata*). Linsdale (1940) reported it abundant in Nevada alfalfa fields.

BEHAVIOR AND ECOLOGY: Of the three former subspecies-groups once included in *C. viridis*, that which now composes the subspecies of *C. oreganus* is the most well researched (except for the Colorado populations of *C. viridis*; see the papers of Duval and his research associates), and ranks only behind *C. horridus* at present as the most ecologically studied North American rattlesnake. Most of the data reported for any of the former subspecies of *C. viridis* can be easily applied to the other subspecies of the original complex. The three northernmost subspecies of *C. oreganus* (*concolor*, *lutosus*, *oreganus*) have more biological data available for them, but, unfortunately, many gaps exist in our knowledge of *abyssus* and *cerberus*.

In northern Idaho, *C. o. oreganus* is surface-active for about 206 days annually (1 April–21 October), and the snake tends to remain at the hibernaculum for up to 3–4 weeks before it disperses in the spring (Diller and Wallace 1996; Wallace and Diller 2001). Hand capture of mature males occurred mostly in the spring and fall, differing significantly from that of mature females, which had three peaks of capture (spring, summer, fall). Immature males were most active from late March to late May, although drift fence captures indicated that they were active over the summer into the fall. Both hand captures and drift fence captures indicated that immature and non-reproductive females had activity peaks only in the spring and fall. Drift fence captures began about a month later than the first hand captures in the spring, but the last fall captures occurred at about the same time for both methods. Most male drift fence captures occurred during the summer, indicating greater movements of them then

Crotalus oreganus cerberus (East of Tucson, Arizona; Arizona-Sonora Desert Museum; John H. Tashjian)

(possibly involving mate-searching). Drift fences also showed that reproductive females are active during the summer months. Wallace and Diller (2001) thought that the hibernaculum was located within the summer feeding range of this population. In southwestern Idaho, based on hand captures, the snake becomes active in mid-April, is most active in early May, and then becomes less so from mid-May to September, and enters overwinter quarters in late September. Drift fence captures showed it initially surface-active in mid-April, becoming more active in late April, remaining active during June and July, and then disappearing in mid-August (Diller and Wallace 1996).

The overwintering biology of the species has been especially well reported. *C. o. oreganus* enters hibernation from 15 September in British Columbia to 16 November in northern Idaho (Carl 1960; Diller and Wallace 1984), but not until October–December in the more southern subspecies. Earliest recorded emergence of *C. o. oreganus* in California is 1 March (Fitch and Twining 1946), but in British Columbia it may not become spring-active until 24 March into April (Carl 1960). The other subspecies, depending on elevation, usually emerge from late March into early May, but some Idaho *C. o. lutosus* do not become active until 26 June (median date, 15 May) (Cobb and Peterson 2008). *C. o. concolor* may spend as long as seven months underground (Houston 2006), and *C. o. lutosus* 189–253 days (median, 206 days) (Cobb and Peterson 2008). There appear to be no sexual or age differences in spring arousal time (Fitch 1949a; Wallace and Diller 2001), but an ET of 16°C may be necessary to bring them out of the den (Woodbury 1951).

At hibernation sites of *C. o. lutosus* in eastern Nevada and western Utah, the mean annual ET was 9°C, that in January and July, −1 °C and 22 °C, respectively. Mean elevation was 1,837 (range, 1,539–2,088) m, and the area experienced about 124 annual frost-free days. Mean insolation at the dens was 1,551 (range, 896–2,178) watts/m², and their mean aspect and slope were 197 (range, 103–342) ° and 20 (range, 6–34) °, respectively (Hamilton and Nowak 2009).

C. o. oreganus and *C. o. lutosus* usually spend the winter in a communal den often involving more than 100 individuals of all ages and sexes, but particularly in the more southern subspecies, individual or small-group hibernation is more normal. At least in the northern populations, hibernacula fidelity occurs. *C. o. oreganus* has homing ability, and can return to its hibernaculum if displaced varying distances at all four compass points from the den (Hirth 1966b). Several northern hibernacula probably once housed more than 500 rattlesnakes each winter, but the numbers have been severely reduced in modern times. Overwintering occurs in desert tortoise (*Gopherus agassizii*) and mammal burrows, dry spring channels, loose cobblestone rocks, gravel banks, talus slopes and landslides, as much as 4.5 m deep rock crevices, many artificial openings in buildings and walls, and garbage dumps (Fitch and Glading 1947; Klauber 1972; Woodbury and Hardy 1948; Woodbury and Parker 1956). Such hibernacula usually face south and are free of water seepage. Northern dens may be shared with lizards (*Eumeces skiltonianus*, *Scloporus graciosus*) and other snakes (*Coluber constrictor*, *Diadophis punctatus*, *Hypsiglena torquata*, *Masticophis taeniatus*, *Pituophis catenifer*, *Rhinocheilus lecontei*, and *Thamnophis elegans*) (Jenkins and Peterson 2008; Parker and Brown 1973; Woodbury and Parker 1956; Woodbury and Smart 1950).

Both ATs and BTs usually decline during the fall (mean BT of Idaho *C. o. lutosus* decreased from approximately 13°C in November to a minimum of 6°C in February through April; Cobb and Peterson 2008), forcing the snakes into hibernacula. Snakes in dens never freeze, and the snakes move about until January, when they become relatively stationary (Cobb and Peterson 2008). They are usually sluggish, but alert at ATs > 3.5°C, and activity resumes when the AT has reached 5.0°C (Vetas 1951). Then the snakes emerge from hibernation with a spike in BT. The lowest BT experienced in Idaho was 4.4°C; the overall mean BT was 8.9°C (Cobb and Peterson 2008).

At a den in British Columbia, the spring thermal gradient was weak to nonexistent during most of the emergence period (Macartney et al. 1989). BM is lost during hibernation, probably due to evaporative water loss, but also to metabolic use of stored fat. Adults lose 4–9% and juveniles 20–50% of their BM (Hirth 1966a; Klauber 1972;

Crotalus oreganus cerberus (Barry Mansell)

Parker and Brown 1974). This is an important factor influencing overwinter surviv-ability of small snakes, particularly neonates. Charland (1989) has reported a survivor-ship rate of only 55% for neonates in their first winter, but that survivorship seemed independent of both BM and condition at birth. Lying in bunches of several individuals with bodies in contact reduces water loss (White and Lasiewski 1971).

Populations occurring at both higher elevations and latitudes seem to be more di-urnally active in the summer than those at lower elevations and more southern lati-tudes. *C. o. oreganus* is more diurnal in the mountains of California than it is in the San Joaquin Valley (Klauber 1972). Nights are usually spent alone under a rock or in some other shelter such as a log or animal burrow (*Geomys bursarius*; Vaughan 1961). Cre-puscular and nocturnal activity does occur in *C. o. lutosus* and *C. o. oreganus*, and espe-cially in desert habitats in southern subspecies (Brooking 1934; C. Ernst, personal observation; Fautin 1946; Klauber 1972). In Idaho, *C. o. oreganus* usually begins to bask at about 0900–1000 hours, retreats into a shelter by early afternoon (1400–1500 hours), re-emerges in late afternoon and evening in the summer (1600–2000 hours, but this period is less predictable and extensive than that in the morning), and is seldom active after dark (2300–2400 hours) (times approximate); it is most active in the morning (0800–1200 hours) in both spring and summer (Diller and Wallace 1984, 1996). Juve-nile *C. o. concolor* are usually active at mid-day and feed diurnally; adults are typically more crepuscular, being more active in the morning and early evening (Houston 2006). Grand Canyon *C. o. abyssus* are usually found resting in the shade of rocks or vegeta-tion in the afternoon, but may be nocturnally active (Young and Miller 1980). If the AT is favorable, the species may emerge shortly after sunrise, usually basks first, then starts its daily movements, and remains active until the evening.

Reported mean BTs of active British Columbia *C. oreganus* are 30–35 (range, 11.9–37.8) °C (Charland and Gregory 1990), in Washington 24.5 (range, 13.5–33.5) °C (Vitt 1974), in southwestern Idaho 28.4°C (Diller and Wallace 1996), and in northwestern Utah 26–30 (range, 16–35) °C (Hirth and King 1969). *C. o. concolor* has been reported foraging in the evening when ATs were as low as 5°C (Porras 2000).

Gravid females have 19–27°C voluntary BTs, and bask most commonly when ST is 15–25°C (Diller and Wallace 1996). Females regulate BT by shifting between warmer and cooler locations, with most periods of inactivity spent near a heat source (Gier et al. 1989). Most maintain a BT of 29–30°C. There is no significant difference in mean and range of BT between gravid and nongravid females before or after birth. Gravid females thermoregulate higher and with less variability before than after birth, 26.9–34.6 (mean, 30.7) °C, and their BTs change significantly after parturition, dropping to 23.3–32.0 (mean, 28.5) °C (Gier et al. 1989). Both gravid and nongravid females have a triphasic diel BT pattern in the summer—BTs rise rapidly in the morning, re-main stable in the afternoon, and decline slowly at night, and during the afternoon stable period, mean BT of gravid females (31.7°C) and those of nongravid females (30.7°C) are not significantly different, but those of gravid snakes are significantly less variable (Charland and Gregory 1990).

Excessive heat will kill the snake; Swift (1933) reported that a 66 cm *C. o. oreganus* succumbed in 20 minutes after exposure to direct sunlight (unfortunately, no BTs were recorded). Blum and Spealman (1933) reported that another exposed 30 min-utes to full sunlight only increased its activity when the AT rose from 25 to 36°C, and, 4 days later when exposed to a quartz mercury arc lamp at 45 cm for 11 minutes, it

Crotalus oreganus
cerberus
(R. D. Bartlett)

showed no effect at ATs of 23–25.5°C. A day later, Blum and Spealman exposed the snake directly to clear noon sunlight for 17 minutes, and as the AT rose from 23 to 36°C the snake increased its activity accordingly, but showed no ill effects after the treatment. The following day they put the snake into an air bath and raised the AT from 38 to 52°C during a period of 14 minutes, at the end of which the snake was dead. As soon as it was introduced to the bath the snake began to move rapidly about. At the end of 7 minutes, when the temperature had reached 49°C, the snake was making undirected writhing movements and gasping. Shortly after, it became rigid. When later examined, its muscles appeared to be in rigor. Blum and Spealman concluded that death was caused by irreversible changes in the tissues produced by excessive temperature, and that heat rigor occurs in the muscles in general after 40°.

Smaller snakes heat and cool faster than larger ones—two neonates had 16–18°C BTs, while adults had BTs above 21°C; in addition, mean BT of adult females were 2.1 and 2.0°C higher than those of males in the spring and summer, respectively, and may be correlated with female viviparity (Hirth and King 1969).

Scarcity of water is a problem in many of the xeric habitats occupied by *C. oreganus*, and the snake loses water through evaporation from its skin and oral cavity to the dry air. When water is available, such as in puddles after rains, the snake often drinks to capacity, or it may coil and lick up the water that accumulates between its body coils. Ashton and Johnson (1998) witnessed a *C. o. concolor* lapping water from its skin during a rain event. At first it emerged from a burrow when the rain began to fall and drank water from a depression in a small rock. It then retreated to the burrow entrance and loosely coiled. After rain drops accumulated on its exposed body, the snake began drinking from its skin, and continued to do so intermittently for 10 minutes. Drinking rain water from its coils has also been observed in *C. o. lutosus* by Aird and Aird (1990).

Horn and Fitch (1942) marked *C. o. oreganus* in California grazing country and observed that during the annual activity period these snakes wander indefinitely with no apparent tendency to return to a given spot, but noted that on average most individuals move less than 3 m a day, and may move only a few meters over a period of months. The mean size of home ranges in California is 12.1 ha for males and 6.5 ha for females (Fitch 1949a). Stark (*in* Macartney et al. 1988) found home ranges to be only 2.4 ha for males and 1.8 ha for females. The heart rate of monitored exploring *C. o. oreganus* is 26–55 beats per minute (Hayes et al. 1994).

Arizona *C. o. abyssus* only engage in rather short meandering movements (Jørgensen et al. 2008). In the Grand Canyon, *C. o. abyssus* tracked by Reed and Douglas (2002) made daily movements of 0–365 m with averages of 45 m per movement and 26 m per day. Males moved 0–365 (mean, 55.2) m, while females moved 0–99 (mean, 28.5) m; both sexes averaged only 0.57 movements per day. Their mean home range areas were minimum convex polygon—males, 15.8 ha; females, 5.2 ha; harmonic mean transformation 50% isopleths—males, 1.9 ha; females, 0.1 ha; and harmonic mean transformation 95% isopleths—males, 16.9 ha; females, 8.0 ha. Overall home range size ranged from less than 4 ha to more than 30 ha.

Males made short, infrequent movements, and Ashton (2003) could find no evidence of migratory behavior or shifts in areas used in the time (mostly during the mating period) he monitored southwestern Wyoming *C. o. concolor*. Eight males located 5–57 times moved 1–20 (mean, 2) times, covered 0–70 (mean, 45) m per movement, and traveled a total distance of 0–876 (mean, 259) m. Adult males moved greater distances than gravid females (mean distance per movement, 4 m; mean total distance traveled, 45 m). Ashton and Patton (2001) studied movements by female *C. o. concolor* in three Wyoming populations. Gravid females emerged from hibernation, moved a short distance to a rookery, then moved very little for the remainder of their GP. From the time of initial capture until just before parturition, the females traveled only 15–108 (mean, 41.2) m; mean distance per movement was only 4.5 (range, 2.0–6.2) m. The mean distance moved per day was 0.74 (range, 0.3–1.9) m.

Jenkins and Peterson (2005, 2008) followed 12 male and 18 female radio-equipped *C. o. lutosus* in the upper Snake River Plain of Idaho during 2003 and 2004 for an average of 64 (range, 30–102) days each. The snakes moved an average total distance of 3,611 m, an average of 1,378 m from their hibernacula, and an average of 53 m per day. Their mean kernel density home range was 10 ha: males, 0.06–11.90 (mean, 5.64) ha; females, 0.04–77.41 (mean, 13.04) ha. No significant differences in movement patterns were noted among disturbance categories, sex, or years, but notable differences occurred between large males and females. Large males were approximately 1 m longer (SVL), had high body condition, and tended to move longer total distances (10,845–12,678 m). The female that moved the greatest distance (5,964 m) did not move half the total distance of most large males; however, all farthest movements from hibernacula were by females (2,200–3,160 m, versus 1,267–1,708 m for males).

The snake is a good climber that ascends into bushes and low trees to either bask or forage (Nussbaum et al. 1983); it has been observed as high as 2.8 m above the ground (Slaugh *in* Klauber 1972). It also readily takes to the water of rivers and lakes, and individuals tested by Klauber (1972) swam quite well.

Ferguson and Thornton (1984) found that *C. oreganus* has a greater O_2 storage capacity, hemoglobin content, and blood volume than the watersnake, *Nerodia sipedon*. *C. oreganus* could remain underwater 30.13 minutes. The hemoglobin of *C. o. oreganus* is functionally controlled by nucleoside triphosphates (NTPs), generally ATP, but the

Crotalus oreganus concolor (Louisiana Purchase Garden and Zoo; John H. Tashjian)

binding affinity is low such that NTP is not saturating at NTP/Hb ratios below 3.5 (Ragsdale et al. 1995). Other blood parameters of *C. oreganus* are listed in Luck and Keeler (1929).

Males engage in combat in both the spring and late summer, corresponding to the mating seasons of *C. oreganus* (Aldridge and Duvall 2002; Fitch 1949a; Fitch and Glading 1947; Gloyd 1947; Hersek et al. 1992 Klauber 1972). The behaviors involved are like those reported for other species of *Crotalus*, and the larger male usually wins (Ashton 2003). Male heart rate at rest (16–46 beats/minute) rises during combat to 43–60 beats/minute and is higher than during courtship (38–50 beats/minute) (Hayes et al. 1994). Feeding behavior also elicits high heart rates (44–60 beats/minute).

REPRODUCTION: Males usually mature earlier than females. Male *C. o. concolor*, *C. o. lutosus*, and *C. o. oreganus* mature in 2–4 years at SVLs of 52.0–56.5 cm with most first reproducing in their third year. Male Idaho *C. o. oreganus* with > 52 cm SVLs and 4+ rattle segments are usually mature; the smallest mature male there had a SVL of 46 cm and 4 rattle segments (Diller and Wallace 1984). Females mature in 3–7 years at SVLs of 55.0–69.3 cm, with most first reproducing in their fourth year, but not until age 6–8 in British Columbia (Diller and Wallace 1984, 1996, 2002; Glissmeyer 1951; Houston 2006; Jenkins et al. 2009; Macartney and Gregory 1988; Macartney et al. 1990). Klauber (1972) listed the following TBLs of the smallest gravid female for each of the 5 subspecies of *C. oreganus*: *abyssus*, 68.4 cm; *cerberus*, 70.1 cm; *concolor*, 52.2 cm; *lutosus*, 55.7 cm; and *oreganus*, 50.3 cm. Female Idaho *oreganus* with SVLs greater than 55 cm and 5+ rattles are usually mature (Diller and Wallace 1984). Ashton and Patton (2001) reported the mean SVL and postpartum BM of reproducing Wyoming female *C. o. concolor* are 53.4 cm and 70 g, respectively.

The stages in the male reproductive cycle of *C. o. oreganus* are the same as those occurring in *C. helleri* (which see for details), and are described and illustrated in Aldridge and Duvall (2002) and Diller and Wallace (1984).

In Idaho, female *C. o. oreganus* have enlarged (> 35 mm) yolked follicles when they emerge from hibernation, then experience secondary vitellogenesis, ovulate from mid-May to mid-June, and undergo parturition over an approximate 30-day period beginning from early September to mid-October (7 September–14 October, followed by a period of secondary vitellogenesis in the fall; Diller and Wallace 1984, 1996). Reproductive females have the greatest percentage of body fat (followed by nonreproductive females and males) and greater BMs, suggesting maternal-fetal exchange of nutrients (Diller 1981; Diller and Wallace 1984, 2002).

The female cycle is usually biennial, but depends on prey availability. Most females reproduce every 2 (range, 2–5) years (Ashton and Patton 2001; Diller and Wallace 2002; Jenkins et al. 2009). In central Idaho, 9.6% of *C. o. oreganus* produced litters for two consecutive years, but only one female had a three-year cycle (although this was probably an underestimate; Diller and Wallace 2002). The mean proportion of female *C. o. oreganus* reproducing in any given year varies from 30.8 to 87.5% in southwest Idaho (Diller and Wallace 1984, 1996, 2002); in contrast, Jenkins et al. (2009) found only 22–27% of females gravid in any given year in southeast Idaho. Glissmeyer (1951) reported that 12.5–66.7% of female Utah *C. o. lutosus* were gravid in different years. Woodbury and Hansen (1950) determined through palpation that only 6 of 30 (20%) adult female *C. o. lutosus* captured at a Utah den in 1950 were gravid and ready to give birth the next summer; they contained a total of 45 large follicles, representing a breeding potential of 64% of the total population that year.

During a 10-year (1994–2003) mark-recapture study of 3 den complexes of *C. oreganus* on the Idaho-National Laboratory on the Upper Snake River Plain of southeast Idaho, Jenkins et al. (2009) noted significant microgeographic differences in the reproductive characteristics of body condition, growth, age of maturity, pregnancy interval, and neonate size, although the 3 populations were clustered within 40 km. Snakes from the most disturbed area were shorter, had poorer body conditions, grew slower, had lower fecundity, and had shorter and poorer-conditioned neonates.

Grand Canyon *C. o. abyssus* breed nocturnally in July, when diurnal ATs are often over 40°C (Reed 2003b). Captive *C. o. cerberus* have courted or mated as early as 2 February after a cooling period ended in late January (Sievert 2002a). The subspecies *concolor* and *lutosus* apparently only mate from late June into at least mid-July (Ashton 2003; Wright and Wright 1957), but *oreganus* breeds in both spring and summer into fall, with sperm being more frequently found in females during May to early August (Aldridge and Duvall 2002; Diller and Wallace 2002; Fitch 1949a; Fitch and Glading 1947; Hersek et al. 1992; Klauber 1972; Macartney and Gregory 1988). In captivity, a cooling period of 6–8 weeks stimulates mating behavior (Houston 2006; Sievert 2002a). Mating may be initiated underground (Fitch and Glading 1947).

Olfaction appears to be the most important cue used by males while searching for a mate. They may locate and court females prior to female ecdysis, but the only successful copulation observed by Ashton (2003) occurred after the females had shed. The courtship and copulatory behaviors (Hayes 1986; Reed 2003b) are essentially like those reported for *C. v. viridis* (which see for details). Copulation may last from 2 to 3 minutes to several hours (Houston 2006; Reed 2003b; Sievert 2002a), and a female may mate several times with more than one male during a single breeding season (Reed 2003b). During the various stages of courtship, the male's heart rate (beats/

minute) is as follows: forward-jerking, 38–43; no movement, 38–43; and chasing the female, 43–50 (Hayes et al. 1994). Mating activity in British Columbia coincided with the peak of the male spermatogenic activity, and up to eight snakes took part (Macartney and Gregory 1988).

Parturition normally occurs in August and September (20 August–18 September for Wyoming *C. o. concolor*; Ashton and Patton 2001), but a captive *C. o. cerberus* has given birth on 26 July (Sievert 2002a). The GP lasts about 90 days in wild Idaho *C. o. oreganus* (Diller and Wallace 1996), but has lasted as long as 143 days in captive *C. o. cerberus* (Sievert 2002a) and 425 days in captive *C. o concolor* (Houston 2006).

Some gravid females may immigrate from their hibernaculum in the spring, but most remain relatively close and form maternity colonies either there or in secondary nearby dens. Little if any feeding may occur during the period of pregnancy, and the females must depend on the fat supply they have built up to carry them through the GP. Following parturition, the emaciated females apparently must double their weight before mating again, and those that cannot in a single summer must delay mating for another year or more (Charland and Gregory 1989; Macartney and Gregory 1988). Basking becomes the prime activity.

Ragsdale and Ingermann (1991) determined the red cell O_2-affinity of gravid female, nongravid female, male, and fetus *C. o. oreganus* at 20 and 34°C. Embryos had significantly higher O_2 affinities (indicated by lower P_{50} values) when compared with maternal red cells at both temperatures. At 20°C, the embryos had a mean P_{50} of 3.4, a mean mMol/L concentration of nucleoside triphosphate (NTP) of 9.6, and an NTP/Hb of 2.47, and at 34°C, a P_{50} of 6.8, compared to the following same measurements in gravid females (6.4, 15.5, and 3.49; and a P_{50} of 8.3), nongravid females (4.9, 10.2, and 2.41; and 8.2), and males (4.8, 9.8, and 2.27; and 7.8).

Litters of 1–15, but normally 3–8, young are produced: *abyssus*, 6–13; *cerberus*, 6–11 (mean, 8); *concolor*, 3–7 (population mean, 4.7); *lutosus*, 3–13 (population means, 5.5–8.3); and *oreganus*, 1–15 (25?) (Diller and Wallace 1996; Jenkins et al. 2009; Nussbaum et al. 1983) (population means, 4.6–9.9) (Ashton 2003; Diller 1981; Diller and Wallace 1984, 2002; C. Ernst, personal observation; Fitch 1949a; Glissmeyer 1951; Kawata 2004; Klauber 1972; Nohavec 1995; Nussbaum et al. 1983; Seigel and Fitch 1984; Shine and Seigel 1996; Sievert 2002a).

Litter size is directly proportional to female SVL (Ashton and Patton 2001; Diller and Wallace 1984, 2002), but neither neonate SVL nor BM is related to female length (Ashton and Patton 2001). Neonate BM is positively correlated with litter size (Diller and Wallace 1984, 2002). Short Wyoming *C. o. concolor* produce litters with a mass 17.9–53.2 (mean, 36) g and RCMs of 25.6–76.0 (mean, 50.0) % (Ashton and Patton 2001). Reported mean RCMs for *C. o. oreganus* have ranged from 29 to 64.37% (Diller and Wallace 1996, 2002; Macartney and Gregory 1988).

Literature measurements of neonates are TBL—19.0–28.0 (mean, 25.2; n = 32) cm, SVL—14.0–29.3 (mean, 20.8; n = 37) cm, and BM—3.4–22.5 (mean, 14.8; n = 41) g. Steehouder (1991) reported that the mean neonate SVL of male *C. o. oreganus* is 26.7 cm, while that of females is 27.0 cm; but mean length and BM of males and females are not significantly different at birth (Diller and Wallace 2002). Sievert (2002a) reported that neonate male *C. o. cerberus* become pitch black with yellow stripes after two weeks, while neonate females retain their birth pattern.

Hybridization of *C. o. concolor* with *C. v. nuntius* has been reported in the upper Colorado River watershed (Tanner 1958), and in captivity between *C. o. oreganus* and *C. scutulatus* in Kern County, California (Cook 1955; Klauber 1972).

GROWTH AND LONGEVITY: *C. oreganus* exhibits intrapopulational, interpopulational, and annual differences in growth. Smaller (younger) individuals grow faster than larger ones.

In British Columbia, male *C. o. oreganus* have mean SVLs of 28.5, 35.8, and 45.3 cm for years 1–3, respectively (mean first year growth, 1.3 cm/month); females average SVLs of 28.2 and 35.2 cm during the first 2 years (Macartney et al. 1990). California *C. o. oreganus* grow only an average of 22 cm, or 2.4 cm/month, during their first year (Fitch 1949a). In Idaho, mean growth of both sexes is not significantly different during the first year of life: 16.2 cm/year, or 2.5 cm/month, during the 190-day active period (Diller and Wallace 2002).

Diller and Wallace (2002) concluded that growth in length of both sexes of *C. o. oreganus* best fits a symptotic growth model. They reported average SVL growths for Idaho males and females of 18.2 cm and 20.5 cm per year and average increases in BM of 49 g and 43 g per year, respectively. These are not significantly different, but adult males in their Idaho population had an asymptotic maximum SVL of 88.8 cm and the females 69.2 cm (actual measurements gave males a maximum SVL of 96 cm and females 79.5 cm). Jenkins et al. (2009) reported growth rates of 0.012–0.022 cm per year in southeastern Idaho. Captive male *C. o. oreganus* have grown in length, on average, about 7% faster than females (Steehouder 1991).

A Utah 40 cm male *C. o. lutosus*, studied by Heyrend and Call (1951), grew an average of 20.7 cm (51%) between captures, while other males 50 cm long grew only 7.5 cm (15%), those 70 cm grew 3.8 cm (5%), and 90 cm males increased only 1.6 cm (1%). Females 40 cm long grew 13.4 cm (33%) between captures, 50 cm females grew 7.1 cm (15%), those 70 cm grew 2.7 cm (4%), and females about 79 cm grew only 1.2 cm (1.5%).

Crotalus oreganus lutosus
(R. D. Bartlett)

As expected, there is a direct correlation between snake weight gain and average prey biomass (Jenkins and Peterson 2008). Captive male and female *C. o. cerberus* fed weekly grew from an average TBL of 25.8 cm to 60 and 50 cm, respectively, in just shy of a year (Sievert 2002a).

Neonates normally undergo ecdysis within their first 10 days. First-year snakes usually shed 2–3 times during late May through September (but only once in British Columbia; Macartney et al. 1990), and 4–5 times their second year (one shed 7 times). After this the ecdysal rate slows to about 2 annual sheds after the fourth year in males, but some gravid females only shed once a year (Diller and Wallace 2002). Jenkins et al. (2009) reported 1.17–1.41 sheds per year. It is likely that adult *C. o. lutosus* and *C. o. oreganus* only shed once per activity season (Heyrend and Call 1951). In June and July *C. o. concolor* and *C. o. oreganus* form ecdysal aggregations, where adults (and possibly juveniles) shed simultaneously (Ashton 1999; Gregory et al. 1987).

Several longevity records were reported by Snider and Bowler (1992) for the various subspecies of *C. oreganus*. A wild-caught male *C. o. lutosus* of unknown age lived for an additional 19 years, 20 days at the Houston Zoo; a wild-caught *C. o. concolor* of unknown age survived 14 years, 6 months, and 28 days at the Fort Worth Zoo; a wild-caught male *C. o. oreganus* lived for 14 years, 3 months, and 10 days at the Zoo Atlanta; and another wild-caught adult *C. o. cerberus* of unreported sex lived 12 years, 5 months, and 4 days at the Phoenix Zoo. Kawata (2004) reported the following survival records at the Staten Island Zoo: *C. o. cerberus*—15 years, 11 months, and 20 days; and *C. o. lutosus*—26 years, 8 months, and 4 days.

DIET AND FEEDING BEHAVIOR: Small reptiles and mammals make up the greatest food bulk of juveniles, and possibly of the smallest subspecies, *C. o. concolor*. However, the only ontogenetic shift in prey by *C. o. oreganus* found by Wallace and Diller (1990) was an increase in mammalian prey size with TBL; snakes less than a year old took shrews, but, as they grew, more mice and voles were captured, and finally prey as large as cottontail rabbits were taken by adults. Yearlings have greater consumption rates (expressed as percent of BM) than do older age classes (Diller and Johnson 1988).

Because of its extensive north-south distribution, *C. oreganus* comes in contact with many types of potential prey, and consequently eats many species within its range. From the following list of reported prey species, it is evident that it will take almost any vertebrate of suitable size (particularly birds and mammals) it encounters: insects—crickets (*Anabrus simplex, Stenopelmatus fuscus*); fish—salmon (Salmonidae); amphibians—salamanders (*Aneides lugubris*), and anurans (*Rana pipiens; Spea hammondii, S. intermontana*); reptiles—lizards (*Cnemidophorous exsanguis, C. tigris; Eumeces gilberti, E. skiltonianus; Gambelia sila; Phrynosoma platyrhinos; Sceloporus graciosus, S. occidentalis, S. undulatus; Urosaurus ornatus; Uta stansburiana*) and snakes (*Charina bottae, Lampropeltis* sp. [captivity], *Phyllorhynchus decurtatus* [captivity]); birds—*Actitis macularia; Callipepla californica; Dendragapus obscures; Eremophila alpestris; Euphagus cyanocephalus; Gallus gallus* (egg); *Junco hyemalis;* grouse; *Mesospiza melodia; Passerculus sandwichensis; Pipilo crissalis, P. erythrophthalmus; Prooecetes gramiineus; Psaltriparus minimus; Spizella passerine; Sturnella* sp.; *Sturnus vulgaris; Zenaida nacroura, Zonotrichia leucophrys*, and a small wren; and mammals—shrews and shrew moles, mostly by juveniles (*Neurotrichus gibbsii; Sorex cinereus, S. vagrans*); rodents (*Amnospermophilus leucurus; Chaetodipus intermedius; Clethrioonomys gapperi; Dipodomys heermanni, D. ordii; Geomys bursarius;*

Lemmiscus curtatus; Marmota flaviventris; Microdipodops megacephalus; Microtus califor-nicus, M. longicaudus, M. montanus, M. pennsylvanicus; Mus musculus [captivity]; *Neo-toma cinerea, N. fuscipes, N. lepida; Ochotona princeps; Ondatra zibethicus; Peromyscus boylii, P. crinitus, P. eremicus, P. maniculatus, P. truei; Perognathus inornatus, P. parvus; Rattus norvegicus* [captivity]; *Reithrodontomys megalotis; Sciurus griseus; Spermophilus co-lumbianus, S. mollis, S. townsendii, S. varigatus; Tamias amoenus, T. merriami, T. minimus; Taimiasciurus hudsonicus; Thomomys bottae; Zapus hudsonius, Z. princeps, Z. trinotatus*); and cottontails and hares (*Lepus americanus; Sylvilagus audubonii, S. nuttallii*). It may possibly also occasionally feed on bats (Whitman *in* Klauber 1972). The production efficiency (proportion of prey BM consumed that is used for growth) is 28%. Both live and scavenged dead prey are eaten (Dornburg and Weaver 2009).

The references for the above prey species are Arno (1969); Arthur and Janke (1986); Ashton (2000); Bullard and Fox (2002); Campbell and Lamar (2004); Carl (1960); Cun-ningham (1959); Diller (1990); Diller and Johnson (1988); Diller and Wallace (1996); Dodge (1938); Dornburg and Weaver (2009); Drake (1921); C. Ernst, personal obser-vation; Evans and Holdenried (1943); Evermann (1915); Fitch (1949a); Fitch and Glad-ing (1947); Fitch and Twining (1946); Germano and Brown (2003); Gilmore (1934); Glading (1938); Hall (1929); Hill (1943); Hubbard (1941); Hulse (1973); Jenkins and Peterson (2008); M. L. Johnson 1995; Kardong (1993); Klauber (1972); Linsdale and Tevis (1951); Lowe et al. (1986); Macartney (1989); Mackessy et al. (2003); B. Martin (1974); Nussbaum et al. (1983); Pack (1930); Pauly and Benard (2002); Reed and Doug-las (2002); Richardson (1915); Schuett et al. (2002b); Stahlecker (2004), Storer and Wil-son (1932); Theodoratus and Chiszar (2000); Wallace and Diller (1990); Weaver and Lahti (2005); Woodbury (1931); Wright and Wright (1957); and Young and Miller (1980).

The diet of *C. o. oreganus* has been well researched. In British Columbia, *C. o. orega-nus* feeds most often from June to August, consuming mostly mammals (overall fre-quency, 95.6%; juveniles, 98.8%; adults, 93.5%) with rodents making up 91%; shrews, 4.9%; and birds, 4.4% of the prey consumed. Juveniles (< 65 cm) prey chiefly on deer mice (*Peromyscus*, 40.5% frequency), voles (*Microtus*, 40.5%), and shrews (*Sorex*, 11.5%), but also pocket gophers (*Thomomys*, 5.1%) and chipmunks (*Tamias*, 1.2%); snakes > 65 cm take the following prey by frequency: voles (50%), red squirrels (*Tamiasciurus*, 12.1%), deer mice (10.5%), pocket mice (*Perognathus*) and gophers (2.4% each); and chipmunks, shrews, and marmots (*Marmota*) (0.8% each) (Macartney 1989). In central Washington, small mammals (*Microtus, Perognathus, Peromyscus*) comprise 74% of the diet; lizards, 20%; and birds, 6% (Weaver and Lahti 2005).

In California, the ground squirrel (*Spermophilus beecheyi*) is its chief prey, followed by kangaroo rats (*Dipodomys*), cottontails (*Sylvilagus*), white-footed mice (*Peromyscus*), pocket mice (*Perognathus*), and pocket gophers (*Thomomys*); mammals predominate, but some birds, lizards (only about 9% of prey; Jaksic and Greene 1984), and spadefoot toads are also eaten (Fitch and Twining, 1946). Juvenile ground squirrels, *Spermophilus townsendii*, comprise 53.3% of prey consumed in southwestern Idaho, followed by *Peromyscus maniculatus* (20.6%), *Dipodomys ordii* and *Perognathus parvus* (6.4% each), and *Sylvilagus nuttallii* (5.7%) (Diller and Johnson 1988). In northern Idaho, voles (*Microtus*, 57.7% occurrence), cottontail rabbits (*Sylvilagus*, 17.8%), and deer mice (*Peromyscus*, 16.2%) make up almost 92% of the biomass ingested and 80% of the total prey taken, followed by shrews (*Sorex*, 2.8%), pocket gophers (*Thomomys*, 2.3%), harvest mice

(*Reithrodontomys*, 1.2%), birds (*Melospiza*, 0.7%), and lizards (*Eumeces*, 0.2%) (Diller and Johnson 1988; Wallace and Diller 1990).

Prey reported for *Crotalus o. oreganus* by Bullard and Fox (2002), Campbell and Lamar (2004), Diller and Wallace (1996), Fitch and Twining (1946), Klauber (1972), and others (see previous list of references) are spadefoot toads (*Spea hammondii*), lizards (*Cnemidophorus tigris, Eumeces gilberti, Sceloporus occidentalis, Uta stansburiana*), snakes (*Charina bottae, Phyllorhynchus decurtatus* [captivity]), birds (*Callipepla californica, Gallus gallus, Melospiza melodia, Passerculus sandwichensis, Pipilo crissalis, Pooecetes gramineus, Zenaida macroura, Zonotrichia leucophrys*), moles and shrews (*Neurotrichus gibbsii, Sorex cinereus*), muskrats (*Ondatra zibethicus*), pika (*Ochotona princeps*), mice (*Mus musculus; Peromyscus boylii, P. maniculatus; Perognathus inornatus, P. parvus; Reithrodontomys megalotis; Zapus hudsonius, Z. princeps, Z. trinotatus*), voles (*Clethrionomys gapperi, Microtus californicus*), rats (*Neotoma fuscipes, Rattus norvegicus*), kangaroo rats (*Dipodomys heermanni, D. ordii*), squirrels (*Marmota flaviventris; Sciurus griseus; Spermophilus beecheyi, S. columbianus, S. townsendii, S. varigatus; Tamias merriami*), gophers (*Thomomys bottae*), and cottontails and hares (*Lepus americanus; Lepus* sp.; *Sylvilagus audubonii, S. nuttallii*).

Young *C. oreganus* are more likely to consume lizards, while adults seem to prefer mammal prey (Wallace and Diller 1990), and an ontogenetic change in venom composition to accommodate the prey switch accompanies growth. They are capable of ingesting rather large prey for their body size; Pauly and Benard (2002) found a 29.65 cm (TBL), 11.91 g (prey removed) juvenile that had swallowed a 15.37 cm (TBL), 9.04 g (relative prey mass, 0.76) *Sceloporus occidentalis*, but the snake was severely injured while swallowing the lizard.

Crotalus o. abyssus is known to feed on fence lizards (*Sceloporus* sp.) and sandpipers (*Actitis macularia*) (Klauber 1972; Stahlecker 2004). It probably also eats small mammals.

Lizards (*Sceloporus undulatus*) and birds (*Dendroica* sp. [?], *Mimus polyglottos, Sialia mexicana*) have been reported as prey of *C. o. cerberus* (Klauber 1972; Schuett et al. 2002b), but it probably also takes small mammals. A 23.5 cm (TBL), 17 g (prey intact) neonate swallowed a 9.2 cm (TBL), 9.05 g *Sceloporus undulatus* (Schuett et al. 2002b).

Crotalus o. concolor is known to prey on lizards (*Cnemidophorus tigris, C. velox; Sceloporus graciosus, S. undulatus; Sceloporus* sp.; *Uta stansburiana*) and rodents (*Peromyscus maniculatus, Reithrodontomys megalotis, Tamias minimus*) (Houston 2006; Klauber 1972; Mackessy et al. 2003).

Crotalus o. lutosus has an extensive prey list (Jenkins and Peterson 2008; Klauber 1972): lizards (*Sceloporus graciosus*), birds (*Callipepla californica, Eremophila alpestris, Euphagus cyanocephalus, Gallus gallus* [egg], grouse, *Pooecetes gramineus*, a small wren, *Spizella passerina*, "woodhouse[?]" jay), mice (*Onchomys leucogaster, Peromyscus maniculatus, Perognathus parvus, Reithrodontomys megalotis*), voles (*Clethrionomys gapperi, Microtus* sp.), woodrats (*Neotoma* sp.), kangaroo mice and rats (*Dipodomys* sp., *Microdipodops megacephalus*), squirrels (*Amnospermophilus leucurus; Spermophilus columbianus, S. mollis, S. townsendii; Tamias minimus; Tamiasciurus hudsonicus*), gophers (*Thomomys bottae*), and young cottontails (*Sylvilagus* sp.) and jackrabbits (*Lepus* sp.).

In most populations of the species, gravid females usually do not feed during gestation or after parturition, but Wallace and Diller (1990) reported that in Idaho reproductive females feed from the spring to early August, and again in the fall after the birth of the young. They thought feeding during the breeding year may explain how some females can reproduce in consecutive years.

C. oreganus is either an active forager, or a sit-and-wait ambusher (Diller 1990). Prey is detected by vision, olfaction, and thermal sensory cues. Hungry snakes that have been deprived of food for some time tongue-flick more often than those that have fed more frequently (Hayes and Hayes 1993). Once prey is detected, visual and thermal cues direct the strike, but the latter are more important for a successful one (Hayes and Hayes 1993); but see below. When the rattlesnake approaches *Spermophilus beecheyi* head on, it behaves as if the ground squirrel has resisted moving away from its advance, and uses the squirrel as the center of a radial search pattern (Hennessy and Owings 1988).

A predatory strike by *C. oreganus* is completed in less than 0.5 second (Kardong and Bels 1998), and can be successful even if the snake is blind as long as its heat sensory pits are not occluded (Kardong and Mackessy 1991). Fang contact lasts only 0.07–0.53 second (Herbert 1998). When striking, the snake's body is formed into several lateral curves and held off the ground. The remainder of the body on the ground serves as an anchor from which the strike is launched. As the strike begins, the lateral curves are straightened to speed the head toward the prey. At the same time, the snake's jaws open by depressing the lower jaw, there is an upward pivoting of the skull to open the upper jaw, and there is a horizontal axis rotation in the neck. Simultaneously, the two fangs are rotated into an erect stabbing position by protraction of the palatomaxillary arches. At contact, the upper jaw immediately closes, plunging the fangs into the prey. The neck is arched upward, apparently allowing the fangs to follow their own curvature deeper into the prey (Kardong 1986a, 1986b). The prey is then released; seldom is it retained if large. The snake does not always make a clean strike, as some strikes are flawed (Kardong 1986b). The head or chest is struck most often, and envenomation of these areas most quickly leads to death (possibly due to mechanical damage caused by the snake's fangs) (Kardong 1996). Movement either forward or backward brings on a strike, but most strikes are directed at forward-moving prey (Schmidt et al. 1993).

Some small prey may simply be swallowed alive without envenomation by larger snakes. Larger prey is usually released immediately, but smaller prey may be retained for a short while. Retention of prey following a bite increases the severity of the envenomation, and a poor strike is usually followed by a second or third. However, holding prey also increases the chance of a retaliatory bite with possible injury to the snake. An envenomated mouse may travel a longer distance than one not bitten (Hayes and Galusha 1984), but the distance traveled does not differ significantly between the mice *Mus musculus* and *Peromyscus maniculatus* (Kuhn et al. 1991).

The snake apparently waits for the venom to take effect, and later follows the wounded prey by its airborne odor trail (Kardong and Smith 2002; Parker 2005; Parker and Kardong 2005). Apparently, the odor of the snake's own venom is the major cue. The snake only needs to inject venom from one fang to gain the necessary information it needs for trailing, and abandonment of a poorly differentiated prey odor trail may represent a snake's response to detection of a poorly envenomated prey (Lavín-Murcio and Kardong 1995; Lavín-Murcio et al. 1993). The odors of the prey's blood or urine contribute little (Smith and Kardong 2005). The vomeronasal organ plays a major role in odor reception during poststrike prey-trailing (Alving and Kardong 1996), and if the organ is occluded trailing behavior becomes modified and there is a lower tongue-flick rate (see *C. viridis*). Hours may be required to find the envenomated prey if it has traveled far before death; if the trail ages significantly, losing its perceptibility, poststrike trailing may be futile (Smith et al. 2000). High prey odor concentrations bring on the most successful trailing behavior (T. L. Smith et al. 2005), but the snake must be able to discriminate between the odor trails of nonenvenomated prey and that which it has struck. The snake may use polarized chemical cues while searching for prey, but not recognize or use these while trailing envenomated prey (Smith and Kardong 2000). The forking of the snake's tongue is apparently an adaptation that helps pick up chemosensory odors by increasing the sampling area (Parker et al. 2008). Captives kept in small quarters may have their ability to trail an odor diminished, but they can regain it if returned to a situation where they must trail their wounded prey (Busch et al. 1996; Chiszar et al. 1999a). Once the dead prey is found, swallowing may begin at either the head or the tail end (Evermann 1915).

Predators and prey evolve together over time. California ground squirrels, *Spermophilus beecheyi*, have evolved several defensive behaviors and physiological responses toward predatory *C. oreganus* (Clucas et al. 2008; Hennessy and Owings 1988; Owings and Coss 2008; Owings et al. 2001; Pain 1999; Rowe and Owings 1990, 1996; Rundus and Owings 2005; Swaisgood et al. 1999a, 1999b). The ground squirrels watch the snake and usually tail-flag during this behavior. Apparently they determine how much danger they are in by assessing the size of the snake (more caution is displayed toward large ones). The squirrel may change its position several times to better observe the predator as it moves about. If the snake is crawling, the squirrel is usually following, and sometimes approaches close to its tail as it passes through the squirrel colony. When the snake coils, it is inspected and probed, sometimes by several squirrels at once (Owings and Coss 1977). Squirrels will repeatedly approach unmoving, coiled snakes more to inspect than to harass them. If the snake can be induced to rattle, the squirrel can assess whether or not it is warm by the rate of rattling; the faster the rattle rate, the warmer the snake, and the more danger to the squirrel (Rowe and Owings 1978; Swaisgood et al. 2003). Smaller or cooler *C. oreganus* are probably in more danger from an attack by the squirrel; however, a nonrattling snake may be mistaken for a nonvenomous one and attacked vigorously, putting the squirrel in greater danger. *S. beecheyi* from areas where rattlesnakes are absent or rare and their burrows are shared with the burrowing owl (*Athene cunnicularia*) may not be able to differentiate the sound made by the snake's rattling from that of the owl's hissing (Pain 1999; Rowe et al. 1986), putting themselves in greater danger from a rattlesnake encounter.

The squirrels become more interactive as the snake approaches a den. They repeatedly move toward and back away from the snake, kick loose substrate at it (females with nearby young), and avoid the snake's strike by launching themselves into the air and away from the strike. A direct correlation occurs between the squirrel's interactions with the snake and how closely the snake approaches the squirrel's burrow. If the snake is moving directly toward the rodent and is within 3 m of its burrow an interaction may develop, but more often it does not. Instead, the squirrel remains stationary in front of the snake's advance, as if to minimize affording the snake information about the burrow location. Hennessy and Owings (1988) thought that the mammal engages the snake before it discovers the burrow because, if it contains unweaned young, they are at greater risk of being predated. *S. beecheyi* is able to differentiate the infrared-sensitive rattlesnake from infrared-insensitive gopher snakes (*Pituophis cateni-*

fer) (Rundus and Owings 2005). It heats up its tail and flags it at the rattlesnake, probably to either lure it away from the burrow or possibly to direct the snake's strike away from the squirrel's body or head. Interestingly, *S. beecheyi* and *S. variegatus* have also evolved an immunological resistance to the venom of *C. oreganus* (Biardi 2000, 2006, 2008; Poran et al. 1987). In addition, the snake's own scent may be used by *S. beecheyi* and *S. variegatus* as a deterrent against predators. The snake's scent is transferred to the rodent's fur by chewing shed rattlesnake skins and then licking the fur; no antiparasite or conspecific function has been identified for this behavior (Clucas et al. 2008).

VENOM DELIVERY SYSTEM: Klauber (1939) reported fang dimensions for 4 subspecies: (1) *abyssus* (n = 5)—FL, 5.3–8.4 (mean, 6.5) mm; mean TBL/FL, 121; mean HL/FL, 5.40; and mean angle of fang curvature, 73°; (2) *concolor* (n = 5)—FL, 4.1–5.2 (mean, 4.7) mm; mean TBL/FL, 132; mean HL/FL, 5.75; and mean angle of fang curvature, 60°; (3) *lutosus* (n = 5)—FL, 6.0–8.8 (mean, 7.3) mm; TBL/FL, 131; mean HL/FL, 5.49; and mean angle of fang curvature, 61°, and (4) *oreganus* (n = 6)—FL, 6.2–7.5 (mean, 6.6) mm; mean TBL/FL, 132; mean HL/FL, 5.96; and mean angle of fang curvature, 60° (probably calculated from a combination of *C. o. oreganus* and *C. helleri*). He also reported measurements for 6 "Arizona" specimens, possibly including *cerberus*, as follows: FL, 6.6–9.6 (mean, 8.4) mm; mean TBL/FL, 115; and HL/FL, 4.93. Penetration by a fang may cause considerable mechanical damage and directly kill struck prey (Kardong 1996).

VENOM AND BITES: Unfortunately, much data concerning the venom of *C. oreganus* has been hidden in papers in which data for the species were either listed as, or combined with, *C. viridis*. Consequently, papers listing only *C. viridis* as the venom donor are discussed under that species, and only those papers specifically listing the donor as *C. oreganus* or one of its five subspecies are discussed here.

The average (maximum or range) of dry venom yield (mg) for the subspecies are as follows: *abyssus*, 60–97 (maximum, 137); *cerberus*, 89–112 (maximum, 150); *concolor*, 9–22 (range, 6–34); *lutosus*, 65–110 (range, 16–240); and *oreganus*, 90 (range, 190–289[?]) (do Amaral 1928; Githens 1933; Glenn and Straight 1977, 1982; Hayes et al. 1992b; Klauber 1972; Russell and Brodie 1974; Russell and Puffer 1970). Klauber (1972) reported an average yield of 112 mg and a maximum of 289 mg for a combined venom sample from *C. helleri* and *C. oreganus*.

The average and range of dry venom (mg) injected during a single defensive bite by *lutosus* and *oreganus* are 33 (range, 16–50) and 64 (range, 41–93), respectively; predatory strikes by *oreganus* average 15 (range, 1–31) mg of dry venom (Hayes et al. 2002; Herbert 1998). Successive venom extractions yield lower quantities and diminished toxicities; the second milking of nine *oreganus* yielded only about 50% of the liquid volume first achieved, and a third milking only 6% of the first (Klauber 1972).

C. oreganus is equipped at birth with enough venom to kill small prey, and a typical adult has enough reserve venom to kill up to four mice in close succession without loss of venom effectiveness, but when presented several mice in succession, the later mice are often struck repeatedly or held in the mouth rather than released (Kardong 1986a, 1986b). This behavior probably results from either the depletion of, or reduced access to, venom reserves (Hayes 1991b; Hayes et al. 1992b).

The venom of *C. o. concolor* is neurotoxic. Venoms of the other subspecies are largely hemotoxic. Enzymes identified in or other activities recorded from, the venoms of the subspecies of *C. oreganus* include a "plethora" of small peptides, azocaseinase, esterase activity (at least six proteases, including arginine esterase, kallikrein-like protease, plasmin-like protease, thrombin-like protease), dipeptidal peptidase IV (with some geographic variation in *C. o. lutosus*), exonuclease, factor X activator, hemorrhagic toxin (HPA metalloprotease), L-amino acid dehydrogenase, Mojave toxins α and β, 5′-nucleotidase, peptide inhibitors ($_p$ENW, $_p$EQW), phosphodiesterase, phospholipase A$_2$ (as *concolor*-toxin, a β-neurotoxin, related to crotoxin) (Pool and Bieber 1981), canebrake-toxin, and *vergrandis*-toxin (Kaiser et al. 1986, in *C. o. concolor*), phosphomonoesterase (alkaline activity), and presynaptic neurotoxin (*C. o. concolor*) (Aird 1985, 2005, 2008; Aird et al. 1988; Bober et al. 1988; Glenn and Straight 1977, 1978, 1985a, 1990; Hayes and Bieber 1986; Henderson and Bieber 1986; Mackessy 1988, 1993a, 1993b, 1996, 1998; Mackessy et al. 2003; Martínez et al. 1990; Mashiko and Takahashi 1998; Meier and Stocker 1995; Munekiyo and Mackessy 2005; Ouyang et al. 1992; Powell et al. 2008; Quinn 1987; Rael et al. 1986; Ramírez et al. 1999; Soto et al. 1989; Tan and Ponnudurai 1991; Weinstein et al. 1985). Metalloprotease activity is approximately five times greater in adults than in younger individuals (Mackessy 1988). Although Rael et al. (1986) reported that the venom of most subspecies does not contain Mojave toxin, Bober et al. (1988) confirmed its presence in varying amounts (mg/mL) in the venoms of *abyssus* (> 2.0), *cerberus* (0.7), *concolor* (> 2.0), and *oreganus* (> 2.0), but not in *lutosus*. Phospholipase A$_2$ activity is absent from the venom of *C. o. abyssus* (Adame et al. 1990), but it is present in the venom of *C. o. concolor*, which has the most virulent venom of the subspecies (Glenn and Straight 1977, 1978).

The electrophoretic protein patterns of the venoms vary considerably between individuals of *C. o. oreganus*, but do not seem to be the result of differences in geographic location or sex. The patterns remain identical for the same individual from the time it emerges from hibernation through the annual activity period (E. K. Johnson 1987).

The ontogenetic change in diet composition previously discussed is associated with corresponding changes in venom composition. Phospholipase A$_2$ activity decreases significantly with size, as does also venom toxicity in *C. o. oreganus*. Protease activity increases significantly with snake length, and is about five times higher in adults than in juveniles; L-amino oxidase and exonuclease activities also tend to increase (Mackessy 1988, 1993a). However, the venom ontogeny is different in *C. o. concolor* and *C. o. oreganus*, probably due to the differences in venom composition. The venom of *C. o. concolor* contains *concolor*-toxin, a phospholipase A$_2$-based ß-neurotoxin. It is absent from the venom of *C. o. oreganus*, which is hemotoxic in effect. Mouse LD$_{50}$ assays indicate no toxicity differences between adult *C. o. concolor* (0.38 μg/g) and juvenile (0.45 μg/g) venoms. Metalloprotease activity is extremely low, but the levels of peptide myotoxins and several serine proteases were positively correlated to snake length. Apparently, the presence of potent neurotoxic components in venom minimizes predigestion metalloproteases (Mackessy et al. 2003).

The specific gravity of the subspecies *concolor* and *oreganus* is usually below 1.08, while that of venom from *abyssus* and *lutosus* is intermediate between 1.07 and 1.08 (Klauber 1972).

Venom lethalities have been determined for several test animals. *C. o. lutosus* is not immune to its own venom; 21 of 25 (84%) given intramuscular injections died. The minimum lethal dose was between 0.024 and 0.05 mL/10 g BM, and one of 0.04 mL/ 10 g BM was lethal; 4 snakes survived injections of 0.025–0.045 mL. Three factors possibly contributed to the lethality of the venom: (1) loss of blood tissue from circulation into tissue space, possibly causing a drop in blood pressure; (2) the effect on the cellular portion of the blood, particularly the erythrocytes, disturbing their normal function and possibly destroying them; and (3) extensive necrosis around the site of injection (Sanders 1951). In contrast, La Rivers (1973) reported that *C. o. lutosus* survived the bites by other individuals of its own subspecies and of *C. o. oreganus*; perhaps a lesser dose of venom was injected in these cases.

The minimum intravenous venom lethal dose from the subspecies *lutosus* and *oreganus* is 0.06 mg for a 350 g pigeon (*Columba livia*), and the estimated median lethal doses for *abyssus* and *cerberus* are 0.06 mg and 0.10 mg, respectively (Githens and George 1931).

Reported means or ranges of LD_{50} values for mice (mg/kg) by subspecies are intramuscular—*cerberus*, 6.0; *concolor*, 0.8–1.3; *lutosus*, 4.0–4.6; and *oreganus*, 3.6; intraperitoneal—*abyssus*, 4.6; *cerberus*, 2.4; *concolor*, 0.13–0.45; *lutosus*, 1.9–6.4; and *oreganus*, 3.2; and intravenous—*concolor*, 0.28–0.48; *lutosus*, 2.01; and *oreganus*, 2.84 (Aird and Kaiser 1985; Githens and Wolff 1939; Glenn and Straight 1977, 1982; Khole 1991; Macht 1937; Russell 1967b, 1983; Russell and Brodie 1974; Russell and Puffer 1970).

Hemorrhagic activity occurs in less than 20 minutes in mice, and its intensity varies between populations of *C. o. lutosus*. Venoms from snakes occupying northern ranges in Utah have high hemorrhagic ability, while snakes from southern Utah and northern Arizona have a lower hemorrhagic capability (Adame et al. 1990). Neurologic symptoms may be paralytic, but are not as severe as those produced from bites by *C. scutulatus*. Venom from *C. v. concolor* has an amino acid sequence similar to that Mojave toxin (Aird et al. 1990a), and is 10–30 times more toxic than that of any other subspecies (Glenn and Straight 1977).

Human envenomations have resulted in altered taste sensations; anxiety; bleeding from the puncture wound (initially); blood coagulation (caused by diminished fibrinogen and platelet destruction); bloody urine; breathing difficulty; cardiac arrest; chills and sweating; diarrhea; discoloration around the bite site; decreases in both red and white blood cells; difficulty in concentrating; cyanosis of lips and face; ecchymosis (hemorrhagic blisters); edema and swelling (including at the bite site and pulmonary); elevation of serum calcium; phosphorus, and potassium; general depression; increased heart rate; infection at bite site; insomnia (from pain); muscle twitching; hysteria (?); necrosis and sloughing around the bite site; numbness (digits, mouth, tongue); pain (burning initially); paralysis of limb bitten; rash (possibly from antivenom); shock; skeletal muscle damage (rhabdomyolysis); slight elevation of BT; swollen and tender lymph nodes; temporary blindness; thirst (possibly severe); tingling sensation over body; vomiting; and weakness with reduced coordination (Butner 1983; Cartwright 1928; Dart and Gustafson 1991; Davidson 1988; Gold and Wingert 1994; Holstege et al. 1997; Kitchens 1992; Klauber 1972; La Grange and Russell 1970; Mackessy 1996; Norris 2004; Norris and Bush 2001, 2007; Russell 1980a, 1980b, 1983; Silvani et al. 1980; Straight and Glenn 1993; Whitlow et al. 2007). Case histories and statistics of human envenomations are found in Klauber (1972), Offerman et al. (2002), Russell and Picchioni (1983), Whitlow et al. (2007), Woodbury (1945), Woodbury and Anderson (1945), and Young and Miller (1980). Envenomation by *C. o. lutosus* has caused death in 4–5 hours (Straight and Glenn 1993); other fatal bites by the species have been reported by Hammerson (1986), Norris and Bush (2007), and Nussbaum et al. (1983). A Utah man was "squirted in the eye" while skinning a dead *C. o. lutosus* (Plowman et al. 1995).

PREDATORS AND DEFENSE: Known predators of *C. oreganus* are fish (*Oncorhynchus mykiss*), bullfrogs (*Rana catesbeiana*), lizards (*Gambelia wislizenii* [captivity]), snakes (*Charina bottae* [captivity]; *Coluber constrictor; Lampropeltis getula; Masticophis lateralis, M. taeniatus; Pituophis catenifer*), eagles (*Aquila chrysaetos, Haliaeetus leucocephalus*), roadrunners (*Geococcyx californicus*), ravens (*Corax corax*), hawks (*Buteo jamaicensis, B. regalis*), great horned owl (*Bubo virginianus*), jays (*Aphelocoma coerulescens*), hogs (*Sus scrofa*), marmots (*Marmota flaviventris*), ground squirrels (*Neotoma cinerea*, during hibernation torpor), bears (*Ursus americanus*), mustelids (*Mephitis mephitis, Taxidea taxus*), canids (*Canis familiaris, C. latrans; Vulpes vulpes*), and bobcats (*Lynx rufus*). (Arno 1969; Ashton 2000; Basey 1988; Campbell and Lamar 2004; Carnie 1954; Cobb and Peterson 1999; Ferrel et al. 1953; Fitch and Glading 1947; Follett 1927; Hornocker et al. 1978; Klauber 1972; Knight and Erickson 1976; Mauldin 1968; Sperry 1941; Steenhof and Kochert 1985; Toweill 1982; Wilcox 2005; Wilcox and Van Vuren 2009; Woodbury 1931, 1952). Possibly the California mountain kingsnake (*Lampropeltis zonata*), feral cat (*Felis catus*), and long-tailed weasel (*Mustela frenata*) are also predators.

In addition, cattle (*Bos taurus*), goats (*Capra hircus*), pronghorns (*Antilocapra americana*), deer (*Odocoileus hemionus, O. virginianus*) and wild horses (*Equus caballus*) trample them (Klauber 1972), and Metter (1963) found one with the quills of a porcupine (*Erethizon dorsatum*) protruding from its head.

When confronted with a kingsnake (*Lampropeltis* sp.), *C. oreganus* will flee, freeze in place, hide its head, body-bridge, or strike and bite (Bogert 1941; Cowles 1938; Weldon and Burghardt 1979; Weldon et al. 1992).

In addition to identifying potential predators by infrared detection, olfaction, and vision, *C. oreganus* has some hearing ability. Young and Harris (2006) recorded the tongue-flick rates of *C. o. oreganus*, and whether or not the snake froze, jerked its head, or rattled in response to several airborne sound frequencies. It did not respond to silence or a sound at 6,000 Hz at 60 dB, but generally did respond to sounds ranging in frequency from about 50 to 4,000 Hz and 55 to 90 dB (one response at 40 dB, but mostly > 60 dB).

The snake's heart rate rises from 16–20 to as high as an average of 47.3 beats per minute during defensive behaviors, and its fear rate is 34–60 beats per minute (Hayes et al. 1994). Its temperament varies from docile to highly nervous and quick to strike (it can strike for a distance of at least 33–40% of its TBL, C. Ernst, personal observation; strike kinematics are discussed by Kardong and Bels 1998); a few respond to an intruder by uncoiling and directly attacking it (Hall 1929).

The snake's ground color and pattern may provide some concealment by matching the habitat background, and it may be mimicked by *Pituophis catenifer* (Harris 2006b; Sweet 1985).

PARASITES AND PATHOGENS: Several parasitic worms have been found associated with *C. oreganus*. Tetrathyridia larvae of the cestode *Mesocestoides variablis* were encysted in the intestinal mesenteries and body wall of *C. o. oreganus* (Goldberg and Bursey 2004; Voge 1953). Goldberg and Bursey (2004) also found cystocanths of oligacanthorhynchid acanthocephalans in *C. o. cerberus*. The digestive tract nematode *Physaloptera obtussima* also infects this *C. o. oreganus* (Morgan 1943). Surprisingly, Woodbury and Parker (1956) found no parasitic mites on the *C. o. lutosus* at a den where the colubrid snakes *Masticophis taeniatus* and *Pituophis catenifer* both harbored the ectoparasites.

The bacteria *Arizona hinshawii* and paracolon type 10 (Pc27) have been isolated from captive *C. o. cerberus* and *C. o. oreganus*, and *C. o. abyssus* and *C. o. concolor* are susceptible to bacterial respiratory diseases (possibly caused by *Aeromonas hydrophila*) (Hinshaw and McNeill 1946; Murphy and Armstrong 1978). Captive *C. o. oreganus* have also died from a paramyxovirus infection (Jacobson and Gaskin 1992). Nieto et al. (2009) have isolated *Anaplasma phagocytophilum*, the cause of the potentially fatal rickettsial disease granulytic anaplasmosis, from *C. o. oreganus* in northern California. The disease is transmitted through bites by the tick *Ixodes pacificus*.

POPULATIONS: Currently, populations of *C. oreganus* may still be fairly large in undisturbed areas, but others have decreased in numbers, and many have been extirpated. Most population data come from the capture of snakes at hibernation dens.

The population of *C. o. lutosus* using a hibernaculum in Tooele County, Utah, was studied from 1939 to the early 1970s by Woodbury (1951), Hirth and King (1968), Parker and Brown (1973, 1974), and Brown and Parker (1982). Woodbury (1951) and his students (see the first issue of volume 7 of *Herpetologica*, 1951) marked 930 *C. o. lutosus* in 1939–1950. Population size was estimated to be 769 snakes in 1940, and the rattlesnakes made up 2,010 (61.2%) of the total 3,285 snake records for the den. However, the numbers of surviving individuals of *C. o. lutosus* taken in successive years were reduced after 1940. The numbers of *C. o. lutosus* using the den had dwindled to only 235 by 1950, in the mid-1960s to only 53–58 (Hirth and King 1968), and by 1969–1972 the population had dropped to only 12–17 individuals (Parker and Brown 1974). A different Tooele County hibernaculum yielded only 4 *C. o. lutosus*, although 1 *Pituo-*

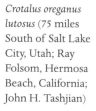

Crotalus oreganus lutosus (75 miles South of Salt Lake City, Utah; Ray Folsom, Hermosa Beach, California; John H. Tashjian)

phis catenifer and 49 Masticophis taeniatus also used it (Woodbury and Parker 1956). At another den used by C. o. lutosus in Juab County, Utah, 69 rattlesnakes were collected in 1949 (54.3% of 127 snakes recorded), where in 1937 an estimated 300 rattlesnakes were killed while the entrance to the den was blasted in an effort to seal it (Woodbury and Hansen 1950). Hall (1929) visited a hibernaculum in White Pine County, Nevada, on 22 May 1929, and collected 25 C. o. lutosus for museum specimens. Hall also reported that persons visited the den several times in the previous 5 years to kill the C. o. lutosus there, and on one occasion shot 139 rattlesnakes; a lesser number had been killed on several other days. Of 258 snakes observed on a 170 km loop in the eastern Snake River Plain of Idaho during 55 trips, C. o. lutosus comprised 19%; males were observed more often than any other age or sex class (Jochimsen and Peterson 2004).

Graf et al. (1939) reported that in the 1930s, C. o. oreganus was the most common snake in the butte country along the western foothills of the Cascade Mountains in Oregon, but its numbers had been much reduced. In 1938, only 3 C. o. oreganus were taken, although shed skins were plentiful at a den where they were told 97 had been killed in April 1936 and 27 in 1937. They also reported that more than 100 rattlesnakes had been killed by a road crew while blasting a rock ledge. Johnson (1942) reported that with the encroachment of civilization the subspecies was becoming much less common in the part of Washington east of the Cascades. At a den in Okanogen County, 201 individuals were collected in 4 days, and almost always 100 or more C. o. oreganus were found per trip (Martin 1930).

In spite of these negative reports, some populations of C. o. oreganus are still fairly large. Fitch (1949a) reported a density of approximately 2.9/ha at his California site (males, 1.3/ha; females, 1.6/ha), where individuals with 25–50 cm SVLs made up 34.2% of the population, and adults longer than 70 cm comprised 52.9% of the snakes. This subspecies comprised 454 (45.9%) of the snakes recorded from 1975 to 1980 within and near the Snake River Birds of Prey Area in southwestern Idaho (Diller and Wallace 1996), and 1,799 at three close hibernacula in southeastern Idaho (Jenkins et al. 2009). At a hibernaculum in north-central Idaho, Wallace and Diller (2001) identified 493 individuals and made 578 additional observations during which the individual identities of the rattlesnakes were not determined. In addition, Diller and Wallace (2002) marked 319 C. o. oreganus during field studies there. Weaver and Lahti (2005) collected 304 of the snakes, either alive or dead, for diet studies on roads in a population's habitat in central Washington. In the south Okanagan Valley of British Columbia, Parsons and Sarell (2005) marked 304 individual C. o. oreganus in 14 months in an important habitat there, but the area is experiencing extensive development that will impact the local snake population; during the same period 21 rattlesnakes were killed and 1 human envenomation occurred.

Three small island populations of C. oreganus have been documented (Klauber 1972). C. o. lutosus is still present on Anaho Island in Pyramid Lake, Washoe County, Nevada, evidenced by eight individuals collected in 1999 (Ashton 2000). However, the C. o. oreganus on Morro Rock, San Luis Obispo County, California, have apparently been extirpated, possibly due to human disturbance and predation by feral cats (Felis catus). The subspecies has also disappeared from Rattlesnake Island in Clear Lake, Lake County, California, because of either predation or habitat destruction by introduced hogs (Sus scrofa) (Ashton 2000; Mauldin 1968).

C. o. abyssus is probably still common in some isolated feeder canyons off the floor of the Grand Canyon, Arizona, but it does not seem to be as plentiful in other areas as in the past. At Indian Gardens, 3–4 individuals could be found in a day during the 1930s and 1940s, but the snake is not as common there today; Young and Miller (1980) contribute this decline to disturbance by tourists. Lowe et al. (1986) reported that in Arizona *C. o. abyssus*, *C. o. cerberus*, and *C. o. concolor* are "neither abundant nor uncommon."

The adult male to female ratio of adult *C. o. oreganus* and *C. o. lutosus* is typically not significantly different from 1:1 (Ashton and Patton 2001; Diller and Wallace 1984, 2002; Jenkins et al. 2009; Klauber 1943a), but exceptions occur. At a north-central Idaho hibernaculum, Wallace and Diller (2001) found the sex ratio of 493 *C. o. oreganus* to be 0.91:1.00; and Julian (1951) reported a 1.47:1.00 sex ratio for *C. o. lutosus* in a Tooele County, Utah, hibernaculum where males always outnumbered females in each of the years from 1940 to 1949. The original sex ratio recorded there was 2.62:1.00 (168 males, 64 females). Later ratios at the den were 0.88:1.00 in 1964, 1:1 in 1965, 0.92:1.00 in 1966, 2.40:1.00 in 1969–1970, and 2:1 in 1970–1971 and 1971–1972 (Hirth and King 1968; Parker and Brown 1973).

The juvenile to adult ratio may vary from 1:1 or 1:2 (Diller and Wallace 1984; Macartney 1985) to 2:1 (Fitch 1949a). The ratios of gravid to nongravid females at Wyoming hibernacula of *C. o. concolor* varied from 2:1 to 5:4, and, if the smallest maturing female was used in the calculation, 2:1 to 5:8 (Ashton and Patton 2001).

More adult Idaho *C. o. oreganus* survive annually than do juveniles or neonates. Diller and Wallace (2002) reported that survival time decreased linearly on a log scale over time, but differed in magnitude in the three age classes. It decreased from a high in 1982 (neonates, 0.666; immature, 0.768; adults, 0.820) to a low in 1990 (neonates, 0.350; immatures, 0.472; adults, 0.552) (survival estimates for neonates only included the time from birth to spring emergence, while survival estimates for the other two age classes was annual survival). Recaptures by Charland (1989) at two hibernacula in British Columbia suggested a neonate minimum survivorship during the winter 1985–1986 of about 55%, and that survivorship was independent of either neonate BM or condition at birth. Charland thought that any survival advantage conferred by size at birth either varies annually or occurs during the first active season when food for small individuals may be difficult to obtain.

Humans and their activities are the greatest threat to *C. oreganus*. Disturbance of its habitat, particularly of sagebrush through burning (both human-induced and natural), development (including road-building), and overgrazing eliminates shelters and reduces availability of prey species (Jenkins and Peterson 2008). Humans also impact populations of *C. oreganus* through collection for the pet and venom trades, wanton killing, and destruction on our highways. Most road mortality occurs during the fall and spring when the snakes are moving to and from their winter quarters (Jochimsen and Peterson 2004).

Chemical contamination may occur at some sites. Three *C. o. lutosus* collected at a radioactive waste storage and disposal area in southeastern Idaho were contaminated with Cs^{137}, possibly gained through predation on *Peromyscus maniculatus* (Arthur and Janke 1986).

To keep encounters of *C. o. oreganus* and humans at a minimum in southern British Columbia, J. R. Brown et al. (2005) have instituted a short-distance translocation project to move the snakes to nearby areas of natural habitat. Twenty *C. o. oreganus* were captured in areas of human activity, fitted with implanted radio-transmitters, and released in natural habitat at distances of 0.2–1.0 km from their initial capture locations. The translocated snakes experienced an increase in summer movements, with movement distance increasing with translocation distance. Observed mortality (none) was significantly lower compared to that in areas of human activity during the same period (8 snakes, all through human interaction). However, 71% of the transplanted snakes returned to their area of initial capture; the mean time of return decreased with increasing translocation distance (average, 19 days at 0.5 km; 16 days at 1.0 km). Brown et al. thought that short-distance translocation does negatively influence the snakes through increased movement distances (also more use of energy reserves?), but has the benefit of a lower mortality rate. They concluded the strategy effective to gain "snake free" time in areas of human interaction, but it does not entirely solve the problem of nuisance rattlesnakes in areas of human activity.

C. o. concolor is considered endangered in Colorado and rare in Wyoming, and *C. o. lutosus* is a controlled reptile in Utah. Jenkins and Peterson (2008) have developed a trophic-based conservation plan for *C. o. lutosus* in Idaho (which see for details).

REMARKS: The systematics and relationships of the taxa in the *Crotalus viridis*-complex have been studied by Ashton and de Queiroz (2001), Douglas et al. (2002), Glenn and Straight (1977), Murphy et al. (2002), Pook et al. (2000), and Quinn (1987); these papers are discussed in that species' account.

The relationships between the subspecies of *C. oreganus* have also been studied. Comparison of venom proteins of *C. o. abyssus* with those of *C. o. concolor, C. o. lutosus*, and *C. viridis nuntius* shows it to be more closely related to *C. o. lutosus* than the other two taxa (Young et al. 1980). Correspondingly, Pook and McEwing (2005) isolated mtDNA in the venom of *C. o. oreganus* and *C. o. lutosus* and found the subspecies closely related.

Crotalus pricei Van Denburgh, 1895
Twin-spotted Rattlesnake
Cascabel de Mamchas-gemelas

RECOGNITION: *Crotalus pricei* is another of the three small montane *Crotalus* occurring within the bounds of the range covered in this book. It has a TBL to 66 cm (Stebbins 2003), but most are shorter than 55 cm. Its slender body is gray, bluish-gray, brownish-gray, to pale brown, or occasionally reddish-brown. Some individuals are patternless, but most have a dorsal pattern consisting of two longitudinal rows of 39–64 (usually 50–52) small gray, dark brown or black, sometimes white-bordered spots, which may be united across the back (Sierra Oriental Madre; less so, Sierra Occidental Madre) or alternate, with those nearest the tail usually forming cross bands. Three additional rows of smaller dark spots may occur along the sides; usually these are obscure in adults. The tail is patterned with 5–11 cross bands, and the proximal segment of the rattle is orange to red. Dorsal body scales are keeled (the lowest row is smooth) with two faint apical pits. Stille (1987) described the microdermoglyphic pattern of the dorsal surface as foveate with raised cell borders or reticulate and shows an electromicrograph; Harris (2005) stated that the dorsal ridges are either straight or

Crotalus pricei pricei
(Chihuahua; San
Diego Zoo;
John H. Tashjian)

swirling in *C. p. miquihuanus*. The dorsal scales normally occur in 23 (21–25) anterior rows, 21 (19–23) midbody rows, and about 17 (15–19) rows near the anal vent. On average, Arizona individuals shed 1.76 times per year (Prival et al. 2002). The throat and venter are gray or light brown; the posteriormost ventrals contain much black mottling and are almost totally black near the tail. Ventrals total 135–171 and subcaudals 18–33, and the anal plate is undivided. The relatively small head has a dark stripe extending diagonally backward from below the orbit to the corner of the mouth, and its light-colored labials are heavily patterned with small dark marks. The rostral is broader than long, and is followed dorsally by 2 large internasals, 1 canthal (rarely 2), 2 supraoculars, 2–3 intersupraoculars, and 4–11 scales in the internasal-prefrontal area (prefrontals are usually absent). Laterally are 2 nasals (the prenasal touches the supralabials, but the postnasal does not touch the upper preocular), 1–2 loreals (separated from the supralabials), 2–3 preoculars, 3–4 postoculars, 3–4 suboculars (the first contacts supralabials 3 and 4, and the second touches supralabials 4 and 5), 9 (8–11) supralabials, and 9–10 (8–12) infralabials. Usually 1–3 small prefoveal scales are present. The chin has a mental scale and a pair of longer than broad chin shields. The species is unique among crotalid rattlesnakes in having the first supralabial curving dorsally behind the postnasal scale to contact a small prefoveal scale lying between the postnasal and loreal scales.

The short, bifurcate hemipenis has a divided sulcus spermaticus, with 25–52 spines and 22 fringes per lobe and numerous spines in the crotch (Campbell and Lamar 2004; Klauber 1972).

The dentition consists of 3 palatines, 6–7 pterygoids, 9–10 dentaries, and a maxillary fang on each side (Klauber 1972).

Males (TBL_{max}, 66 cm) have 135–164 (mean, 156) ventrals, 21–33 (mean, 25) subcaudals, and tails 6.6–12.0 (mean, 8.8) % of TBL. Females (TBL_{max} to about 57 cm; C. Ernst, personal observation) have 143–171 (mean, 163) ventrals, 18–27 (mean, 22) subcaudals, and tails 5.2–8.6 (mean, 7.5) % of TBL.

GEOGRAPHIC VARIATION: Two subspecies have been described. *Crotalus pricei pricei* Van Denburgh, 1895, the western twin-spotted rattlesnake or cascabel manchas-gemelas de Price, occurs in the Sierra Madre Occidental from the Chirchuahua, Huachuca, Pinaleño, and Santa Rita mountains of southern Arizona southward through northeastern Sonora, western Chihuahua, and western Durango in Mexico. An isolated population occurs west of Rincón de Ramos in the Santa Fria of Aquascalientes. It is the larger subspecies with a TBL_{max} of 66 cm; it has a grayish body; an average of 52–55 more often divided mid-dorsal body blotches; usually 1 canthal; 2–3 preoculars and 3–4 postoculars on each side, and 149–171 ventrals. *Crotalus p. miquihuanus* Gloyd, 1940, the Miquihuanan rattlesnake or cascabel de Miquihuana, is found in the Sierra Madre Oriental from southeastern Coahuila and southern Nuevo León into southwestern Tamaulipas and north-central San Luis Potosí. It is smaller with a TBL_{max} of about 55 cm; it has a brownish body, about 48–49 divided mid-dorsal blotches; 2 canthals; 2 preoculars and 3 postoculars on each side; and 135–145 ventrals.

CONFUSING SPECIES: *Crotalus lepidus* and *C. willardi*, which occur within the range of *C. pricei*, can be identified by the key presented above. The small Mexican species, *C. intermedius*, which occupies a nearby allopatric range, has a single row of dorsal body blotches, loreal-supralabial contact, contact between the anterior subocular and supralabials 4 and 5, the second subocular rarely contacting a supralabial, and usually no prefoveals (other differences are listed in McCranie 1980).

Crotalus pricei pricei
(R. D. Bartlett)

Distribution of
Crotalus pricei.

KARYOTYPE AND FOSSIL RECORD: Unreported.

DISTRIBUTION: In the west, *C. pricei* is found from southeastern Arizona (the iso-lated Santa Rita, Huachuca, Pinaleño [Mt. Graham], Dos Cabezus, and Chiricahua mountains; Johnson and Mills 1982) south into Mexico, where its main range is from northeastern Sonora and western Chihuahua southward through western Durango (Chrapliwy and Fugler 1955), with an isolated population in Aguascalientes (Camp-bell and Lamar 1989, 2004). A second eastern Mexican range occurs from southeast-ern Coahuila (Bryson and Lazcano 2003) and southwestern Nuevo León to southern Tamaulipas and San Luis Potosí (Campbell and Lamar 2004). The entire range lies between about 1,800 and 3,200 m elevation.

These ranges were probably continuous during the Pleistocene (Rancholabrean) glaciation, but later drying of the intervening lowlands caused restriction of the scrub pine-oak woodlands to high altitudes, and isolated the Arizona populations from each other and those in Mexico.

HABITAT: This snake is a denizen of rocky areas. It particularly favors the limestone outcrops, rocky ledges, talus slopes, or volcanic rocks of canyons and ridges. Over the range, such areas are vegetated with bracket fern (*Pteridium aquilinim*); manzanita (*Arctosaphylos* sp.); maquey (*Agave* sp.); buckwheat (*Eriogonum jamesii*); scrub-brush, pine (*Pinus*, often *P. ponderosa*)-oak (*Quercus* sp.), or coniferous (*Pinus*) woodlands. Lowe et al. (1986) suggested that high-mountain hardwoods such as alders (*Alnus oblongifolia*), aspen (*Populus tremuloides*), box elders and other maples (*Acer negundo, Acer* sp.), cherry trees (*Prunus* sp.), Gambel oaks (*Quercus gambelii*), and New Mexico locust (*Robinia neomexicana*) may be part of the snake's habitat in Arizona. South-facing slopes seem to be preferred (Axtell and Sabath 1963). When available, *C. pricei* will use wooded shelters, and has been found in logged (*Pinus*) areas in Chihuahua sheltering under piles of branches, in hollow stumps, or under logs (Bryson et al. 2002b). In Durango, grassy areas and gently rolling hillsides covered with manzanita and scrub oaks are also inhabited. Water is probably obtained by drinking that trickling over rocks (Kauffeld 1943b).

BEHAVIOR AND ECOLOGY: In the past, available life history data on this species have been scare and anecdotal, but recent studies in southeastern Arizona by Prival and his associates have increased our knowledge of this animal considerably.

Armstrong and Murphy (1979) collected active *C. p. pricei* in southeastern Arizona as early as 17 March (AT 17°C), but they most often first become surface-active in April or early May and remain so into October. Those in the Chiricahua Mountains are most active in July and August during the late summer monsoon season. Drier conditions in August and September 1998 caused a drop in snake activity from the wetter monsoon season of 1997 (Prival 2000a; Prival et al. 2002).

When the daily ETs increased, the snakes moved from the talus slopes to the cooler, more favorable microclimates in nearby wooded areas. Gravid females typically made moderate movements from mid-May to mid-July, but almost no movement from mid-July to the end of August, and with more movement (returning to hibernacula?) in September and October (Prival 2000a; Prival et al. 2003). Males were found moving from one talus slope to another in mid- to late October (Prival et al. 2003). If the AT is favorable, some basking and surface movement may occur during December and early January (Prival et al. 2003).

In Mexico, *C. p. pricei* is active from May through September, sometimes under adverse weather conditions; in its eastern Mexican range, *C. p. miquihuanus* probably follows a similar annual cycle, but is only found in numbers during the July–September rainy season (Bryson and Lazcano 2003).

At least in Arizona, *C. pricei* must spend some of the colder months hibernating.

Crotalus pricei pricei (Cochise County, Arizona; Steve W. Gotte)

Typical hibernacula are talus slopes or under large rocks, but animal burrows and hollow logs or stumps are probably also used. The snakes seem to overwinter singly. Two males monitored by Prival et al. (2003) spent the winter at the edge of a talus slope where snow melt was rapid. At one of these sites, the ST (talus) was 15.6°C on 15 December. A female tracked by them overwintered off the talus. She remained in dense buckwheat (*Eriogonum jamesii*) until at least mid-December, but from mid-January to mid-March she was found under a surface rock.

Captives are most active in the late afternoon and at night (Kauffeld 1943b), but in nature the nights at higher elevations are usually cold (the high-elevation habitats of *C. p. miquihuanus* may only reach a high daily AT of 26°C; Bryson and Lazcano 2003), and although *C. pricei* may be nocturnally active on the warmest nights, most activity seems diurnal, particularly after 1100 hours. This is especially true immediately following precipitation, when this snake is commonly found basking, sometimes in pairs (possibly a prelude to mating during the late summer monsoon season; see below). Humid days also bring it out from its daily retreats under rocks and in rock crevices.

C. pricei is adapted to rather low ETs. Active individuals have been found at ATs of 11°C (Armstrong and Murphy 1979) to 27°C (Wright and Wright 1957), and BTs of 5 sent to Brattstrom (1965) by Frederick Gehlbach averaged 21.1 (range, 18.0–23.8) °C.

A total of 198 Arizona BTs (160 during the monsoon season) recorded by Prival et al. (2002) averaged 26.1 (range, 12.0–39.0) °C, and only rarely dropped below 10°C or exceeded 35°C. The BT during the monsoon months averaged 1.59°C above the ST, and mean BT of gravid females averaged 4.64°C higher than the ST. Mean BTs of males and nongravid females were closer (1.11°C and 0.97°C, respectively) to the ST. Gravid females probably require higher BTs for embryonic development, and either bask or seek warmer sites. The winter BTs between November and April of 13 *C. p. pricei* were 5.0–29.8 (mean, 18.5) °C, substantially warmer than the AT, but not significantly different from the ST (Prival et al. 2003).

Bryson et al. (2008) reported the following body and environmental temperatures for Mexican *C. pricei*: *C. p. miquihuanus*—BT, 19.7 (range, 10.1–29.6) °C; AT, 19.3 (range, 12.8–26.1) °C; and ST, 18.6 (range, 9.7–28.2) °C; and *C. p. pricei*—BT, 21.0 (range, 16.8–23.4) °C; AT, 20.2 (range, 18.5–21.9) °C; and ST, 21.3 (range, 18.7–23.0) °C.

Prival et al. (2002) studied *C. p. pricei* in 1997 and 1998 in Arizona. During 1997, 4 males moved an average of 14.0 m per week, and 1 female moved an average of 42.5 m per week; in 1998, 5 males moved an average of 87.8 m per week, and 4 females averaged 19.7 m per week. In 1997, the mean home range of 4 males was 0.16 ha; the lone female had a home range of 0.83 ha; but in 1998 the 5 males and 4 females tracked had mean home ranges of 2.29 ha and 0.20 ha, respectively. During the monsoon months, the home ranges of 9 males were 0.004–5.34 (mean, 1.37) ha. That of the lone nongravid female monitored in 1997 was much larger during the monsoon period than were the home ranges of gravid females tracked in 1998. The rainy season was more intense in 1997 than in 1998 (see above), and the snakes were more active in 1998. On average, males moved 73.8 m per week farther in 1998 than in 1997 during the monsoon months; 1 male monitored both years moved almost 4 times as much in 1998 than in 1997. Nongravid females moved an average of 53.5 m per week between 24 July and 29 August 1997, and 30 m per week between 29 August and 30 September. Three gravid females tracked before and after parturition in 1998 only moved an average of 3.2 m per week between mid-July and the end of August when they gave birth, but

Crotalus adamanteus (Collier County, Florida; Roger W. Barbour)

Crotalus atrox (Pima County, Arizona; Brad Moon)

Crotalus basiliscus (Minas Nuevas, Sonora; Eric Dugan)

Crotalus catalinensis, juvenile, dark morph (Santa Catelina Island; Brad Moon and Ali M. Rabatsky)

Crotalus mitchellii mitchellii (Monseratte Island, Baja California Sur; Ali M. Rabatsky)

Crotalus mitchellii angelensis (Angel de la Guarda Island, Gulf of California; Ali M. Rabatsky)

Crotalus mitchellii pyrrhus (R. D. Bartlett)

Crotalus molosssus molossus, yellow morph (Santa Cruz, County, Arizona; Brad Moon)

Crotalus molossus nigrescens (Southeast of Yerbanis, Durango; University of Texas El Paso; John H. Tashjian)

Crotalus oreganus oreganus (Calaveras County, California; John H. Tashjian)

Crotalus oreganus cerberus (R. D. Bartlett)

Crotalus oreganus concolor (R. D. Bartlett)

Crotalus oreganus lutosus (R. D. Bartlett)

Crotalus pricei pricei (Cochise County, Arizona; Paul Hampton)

Crotalus pricei miquihuanus (Cerro Potosi, Nuevo León; Arizona-Sierra Desert Museum; John H. Tashjian)

Crotalus ruber ruber
(John H. Tashjian)

Crotalus ruber exsul
(Cedros Island,
Pacific Ocean, Baja
California Norte;
United States
International
University; John H.
Tashjian)

*Crotalus ruber
lorenzoensis* (San
Lorenzo Sur Island,
Ali M. Rabatsky)

*Crotalus ruber
lucasensis*
(R. D. Bartlett)

Crotalus scutulatus scutulatus (Cochise County, Arizona; Carl H. Ernst)

Crotalus scutulatus salvini (El Limon, Vera Cruz; Terry Basey, Arcadia, California; John H. Tashjian)

Crotalus stephensi (R. D. Bartlett)

Crotalus tigris
(Barry Mansell)

Crotalus tigris (Pima
County, Arizona;
R. D. Bartlett)

Crotalus tortugensis
(Tortuga Island,
Baja California Sur;
R. D. Bartlett)

Crotalus totonacus
(Soto la Marina,
Tamaulipas; Gladys
Porter Zoo;
John H. Tashjian)

Crotalus viridis
viridis
(R. D. Bartlett)

Crotalus viridis
nuntius
(R. D. Bartlett)

Crotalus willardi
willardi
(R. D. Bartlett)

Crotalus willardi
amabilis (Sierra del
Nido, Chihuahua;
Dallas Zoo;
John H. Tashjian)

Crotalus willardi
meridionalis
(R. D. Bartlett)

Crotalus willardi obscurus (R. D. Bartlett)

Crotalus willardi silus (Gladys Porter Zoo; John H. Tashjian)

increased their movements to an average of 73.1 (range, 38.5–133.6) m per week afterward.

C. pricei is a good swimmer, although seldom required to do so. It swims with the same typical lateral undulations used in crawling, but does not elevate the tail rattle above the water (Klauber 1972). It climbs into shrubs to a height of at least 20 cm to forage or possibly bask (Gumbart and Sullivan 1990).

Mahaney (1997a) observed combat between two captive males on 10 July, and the winner of the bout later copulated with a female.

REPRODUCTION: Stebbins (2003) reported a mature length of 30.5 cm for both sexes of *C. pricei*, and the smallest gravid female measured by Klauber (1972) was 30.1 cm. Goldberg (2000b) reported mature males with 32.2 cm and 33.3 cm SVLs, and a mature female with a 33 cm SVL. Prival et al. (2002) noted that gravid female Arizona *C. p. pricei* had SVLs of 36.4–50.0 (mean, 43.0) cm. Such lengths are probably attained during the third year (Prival et al. 2002).

Goldberg (2000b) studied the gametic cycles of both sexes. No males had regressed (inactive) testes. Recrudescence with primary and secondary spermatocytes and some spermatids occurred from June to August, and spermiogenesis with metamorphosing spermatids and mature sperm present began in June and lasted into October. Sperm were present in the vas deferens of 93% of the males, including all from June to September. Mating coincides with hypertrophy of the kidney sexual segments (Saint Girons 1982), which were enlarged and contained secretory granules in 89% of the males examined from June to October. Apparently, females of both subspecies store sperm over winter for spring fertilization (Bryson and Lazcano 2003; Goldberg 2000b; Prival et al. 2002), and the young may be born more than a year later (Campbell and Lamar 2004; Mahaney 1997a).

Three females examined by Goldberg (2000b) on 7 May, 11 June, and 12 August, respectively, were not undergoing vitellogenesis; two females had started yolk deposition by 6 July and 27 September and would probably have ovulated the following spring. Another female that would probably have also ovulated the following spring had enlarged follicles (> 10 mm) on 23 January, and four females had ovulated by 18 May, 7 June, 29 June, and August and probably would have given birth later that year. Prival (2000a) palpated 28 female Arizona *C. p. pricei* between mid-August and mid-September and found 46.4% were gravid; Prival et al. (2002) detected no embryos before 9 June or after 1 September. Most females carrying young were found in July and August, when 50% of 36 females were gravid, and no female was gravid in consecutive years. Apparently, the female reproductive cycle is biennial.

Adult Arizona *C. p. pricei* of both sexes have been found within 1 m of each other between 9 August and 21 September, and courtship was observed between 0851 and 0949 hours on 21 August by Prival et al. (2002). During courtship the 2 snakes constantly intertwined their tails, and the male performed about 2.5-minute cycles of chin-pressing and tongue-flicking, followed by 4 seconds of vigorous tail movement and 20 seconds of motionlessness; but it could not be determined if mating actually occurred. However, Ball et al. (2003) witnessed copulation by another pair in Arizona on 1 September, during which the snakes had intertwined tails, the male was on top of the female's back, and he made pulsating movements while crawling over the female. Copulation continued for about an hour, but the female broke off contact and retreated under a rock when a flash photograph was taken. A captive mating witnessed by Armstrong and Murphy (1979) occurred on 9 July.

C. pricei is ovoviviparous and bears live young in July or August (*C. p. miquihuanus*, 1 July-19 August; *C. p. pricei*, 9 July–17 August) in the wild (Armstrong and Murphy 1979; Bryson and Lazcano 2003; Liner and Chaney 1986). Captives have given birth as early as 6 June (Mahaney 1997a). Litters contain 3–9 (mean, 5.55; n = 20) young; *C. p. miquihuanus*—3–6 (mean, 5; n = 4) young (Armstrong and Murphy 1979; Bryson and Lazcano 2003; Liner and Chaney 1986); *C. p. pricei*—3–9 (mean, 5.7; n = 16) young. Short females probably produce the smallest litters. Bryson and Lazcano (2003) reported RCMs of 43.7% and 57.3% for litters by 2 Coahuila *C. p. miquihuanus*; RCMs of Arizona *C. p. pricei* reported by Mahaney (1997a) and Prival et al. (2002) were 53% and 31.7–47.3 (mean, 39.4) %, respectively.

Neonate *C. p. miquihuanus* have 11.4–14.3 (mean, 12.6; n = 6) cm SVLs and BMs of 2.3–4.0 (mean, 3.2; n = 5) g. Neonate *C. p. pricei* have 12.8–22.3 (mean, 16.3; n = 64) cm TBLs and BMs of 2.4–10.3 (mean, 4.85; n = 36) g.

GROWTH AND LONGEVITY: During an Arizona study, the mean SVL growth rate for all snakes was 0.063 mm per day. Juveniles (mean, 0.254 mm per day) grew faster than adults (mean, 0.015 mm per day). The mean gain in BM/day for all snakes was −0.019 g (juveniles, −0.085 g; adults, −0.045 g). The mean SVL growth rate for all snakes between ecdysis events was 7.26 mm (juveniles, 25.1 mm; adults, 2.24 mm). The mean gain in BM between sheds for all snakes was 2.06 g (juveniles, 5.72 g; adults, 1.03 g) (Prival et al. 2002).

Five captive-born young grew an average of 9.8 cm and increased their BM an average of 11.5 g during their first 125 days (Kauffeld 1943b). The greatest increase (17 to 28 cm) was by the largest neonate.

A captive-born male *C. p. pricei* was still alive after 15 years, 8 months, and 9 days at Zoo Atlanta; another wild-caught adult lived more than 10 years in captivity (Kauffeld

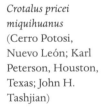

Crotalus pricei miquihuanus (Cerro Potosi, Nuevo León; Karl Peterson, Houston, Texas; John H. Tashjian)

1969). The record survivorship in captivity for *C. p. miquihuanus* (a wild-caught adult male) is only 5 years, 9 months, and 6 days at the Columbus Zoo in Ohio (Snider and Bowler 1992). Longevity in the wild is unknown.

DIET AND FEEDING BEHAVIOR: Reported prey of *C. pricei* include lizards (*Barisia imbricata, B. levicollis; Sceloporus grammicus, S. jarrovi, S. poinsetti, S. torquatus*), neonate snakes (*C. pricei*), small birds (*Catherpes mexicanus, Junco phaeonotus*), and rodents (*Neotoma [mexicana ?], Peromyscus [boylii ?]*) (Armstrong and Murphy 1979; Bryson et al. 2003; Bryson and Lazcano 2003; Campbell and Lamar 2004; Gumbart and Sullivan 1990; Klauber 1972; Prival 2000a; Prival et al. 2002; Woodin 1953). Captives have readily taken *S. undulatus* and *Anolis carolinensis* (Kauffeld 1943b). Klauber (1972) found a mouse in a wild adult, and Wright and Wright (1957) listed both mice and a "shed skin" as foods of captives. B. Martin (1974) stated they eat invertebrates, but this has not been verified.

Prival et al. (2002) detected food boli in 36% of the 134 *C. p. pricei* they palpated: juveniles, 54%; adults, 29% (males, 44%; females, 25%). Feeding occurred most often during the monsoon season. Lizards were the primary prey, and *Sceloporus* scales occurred in 74.2% of the 31 fecal samples examined in which the remains could be identified. Snake and bird remains were found in only 1 (3.2%) fecal sample each. Twenty-nine percent of the samples contained mammal hair, and mammal bones were found in 2 (6.4%) others. *Sceloporus* sp. were found in 75% of the 24 snakes captured on talus; 25% contained mammal remains. *Sceloporus* accounted for 57% of the prey of 7 snakes producing samples, while mammals made up the other 43%. Both the bird and snake prey were from talus snakes. It is not known whether the conspecific neonate was captured alive or consumed as carrion.

Lizards are usually struck in the thorax and then held in the mouth following a strike, but mice are struck, released, and later trailed with much tongue-flicking (Cruz et al. 1987). Mice are normally struck multiple times during a predatory attack (Chiszar et al. 1986c). *C. pricei* does not wave its tail to lure lizards, as do some other small pit-vipers (Kauffeld 1943b). Digestion is rapid after a lizard-meal, with defecation usually taking place on day 2–3 after feeding, and often again on day 5 (Kauffeld 1943b).

VENOM DELIVERY SYSTEM: The fangs of *Crotalus pricei* are among the shortest of those of all rattlesnakes. Twelve adult *C. p. pricei* with 40.2–51.6 cm TBLs that we measured had 2.0–3.3 (mean, 3.1) mm FLs, and a 21.5 cm neonate had 1.0 mm fangs. Klauber (1939) reported that the FLs of 5 *C. triseriatus pricei* (= *C. p. pricei*) with TBLs of 52.5–56.5 cm were 3.3–3.7 (mean, 3.5) mm. Their TBL/FL and HL/FL ratios were 155% and 7.5%, respectively, and the fangs had only a 48° angle of curvature. In respect to fang curvature, the fangs of *C. p. pricei* are "flatter" than those of other rattlesnakes (Klauber 1972). Because its fangs are short, this is not surprising, as the fang becomes more curved with increased length (Klauber 1939, 1972).

VENOM AND BITES: The venom is probably highly toxic, but yields are low. Thirty *C. p. pricei* produced a total of 182 mg of dried venom, or approximately only 8 mg per adult (Klauber 1972). Minton and Weinstein (1984) recorded a yield of only 4.1 mg per snake. The chemical composition of the species' venom has not been adequately studied. Minton and Weinstein (1984) recorded no protease activity.

The LD_{50} (mg/kg) for mice is 0.95 for intravenous and 11.50 for intramuscular injections (Glenn and Straight 1985b); Mori et al. (1987b) reported an intravenous LD_{50} of 5.2 μg/g. The minimum intravenous lethal dose for a 350 g pigeon is 0.2 mg of dried venom (Githens and George 1931). A nestling yellow-eyed junco (*Junco phaeonotus*) displayed extreme edema around the fang marks left by a *C. p. pricei* (Gumbart and Sullivan 1990). The human LD_{50} has not been calculated.

Four human envenomations resulted in both local and systemic symptoms more serious than expected from such a small rattlesnake (Minton and Weinstein 1984), and Bernstein et al. (1992) reported secondary infection at the bite site.

PREDATORS AND DEFENSE: Other than the case of cannibalism reported above, data on predation are lacking. Juveniles, at least, probably occasionally fall victim to great horned owls (*Bubo virginianus*), various large hawks (*Buteo* sp.), skunks (*Conepatus mesoleucus, Mephitis macroura, M. mephitis; Spilogale gracilis*), badgers (*Taxidea taxus*), coyotes (*Canis latrans*), and ophiophagus snakes (*Lampropeltis getula, L. pyromelana, L. triangulum*). Humans probably kill more than all natural predators combined (see below). *C. pricei* may be attacked by parental passerine birds defending their nestlings (Gumbart and Sullivan 1990). When confronted with the snake *Lampropeltis triangulum*, *C. p. pricei* displayed a body-bridge defensive posture (Weldon and Burghardt 1979; Weldon et al. 1992).

C. pricei is shy, and quickly crawls to shelter if possible. Its rattle is soft and at times barely audible above accompanying insect noises; the total range of frequency is 4.9–24.1 Hz (Young and Brown 1993), and the SVL-adjusted loudness is about 70 dB (Cook et al. 1994). Although usually mild-mannered, it will strike if provoked. In spite of its small size, this rattlesnake is dangerous, and should be respected; it should not be handled.

PARASITES AND PATHOGENS: The twin-spotted rattlesnake is known to harbor the tetrathyridia larvae of *Mesocestoides* cestodes and unidentified nematodes (Ernst and Ernst 2006; Goldberg and Bursey 1999a; Klauber 1972).

POPULATIONS: *C. pricei* is often the most frequently encountered rattlesnake at higher elevations within its range (Armstrong and Murphy 1979), and several may occur in a small space of only a few square meters; Kauffeld (1943a) collected three within an hour in a small area of exposed boulders. Prival (2000a) reported that 90–100 *C. p. pricei* lived in the talus slopes within his Arizona study area, and in 1997–2000, 127 individual snakes were captured, measured, and marked there (Prival et al. 2002). Johnson and Mills (1982) found it present at 21 sites during their survey.

Prival et al. (2002) reported some dynamics for their Arizona population of *C. p. pricei*. SVLs ranged from 16.8–57.2 cm, with most individuals between 30.0 and 49.9 cm; the majority of females were 40.0–49.9 cm, and males dominated in the size classes 30.0–39.9 cm and those > 45.0 cm. Juveniles' SVLs were < 29.9 cm. The total BM of the 127 snakes was approximately 6.8 kg (extrapolated from statistics presented in the paper). The male to female sex ratio of snakes > 30.1 cm SVL was skewed toward males (58.3%), but did not differ significantly from 1:1. Adults composed 63% of the snakes captured, juveniles 37%. The sex ratio of a litter reported by Mahaney 1997 was 2 males to 5 females, or 0.40:1.00.

The twin-spotted rattlesnake was formerly quite common in the Chiricahuas of southeastern Arizona, but somewhat less so in the four other mountains it occupies in that state (Klauber 1972), but now it is not as common as before 1982. Humans have destroyed many habitats, numerous *C. p. pricei* have been removed for the pet trade (Prival 2000a), and a few others have died on roads, which now experience more traffic. Today, the major potential threats in Arizona include mining, grazing, logging, and recreational or other development; illegal collection for the pet trade is the greatest current threat (Johnson and Mills 1982; Prival 2000a). The species has been given legal protection by the state of Arizona.

REMARKS: Electrophoretic studies of venom proteins by Foote and MacMahon (1977) indicate *C. pricei* is most closely related to *C. cerastes* and *C. willardi*; in contrast, a study of its dorsal scale microdermatoglyphics by Stille (1987) shows its nearest relatives to be *C. mitchelli* and *C. cerastes*. More recent studies of the mtDNA and RNA of *C. pricei* by Murphy et al. (2002) have indicated that the species belongs to the *Crotalus polystictus*-group of mostly small Mexican rattlesnakes, which includes, along with those two species, *C. cerastes*, *C. intermedius*, *C. transversus*, and *C. willardi*. *C. intermedius* appears to be its closest relative.

The species was reviewed by McCranie (1980b). An interesting paper by Moll (2003) supplies information regarding the collecting trip during which *C. pricei* was first discovered, and the identity of William ("Billy") Wightman Price, who helped collect the first specimen.

Crotalus ruber Cope, 1892
Red Diamond Rattlesnake
Cascabel Diamante Rojo

RECOGNITION: A freshly shed red-phase northern *Crotalus ruber* is among the most beautiful of rattlesnakes. The species' body color ranges from brick-red to reddish-gray, pinkish-brown, or tan. *C. ruber* from the southern half of Baja California are somewhat darker, being yellowish-brown to olive-brown (see "Geographic Variation"). A series of 20–42 (mean, 33–35) light-bordered, diamond-shaped, reddish-brown dorsal blotches are present. These are clearly separated by white or tan on the anterior 50–67% of the body. Some individuals have the diamonds poorly outlined, with the pale border reduced or incomplete along the lateral edges. A lateral row of smaller diamond-shaped dark blotches may also be present along the body. The tail bears 2–7, normally 3–5, conspicuous black rings; the proximal rattle segment is black. The snake's TBL_{max} is 162.5 cm (Klauber 1972), making it one of the largest rattlesnakes; Sibley (1951) reported a maximum length of 190.5 cm, but this has not been confirmed. Mainland individuals commonly exceed 100 cm, while those living on islands rarely reach 90 cm. The dorsal body scales are keeled (except on lateral rows 1–2) and pitted; according to Stille (1987) and Harris (2005), the linear microdermoglyphic ridges on the dorsal surface are mostly swirled. The dorsal body scales lie in 29 (25–30) anterior rows, 25–33 (see "Geographic Variation" for subspecific means) rows at midbody, and 27–29 (25–30) posterior rows. The unmarked white to cream venter has 179–206 ventrals, 15–29 subcaudals, and an undivided anal plate. The head is unicolored reddish with no dorsal pattern. A diagonal dark, light-bordered stripe extends from the lower edge of the orbit to the corner of the mouth on each side. Its posterior light border may be incomplete, indistinct, or totally absent. A small, somewhat indistinct, lateral blotch may also be present. Dorsally, the head scales consist of a rostral (broader than long in the north, longer than broad in the south), 2–4 internasals (small, and touching the rostral), 2 (3) canthals on each side (the posterior is large, prefrontals are absent), 8 or more scales in the internasal-prefrontal area, 2 supraoculars, and 4–10 (usually more than 6) intersupraoculars. The lateral scales include 2 nasals (the prenasal usually touches the first supralabial on mainland snakes, but may be prevented from contact by prefoveals on individuals from Cedras Island; the postnasal is usually prevented from contacting the upper preocular by the posterior canthal), 1–2 (3) loreals, 1–2 (3) preoculars (normally not touching the postnasal), 2–3 postoculars, 4–5 interoculabials (1–3 separating the anterior subocular from the supralabials), several suboculars, 15–17 (12–19) supralabials, and 16–18 (13–21) infralabials (the first is usually transversely divided). The chin has a single mental scale and 2 large chin shields, but normally no interchin shields.

The long, bifurcate hemipenis has attenuated lobes, a divided sulcus spermaticus, about 62 spines and 50 fringes per lobe, and 1–3 spines in the crotch between the lobes (Campbell and Lamar 2004; Klauber 1972).

Besides the fang, the dental formula is 3 palatine, 8 (6–9) pterygoid, and 9–10 (8–11) dentary teeth on each side (Klauber 1972).

Males (TBL_{max}, 162.5 cm) have 179–203 (mean, 194) ventrals, 20–29 (mean, 26) subcaudals, and 3–7 (mean, 5) dark tail rings. Females (TBL_{max}, >100 cm) have 183–206 (mean, 197) ventrals, 15–26 (mean, 21) subcaudals, and 2–5 (mean, 4) dark tail rings.

GEOGRAPHIC VARIATION: We follow Campbell and Lamar (2004) in recognizing four subspecies. *Crotalus ruber ruber* Cope, 1892, the red diamond rattlesnake or cascabel de diamante rojo, ranges from southwestern San Bernardino, southeastern Los Angeles, Riverside (west of the Mojave Desert in Riverside), Orange, extreme southwestern Imperial, and San Diego counties in the United States; southward to Loreto, Baja California Sur, but excluding the deserts of Baja California Norte east of Sierra Juárez, and also occurs on the islands of Angel de la Guarda, Monserrate, Pond, and San Marcos in the Gulf of California. It is the largest subspecies with a TBL_{max} of 162.5 cm; it is reddish in color, with 29 (25–31) midbody scale rows, 185–203 (mean, 194) ventrals in males and 188–206 (mean, 194–197) in females, 22–29 (mean, 25–26) subcaudals in males and 16–25 (mean, 20–21) in females, 29–42 (mean, 36) dorsal body blotches, and the rattle matrix well developed with a long string of segments (see Rowe et al. 2002). *Crotalus r. exsul* Garman, 1884, the Cedros Island diamond rattlesnake or cascabel de la Isla Cedros, is restricted to that island off the Pacific Coast of Baja California Norte. It is pinkish to pale reddish-gray, and has a TBL_{max} of 94 cm, 27 (27–30) midbody scale rows, 188–196 (mean, 191–192) ventrals in males and 192–201 (mean, 195–196) in females, 18–26 (mean, 22) subcaudals in males and 17–23 (mean, 19–20) in females, 30–37 (mean, 32) dorsal body blotches, and a typical rattle. *Crotalus r. lorenzoensis* Radcliffe and Maslin, 1975, the San Lorenzo Island Rattlesnake or cascabel de Isla San Lorenzo de Sur, is found only on that island in the Gulf of California. It is the smallest subspecies with a TBL_{max} of 87.2 cm and is pale yellowish-brown to reddish-brown, typically with only 25 (25–27) midbody scale rows, 189–195 (mean, 192) ventrals in males and 186–193 (mean, 190) in females, 21–23 (mean, 22) subcaudals in males and 15–19 mean, 16–17) in females, 31–39 (mean, 35) dorsal body blotches, and a rattle matrix that is inwardly shrunken with about 50% of adults lacking rattle segments (Rabatsky 2006). Grismer (1999a) would elevate this taxon to full species status. *Crotalus r. lucasensis* Van Denburgh, 1920, the San Lucan rattlesnake or cascabel de San Lucan, is found on the mainland of Baja California Sur from Loreto southward to Cape San Lucas, and on the islands of Danzante and San Jose in the Gulf of California and Santa Margarita off the Pacific Coast. This large snake (TBL$_{max}$, 130.6 cm; Klauber 1972) is distinguished by its darker brownish or olive-brown body color, 27 (25–33) midbody scale rows, 181–195 (mean, 189) ventrals in males and 183–203 (mean, 192–193) in females, 22–29 (mean, 25) subcaudals in males and 17–23 (mean, 20) in females, 20–39 (mean, 30) dorsal body blotches, and a normal rattle. *C. r. ruber* and *C. r. lucasensis* intergrade in the vicinities of Bahía Concepción and Loreto, Baja California Sur (Klauber 1972; Murray 1955).

An evaluation of mtDNA variation in the subspecies of *C. ruber*, excluding *C. r. lorenzoensis*, revealed three clades with only shallow genetic divergence: (1) a *C. r. exsul* clade (closely related to that of *C. r. ruber*), (2) a well-defined *C. r. ruber* clade occupying southern California and Baja California Norte, and (3) a sister clade in Baja California Sur represented by the only specimen of *C. r. lucasensis* tested. Such shallow genetic diversity probably resulted from Pleistocene climatic events (Douglas et al. 2006).

Harris and Simmons (1978) named two other Mexican subspecies, *C. r. elegans* from Isla Angel de la Guarda and *C. r. monserratensis* from Isla Monserrate, but these are not considered valid for several reasons (Campbell and Lamar 2004; McCranie and Wilson 1979; Murphy and Ottley 1984), and they have subsequently been included in the synonomy of the species by Beaman and Dugan (2006).

Case (1978) compared the smaller-sized *Crotalus*, including *C. ruber*, on Mexican islands to individuals of the same species from adjacent mainland populations, and developed a model to explain the optimal body size based on predator-prey relationships. We refer the reader to his paper for details.

Crotalus ruber ruber
(R. D. Bartlett)

Crotalus ruber ruber
(San Marcos,
California;
John H. Tashjian)

CONFUSING SPECIES: *C. ruber* can be recognized by usually having its first infralabials laterally subdivided. Other species of *Crotalus* within its range can be distinguished from it by the above key.

KARYOTYPE: Stewart and Morafka (1989) and Stewart et al. (1990) reported a diploid chromosome total of 36, consisting of 16 macrochromosomes (2 pairs of metacentrics, 4 pairs of submetacentrics, and 2 pairs of telocentrics), and 20 microchromosomes. Sex inheritance is ZW for females and ZZ for males.

FOSSIL RECORD: A Late Pliocene (Blancan) vertebra from a *Crotalus* discovered in southernmost Baja California Sur seems closest to those from a moderate-sized *C. ruber* (Miller 1980).

DISTRIBUTION: *C. ruber* occurs from San Bernardino, Los Angeles, Riverside, Orange, Imperial, and San Diego counties, California, southward through peninsular Baja California, Mexico (except the northeastern desert area east of Sierra de Juárez). It also is found on the islands of Angel de la Guarda, Danzante (Goodrich et al. 1978; Vaughan and Schwartz 1980), Monserrate, Pond, San Jose, San Lorenzo de Sur, and San Marcos in the Gulf of California; and on Isla Cedros and Isla de Santa Margarita (Klauber 1972; Wong 1997). It possibly also occurs on Isla Magdalena off the Pacific Coast of Baja California Sur (Bostic 1975).

HABITAT: *C. ruber* is an inhabitant of rocky habitats with thick vegetation: desert scrub, thornscrub, cacti (*Fouquieria* sp., *Machaerocereus gummosus*, *Opuntia cholla*), chaparral (*Arctostaphylos* sp., *Quercus dumosa*), pine (*Pinus*)-oak (*Quercus*) woods, and tropical deciduous forest. Dugan et al. (2004) reported that telemetered *C. r. ruber* preferred cactus patches and coastal sage scrub habitats. Some other common plants in its habitat are aster (*Arctotheca nivia*), bunch grass (*Bouteloua curtipendula*), dalea (*Dalea* sp.), elephant trees (*Bursera hindsiana*, *B. microphylla*), ironweed (*Olneya tesota*), jojoba grass (*Simmondsia chinensis*), lengua de gato (?) (*Bourreria sonorae*), limberbush (*Jatropha cu-*

neata), mimosa (*Desmanthus fruticosus*), morning glory (?) (*Jacquemontia abutiloides*), nightshade (*Solanum hindsianum*), paloblanco (*Lysiloma candida*) and palo verde (*Cercidium peninsulare*) trees, sagebush (*Artemisia tridentata*), saltbush (*Atriplex barclayana*), spurges (*Euphorbia madalenae*), and sumac (*Rhus* sp.). *C. ruber* occasionally ventures into cultivated areas and grasslands. In California, it is most common in the western foothills of the Coast Ranges, but also lives in the dry, rocky inland valleys; it seems to prefer cactus patches and coastal sage scrub (Dugan and Hayes 2005). Its total elevation ranges from near sea level to at least 1,520 m.

A model developed by Halama et al. (2008) to test for landscape niche characteristics indicated that the variables that seem most closely associated with the distribution of the species over its range in Southern California are low elevation, high winter precipitation, and the presence of chaparral and sage scrub habitat.

BEHAVIOR AND ECOLOGY: Klauber (1926) collected active *C. ruber* in San Diego County from February to November in 1923. Dugan and Hayes (2005) reported the active season of telemetered *C. r. ruber* as March–November, but Armstrong and Murphy (1979) and Klauber (1972) stated that they collected it in every month of the year. Most surface activity occurs from April to June, corresponding to the mating season (Dugan and Hayes 2005). Eighty-nine of the 151 individuals (59%) collected by Klauber (1926) were from that period, and 55 (36%) more were from July to September; only 2 (1.3%) were collected in October, and only 5 (3.3%) were found in February (3) and March (2). The earliest and latest actual recorded dates of surface activity are 20 February and 9 October (Wright and Wright 1957).

From late October or early November to the end of March, cool ETs cause *C. ruber* to spend more time in rodent middens or underground in rock crevices, outcrops, and

Distribution of
Crotalus ruber.

animal burrows. Greenberg (2005) found Southern California *C. r. ruber* chose winter hibernacula that receive less solar radiation than their summer retreats. He thought it possible in regions with only mild winters, where low ETs never become lethal, that it may be best for the snake to be as cold as possible while hibernating, such as at a site with low irradiance. This would keep the snake's basal metabolic rate low during the several months of fasting hibernation and help preserve its stored energy reserves. Several may congregate in a hibernaculum; Klauber (1972) mentioned one such den where 24 *C. r. ruber* were blasted out of a rock crevice during road construction, and Brown et al. (2008) found up to 7 individuals using a single den in San Diego County.

C. ruber can be found abroad during the day or at dusk, especially on warm days following rains, but is more diurnal during the cooler spring and fall months than in the hot summer months when it is almost entirely nocturnal. All *C. r. ruber* captured in the spring by Dugan et al. (2004) were found within or near cactus patches on south-facing slopes. It must have an afternoon retreat in the summer; animal burrows, brush piles, cactus clumps, and rock crevices are used. These retreats receive high radiation, and may be important for keeping the snake warmer overnight (underground retreats are usually deep enough for the snake to maintain a fairly stable BT) and to warm it quicker in the morning (C. Ernst, personal observation; Greenberg 2005). It has been found active from 1840 to 0325 hours at ATs of 18–30°C (Klauber *in* Wright and Wright 1957). BT of an active snake recorded by Brattstrom (1965) was 24°C.

Brown et al. (2008) found that resident male *C. r. ruber* in San Diego County had home ranges nearly triple the size of those of resident females, and that they typically moved twice as far as the females during the activity season. Twelve resident snakes had minimum kernel home ranges of 0.3–6.2 ha. In contrast, 5 snakes relocated by them only established home range areas of 1.6–1.8 ha. The home ranges and distances moved by relocated snakes decreased with time. Two of the snakes relocated more than 1,000 m were able to find hibernacula used by other snakes in their new areas. In another California study, Dugan et al. (2008) determined that the maximum convex polygon (MCP) and fixed kernel (FK) mean home ranges of male *C. r. ruber* over 2 active seasons and the winter between were active season 2003—MCP, 2.23 ha; FK, 2.06 ha; winter 2003–2004—MCP, 0.07 ha; FK, 0.07 ha; and active season 2004—MCP, 2.55 ha; FK, 1.00 ha; the mean daily movements (m per day) for these periods were 5.97, 1.35, and 12.46, respectively. Greenberg and McClintock (2008) compared the home ranges of nine *C. r. ruber* on the south side of the Coachella Valley in Riverside County, using planimetric (two-dimensional) and topographic (three-dimensional) estimates. The planimetric calculations underestimated the home range on average by 9% (10% for females, 8% for males).

Tevis (1943) tracked a *C. r. lucasensis* in Baja California Sur for 2 nights and a day, during which the snake made a successful predatory strike but did not move outside an estimated circular area 7.6 m in diameter.

C. r. ruber is an accomplished swimmer that occasionally swims in reservoirs, and even in the Pacific Ocean (Klauber 1972). It is also a good climber, and often ascends cacti, bushes, and small trees (Klauber 1972). Hollingsworth and Mellink (1996) saw a *C. r. lorenzoensis* climb 2 m into an elephant tree (*Bursera microphylla*), where it eventually stopped and coiled in the upper branches. They thought that its climbing, coupled with the reduction of that subspecies' rattle, could be related to predation on arboreal mammals or birds.

Males participate in dominance combat bouts, especially during the breeding season. Two males face each other with one lying on top of the other. Their heads and necks are raised and swayed from side to side while their tongues continually flick. One male rises above the other and bends his head toward the other snake until their heads touch. The head of the ventral male may only be raised 3–4 cm above his body, but he usually also elevates the anterior portion of his body an additional 30–40 cm. The dorsal male rises at the same time to maintain a superior position, and then both snakes push their raised bodies and necks against each other. They may separate, then rejoin to repeat the process until one pushes the other to the ground and pins it there,

Crotalus ruber exsul (Cedros Island, Pacific Ocean, Baja California Norte; United States International University; John H. Tashjian)

after which the pinned male usually crawls away (Shaw 1948; includes sequential photographs of the act).

REPRODUCTION: Most available reproductive data pertaining to the red diamond rattlesnake are from the subspecies *C. r. ruber*. Complementary data are almost totally absent for the Mexican subspecies.

The TBL of *C. r. ruber* at sexual maturity has been listed as 60–76 cm (Stebbins 2003; Wright and Wright 1957). The smallest gravid female examined by Klauber (1972) was 73.3 cm. The smallest male with sperm and the smallest female with enlarged follicles (> 6 mm) reported by Goldberg (1999b) had 63.7 cm and 66.2 cm SVLs, respectively. Attainment of such lengths probably takes 1–2 years of growth. The smallest adult female *C. r. lucasensis* measured by Klauber (1972) was 73.6 cm.

Goldberg (1999b) found the testes of male *C. r. ruber* were regressed with spermatogonia and Sertoli cells from February to May; recrudescent with dividing spermatogonia, primary and secondary spermatocytes, and a few spermatids in April–August; and undergoing spermiogenesis with metamorphosing spermatids and mature sperm in August. Spermiogenesis probably ends in September–October, and mature sperm is stored over the winter in the vas deferens. He also found secretory granules in the hypertrophied renal sexual segments of 100% of the males examined in February–July, but only in 33% in August.

Females of *C. r. ruber* have enlarged (> 10 mm) ovarian follicles from March to June and in September, but some females have inactive ovaries in April–August (Goldberg 1999b). Ovulation occurs sometime between March and June. Females of this subspecies typically begin vitellogenesis in the fall of one reproductive season and complete it the following spring. The reproductive cycle is probably biennial, as only 47% of the females examined by Goldberg were reproductive. Sperm from spring matings is apparently stored and used to fertilize the following spring's ovulated eggs.

Apparently courtship and mating by wild *C. r. ruber* occur exclusively in the spring, and have been observed from early March into May (Armstrong and Murphy 1979; Dugan et al. 2004, 2008; Klauber 1972). However, captives have mated at almost any time of the year, but most often from February to June. Captive copulations witnessed by Perkins (1943) lasted from more than 2 hours to almost 23 hours. During courtship, the male nudges and tongue-flicks the female's back and head, and may spastically jerk. During a lengthy mating the male may change body positions several times, and the attached pair may move about, one crawling and dragging the other along with it. While copulating, the male sometimes pulsates his tail (C. Ernst, personal observation).

The young of *C. r. ruber* are born after a GP of 141–190 (mean, 165; n = 4) days from July to September, but usually in August and September. Litters of this subspecies contain 3–20 (mean, 8; n = 40) young. Neonates are more gray than adults, but later turn reddish. Their TBLs are 28.0–35.0 (mean, 31; n = 23) cm.

In Baja California Sur, *C. r. lucasensis* courts and mates in March–May, with parturition taking place in August–December (Grismer 2002). Klauber (1972) reported that the smallest neonate he measured was 27.3 cm, and that newborns average 29 cm at birth.

C. r. ruber has hybridized with *C. helleri* in both the wild and captivity (Armstrong and Murphy 1979; Klauber 1972; Perkins 1951).

GROWTH AND LONGEVITY: The growth patterns of 249 *C. ruber* from San Diego County, California, were analyzed by Klauber (1972). Neonates, averaging about 30 cm in TBL, appeared in September, grew to 39 cm, and had a button rattle by November. After emerging from hibernation the following March, the snakes grew to nearly 50 cm and had 2 rattle segments by mid-April. By the end of that month, most had 3 rattle segments and some had grown to 60 cm; few were shorter than 45 cm. They had 45–65 cm TBLs in May. The first 4 rattle segments appeared in June when the snakes were 46–70 cm. Those measured in July were 50–70 cm, and some had 5 rattle segments. By September, the yearlings averaged 67 mm. Two-year snakes probably averaged about 87 cm with 7–8 rattle segments and the first indications of sexual dimorphism, and 3-year-old males had TBLs of about 94 cm and females about 84 cm. By the following spring, an additional 2.5 cm had been added to their length and the tail had as many as 9–11 rattles. The male TBL exceeded that of the female by about 10%.

A male *C. r. ruber*, wild-caught as an adult, survived an additional 19 years, 2 months, and 27 days in captivity in the private collection of James E. Gerholdt of Webster, Minnesota (Snider and Bowler 1992). At the San Diego Zoo, another *C. r. ruber* survived 14 years, 6 months, and a hybrid *C. r. ruber* × *C. helleri* was still alive after 13 years, 8 months at the time the report was submitted (Shaw 1969).

DIET AND FEEDING BEHAVIOR: Reported prey of wild *C. ruber* include lizards (*Cnemidophorus tigris*, a typical individual contains a live weight caloric value of 1.3–1.7 kcal/g, Essghaier and Johnson 1975; *Ctenosaura hemilopha*); small snakes; birds; and mammals—rabbits (*Sylvilagus auduboni, S. bachmani*), ground squirrels (*Amnospermophilus leucurus, Spermophilus beecheyi*), kangaroo rats (*Dipodomys deserti, D. merriami*), wood rats (*Neotoma fuscipes, N. lepida*), mice (*Mus musculus, Peromyscus maniculatus*), voles (*Microtus* sp.), and spotted skunks (*Spilogale gracilis*) (Armstrong and Murphy 1979; Campbell and Lamar 2004; Dugan and Hayes 2005; C. Ernst, personal observation; Grismer 2002; Hammerson 1981; Klauber 1972; Patten and Banta 1980; Stebbins 1954, 2003; Tevis 1943; Van Denburgh and Slevin 1921a; Vaughan and Schwartz 1980; Wright and Wright 1957). *Sceloporus orcutti* may also be a prey item, but Mayhew

Crotalus ruber lorenzoensis (Arizona-Sonora Desert Museum; John H. Tashjian)

(1963) reported that he found the lizard and *C. r. ruber* within 60 cm of each other on 2 occasions without witnessing any interaction between them.

Young *C. ruber* feed on mice and lizards, but adults prefer larger mammalian prey. Of 57 snakes containing food items examined by Klauber (1972), 53 (93%) had eaten mammals, 3 (5.3%) lizards, and 1 (1.8%) a bird. Captives have consumed rodents (*Mus musculus*, *Microtus* sp.), lizards (*Anolis carolinensis*, *Eumeces fasciatus*), and a western rattlesnake (*Crotalus viridis*) (Cunningham 1959; C. Ernst, personal observation; Klauber, 1972). Some prey may possibly be consumed as carrion (Cowles and Phelan 1958; Hammerson 1981; Patten 1981; Patten and Banta 1980). It took more than a year for one *C. ruber* to defecate the remains of its last meal (Marsh *in* Lillywhite et al. 2002).

Occasionally plant material may be ingested accidentally; Klauber (1972) found a sumac (*Rhus* sp.) leaf in the intestines of a *C. r. ruber*.

Prey is located by sight, olfaction, and BT-sensing by the loreal pit. The normal method of prey capture is by ambush, but *C. ruber* will also actively follow scent trails (Dullemeijer 1961; C. Ernst, personal observation). After initial detection, usually indicated by a raising of its head and neck in an S-shaped curve and an increase in tongue-flicking, the snake remains almost motionless, only slowly moving its head in the direction of the prey. Most prey is bitten only once when it enters the snake's striking range. It is then released, followed later by odor-trailing, and finally swallowed after it is dead (Chiszar et al. 1986a; Dullemeijer 1961; C. Ernst, personal observation). Dullemeijer and Povel (1972) described the coordinated muscle and bone movements that occur in the skull during prey ingestion, and we refer the reader to their paper for the intricate details.

VENOM DELIVERY SYSTEM: *C. ruber* has among the longest FLs of the species *Crotalus*. Klauber (1939) interpreted the following data from regression lines for 270 *C. r. lucasensis* with TBLs of 90–130 cm: FL, 9.9–13.2 (mean, 11.6) mm; mean TBL/FL, 95; and mean HL/FL, 4.38. Similarly, these calculations for 100 *C. r. ruber* with TBLs of 90–130 cm were FL, 9.5–12.9 (mean, 11.3) mm; mean TBL/FL, 97; and mean HL/FL, 4.51; and those of seven *C. r. exsul* with TBLs of 71.3–94.0 cm were FL, 7.5–9.6 (mean, 8.4) mm; mean TBL/FL, 94; and mean HL/FL, 4.37. He also reported that the mean fang curvature for the species was 70°.

VENOM AND BITES: *C. ruber* has high venom yields and hemorrhagic and enzyme activities similar to those of the *C. atrox*-group (Glenn and Straight 1985b). Although Minton (1956) thought the toxicity low compared to some other species of *Crotalus*, the snake's large size allows a greater volume of venom to be injected during a strike, and it is capable of killing a human (Shaw and Campbell 1974).

Venom yields from individual snakes and populations are variable. A single extraction from a *C. r. exsul* yielded only 54 mg, but this is more than the yield of several similar short rattlesnake species (Klauber 1972), while the maximum yield from several *C. r. lucasensis*, a much larger snake, ranged from 669 to 710 (means, 230–234) mg (Glenn and Straight 1982; Klauber 1972). *C. r. ruber*, the most studied subspecies, has produced maximum yields of 668–670 (means, 350–364) mg (Glenn and Straight 1982; Klauber 1972; Russell and Puffer 1970). Other yields (some approximate) from *C. r. ruber* are 49–203 mg (Glenn and Straight 1985b), 120–350 mg (Russell and Brodie 1974; Russell and Puffer 1970), 120–450 mg (Ernst and Zug 1996), 150–350 mg (Dowling 1975), 300–350 mg (Mackessy 1985), and 550 mg (do Amaral 1928). The larger the snake, the greater the yield; adults contain about 6–15 times as much venom as do juveniles (Norris 2004). Klauber (1972) reported the following high liquid yields by snake length: 60 cm, 0.24 mL; 70 cm, 0.43 mL; 80 cm, 0.70 mL; 90 cm, 1.07 mL; 100 cm, 1.57 mL; 110 cm, 2.20 mL; and 120 cm, 3.04 mL. Adults typically secrete about 0.72 mL (0.24 g dried) of venom per strike, and an exceptional one may inject 1.65 mL (0.55 g of dried venom; do Amaral 1928). Mackessy (1985) reported total yields of 28–35 mg from juvenile *C. r. ruber*; a typical juvenile bite yield is only 1.2–2.3 mg (do Amaral 1928; Glenn and Straight 1985b).

The yellowish-apricot venom has a specific gravity in adults of 1.095–1.100, but only 1.042 in neonates (Klauber 1972). The pH when freshly extracted is 5.25–5.70 (Mackessy and Baxter 2006). According to Rael et al. (1986), no Mojave toxin is present; but Bober et al. (1988) recorded positive immunodiffusion reactions of > 2.0 mg/mL for the toxin in the venom of *C. r. ruber* from near San Ignacio, Baja California, and from San Marcos Island. Perhaps venom races exist in *C. ruber* (see also Straight et al. 1992). Other enzymes (both purine and pyrimidine) and toxins identified in this species' venom are argine E-I and E-II ester hydrolases, esterase (BAEE; adult activity range, 17–189), kallikrein-like enzyme, L-amino acid oxidase, galactoside-binding (C-type) lectin, peptide inhibitors, phosphodiesterase (adult activity range, 44–278), phospholipase A_2 (which is clinal from north to south on the mainland; Straight et al. 1992), and proteases, including three hemorrhagic metalloproteinases (HT-1, HT-2, and HT-3 containing Zn, with an adult activity range of 72–163) (Aird 2005; Glenn and Straight 1985b; Hamako et al. 2007; Mackessy 1985, 1998; Mashiko and Takahashi 1998; Meier and Stocker 1995; Mori et al. 1987a, 1987b; Mori and Sugihara 1988, 1989a, 1989b; Munekiyo and Mackessy 2005; Norris 2004; Straight et al. 1992; Takahashi and Mashiko 1998; Takeya et al. 1990, 1993; Tan and Ponnudurai 1991). Protease activity is low to nonexistent in venom from juvenile *C. r. ruber* (Glenn and Straight 1985b).

McCue (2005) reported a relative venom toxicity of approximately 0.50 on a $\log_{10}$ scale. The LD_{50} values of venom from *C. r. ruber* for mice are intravenous, 3.48–3.97; intraperitoneal, 4.22–11.0 mg/kg; and subcutaneous, 21.25 mg/kg (Githens and Wolff 1939; Glenn and Straight 1982; Macht 1937; Russell and Brodie 1974; Russell and Puffer 1970). The venom of neonates has a lower mouse LD_{50} (mean 2.8 intraperitoneal and > 5.0 intramuscular) than that of adults (Glenn and Straight 1985b). The minimum mouse intraperitoneal LD_{50} for *C. r. lucasensis* venom is 4.0 mg (Githens and Wolff 1939). A 350 g pigeon has an LD_{50} of 0.5–0.6 mg dried venom from *C. r. ruber* (Githens and Butz 1929; Githens and George 1931). After 26–27 years of storage, the toxicity of the venom for cats decreased only slightly; before death the injected cats experienced an initial lowering of the rate of breathing; a drastic lowering of arterial, cisternal, and venous blood pressures; a more erratic EEG; and finally, a complete neuromuscular block of the diaphragm in 22 minutes (Russell et al. 1960). A kangaroo rat (*Dipodomys* sp.) observed by Tevis (1943) that was struck by a *C. r. lucasensis*, traveled less than a meter, stopped to lick the wound, straightened out and wobbled a few more centimeters, its hind quarters apparently paralyzed, and fell over on its left side, convulsed, and died in 2.5 minutes.

The estimated lethal venom dose for a human is about 100 mg (Dowling 1975; Ernst and Zug 1996; Klauber 1972). Human envenomations have resulted in great pain, edema and swelling, discoloration in the area around the bite, hemorrhage blebs (ecchymosis), a slight increase in BT, an initial increase followed by a lowering of blood pressure, a decrease in blood platelets (bone marrow production of cells does not seem affected), necrosis, and bloody diarrhea (Bush 2006; Clarke 1961; Klauber 1972; Lyons 1971; Mackessy 1985; Norris 2004; Norris and Bush 2001, 2007; Russell 1983). The venom causes local hemorrhaging by direct action on blood vessel walls. The two metalloproteinases HT-1 and HT-2 are responsible for this, and three proteolytic hemorrhagins degrade fibrinogen and cause myonecrosis. Case histories of bites are in Clarke (1961), Klauber (1972), and Lyons (1971). The venom has some bactericidal properties (Glaser 1948).

Venom is usually injected during a bite, but Klauber (1972) received a spray of venom or saliva droplets from a strike by a snake with a previously injured jaw when he was beyond the snake's range, and thought that the injury may have caused the venom to accumulate on the snake's lower jaw, which was then thrown onto him by the force of the strike.

PREDATORS AND DEFENSE: According to Dugan and Hayes (2005), adult *C. r. ruber* experience a low rate of predation. Coyotes (*Canis latrans*) are known predators (Grajales-Tam et al. 2003); but kingsnakes (*Lampropeltis* sp.) and other ophiophagus snakes, birds of prey (*Bubo virginianus*, *Buteo* sp.), and carnivorous wild mammals, such as raccoons (*Procyon lotor*), badgers (*Taxidea taxus*), skunks (*Mephitis mephitis*, *Spilogale gracilis*), felids (*Lynx rufus*, *Puma concolor*), and other canids (*Canis familiaris*, *Urocyon cinereoargenteus*, *Vulpes macrotis*), probably also at least occasionally take juveniles and small adults. However, larger adults have few enemies other than humans.

C. ruber is mild-mannered for such a large snake, and aggressive behavior seems rare. Some will lie quietly without rattling when closely approached (Ritter 1921), and may even hide their heads (Medica 2009); others rattle excessively. However, it can bite at the least expected moment, and if cornered or provoked can put up a spirited fight—coiling, raising up, and striking viciously. *C. r. lucasensis* performs a body-bridge when introduced to the odor or sight of *Lampropeltis getula* or *L. mexicana* (Bogert 1941; Rubio 1998; Weldon and Burghardt 1979; Weldon et al. 1992). During laboratory tests by Cowles and Phelan (1958), the heart rate of *C. r. ruber* increased 42% when presented with the odor of *L. getula* and 32–38% when tested with methyl mercaptan from the skunk, *Spilogale putorius*. In addition, Clarke and Marx (1960) reported that the approach of a potential enemy also causes the snake's heart rate to increase.

PARASITES AND PATHOGENS: The red diamond rattlesnake is a host for the following parasites: cestode, *Mesocestoides* sp.; nematode, *Ophidascaris labiatopapillosa*; pentastomids, *Porocephalus* sp.; and mites, *Ophionyssus serpentium* (Ernst and Ernst 2006; Goldberg and Bursey 1999b; Rego 1980/1981, Riley and Self 1979; Schroeder 1934). Captives have tested positive for various *Arizona* and *Salmonella* bacteria (Ghoniem and Refai 1969; Murphy and Armstrong 1978; Zwart et al. 1970). Harshbarger (1974) reported a sarcoma in one.

POPULATIONS: No quantitative data on the population dynamics of *C. ruber* have been published. Locally, it may be common, especially around a hibernaculum. Of more than 12,000 snakes collected or observed by Klauber (1972) from 1923 to 1938 in San Diego County, *C. r. ruber* made up 1.1–10.6% in any zone, but was most numerous in the inland valleys (10.6%), desert foothills (9.1%), and foothills (8.3%) and along the coast (6.6%). It comprised 7.8% of all the snakes collected. Murphy (1990) reported

Crotalus ruber lucasensis (Monserrate Island, Baja California Sur; John H. Tashjian)

that *C. r. ruber* made up 12.6% of the 111 snakes of 15 species he collected during a 4-day (28–31 May 1987) study in San Diego and Imperial counties.

The automobile (C. H. Ernst, personal observation; Klauber 1972; Murphy 1990), rifle, and collection for the pet trade have taken their toll on populations of *C. r. ruber*, but habitat destruction probably has caused the greatest loss. Reserves should be established in Southern California for this snake, and any conservation plan must include preservation of its chaparral and sage scrub habitat (Halama et al. 2008). The subspecies *C. r. exsul* and *C. r. lorenzoensis* are restricted to only one island each, and *C. r. lucasensis* has a highly fragmented distribution; all three are considered at least vulnerable to extinction (Greene and Campbell 1992).

Using data from Tracey (2000), Tracey et al. (2005) developed a nonlinear regression model for movement by *C. r. ruber* in response to a single landscape feature, particularly the effects of roads and housing developments, on telemetered snakes in San Diego County. This model should prove helpful in planning future conservation strategies for suburban *C. ruber*. We refer the reader to their paper for details.

REMARKS: Murphy et al. (1995) compared the genetic and morphological diversity of *Crotalus exsul* Garman, 1884 from Isla de la Cedros, Mexico to *C. ruber* from mainland Baja California, Southern California, and several islands in the Gulf of California, and, because of the closeness of the two snakes, proposed that they both be included in the single species *C. exsul* by priority. However, the International Commission on Zoological Nomenclature (2000) gave the name *ruber* precedence for the species because of its extensive use since 1892 for populations over the mainland and on the islands in the Gulf of California, as *exsul* has been used almost solely for the island populations off the southern Pacific coast of Baja California. Thus we use the specific name *Crotalus ruber*.

Klauber (1972) thought *C. ruber* was derived from *C. atrox*, and Foote and MacMahon (1977) and Gloyd (1940) thought it closely related to the other two diamondback rat-

Crotalus ruber lucasensis (Monserrate Island, Baja California Sur; John H. Tashjian)

tlesnakes, *C. atox* and *C. adamanteus*, and the Mexican *C. tortugensis*. However, recent mtDNA and RNA studies by Murphy et al. (2002) show it to be most closely related to *C. catalinensis*. *C. ruber* is a member of the *C. atrox*-group of rattlesnakes, which also includes *C. adamanteus*, *C. catalinensis*, and *C. tortugensis*.

The taxonomy and literature of the species are reviewed in Beaman and Dugan (2006).

Crotalus scutulatus (Kennicott, 1861)
Mojave Rattlesnake
Chiauhcoatl

RECOGNITION: The following description of the Mojave rattlesnake is based on those presented in Bush and Cardwell (1999), Campbell and Lamar (2004), Ernst and Ernst (2003), Gloyd (1940), and Klauber (1972).

The TBL_{max} of *C. scutulatus* is 137.3 cm (Tennant 1984), but most adults are less than 100 cm long. Its body is greenish-gray, olive-gray, olive, yellowish-green, greenish-brown, brown, or even yellowish, and patterned with 27–44 (mean, 37) light-bordered (sometimes interrupted), yellowish-olive, dark gray or brown, diamond-shaped, or oval to hexagonal, dorsal blotches. The light scales separating these blotches usually lack dark pigment. Cardwell and Alexander (2006) and Nickerson and Mays (1968) described albinistic individuals and variations to the blotch pattern. Some individuals may contain one or two dorsal stripes, particularly those from Mexican populations. The tail has 2–8 alternating light gray and dark brown to black bands crossing its base, with the dark ones narrower than the light ones. The proximal rattle segment has some black pigment on its dorsal surface (occasionally the rattle is absent; Cardwell and Banashek 2006). Dorsal body scales are keeled and pitted; the dorsal surface's microdermoglyphic pattern is of swirling linear ridges (*C. s. salvini*; Harris 2005; Stille 1987). The scales occur in 23 (21–25) anterior rows, 25 (21–29) rows at midbody, and 21–23 (19–25) posterior rows. The venter is white to cream with only slight pigment encroachment along the sides, and some small dark spots under the tail. It contains 165–192 ventrals, 15–29 subcaudals, and an undivided anal plate. The head is patterned with a dark, light-bordered stripe that extends from the orbit downward to above the angle of the mouth, some dark pigment occurs on the posterior borders of the supraoculars, and a pair of dark blotches is usually present on the occiput. The rostral contacts two small internasals and is usually higher than it is wide. No prefrontals are present, but there are two pairs of canthals with the posterior pair the largest. Between the internasals and intersupraocular scales lie 6–21 scales, but normally only 2 (occasionally 3) rows of intersupraoculars occur between the large supraocular scales (those most posterior are smaller). Commonly, a large, sometimes subdivided crescentic scale is positioned along the posterior-medial edge of each supraocular. Laterally on the face are 2 nasals (the prenasal contacts the first supralabial; the postnasal rarely contacts the upper preocular), 1 (rarely 2) loreal, 2 preoculars, 2 (3) postoculars, several suboculars, normally only 1 interoculabial, 13–15 (12–18) supralabials, and 14–16 (12–18) infralabials. On the underside of the head are a small mental scale, 2 pairs of chin shields (the second is the largest), but no small scales lying between the chin shields.

The bifurcate hemipenis has a divided sulcus spermaticus, each lobe with approximately 49 spines and 40 fringes, and 0–2 spines lying in the crotch (Klauber 1972).

C. scutulatus has 3 palatine teeth, 7 (6–8) pterygoid teeth, and 9–10 dentaries (Klauber 1972).

Males (TBL$_{max}$, 137.3 cm) have 155–190 (mean, 177) ventrals, 21–29 (mean, 25) subcaudals, and 5 (3–8) dark tail rings. Females (TBL$_{max}$ probably about 100 cm) have 165–192 (mean, 181) ventrals, 15–25 (mean, 19) subcaudals, and 3–4 (2–6) dark tail rings.

GEOGRAPHIC VARIATION: Two subspecies have been described. *Crotalus scutulatus scutulatus* (Kennicott, 1861), the Mojave green rattlesnake or chiauhcoatl, ranges from extreme southwestern Utah, adjacent southern Nevada, and southeastern California

Crotalus scutulatus scutulatus (Clark County, Nevada; John H. Tashjian)

*Crotalus scutulatus
scutulatus*
(R. D. Bartlett)

southwestward through western, central, and southern Arizona, northern Sonora, and Chihuahua into Hildalgo and Otero counties in New Mexico and western Texas; and south in Mexico to the state of Querétaro. Its TBL_{max} is 137.3 cm, but most individuals are less than 100 cm long; it is usually greenish, but may be brown or yellow with 36–37 (27–44) normally light-bordered body blotches; 3–5 (2–8) dark tail bands (the light ones are gray or white); 2 unobscured occipital marks; usually a complete angular stripe on the side of the face; a black dorsal surface on the proximal rattle segment; 178–181 (166–192) ventrals; 19–24 (15–29) subcaudals; 14–15 (12–18) supralabials; and 15–16 (12–18) infralabials. Within the United States, *C. s. scutulatus* has two distinct venom types (venoms A and B), but also populations with venom intergrade between the two types (see "Venom and Bites"). *Crotalus s. salvini* Günther, 1895, the Huamantlan rattlesnake or chiauhcoatl de Salvin, resides in Querétaro (Domínguez-Laso et al. 2007), Hidalgo, Tlaxcala, Pueblo, and southwestern Veracruz, Mexico. Its TBL_{max} is about 100 cm (Klauber 1972), but most individuals are 60–70 cm long; it is olive-gray to yellow with 32–33 (30–35) brownish-olive to black, borderless, body blotches; 4–6 dark tail bands (the light bands are the same color as the body ground color); the 2 occipital marks sometimes obscured by dark pigment; a reduced, angular, dark (often black) stripe on the side of the face; the dorsal portion of the proximal tail segment with slight brownish pigment; 168–172 (165–175) ventrals; 19–25 (18–26) subcaudals; 13–14 (12–15) supralabials; and 14 (13–15) infralabials.

CONFUSING SPECIES: Other species of *Crotalus* that occur within the range of *C. sculatatus* can be identified by the key presented above and Bush and Cardwell (1999). *C. atrox* has broader dark tail bands (3+ scales wide versus 2 or less in *C. scutulatus*), more numerous dorsal head scales, and a light-bordered lateral dark facial stripe that ends in front of the corner of the mouth.

KARYOTYPE: The Mojave rattlesnake has 36 diploid chromosomes—16 macrochromosomes (including 4 metacentric, 6 submetacentric, and 4 subtelocentric), and 20 microchromosomes. Females are ZW; males ZZ (the Z macrochromosome is meta-centric-submetacentric; the W macrochromosome is submetacentric-subtelocentric) (Baker et al. 1972; Cole 1990; Zimmerman and Kilpatrick 1973). In contrast, Stewart et al. (1990) reported the species has 4 metacentric, 8 submetacentric, and 4 telocentric macrochromosomes.

FOSSIL RECORD: Pleistocene (Rancholabrean) fossils have been found in Distrito de Zumpango, Mexico (Brattstrom 1954b, 1955b), and in Deadman Cave, southern Arizona (Mead et al. 1984).

DISTRIBUTION: *C. scutulatus* ranges from southwestern Washington County, Utah (Klauber 1932; Woodbury and Hardy 1947a); Lincoln and Clark counties, Nevada; and Kern, Los Angeles, and San Bernardino counties, California, southeast through Arizona and southward from Trans-Pecos, Texas, and southwestern Hildalgo and Otero counties, New Mexico, to the southern edge of the Mexican Plateau in Puebla and adjacent Veracruz. The species' total distribution encompasses elevations from near sea level to about 2,500 m.

HABITAT: *Crotalus scutulatus* is mostly a flatland species, but it can be found on rocky or wooded hillsides and on lava flows. It inhabits prairie valleys, riparian stretches, semiarid grasslands, open brush areas, and deserts where the soil is not particularly rocky. Both mammal (*Ammospermophilus leucurus*, *Dipodomys* sp., *Spermaphilus beecheyi*, and *Vulpes macrotis*; discussed below) and tortoise (*Gopherus agassizii*) burrows are used for retreats. Some typical vegetation in its various habitats includes agave (*Agave lecheguilla*), allthorn (*Koeberlinia spinosa*), cacti (*Neobuxbaumia* sp., *Opuntia* sp.), creosote (*Larrea tridenta*), grasses (*Bouteloua* sp., *Hilaria* sp., *Sporobolus* sp.), hackberry (*Celtis*

Distribution of
Crotalus scutulatus.

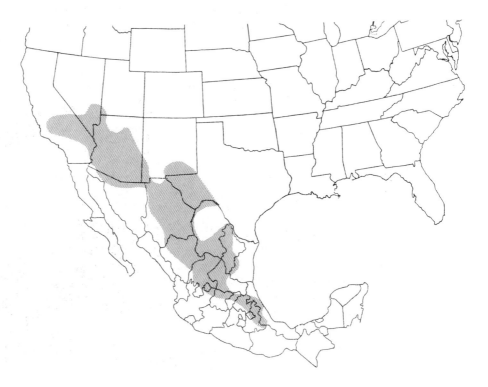

pallida), ironwood trees (*Olneya tesota*), limberbush (*Jatropha* sp.), manzanita (*Arctosa-phylos* sp.), mariola (*Parthenium incanum*), mesquite (*Prosopis* sp.), mimosa (*Mimosa bi-uncifera*), Morman tea (*Ephedra nevadensis*), ocotillo (*Fouquieria* sp.), palmetto (*Sabal mexicana*), palo verde (*Cercidium floridum, C. microphyllum*), tarbush (*Flourensia cernua*), and various yuccas (*Yucca* sp., including the Joshua tree *Y. brevifolia*). In its wooded hillside habitats are various aspens and poplars (*Populus* sp.), junipers (*Juniperus* sp.), oaks (*Quercus* sp.), and pines (*Pinus* sp.). In Chihuahua, the density of plants more than 1 m tall in 57 plots within the habitat of *C. s. scutulatus* was 41.1 (Reynolds and Scott 1982).

BEHAVIOR AND ECOLOGY: Most data on the biology of the species concerns the northern subspecies *C. s. scutulatus*; information on the southern subspecies *C. s. sal-vini* is sparse.

The annual activity cycle of *C. s. scutulatus* in the United States lasts from mid-March to November. Most individuals are active from late April or early May through September, with May and July–August the months of peak surface activity (particularly August—Degenhardt et al. 1996; Reynolds 1982).

Much time is spent in animal burrows or under rocks. Such burrows help the snake in its thermoregulation by providing a cooler site during the hot, summer days, and a warmer retreat when the air is cool in the spring and fall. The snakes stop surface activity when ATs turn cool, and remain in subsurface retreats during the winter. Wright and Wright (1957) reported that wild honeybees are often found in the same hole (double trouble for the would-be predator or the inattentive herpetologist).

During a study in the Mojave Desert of California during March–October, underground *C. s. scutulatus* were found to occupy *Ammnospermophilus leucurus* and/or *Dipodomys* sp. burrows 92% of the time; burrows of *Vulpes macrotis*, 5% of the time; and those of *Spermophilus beecheyi*, 1% of the time (Cardwell 2005a). Gates (1957) found one apparently estivating in early June within an Arizona cave.

Crotalus scutulatus salvini (Pueblo; Staten Island Zoo; John H. Tashjian)

The snakes may hibernate at these sites if the burrows are deep enough, but, if too shallow, they must seek retreats below the frost-line. Cardwell (2005a) reported that of 37 winter documented refuge selections by 17 *C. s. scutulatus* in California's Mojave Desert 27 (73%) were in *Ammospermophilus leucurus* or *Dipodomys* sp. burrows, 7 (19%) were in *Vulpes macrotis* burrows, 2 (5%) were located in unidentified burrows under a collapsed Joshua tree (*Yucca brevifolia*), and 1 (3%) was in a *Spermophilus beecheyi* burrow. A male spent the winters of 2003–2004 and 2004–2005 in the same *Vulpes* burrow, but this was exceptional, as the norm is not to use a hibernaculum more than once (Cardwell 2005a, 2008). One snake apparently shared its hibernaculum in a *Vulpes* burrow with a burrowing owl (*Athene cunicularia*) (Cardwell 2005b). Baxter and Stewart (1990) found one hibernating in a desert tortoise (*Gopherus agassizii*) burrow. *C. s. scutulatus* apparently does not aggregate for hibernation, but instead spends the winter alone or in pairs (Lowe et al. 1986).

According to Campbell and Lamar (2004), this snake is mostly nocturnal at lower elevations, but frequently diurnal at high elevations during the cooler months. It is commonly crepuscular or nocturnal when ATs rise to 32–43°C (Rader 1995). Although primarily nocturnal in the summer in the United States (1959–2337 hours; Klauber *in* Stebbins 1954), the snake forages or basks (often on or along the side of a road) in early morning, and may forage during the late afternoon in the spring and fall. At higher elevations it is often abroad during the day, even in summer (particularly *C. s. salvini*). During August 2001 through April 2004, Cardwell (2006) made more than 2,700 observations of *C. s. scutulatus* in San Bernardino County, California, during which 1,097 (67%) of the snakes were found motionless in a resting/ambush posture.

Klauber (1972) found *C. s. scutulatus* active at ATs of 17–27°C. Brattstrom (1965) recorded BTs of 22.2–34.0 (mean, 30.0) °C from wild *C. s. scutulatus*, and Pough (1966) reported a mean BT of 26.5 (range, 21.9–29.8) °C. According to Brattstrom (1965), the CT_{max} of the species is 42°C. Plummer (2000) observed an individual apparently become heat-stressed while feeding on a dirt road at 1047 hours; it dragged its prey off the road to a shaded spot before resuming swallowing it.

Cardwell (2008) reported that reproductive pairs of *C. s. scutulatus* had the following mean BTs: all males, 24 (range, 13–32; n = 23) °C; all females, 27 (range, 15–36; n = 15) °C; copulating males, 22 (range, 13–32; n = 7) °C; copulating females, 20 (range, 15–24; n = 2) °C; males on surface, 23 (range, 13–32; n = 13) °C; females on surface, 28 (range, 22–36; n = 7) °C; males underground, 25 (range, 15–32; n = 9) °C; and females underground, 25 (range, 15–31) °C.

Obtaining free water is a problem for an arid habitat–dwelling snake such as the Mojave rattlesnake. It possibly obtains water from pools or brooks when they are available, but Cardwell (2006) did not see any snake drink from available free habitat rainwater, but reported another method that this animal uses to drink. He observed *C. s. scutulatus* in San Bernardino County, California, harvest rain drops from the surface of their body skin (a behavior also known for *C. oreganus*; Aird and Aird 1990; Ashton and Johnson 1998). In 12 of 16 (75%) encounters during rainfall, the snakes were coiled with the body spiraling out from the tail in the center. The coils of the anterior body were adjacent to, rather than on top of, each other. The venter was very flattened dorsoventrally, maximizing the surface area exposed to the rain. Adjacent posterior ventral coils were touching, forming a shallow trough between them where rainwater could collect. The snakes' heads were facing nose downward with the rostral area in the trough. Six additional observations of this same posture were observed after rain events. When drinking from the coil trough, the snake positioned its head so that the rostrum was within 2 mm of, or actually in contact with, its body scales. Slight rhythmic movements occurred in its temporal-mandibular muscles and the rostral area was moved every few seconds to different points along the trough. Although no actual jaw movement was observed, water was seen to disappear as the snake's rostral area came in contact with it. Mean BTs and ATs recorded during rain-harvesting events were 12–32 (mean, 18.2) °C and 11–30 (mean, 17.6) °C, respectively.

Males are more surface-motile than are females. During March–May and August–November 2003 and 2004, mean daily movement of male California *C. s. scutulatus* was 38 (range, 1–89) m per day, which was more than 3 times greater than that of females, 12 (range, 2–29) m per day. The mean male home range was 31.6 (range, 14.6–52.6) ha; that of females, 5.6 (range, 2.0–10.5) ha (Cardwell 2008). *C. s. scutulatus* uses a typical undulatory movement during flight (Bartholomew and Nohavec 1995).

A possible incident of climbing by this snake was reported by Boone (1937), who related that a "yellow Pacific rattlesnake" (thought to have been a Mojave green rattlesnake by Klauber 1972) climbed 1.5 m into a mesquite tree. The species is quite capable of swimming when placed in water, moving with lateral undulations, with its head held high out of the water (Klauber 1972).

Arizona male *C. s. scutulatus* are agonistic toward each other from late summer to early autumn (July to September, the monsoon season), and in early to midspring (March to May), probably corresponding to the mating seasons of the species (Reiserer *in* Cardwell 2008; Schuett et al. 2002a). Cardwell (2008) observed male combat on 21 August, but never encountered multiple male *C. s. scutulatus* together. In addition, 65 of 78 (83%) rattlesnake envenomations of humans in Southern California in 2003 and 2004 occurred during these periods of male aggression (Cardwell et al. 2005).

REPRODUCTION: Respectable reproductive data are available for the northern subspecies, *C. s. scutulatus*, particularly involving the recent radio-telemetry study by Cardwell (2008) in San Bernardino County, California. Unfortunately, comparable data are unknown for the Mexican subspecies *C. s. salvini*.

In the western Mojave Desert in California, male *C. s. scutulatus* mature at an SVL of 40 cm and 1.5 years of age, and females mature at an SVL of 60 cm at 2 years of age (Cardwell 2008). The shortest vitellogenic female and male with sperm examined by Cardwell (2008) had SVLs of 61.1 cm and 41.1 cm, respectively. The smallest reproductive female and male *C. s. scutulatus* from Arizona, New Mexico, Texas, and Chihuahua, Mexico, examined by Goldberg and Rosen (2000) had SVLs of 60.0–69.9 cm and 41.1–47.8 cm, respectively; another female with a SVL of 54.3 cm was undergoing secondary vitellogenesis, but it is not known if these follicles would have completed development. The smallest gravid female examined by Klauber (1972) was 63 cm in TBL.

Goldberg and Rosen (2000) found females with inactive ovaries from May to September, some undergoing early yolk deposition in June and August–November, and others with enlarged follicles (> 12 mm) or embryos in March–April and June–July. Twenty-eight percent of the March–August females contained > 12 mm follicles or developing embryos, and would have probably produced young that year. Minton (1959) found a gravid Texas female in early July, and Van Devender and Lowe (1977) collected a gravid female on 18 July in Chihuahua. Cardwell (2008) palpated nine females with embryos between 2 May and 21 August, and observed seven postpartum ones between 18 August and 9 November in California. Females probably reproduce biennially or triennially. Although drought curtailed courtship and copulation during the 2002 activity season in California, the gravid rate was not significantly affected in 2002 and 2003 (Cardwell 2008), so viable sperm apparently can be stored over the winter or longer.

Male *C. s. scutulatus* examined by Goldberg and Rosen (2000) either had regressed testes in March–May and October, were undergoing recrudescence in March–July, or were going through spermiogenesis in June–September; mature sperm was present in the vas deferens throughout the annual activity season. The kidney sexual segment was enlarged and contained secretory granules in 100% of the males examined in March–May, 89% in June, 82% in July, and 100% again in August–October. Schuett et al. (2002a) described a similar testicular histological cycle.

Testicular activity of male Chihuahuan *C. s. scutulatus* peaks in August, when the seminiferous tubule diameter is significantly greater than in other months. Testicular length does not vary significantly from month to month, but testicular mass does, peaking in September. Spermiogenesis occurs from July through August, and mature sperm are present in the accessory ducts throughout the summer (Jacob et al. 1987).

The levels of plasma testosterone, 5α-dihydrosterone, and 17β estradiol in males undergo seasonal cycles. The lowest levels occur in June; the highest levels, in August–September and again in March. Peak levels of the three steroid hormones correspond to the bimodal mating seasons, spermiogenesis in summer, and hypertrophy of the kidney sexual segment. Regression analysis shows that they are both positively and significantly correlated with each other (Schuett et al. 2002a).

The natural mating season of *C. s. scutulatus* is bimodal (Lowe et al. 1986; Reiserer 2001; Schuett 1992; Schuett et al. 2002a). The spring mating season lasts from February to May in Arizona (Lowe et al. 1986). Thirty-one pairs were observed engaged in reproductive activity in the western Mojave Desert of California by Cardwell (2008). Of the 25 in which the specific behavior could be identified, 3 involved a male trailing a female, 15 courtship, and 7 copulation. The encounters occurred during the periods of 16 March to 16 May (7 observations in March, 7 in April, and 4 in May) at 0849–2302 hours (March, 0849–2035 hours; April, 1106–1814 hours; May, 1040–2302 hours) and from 21 August to 7 October (2 in August, 10 in September, and 1 in October) at 1834–2119 hours (August, 2024–2055 hours; September, 1833–2022 hours; and October, 2119 hours). ATs during these events ranged from 7.0°C to 36.0°C. Cardwell classified the overhead at the sites as clear (n = 6), clear/foliage (1), clear/burrow (1), foliage (11), foliage/burrow (1), burrow (10), and various (1); the distance to the nearest shrub varied from 0 to 4 m.

In August 1975 (exact date not given), Jacob et al. (1987) found a heterosexual pair of *C. s. scutulatus* lying on a road in Chihuahua. Soon a second male emerged onto the highway directly in line with the original pair and less than 2 m behind them; however, the snakes did not copulate.

The sequence of courtship/mating behaviors by *C. s. scutulatus* has been listed by Cardwell (2008): trailing, accompaniment, courtship, and copulation. Male courtship involves lying on top of the female, head-jerking, rapid tongue-flicking, chin-rubbing, and tail-searching for her vent. Hardy (1998) has reported male-male copulations by captive *C. s. scutulatus*.

Parturition by *C. s. scutulatus* takes place in July–September after a GP of about 170 days, with most newborn young being born in mid- to late August (Greene et al. 2002; Klauber 1972). It results in an abrupt loss of an estimated 30–50% of female BM (Cardwell 2008). Van Devender and Lowe (1977) reported a birth on 18 July in Chihuahua. Cardwell (2008) first encountered neonates in California's Mojave Desert from 14 August to 9 October, and postpartum radio-telemetered females were recorded on 25 August and 3 September (n = 2). Observations indicate that the female may remain with her neonates for a period of time after parturition, and possibly defend them (see Greene et al. 2002 for a discussion).

Litters of *C. s. scutulatus* contain 2 (Stebbins 2003) to 17 (Mellink 1990) young (mean, 8.2; n = 57). Neonate *C. s. scutulatus* are 17.5–28.3 (mean, 25.1; n = 23) cm in TBL and 9.5–13.9 (mean, 11.3; n = 25) g in BM. Klauber (1972) reported an average TBL of 25 cm for newborn *C. s. salvini*.

A natural hybrid *C. s. scutulatus* × *C. v. viridis* has been found in Hudspeth County, Texas (head illustrated in Murphy and Crabtree 1988), and captive hybridizations between *C. scutulatus* and the rattlesnakes *C. atrox*, *C. cerastes laterorepens* (illustrated in Powell et al. 1990), *C. oreganus oreganus* and *C. unicolor* (illustrated in Klauber 1972) have also been reported (Cook 1955; Jacob 1977; Klauber 1972; Perkins 1951; Powell et al. 1990). Venom characteristics indicate that *C. s. scutulatus* and *C. v. viridis* regularly hybridize in New Mexico (Glenn and Straight 1990).

GROWTH AND LONGEVITY: Few growth data are available. Van Devender and Lowe (1977) reported that 2 first-year Chihuahua juveniles had 32.1 cm and 33.7 cm SVLs. BM is positively coordinated with the SVL (Schuett et al. 2002a).

A wild-caught male of unknown age, still alive at the time of the report, had survived 14 years and 5 days in captivity at the Central Texas Zoo, Waco (Snider and Bowler 1992).

DIET AND FEEDING BEHAVIOR: *C. scutulatus* is predominantly a mammalian predator, although several other types of prey have been reported. Reynolds and Scott (1982) found mammals in 91.7% of the Chihuahuan *C. s. scutulatus* they examined— kangaroo rats occurred in 39.6% of the stomachs containing food items, pocket mice in 20.8%, white-footed mice in 16.5%, ground squirrels in 10.4%, jackrabbits in 4.2%, and cottontails in 2.1%. Klauber (1972) found mammal remains in 21 specimens of *C. s. scutulatus* but lizard remains in only 2. Prey is selected on the basis of size. Prey that is either too large or too small is rejected, and potential prey that could possibly harm the snake is not accepted (Reynolds and Scott 1982). Klauber (1972) found mammal remains in two *C. s. salvini*. Both live prey and carrion are consumed.

Known natural prey include mammals—kangaroo rats (*Dipodomys merriami, D. panamintinus, D. spectabilis*), pocket mice (*Perognathus flavus, P. intermedius, P. longimembris, P. pencillatus*), white-footed mice (*Peromyscus eremicus, P. maniculatus*), ground squirrels (*Ammospermophilus leucurus; Spermophilus spilosoma, S. tereticaudus*), and hares and rabbits (*Lepus californicus, Sylvilagus audubonii*); birds' eggs; reptiles—lizards (*Colenyx brevis, Cnemidophorus tigris, Holbrookia* sp., *Phrynosoma platyrhinos, Sceloporus* sp., *Uta stansburiana*) and snakes (*Phyllorhynchus decurtatus*); amphibians—toads (*Bufo* sp.), spadefoots (*Scaphiopus* sp., *Spea* sp.) and frogs; centipedes; and insects (Bogert *in* Klauber 1972; Boone 1937; Cardwell 2008; Cardwell et al. 2005; Cromwell 1982; Dammann 1961; Huey 1942; Jennings *in* Brown 1997; Johnson et al. 1948; Kauffeld 1943a; Klauber

1972; Lowe et al. 1986; Parker 1974; Plummer 2000; Reynolds 1978; Reynolds and Scott 1982; Turner *in* Tennant 1984). In addition, captives have taken brown rats (*Rattus norvegicus*), house mice (*Mus musculus*), woodrats (*Neotoma albigula*), lizards (*Anolis carolinensis, Eumeces fasciatus, Holbrookia* sp., *Uta stansburiana*), and a snake (*Crotalus cerastes*) (Brown and Lillywhite 1992; C. Ernst, personal observation; Klauber 1972; Strimple 1993b; Vorhies and Taylor 1940).

Young snakes probably take lizards more frequently than do adults (Dammann 1961; Reynolds 1978; Turner *in* Tennant 1984).

C. scutulatus mostly captures prey by ambushing it, but the snake probably takes some while actively foraging. Prey are found by infrared detection, sight, and olfaction (prey odor trails are followed as the snake rapidly tongue-flicks; C. Ernst, personal observation). When lying in ambush, the snake assumes a rounded coil-posture with its head on top facing and within striking distance of a prey trail or close to the entrance of a rodent burrow. Bitten mammals are usually released immediately and allowed to wander off until quickly succumbing to the venom. The snake then follows the animal's scent trail and swallows it after it is dead. Because it uses rodent burrows extensively as retreats, underground feeding probably also occurs.

VENOM DELIVERY SYSTEM: The Mojave rattlesnake does not have particularly long fangs for its size, but long fangs are not needed when a snake has extremely virulent venom. Seven *C. s. scutulatus* (most probably from San Bernardino County, California) examined by Klauber (1939) with 81.2–108.5 cm TBLs had 6.7–8.8 (mean, 7.6) mm FLs, a mean TBL/FL of 126%, and a mean HL/FL of 5.14%. The mean angle of fang curvature was 64°. In compliance, Rader (1995) reported that 83.8–109.2 mm *C. s. scutulatus* from the eastern Mojave Desert of California only have 7–9 mm fangs. Klauber (1927) reported that the smaller *C. s. salvini* had a mean TBL/FL of 97% and an HL/FL of 4.4%.

VENOM AND BITES: *C. s. scutulatus* from the southern portion of the subspecies' range is probably the most dangerous snake in the United States; coupled with *C. s. salvini* they are among the most virulent snakes in the Americas. McCue (2005) has reported that it has the highest relative toxicity, 2.00 on $\log_{10}$ scale, of any American snake that he tested, comparable only to that of the cobra, *Naja naja*. The venom found in these populations is highly neurotoxic, but in other more northern and western populations of *C. s. scutulatus* it contains hemotoxic elements (discussed later).

The venom of southern individuals affects the heart, skeletal muscles, and neuromuscular junctions (Castilonia et al. 1980). The lethal portion is a presynaptic-acting acidic protein chain composed of 88 amino acids termed Mojave toxin α, which belongs to the phospholipase A_2-complex and has a molecular weight of approximately 20,000 daltons. It blocks transmission of impulses from nerve to muscle and alters uptake and release of several neurotransmitters (Harwood 1982; Wooldridge et al. 2001). Its amino acid sequence is very similar to that of related toxins from the venom of South American *C. durissus* and some populations of *C. viridis*.

In addition, a second virulent fibrinogenolytic hemorrhagic toxin (Martínez et al. 1990), Mojave toxin β (a basic protein chain composed of 122 amino acids), is also present in some populations of *C. scutulatus* (see below). Surprisingly, pooled venom from 12 *C. s. scutulatus* (localities unknown) tested negatively for Mojave toxin in cross-activity tests using rabbit antimyotoxin by Li and Ownby (1994).

Protein concentrations, presumably of these different toxins, vary between individual snakes (Johnson et al. 1968b) so that in some cases smaller *C. scutulatus* may have more potent venom than larger individuals. However, larger snakes with less potent venom can be just as deadly because of their potential for injecting greater amounts of venom per bite. The seriousness of the envenomation is dependent upon both the quantity and quality of the venom injected.

Electrophoretic studies have revealed variation in the venoms from the different populations of *C. s. scutulatus* in Big Bend, Texas, and southeastern Arizona, suggesting that several genetically diverse groups occur in the United States (Rael et al. 1984, 1986, 1993). Additional testing has shown that two distinct venom-type populations and a zone of intergradation occur in Arizona (Glenn and Straight 1989, 1990; Glenn et al. 1983; Huang et al. 1992). The venom of the Chihuahuan population (venom A) contains the neurotoxic Mojave toxin α and is lacking in hemorrhagic and specific proteolytic activities. The Sonoran population (venom B) lacks Mojave toxin α, but does contain Mojave toxin β and produces hemorrhagic and proteolytic symptoms. Venoms of *C. s. scutulatus* from regions between the venom A and venom B populations in Arizona contain both Mojave toxins α and β, and produce the proteolytic and hemorrhagic activities of venom B. The interperitoneal LD_{50} values of the A+B venoms are 0.4–2.6 mg/kg, compared to 0.2–0.5 mg/kg for venom A individuals and 2.1–5.3 mg/kg for venom B individuals. High-pressure liquid chromotography shows that A+B venoms exhibit a combined protein profile of venoms A and B. These data indicate that an intergrade zone exists between the two venom types that arcs around the western and southern regions of the venom B population. Within these regions, *C. s. scutulatus* can have three major venom types.

Wilkinson et al. (1991) conducted starch gel electrophoretic studies of 55 enzymes from tissues of *C. s. scutulatus* within the venom intergrade zone, and found high gene flow between venoms A and B with low genetic divergence, indicating the two venom populations are conspecific.

Studies of the venom yield and toxicity of *C. s. scutulatus* indicate that the amount of venom injected in one bite is sufficient to cause the death of its natural prey as well as a human. The neurotoxic venom A is about 10 times more toxic than that of most other North American crotalid snakes (Ernst and Zug 1996).

Several mean dried venom yields have been reported. Klauber (1972) made 228 milkings of *C. s. scutulatus* (217 adults, 11 juveniles) that yielded a total of 10,498 mg of venom, an average yield of 77 mg per extraction, and a maximum yield of 141 mg. Other reported yields are 50–90 mg (Minton and Minton 1969; Russell and Brodie 1974; Russell and Puffer 1970), 77 (50–150) mg per fresh adult snake (Ernst and Zug 1996), 75–150 mg (Russell 1983), and 150 mg (Altimari 1998). Size-related differences between the sexes of *C. s. scutulatus* occur in the mean total dry venom yield: 13 females with 52.0–87.5 cm SVLs, 22.9 (range, 8–45) mg; 16 males with 60.5–110.0 cm SVLs, 61.6 (range, 15–139) mg (Glenn and Straight 1978).

The raw venom has a 79.5–198.0 mg/mL protein and other constituents concentration (Johnson et al. 1968b). Brown (1973) and McCue (2005) reported that proteins made up 70% of the solid content of the venom. Tests by Glenn et al. (1983) revealed mean proteolytic activities ($\Delta OD/100$ $\mu g/15$ minutes) of 0.005 (range, 0–0.013) and 0.192 (range, 0.114–0.246); and mean esterase activities ($\Delta OD/20$ $\mu g/$minute) of 0.90

(range, 0.37–1.78) and 0.88 (range, 0.35–1.59) for venoms A and B of *C. s. scutulatus*, respectively.

These same activity parameters in *C. s. salvini* were 0.001 (0–0.003) and 1.04 (0.92–11.8), respectively.

Venom from Arizona *C. s. scutulatus* with type A neurotoxic venom exhibited proteolytic but no hemorrhagic activity, but those from Arizona with either type B hemotoxic venom or a combination of A and B venoms showed both proteolytic and hemorrhagic activities. In contrast, venom from Texas snakes exhibited neither proteolytic nor hemorrhagic activity (Glenn and Straight 1989).

A more recent study of the venoms of this subspecies from Arizona and Texas by Sánchez et al. (2005a, 2005b) revealed that the venom of the Texas snakes lacked disintegrin genes, but did contain the Mojave toxin gene Mta/Mtb. The venom was not hemorrhagic, was generally not proteolytic, and did generally show fibrinogenolytic activity (with the exception of samples from Jeff Davis County). Those Arizona venoms with intermediate toxicity (intravenous LD_{50}, 0.84–1.05) were not hemorrhagic, were generally not proteolytic, did not contain disintegrin activity, but did possess disintegrin genes. Venoms with an intermediate LD_{50} of 1.05 demonstrated fibrinogenolytic activity. Snakes with the least toxic venom (intravenous LD_{50}, 2.9–5.5 mg/kg) were hemorrhagic, were proteolytic, had disintegrin activity (caused platelet aggregation and retraction), contained disintegrin genes, contained Mtb gene subunits, but lacked Mta subunits.

The various venom types of the Mojave rattlesnake are among some of the most studied. Important publications discussing their chemistry are those of Aird et al. (1990a, 1990b), Al-Joufi and Bailey (1994), Beasley et al. (1993), Bieber et al. (1975, 1990a, 1990b), Bober et al. (1988), Brown (1973), Castilonia et al. (1981), de Roodt et al. (2003), Faure (1999), Gawade (2004), Glenn and Straight (1985a), Hawgood (1982), John et al. (1994), Johnson and Bieber (1988), Lynch (2007), Mashiko and Takahashi (1998), McCue (2005), Meier and Stocker (1995), Nair et al. (1979), Powell et al. (2008), Rathbun and Heim (1982), Soto et al. (1989, 2007), Stegall et al. (1994), Tu (1977), Tubbs et al. (2000), Weinstein et al. (1985), and Wooldridge et al. (2001). These publications contain more detailed information than can be included in this book, so we refer the reader to them for details.

The mean lethal dose for a 350 g pigeon of *C. s. scutulatus* venom is 0.013 (0.005–0.020) mg (Githens and George 1931). Githens and Wolff (1939, intravenous) and Macht (1937, intraperitoneal) published LD_{50} values of 0.012–0.014 and 0.24 mg, respectively, for mice. Arce et al. (2003) reported a mean intraperitoneal LD_{50} for 16–18 g mice of 0.04 (range, 0.04–0.05) mg and a mean intravenous one of 0.15 (range, 0.12–0.19) mg. Minton and Minton (1969) reported the following mean LD_{50} values for standard mice: intravenous, 4.2 mg; intraperitoneal, 4.6 mg; and subcutaneous, 6.2 mg (probably *C. s. scutulatus*, venom types not differentiated); and Russell and Puffer (1970) listed 0.23 mg/kg intraperitoneal and 0.21 mg/kg intravenous LD_{50} values. Russell (1983) reported an intravenous mouse LD_{50} of 0.23 mg/kg (subspecies not listed, but presumably *C. s. scutulatus*). For *C. s. scutulatus* from Arizona-California, and Utah with venom A the mean intraperitoneal LD_{50} values for mice are 0.24 (range, 0.13–0.54) mg/kg and 0.11 (range, 0.09–0.12) mg/kg, respectively; and for those with venom B from northern Arizona are 2.8 (range, 2.3–3.8) mg/kg (Glenn and Straight

1978). Johnson et al. (1966) reported a mouse intraperitoneal LD_{50} of 1.2–2.5 (mean, 1.8) mg/kg.

Glenn et al. (1983) after further tests on *C. s. scutulatus* venom reported the mean intraperitoneal LD_{50} values to be 0.28 (range, 0.22–0.46) mg/kg for venom A snakes from near Tucson, Arizona, and 3.3 (range, 2.0–6.0) mg/kg for venom B snakes from north of Tucson. Bieber et al. (1975) reported intravenous LD_{50} values of 0.18 μg/g and 0.21 μg/g and an intraperitoneal LD_{50} of 0.23 μg/g for mice. An intraperitoneal LD_{50} of 0.23 mg/kg was reported by Russell and Brodie (1974).

Venom from *C. s. salvini* has an intraperitoneal LD_{50} of 0.30 (range, 0.18–0.40) mg/kg for mice (Glenn and Straight 1978; Glenn et al. 1983).

Paramecium multimicronucleatum has an LD_{50} of 0.49–0.55 (mean, 0.52) mg/mL when the venom is introduced into its medium (Johnson et al. 1966). The protozoan reacts to the venom with avoiding reactions, followed by cytolysis and cell fragmentation.

The estimated dry venom dosage needed to kill an adult human is relatively small, probably only 10–15 mg (Minton and Minton 1969), so a typical *C. s. scutulatus* contains 5–15 times more venom than is needed for a fatal envenomation of a human.

Human neurotoxic envenomation (venom A) by *C. s. scutulatus* is usually severe with high mortality occurring in untreated bites (Dart et al. 1992; Hardy 1983, 1986; Russell 1967a). A notable death from the bite of an Arizona *C. scutulatus* was that of Dr. Frederick A. Shannon, a well-known snakebite expert (Pinney 1981). In Arizona, bites by this snake and *C. atrox* are involved in most fatal cases. In Southern California, most human envenomations occur during bimodal mating periods (see "Reproduction"), when males are very motile and come into contact with humans more often (Cardwell et al. 2005).

Recent articles in the popular press have suggested that the venoms of several southwestern species of *Crotalus,* especially *C. helleri* and *C. scutulatus,* may be becoming more virulent. There is no scientific basis for this speculation (Hayes and McKessy 2010).

Symptoms of envenomation by *C. s. scutulatus* (combining those produced by both venom A and B types because of the overlap in some symptoms produced) include local pain and swelling (edema), blood blisters, ecchymosis, necrosis, fragility of veins, cardiopulmonary arrest, elevated heart rate, lowered blood pressure, double vision, difficulty in speaking and swallowing water, decrease in blood platelets and fibrinogen levels, increase in fibrinolytic split products, prevention of fusion of primary and C_2 myoblasts to multinucleate myotubes, shock, renal failure, drooping eyelids, depression, and diarrhea (Bush and Jansen 1995; Castilonia et al. 1980; Clark et al. 1997; Dart et al. 1992; Farstad et al. 1997; Hardy 1983, 1986; Jansen et al. 1992; Klauber 1972; Norris 2004; Norris and Bush 2001, 2007; Smith 1990; Ziolkowski and Bieber 1992).

Farstad et al. (1997) reported that the following symptoms were displayed in envenomations by *C. s. scutulatus* with Mojave toxin: edema, 78–100%; ecchymosis, 25–33%; pain, 0–1%; and neurotropic symptoms, 50–78%. In contrast, they reported that bites by *C. s. scutulatus* lacking Mojave toxin resulted in 100% edema, 73% ecchymosis, 67% pain, and only 7% neurotropic symptoms. Fatalities usually occur from respiratory failure. Hardy (1983) discussed 15 case histories of human envenomations.

In the United States, *C. s. scutulatus* is often very common in some heavily populated areas. Rapid urban expansion of some Arizona cities where this snake is common has brought it into more frequent contact with humans, creating a potentially dangerous scenario, particularly regarding children and pets.

PREDATORS AND DEFENSE: Predation reports are practically nonexistent. *Masticophis lateralis* ate one (Klauber 1972), and Cardwell (2008) suspected that coyotes (*Canis latrans*) had probably eaten three of his telemetered rattlesnakes. However, humans are the worst enemy—slaughtering many each year, particularly on the highways or through habitat destruction.

The sight or odor of ophiophagous snakes, such as *Coluber constrictor* (C. Ernst, personal observation), *Drymarchon corais*, and various species of *Lampropeltis* brings about striking, body-bridging, trunk inflation, and even body-flipping by *C. s. scutulatus* (Bogert 1941; Weldon and Burghardt 1979, Weldon et al. 1992). It can strike while moving (Medica 2009). *C. s. salvini* has also performed body-bridging when confronted with *Drymarchon* (Weldon and Burghardt 1979). When tested with the odor of methyl mercaptan from the skunk, *Spilogale putorius*, the heart rate of *C. scutulatus* increased 27% (Cowles and Phelan 1958). Medica (2009) has described a sequence of head-hiding in the species.

Individual temperaments vary, and, although some *C. scutulatus* are calm, in our experience, others are extremely nervous, excitable, and aggressive, particularly *C. s. salvini*, but also some individual *C. s. scutulatus*. If the snake cannot escape, it will coil, raise up, continuously rattle (the SVL-adjusted rattle loudness is about 70 dB, Cook et al. 1994; and the dominant frequency and frequency range are 10.5 kHz and 3.3–20.3 kHz, respectively, Young and Brown 1993), and strike viciously. Occasionally, one will even advance on the disturber (Bartholomew and Nohavec 1995), and sometimes flatten the neck in a hood-like display (Brown et al. 2000; Glenn and Lawler 1987), or even the head and trunk (Armstrong and Murphy 1979). Occasionally, one will strike so violently that its entire body becomes momentarily airborne. This is particularly true of *C. s. salvini*. It is also common for this species to strike through collecting bags, so one must be very careful with this extremely dangerous snake.

PARASITES AND PATHOGENS: *C. scutulatus* is the host of several parasites: protozoans—*Sarcocystis crotali* (Enzeroth et al. 1985); and worms—unidentified oligacanthorhynchid larval cytacanth acanthocephalans, and the digestive tract nematodes *Kalicephalus inermis*, *Physocephalus sexalatus*, and *Thubunaea cnemidophorus* (Babero and Emmerson 1974; Bolette 1997a; Ernst and Ernst 2006; McAllister et al. 2004; Widmer 1967). In addition, captives of this snake have succumbed to paramyxovirus infections (Jacobson and Gaskin 1992).

POPULATIONS: The Mojave green rattlesnake appears to be at least locally common in some areas of Arizona (Strimple 1996); it is the most common rattlesnake around Wickenburg (Gates 1957). *C. s. scutulatus* is more abundant in grassland habitats than is *C. atrox*. Wherever kangaroo rats (*Dipodomys*) are common, this snake may also be fairly abundant. Elsewhere, its numbers seem lower and its populations more scattered or spread out.

Of 425 rattlesnakes collected in central Arizona, 147 (34.6%) were *C. s. scutulatus* (Klauber 1972); and of 368 snakes observed on Arizona Rt. 85 in 1988–1991, 18 (4.9%) were this species (Rosen and Lowe 1994). *C. s. scutulatus* was also the most observed snake found on Chihuahuan highways in 1975–1977—104 / 418 (24.9%) (Reynolds 1982;

Reynolds and Scott 1982). In contrast, Klauber (1932) only collected one individual on the road in southwestern Nevada to Hoover Dam in 1931 and 1932 while reporting finding a number of nonvenomous snakes, 1 *C. helleri*, 1 *C. stephensi*, 3 *C. mitchellii*, and 9 *C. cerastes*.

In late summer and the fall, neonates increase the population considerably.

The numbers of *C. scutulatus* seem to be declining in some areas. Plant successional changes have caused a decline in *C. s. scutulatus* along two Arizona roads—in 1959–1961, 125 were found on Portal Road and U.S. 80 in Arizona, but in 1987 and 1989 a total of only 56 were observed (Mendelson and Jennings 1992). Campbell and Lamar (2004) have also remarked on its decline due to the succession of grasslands to shrub habitat, and Cruz-Sáenz et al. (2009) thought it endangered in Jalisco.

The adult sex ratio in Cardwell's (2008) California population was 19 males to 31 females, or 0.61:1.00; the juvenile to adult ratio was 30 to 50, or 0.60:1.00.

C. s. scutulatus is protected as a species of special concern in Utah, but it is not protected anywhere else within its range.

REMARKS: On morphological grounds, *C. scutulatus* has almost always been linked to the *C. viridis*-complex of rattlesnakes (Brattstrom 1964a; Klauber 1972), but some other relationships have been proposed. Do Amaral (1929d) suggested that *C. molossus* was the closest ancestor to the species. Gloyd (1940) thought *C. scutulatus* close to the ancestral *Crotalus* stock, and somewhat intermediate to the *C. atrox*- and the *C. viridis*-groups of North American rattlesnake evolution, but somewhat closer to *C. viridis*. More recent molecular studies by Murphy et al. (2002) place *C. scutulatus* in the *C. viridis*-group, along with *C. horridus*, with its closest relative *C. viridis*. Strengthening this hypothesis, Glenn and Straight (1990) found that venom toxins of *C. viridis* from southwestern New Mexico are more similar to the Mojave toxin of *C. scutulatus* than to similar toxins from *C. oreganus concolor*, indicating natural hybridization between the two species in that region.

In contrast, based on venom protein electrophoretic comparisons, Foote and Mac-Mahon (1977) thought the species closer to *C. horridus*, *C. molossus*, and *C. tigris*.

Minton and Minton (1991) have discussed the use of dried rattlesnake carcasses in Mexican folk medicines against cancer, kidney, and skin diseases in the rural areas of Querétaro; possibly *C. scutulatus* is one of those used.

The taxonomy and literature concerning *C. scutulatus* were reviewed by Price (1982).

Crotalus stephensi Klauber, 1930
Panamint Rattlesnake

RECOGNITION: This northern relative of *C. mitchellii* has a TBL_{max} of 94.3 cm (Klauber 1972), but most adults are 65–80 cm. Its body is gray, bluish-gray, butterscotch, yellowish-brown, tan, or reddish-brown. Present are 36–37 (27–43) boldly light-bordered (sometimes irregular), normally subhexagonal but sometimes rectangular, square, or even diamond-shaped, buff-gray, reddish, or brown dorsal blotches (which may not be very distinguishable from the body color). A series of smaller lateral blotches lies between the dorsal blotches, and these sometimes fuse to form a dark stripe-like area near the ventrals. The tail has 2–9 dark brown or black bands, of which the distalmost 15% of the bands (usually 2–3) are usually black and often coalesce. The proximal segment of the rattle is dark brown or, usually, black. The keeled and pitted body scales have swirl-like surface ridges (Harris 2005), and Stille (1987) describes the microdermoglyphic pattern as vermiculate. The scales may be black-tipped, particularly at the anterior edges of the body blotches. The body scales lie in 23 (21–23) anterior rows, 23 (21–25) rows at midbody, and 23 (21–24) posterior rows. The venter is cream, yellowish-brown, or tan and is marked with numerous darker spots. Present on it are 162–185 ventrals, 12–28 subcaudals, and an undivided anal plate. The head pattern is poorly

Crotalus stephensi (Mercury, Nevada; Ray Folsom, Hermosa Beach, California; John H. Tashjian)

Crotalus stephensi (R. D. Bartlett)

defined, but some anterior dark pigment may be present (no vertical light line occurs on the posterior border of the prenasals and first supralabials). Dorsal cephalic scales consist of a broader than long rostral, 2 prenasals that touch the rostral (no scales occur between the prenasals and the rostral), usually 2 internasals, 2 pairs of canthals, numerous small scales in the prefrontal area, 2 pairs of creased, pitted, or subdivided supraoculars usually with irregular borders, and 5 (4–7) intersupraoculars. Laterally are 2 nasals, 1 loreal (longer than high), 2 (3) preoculars (the upper is only occasionally subdivided, and is separated from the postnasal by loreal or loreal-postcanthal contact), 2 (3) postoculars, 2–3 suboculars, 12–17 (mean, 14) supralabials, and 12–18 (mean, 14–15) infralabials.

The sulcus spermaticus of the two-lobed hemipenis is forked with a branch extending to the tip of each lobe. The proximal area is naked, but the crotch between the lobe has 1–3 spines and each lobe has about 60 spines and 31 fringes (see photograph in Klauber 1972).

The lateral dentition of *C. stephensi* has not been described, but is probably similar to that of *C. mitchellii*, which consists of a maxillary fang, 2–3 palatine, 7–10 pterygoid, and 7–10 dentary teeth (Klauber 1972).

Males (TBL_{max}, 94.3 cm) have 162–181 (mean, 174) ventrals, 21–28 (mean, 25) subcaudals, 6–7 (5–9) dark tail bands, and a mean TL/TBL of 8.1%. Females (TBL_{max} unreported, but probably at least 80 cm) have 173–185 (mean, 179) ventrals, 17–24 (mean, 20) subcaudals, 4–5 (2–6) dark tail bands, and a mean TL/TBL of 6.0%.

GEOGRAPHIC VARIATION: None reported.

CONFUSING SPECIES: *C. mitchellii* has small scales between the rostral and the internasals, smooth supraoculars, and the black tail pigment not restricted to the distal 15% of the tail.

KARYOTYPE: The karyotype has not been described, but is probably similar to that of *C. mitchellii* (which see for details).

FOSSIL RECORD: The Pleistocene (Rancholabrean) fossil *C. mitchellii* reported from Gypsum Cave, Clark County, Nevada, by Brattstrom (1954b) was most likely this species.

DISTRIBUTION: *C. stephensi* ranges from southwestern Nevada (Clark, Esmeralda, Mineral, and Nye counties) to east-central California (Inyo, Kern, Mono, and San Bernardino counties).

HABITAT: *C. stephensi* is usually found in rocky habitats on buttes, in deserts, or on mountains to a maximum elevation of 2,438 m (Sierra Nevada Mountains, Nevada; Meek 1906). Most, however, are found at elevations below 2,000 m.

Being naturally adapted to very dry conditions, captives decline in health if the relative humidity in their cage is too high, resulting merely from evaporation from a water dish, and the snake should be offered water only for a short time once a week (Murphy and Armstrong 1978).

BEHAVIOR AND ECOLOGY: Knowledge of the biology of the panamint rattlesnake is spotty at best, and most has been summarized by Klauber (1972) and Wright and Wright (1957). It may be at least surface-active from late March to late September

Distribution of
Crotalus stephensi.

(Douglas et al. 2007; Stejneger 1893; Wright and Wright 1957). The winter is spent underground, often with other individuals of its own species; two winter dens contained 104 and 180 *C. stephensi*, respectively (Klauber 1972). It is mostly nocturnal in the summer, but nighttime activity declines as the desert cools. Some individuals may be active in the morning or late afternoon, especially in the spring or fall.

It will climb as high as a meter into vegetation (Gander *in* Klauber 1972). The snake can also swim (Baldwin *in* Klauber 1972), although probably seldom if ever is required to do so. A male combat dance was observed in the Panamint Mountains, California, on 16 April (Klauber 1972).

Merriam (*in* Stejneger 1893) reported male combat in this species (Klauber 1972).

REPRODUCTION: Females are ovoviviparous. The smallest gravid female examined by Klauber (1972) was 67.4 cm TBL. A female contained small eggs on 10 June (Cunningham 1959), and Klauber (1972) reported a female had 1 egg, 2 females contained 8 eggs, and 3 others contained 10 eggs.

Hybridization with *C. mitchellii pyrrhus* has occurred in captivity (Porter 1983; but see "Remarks"). Mean neonate TBL is 23.0 cm and the maximum is 23.4 cm (Klauber 1972).

GROWTH AND LONGEVITY: No growth data are available. A wild-caught, adult male *C. stephensi* survived an additional 20 years, 3 months, and 17 days at the Los Angeles Zoo (Snider and Bowler 1992).

DIET AND FEEDING BEHAVIOR: The diet of this rattlesnake consists of lizards (*Callisaurus draconoides, Cnemidophorus tigris, Phrynosoma platirhinos*), birds (*Toxostoma* sp.?), and rodents—kangaroo rats (*Dipodomys* sp.), pocket mice (*Perognathus* sp.), wood rats (*Neotoma* sp.), and ground squirrels (*Ammospermophilus leucurus*) (Klauber 1972; Meik 2005; Stejneger 1893; Wright and Wright 1957). Captives will take live or dead mice (*Mus musculus*), and records exist of cannibalism followed by regurgitation in captivity (Porter 1983).

VENOM DELIVERY SYSTEM: Five 74.6–88.5 cm TBL, *C. stephensi* examined by Klauber (1939) had FLs of 5.6–7.1 (mean, 6.2) mm, a mean TBL/FL of 128%, and a mean HL/FL of 5.59%. The angle of fang curvature was 68°.

VENOM AND BITES: Thirteen fresh *C. stephensi* milked by Klauber (1972) yielded a total of 837 mg of dried venom, or an average of 73 mg for an equivalent of 11.5 fresh snakes; the maximum individual yield was 129 mg. Glenn and Straight (1982) reported the average and maximum venom yields of 18 individuals were 72–73 and 80–129 mg, respectively. Bober et al. (1988) and Henderson and Bieber (1986) found no myotoxin-like enzymes in the species' venom.

The minimum lethal dose of dried venom for a 350 g pigeon (*Columba livia*) ranges from 0.16 to 0.20 (median, 0.18) mg (Githens and George 1931). Macht (1937) reported that the minimum lethal dose of dried venom for a 22 g mouse (*Mus musculus*) is 0.075 mg or 5.46 mg/kg.

PREDATORS AND DEFENSE: *C. stephensi* has preyed on its own species in captivity (Porter 1983); perhaps this occasionally occurs in the wild.

In contrast to *C. mitchellii*, *C. stephensi* is more mild mannered and less excitable. It often remains quietly coiled when approached, and seems reluctant to either rattle or strike (C. Ernst, personal observation). In the past some foolish herpetologists have picked up and carried this snake with their bare hands (see Klauber 1972; Wright and Wright 1957). This is not recommended; the snake is dangerous!

PARASITES AND PATHOGENS: Babero and Emerson (1974) found the nematode *Thubinea cnemidophorus* in the digestive tract of this snake.

POPULATIONS: Klauber (1932) reported seeing a skin of *C. stephensi* at Las Vegas Landing (Nevada) on the Colorado River, and that the Death Valley Expedition (Stejneger 1893) had collected three in Las Vegas Wash. Klauber was told the snake was not uncommon in the area.

REMARKS: Douglas et al. (2006) found large mtDNA sequence divergence in the then mainland subspecies of *C. mitchellii* (*C. m. mitchellii*, *C. m. pyrrhus*, and *C. m. stephensi* [sensu lato]). They then conducted further a molecular study comparing these subspecies using two marking systems: mtDNA ATPase 6 and 8 genes, and introns 5 and 6 of the nuclear DNA ribosomal protein gene (Douglas et al. 2007). The results of the later study indicate that *C. stephensi* is 6.4–7.3% divergent from *C. m. mitchellii*, and 5.2–6.7% divergent from *C. m. pyrrhus*; and also differs from the other two taxa at a single nucleotide polymorphism. It appears that *C. stephensi* is a sister group to the other two taxa. In addition, no molecular evidence of interbreeding between *pyrrhus* and *stephensi* was found (but see "Reproduction"). Consequently, Douglas et al. (2007) recommended that *C. stephensi* be elevated to full species rank. In agreement, a morphological analysis of 30 external characters of snakes from the contact zone between *C. stephensi* and *C. m. pyrrhus* by Meik (2008) contradicted the hypothesis that these 2 taxa interbreed, indicating that *C. stephensi* is reproductively isolated from *C. mitchellii*. The separation of *C. stephensi* from *C. mitchellii* probably took place 3.26–3.69 million years ago (Douglas et al. 2006).

C. stephensi, as a former subspecies of *C. mitchellii*, is a member of the *mitchellii*-group of *Crotalus*, which also includes *C. tigris* (Murphy et al. 2002).

The species was reviewed by Campbell and Lamar (2004), Gloyd (1940), Klauber (1972), and McCrystal and McCord (in part, 1986).

Crotalus tigris Kennicott, *in* Baird 1859
Tiger Rattlesnake
Cascabel Tigre

RECOGNITION: *C. tigris* has a possible TBL_{max} of 91.4 cm (Wright and Wright 1957), but the largest specimen measured by Klauber (1972) was 88.5 cm. Most individuals are shorter than 75 cm. It has a comparatively small head (TBL/HL, 27%), and a proportionately long rattle (mean number of rattle segments, 8.13; mean depth of exposed lobe of basal segment, 12.2 mm; Rowe et al. 2002). Body color varies from gray or lavender to pink, yellowish-brown, or orange. A series of 42–43 (34–52) faint, irregularly shaped (broader mid-dorsally than the area between the bands), gray, olive, or brown bands cross the back. The blotches are more pronounced and darker posteriorly, and no other species of *Crotalus* has cross bands on the anterior portion of the body. Dixon et al. (1962) reported a Sonoran individual with unusually rounded, more broad, dorsal blotches. A series of small blotches alternates with the dorsal bands along the lower sides. Four to 10 (mean, 7) indistinct, speckled bands are present on the tail. Dorsal pitted body scales are keeled and linearly ridged on the surface (Harris 2005; Stille 1987); electromicrographs of the scale surface are in Harris (2005). The scales occur in 24–25 anterior rows, 23–24 (20–31) rows at midbody, and 20 (19–21) posterior rows. The venter is cream, buff, or greenish-brown with gray, lavender, or brown, and mottled with small spots on each ventral scute. The venter bears 146–183 ventrals, 16–27 subcaudals, and an undivided anal plate. Head markings are poorly developed, but a dark cheek stripe that extends diagonally backward from the orbit to

Crotalus tigris
(R. D. Bartlett)

the corner of the mouth is normally present, dark streaks may occur on the temporals, and a pair of dark marks occurs in the occipital area. Enlarged dorsal head scales include a broader than long rostral, 2 small internasals in contact with the rostral, 11–37 smaller scales in the internasal-prefrontal area (prefrontals are absent), 4 canthals, 2 supraoculars, and 3–8 intersupraoculars. On each side of the face are 2 nasals (the prenasal usually touches both the rostral and first supralabial, but the postnasal rarely contacts the upper preocular because of canthal-loreal contact), 1 (2) loreal, 2–3 preoculars, 1 (2) postocular, several suboculars, 1 interoculabial, 12–14 (11–16) supralabials, and 13–15 (11–16) infralabials.

The short, bifurcated hemipenis has a divided sulcus spermaticus with a branch extending to the tip of each lobe, about 64 spines and 40 fringes on each lobe, and only 1–2 spines in the crotch between the lobes (Klauber 1972).

The tooth counts on each side of the head are 1 maxillary fang, 3 palatines, 8 (7–9) pterygoids, and 9–10 dentaries (Klauber 1972).

Males (TBL_{max} at least to 88.5 cm, see above) have 146–172 (mean, 163) ventrals, 24–25 (23–27) subcaudals, 8 (6–10) dark tail bands, and a TBL/TL of 8.4%. Females (TBL_{max} unreported, but Goldberg 1999a reported a maximum SVL of 64.7 cm) have 164–183 (mean, 170) ventrals, 19–20 (16–27) subcaudals, 6 (4–7) dark tail bands, and a mean TBL/TL of 6.4%.

GEOGRAPHIC VARIATION: Although Douglas et al. (2006) reported three distinct mtDNA clades exist among the populations of *C. tigris*, and differences in ground color and patterns occur between populations, no subspecies have been described.

CONFUSING SPECIES: Other species of *Crotalus* can be identified by using the key presented above; no other North American *Crotalus* has dorsal bands on its anterior body.

KARYOTYPE AND FOSSIL RECORD: Unreported.

DISTRIBUTION: *C. tigris* is found in isolated populations in the Sonoran Desert from southern Yavapapai County in south-central Arizona (Fowlie 1965; Lowe et al. 1986) southward through almost all of Sonora, Mexico, to near 110°N (map, Campbell and Lamar 2004), and the Isla Tiburon in the Gulf of California. An isolated population is present in southeastern Cochise County, Arizona (Stebbins 2003). It occurs neither in the high mountains of eastern Sonora nor in the northwestern panhandle. Fowlie (1965) thought that its present isolated foothill distribution may have resulted from the displacement of the tiger rattlesnake in the intervening areas by *C. atrox* and *C. cerastes*.

HABITAT: *C. tigris* is found exclusively in local colonies living in rocky habitats (canyons, ravines, hillsides, or talus slopes in deserts), sandy areas, the chaparral biome, or mesquite grasslands at elevations of about 30 m to at least 2,440 m, but usually below 1,650 m. Typical plants in its microhabitat are brittlebush (*Encelia farinosa*); various cacti, particularly ocotillo (*Fouquieria splendens*) and saguaro (*Carnegiea gigantea*); creosote bush (*Larrea tridentata*); mesquite (*Prosopis glandulosa, P. velutina*); and palo verde (*Cercidium microphyllum*). During the winter and spring, Arizona *C. tigris* are strictly rock-dwellers, but in the summer they switch their microhabitat to the edges of arroyos (Beck 1995).

BEHAVIOR AND ECOLOGY: As can be seen from the lack of behavioral, ecological, and reproductive data presented below, such studies of *C. tigris* are needed.

C. *tigris* becomes active in late spring or early summer, when it often emerges after rain events, and remains surface-active into October or early November. It overwinters in rock crevices or animal burrows.

According to Campbell and Lamar (2004), this snake may be active at any time of

Distribution of
Crotalus tigris.

Crotalus tigris (Pima
County, Arizona;
Brad Moon)

the day, and this is true, overall. However, it is chiefly nocturnal (Armstrong and Murphy 1979; Condon 2005; Pavlik 2007; Taylor 1936; Van Denburgh 1922) or crepuscular (Zweifel and Norris 1955), beginning daily foraging at dusk and continuing into the night until its environment becomes too cold, but some individuals are occasionally found basking or foraging in the morning. Normally it seeks shelter when the AT becomes too hot. Beck (1995) reported the following mean durations (in hours) for surface activity by the various seasons: winter, 1.5; spring, 9.6; dry summer, 18.0; wet summer, 21.6; and fall, 7.0.

Few data are available concerning its thermal requirements. The mean BT of active *C. tigris* taken by Beck (1995) was 29.5 (approximately, 25.9–33.1) °C. Kauffeld (1969) reported that one had a cloacal temperature of about 30.5 °C when captured. The snake's mean BTs (°C) when at rest seasonally were winter, 15.3; spring, 24.8; dry summer, 31.3; wet summer, 27.1; and fall, 23.4. Beck reported that *C. tigris* prefers elevated ETs after feeding. The BT of one that had fed was significantly higher than those of unfed rattlesnakes of other species during 1200–1600 hours, and it moved 290 m to its hibernaculum during the 9 days after it had fed. Beck (1995) calculated the seasonal maintenance energy budgets for a 300 g *C. tigris*, which see for details. The O_2 consumption rates while alert were twice the resting values. Pavlik (2007) and Condon (2005) reported the snake active at ATs of 23.9°C and 27.0 °C, respectively.

The snakes studied by Beck (1995) had a mean home range of 3.48 (approximately, 2.43–4.53) ha. The mean distance traveled during the year was 9.29 km, the average distance traveled per movement bout was 9.29 m, and the mean distance traveled per day was 70.8 m. The mean distances (m) moved per day during the various seasons were winter, 0.3; spring, 9.9; dry summer, 36.7; wet summer, 44.4; and fall, 38.1. One male traveled 290 m to its overwintering rock outcrop during the 9 days after it had fed (Beck 1996).

However, the home range may be larger than that reported by Beck (1996). During a 5-year study, 1997–2001, Goode et al. (2008) implanted radio-transmitters into 31 adult Arizona *C. tigris* (15 females, 16 males), and an additional 58 individuals in 2002–2005, to study their movements and determine their home range sizes. Fourteen males had larger home ranges (mean, 13.1 ha; range, 4.5–125.3 ha) than 8 females (mean, 3.9 ha; range, 0.8–7.6 ha) with 30 or more locations. Males seemed to also have greater annual variation in home range location than females, although only 3 females had at least 30 locations per year over multiple years for comparison. Males moved more per day in summer than did females, but not in either spring or fall. The maximum mean distance moved from the den was similar: males, 490 m; females, 447 m. One male made an atypical movement of 2.7 km to a new den. Males transversed their home range throughout the active season, sometimes passing within less than 50 m of their dens. Females, however, once they had moved away from their den tended to maintain their distance from it until they returned in the fall.

Activity is not restricted to the ground, and *C. tigris* has been found in bushes 60 cm above the desert floor (Klauber 1972; Stebbins 1954); Pavlik (2007) observed one coiled on the branch of a tree 3.4 m above ground. In spite of being an inhabitant of the desert, the tiger rattlesnake swims well with typical lateral undulations when placed in water (Klauber 1972).

REPRODUCTION: Little is known of the reproductive biology of this snake. Stebbins (2003) reported that the shortest TBL of mature males and females is 46 cm, a length probably reached in 2 to 3 years. The SVLs of the smallest male with sperm and the smallest female with enlarged ovarian follicles examined by Goldberg (1999a) were 51.2 cm and 54.1 cm, respectively. The shortest TBL of a pregnant female examined by Klauber (1972) was 61.6 cm.

Spermiogenesis, with both metamorphosing spermatids and mature sperm, occurs in males during June–October, and sperm is present in the vas deferens in May–October, making males ripe for mating (Goldberg 1999a). However, recrudescence, with spermatogonial divisions, primary and secondary spermatocytes, and some spermatids, also occurs from May to August. Other males have regressed testes with spermatogenesis and Sertoli cells from October through July.

Females have enlarged follicles (> 6 mm) from August to November (Goldberg 1999a). Stebbins (1954) reported that two females taken in October contained "eggs" (presumably enlarged follicles). Vitellogenesis begins in the fall and is completed the following spring, when the eggs are ovulated. Females apparently have a biennial reproductive cycle.

Mating possibly takes place in both the spring (April, Van Denburgh 1922) and late summer to early fall, although most observations are from late summer or the fall. Lowe et al. (1986) reported finding males and females together in late May and mid-August. Copulation may last several hours. The young are born from late June through September (Lowe et al. 1986), but most from late July to September, after a GP of about 150 days (Rossi and Rossi 1995). Litters contain 2–6 (mean, 4; n = 16) young. Neonates have 21.0–25.8 (mean, 22.8; n = 8) cm TBLs and a 2–9 (mean, 6.8; n = 3) g BM.

GROWTH AND LONGEVITY: The growth rate is unknown. A wild-caught adult survived an additional 15 years, 3 months, and 3 days at the Staten Island Zoo (Snider and Bowler 1992).

DIET AND FEEDING BEHAVIOR: Wild *C. tigris* are known to have eaten lizards (*Cnemidophorus* sp.; *Crotaphytus collaris*; *Sceloporus clarkii*, *S. magister*) and small rodents (*Dipodomys* sp., *Neotoma* sp., *Perognathus hispidus*, *Peromyscus eremicus*, *Thomomys* sp.) (Amarello and Goode 2004; Armstrong and Murphy 1979; Campbell and Lamar 2004; Fowlie 1965; Lowe et al. 1986; B. Martin 1974; Ortenburger and Ortenburger 1927 [1926]; Pavlik 2007; Stebbins 1954). Captives have taken rodents (*Mus musculus*, *Neotoma albigula*, *Perognathus intermedius*, *Rattus norvegicus*) and various lizards (Beck 1996; C. Ernst, personal observation; Kauffeld 1943a).

Although it probably catches much of its prey through ambush, *C. tigris* must first search for probable successful ambush sites (Berg and Bartz 2005). Van Denburgh (1922) noted that in sandy microhabitats *C. tigris* has a habit of "worming" out shallow depressions in which it lies flush with the surface and is difficult to see, especially when its coloration blends in with the soil; this is probably related to ambushing prey. The snake also actively forages, as evidenced by Armstrong and Murphy (1979) collecting one as it was investigating a woodrat (*Neotoma*) nest and Lowe et al. (1986) observing another snake removing a dead cactus deermouse (*Peromyscus eremicus*) from a rock crevice.

The snake's small head may restrict it from feeding on too large an animal; but contrary to the belief of Vorhies (1948) it is capable of ingesting full-grown mice and smaller kangaroo mice and woodrats. Possibly young individuals rely more heavily on lizards for food, while adults depend mostly on rodents. On the other hand, the width of the snake's rattle segments may indicate its nutrient level: wide, well fed; more narrow, less well fed (Smith and Goode 2005).

VENOM DELIVERY SYSTEM: In compliance with its small head size the tiger rattlesnake also has short fangs. Five 74.6–88.5 cm TBL, *C. tigris* had 4.0–4.6 (mean, 4.4) mm FLs, a mean TBL/FL of 165%, a mean HL/FL of 6.11%, and a mean angle of fang curvature of 69° (Klauber 1939).

VENOM AND BITES: *C. tigris* produces a potent whitish venom with neurotoxic elements (Klauber 1972), but because of its smaller head and shorter fangs, the adult venom yield is low: 6.4–17 mg of dried venom per fresh adult (Altimari 1998; Brown 1973; Klauber 1972; Minton and Weinstein 1984; Weinstein and Smith 1990), and 0.18 mL wet (do Amaral 1928). Lowe et al. (1986) reported a maximum individual yield of 60 mg.

The venom seems to have low protease activity and no hemolytic activity; metalloproteases are apparently absent (Minton and Weinstein 1984; Weinstein and Smith 1990). Weinstein and Smith (1990) isolated 4 toxins; one made up about 10% of the total venom protein and had an interperitoneal LD_{50} of 0.05 mg/kg in mice. This particular toxin showed complete immunoidentity with crotoxin and Mojave toxin, indicating the presence of isoforms of crotoxin and/or Mojave toxin exist in the venom. Mojave toxin activity or Mojave toxin α-like proteins have been also been identified in the venom by Bober et al. (1988), Glenn and Straight (1985a), and Henderson and Bieber (1986); Powell et al. (2004a) concluded that the venom's presynaptic neurotoxic subunit is indeed Mojave toxin.

The LD_{50} values for mice are 0.060 mg/kg interperitoneal, 0.056 mg/kg intravenous, and 0.21 mg/kg subcutaneous; and for a 350 g pigeon, 0.004 mg (Githens and Wolff 1939; Glenn and Straight 1982). The lethal venom dose for a human has not been determined. The several recorded human envenomations by tiger rattlesnakes produced little local reaction and no significant systemic symptoms.

However, a bite of a 14-year-old boy by a supposed *C. tigris* produced the following symptoms (Riffer et al. 1987). The boy was bitten on the right middle finger while attempting to capture the snake, and arrived at an outlying hospital 30 minutes after the bite. At that time the finger was swollen, with a single puncture wound visibly evident; the boy was alert. A routine blood examination was performed at that time, and all counts seemed within normal range. He was given an IV of Ringer's solution, and 13 vials of Wyeth crotalid polyvalent antivenom over the next 4 hours with an increase in oral temperature to 38°C as the only complication. He was then transported to the Arizona Regional Poison Management Center in Phoenix. When he arrived, 6.5 hours after the envenomation, he was in mild to moderate distress, complaining of pain in his right hand. The right arm was swollen to the elbow, and his blood pressure, pulse rate, and breathing rate had all increased, but a noninvasive arterial vascular examination comprising skin temperatures, blood pressures, and pulse volumes indicated no vascular compromise. His leukocyte and platelet count, hemoglobin index, hematocrit, fibrinogen level, red cell morphology, and urinalysis were normal, although the white cell and platelet counts had risen and the hematocrit was slightly depressed. The pain was controlled with morphine sulfate IVs every 2–3 hours, and the slight fever quickly brought under control with a single dose of acetaminophen. Twelve hours after the envenomation, the fibrinogen had risen to 230 mg/dL and the platelet count was even higher, but near the normal range. The hand was still swollen three days later when the patient was discharged from the center, but the swelling was completely gone two weeks later at a follow-up examination. No residual impairment occurred. Because the rattlesnake was not satisfactorily identified, a species other than *C. tigris* may have been involved, but the lack of severe circulatory effects seems to indicate that a tiger rattlesnake was the culprit.

PREDATORS AND DEFENSE: No records of natural predation exist.

C. tigris is behaviorally unpredictable, sometimes rattling when approached but at other times remaining silent. It is generally inoffensive, but will strike if bothered too much. Its rattling occurs at a frequency range of 0.7–19.0 Hz (Young and Brown 1993) and at an SVL-adjusted loudness of about 77 dB (Cook et al. 1994). When confronted with the odor of ophiophagous snakes, it reacts by bridging its body (Weldon and Burghardt 1979; Weldon et al. 1992).

PARASITES AND PATHOGENS: Acanthocephalan oligacanthorynchid cystocanth larvae (not identified to taxon) were found in *C. tigris* by Goldberg and Bursey (1999a). Condon (2005) reported a tiger rattlesnake ectoparasitized by blood-sucking insects, possibly biting midges (Ceratopogonidae). Serotypes of potentially dangerous *Salmonella* bacteria have been isolated from zoo captives (Grupka et al. 2006).

POPULATIONS: In Arizona, *C. tigris* is now restricted to isolated, discontinuous (3–25 km apart; Amarello and Goode 2004), in some cases small, populations. Only 6 of 368 (1.6%) snakes recorded along Arizona State Road 85 in Pima County between 1988 and 1991 were *C. tigris* (Rosen and Lowe 1994). In 2 more recent studies in the Tucson area, Amarello and Goode (2004) implanted radio-telemeters in 67 *C. tigris* in 5 populations between 1997 and 2004, and Goode and Smith (2005) captured 228 individuals for a long-term study. Some local populations seem small, but nowhere has there been a concentrated study of its population dynamics. Habitat destruction, especially as suburbs expand around Tucson (Goode and Smith 2005) and Phoenix, automobiles (Rosen and Lowe 1994; Zweifel and Norris 1955), collection for the pet trade, and the wanton killing of any rattlesnake discovered (Goode and Smith 2005) have reduced local populations. Goode and Smith (2005) reported that at a newly constructed golf course near Tucson, the snake takes refuge in tee boxes and irrigated vegetation along the fairways out of proportion to its availability.

REMARKS: Brattstrom (1964a) proposed that *C. viridis* was the probable ancestor of *C. tigris*; however, electrophoretic studies of venom proteins by Foote and MacMahon (1977) suggest that *C. tigris* is more closely related to *C. mitchellii*, *C. scutulatus*, and the tropical *C. durissus* than to the *C. viridis*-group, and that it and *C. scutulatus* are closer to *C. molossus* and *C. horridus* than to either the *C. viridis*- or *C. atrox*-groups. *C. tigris* has long been thought to be closely related to *C. mitchellii* (do Amaral 1929a, 1929d), and molecular studies by Murphy et al. (2002) confirm this; thus it is a member of the *C. mitchellii*-group, along with *C. stephensi* (Murphy et al. 2002). Therefore, it was surprising that a study of dorsal scale microdermatoglyphic patterns shows it and *C. mitchellii* to be widely separated (Stille 1987). Perhaps scale-sculpturing is a poor taxonomic character and more indicative of habitat use than of evolutionary relationships.

A molecular study of the population of *C. tigris* around Tucson, Arizona, revealed 6 novel polymorphic microsatellite loci for the species, but all loci exhibited high variability (5–41 alleles) (C. S. Goldberg et al. 2003). Munguia-Vega et al. (2009) discovered 11 additional microsatellite loci in *C. tigris* from the Rincon and Tucson mountains in southern Arizona which they plan to use in tracking the long range potential effects of urbanization in populations in the Tucson area. In addition, Douglas et al. (2006) found lower levels of molecular diversification in *C. tigris*, which has a relatively isolated range, than in other species of rattlesnakes with more extensive ranges, and related this to Pleistocene climatic events.

Crotalus tortugensis Van Denburgh and Slevin, 1921
Tortuga Island Rattlesnake
Cascabel de Isla Tortuga

RECOGNITION: This endemic rattlesnake has a TBL_{max} of 105.8 cm (Klauber 1972), and most adults are larger than 75 cm. The body is gray, gray-brown, pinkish-tan, brown, or purplish-brown; 32–41 (mean, 37) dark, hexagonal, or diamond-shaped, dorsal blotches are present along the body. Those most anterior are well defined, with white to cream, incomplete borders and lighter areas on each side of the median. The lateral edges of the anterior blotches usually lack the light border, and posteriorly the blotches fade and all borders tend to disappear. Small black, irregular, marks are often present within, at least, the anterior dorsal blotches, and a longitudinal row of more faint, smaller, unbordered, lateral spots may be present. The tail has 5–6 (3–7) black bands that are as broad as or broader than the white or cream bands separating them; the proximal rattle segment is black. Dorsal body scales are keeled and pitted; Harris

(2005) and Stille (1987) describe the scale microdermoglyphics, and Harris includes an electron microscope image of the scale surface. The body scales lie in 25 anterior rows, 23–26 (25–27) midbody rows, and 23 posterior rows. The venter is gray, whitish, or cream with 183–184 (174–190) ventrals, 21–22 (16–25) subcaudals, and an undivided anal plate. The head (mean HL/TBL, 3.8% in males) is smaller relative to that of its close relative *C. atrox* (mean HL/TBL in males, 4.5%; Spencer 2003). The head pattern is uncomplicated and somewhat faded. The dorsal surface of the head is the same color as the body, and is patterned with irregular black marks. A dark, light-bordered, stripe runs diagonally backward from the orbit to the corner of the mouth. The labials are often of a lighter color than the rest of the head. Dorsal head scales consist of a broader than long rostral scale, followed by 2 internasals (touching the rostral), 2 pairs of canthals (the second pair is the larger), 1–2 (3) intercanthals, 15 (8–21) scales before the supraoculars, 2 large supraoculars, and 4–5 intersupraoculars. Laterally are 2 nasals (the prenasal is largest and touches the first supralabial), 1 loreal (rarely a second smaller upper loreal is present) that usually touches the postcanthal and separates the postnasal from the preocular, 2 preoculars, 2 (3) postoculars, 3–4 suboculars, 2–3 interoculabials separating the anterior subocular from the supralabials, 15–16 (14–18) supralabials, and 16–17 (14–19) infralabials (the first pair is not transversely divided). A mental scale, 2 pairs of chin shields, and 1–2 smaller scales separating the chin shields are present on the chin.

The hemipenis is deeply divided into two attenuated lobes, with a branch of the bifurcated sulcus spermaticus extending to the end of each lobe. Only a few large spines adorn the basal area, and each lobe contains approximately 95 spines and an average of 50 fringes (Campbell and Lamar 2004; Klauber 1972).

Three *C. tortugensis* examined by Klauber (1972) had 3 palatine, 9 pterygoid, and 10 dentary teeth in addition to the maxillary fang on each side.

Males (TBL_{max}, 105.8 cm) have 180–190 (mean, 184) ventrals, 22–25 (mean, 24) subcaudals, 4–7 (mean, 5.3) dark tail rings, and 4.1–7.7 (mean, 6.1) cm TLs. Females

Crotalus tortugensis (Tortuga Island, Baja California Sur; Charles Hanson, Tucson, Arizona; John H. Tashjian)

Crotalus tortugensis
(Tortuga Island,
Baja California Sur;
San Diego Zoo;
John H. Tashjian)

(TBL$_{max}$, 88.1 cm; Beaman and Spencer 2004; Klauber 1972) have 183–190 (mean, 186) ventrals, 16–20 (mean, 18) subcaudals, 3–4 (mean, 3.4) dark tail rings, and 3.7–6.0 (mean, 4.8) TLs.

GEOGRAPHIC VARIATION: None.

CONFUSING SPECIES: Other insular Mexican rattlesnakes can be distinguished by using the above key, and those in Campbell and Lamar (2004) and Klauber (1972). *Crotalus atrox* has a larger head (mean HL/TBL, 4.5%), normally two loreal scales, no scales between the chin shields, and more distinct, completely light-bordered, diamond-blotches.

KARYOTYPE: Thirty-six chromosomes make up the karyotype: 16 macrochromosomes (4 metacentrics, 8 submetacentrics, and 4 telocentrics), and 20 microchromosomes. Sex determination is by ZZ in males and ZW in females (Stewart and Morafka 1989; Stewart et al. 1990). Photographs of the male karyotype are included in Stewart et al. (1990).

FOSSIL RECORD: No fossils have been reported.

DISTRIBUTION: *C. tortugensis* is only found on Tortuga Island in the Gulf of California, Baja California Sur, Mexico.

HABITAT: Tortuga Island is a small, low (311 m elevation), volcanic peak thought to be the youngest island in the area. It is very xeric, totally lacking freshwater except during rain events, rocky, barren, and covered with scattered vegetation and cacti. The dominant plants are the small torchwood tree *Bursaria hindsiana* (along temporary watercourses), and the three small shrubs *Encelia farinosa* (brittlebush, rama blanca), *Jatropha cuneata* (limberbush), and *Sphaeralcea hainessi* (globemallow) (Ruth 1974). Lindsay (1962) reported the discovery of lush green moss growing under a rock kept moist by warm, damp air from a small volcanic fumarole.

BEHAVIOR AND ECOLOGY: The few behavioral data available are that it has been collected in March (Lindsay 1962; Ruth 1974), in an area of lava boulders in October (Brown *in* Armstrong and Murphy 1979), and in June (Van Denburgh and Slevin 1921b); and that the snake is diurnally active (Armstrong and Murphy 1979).

REPRODUCTION: The only reproductive data are from Klauber (1972), who reported the smallest gravid female he examined was 76.2 cm, 2 litters contained 5 and 6 young, and the mean TBL at birth was 25 cm. Reproductive data for this species are sorely needed.

GROWTH AND LONGEVITY: Information concerning growth rates is unavailable. A male of unknown age when caught survived 18 years, 3 months, and 21 days at the San Diego Zoo (Snider and Bowler 1992); Shaw (1969) listed it as 18 years and 4 months.

DIET AND FEEDING BEHAVIOR: Van Denburgh and Slevin (1921b) reported that *C. tortugensis* feeds on mice (probably *Peromyscus dickey*, which occurs on Tortuga Island), and Klauber (1972) found mammal hair in several specimens he examined. Captives readily eat both live and dead house mice (*Mus musculus*) and dead brown rats (*Rattus norvegicus*). Klauber (1972) recorded a mean duration from ingestion to defecation of 9 (3–6) days for 8 meals (probably either of *Mus* or *Rattus*) by an adult *C. tortugensis*.

VENOM DELIVERY SYSTEM: Ten *C. tortugenesis* with TBLs of 74.5–103.8 cm had 6.8–9.4 (mean, 8.5) mm FLs; the mean TBL/FL was 112%, the mean HL/FL was 4.57%, and the mean angle of fang curvature was 72° (Klauber 1939, 1972).

VENOM AND BITES: Seventeen (6 large, 11 medium; 11.5 equivalent adults) fresh *C. tortugensis* produced 641 mg of venom, or 56 mg of venom per fresh adult when

Distribution of
Crotalus tortugensis.

milked (Klauber 1972). Glenn and Straight (1982) obtained yields of 67 mg and 103 mg from a 87.5 cm female and a 98 cm male, respectively, following 26 days of fasting. Whether centrifuged or left standing, the liquid venom will separate into an upper yellow portion and a lower colorless portion; the specific gravities of these two portions are 1.064 and 1.080, respectively (Klauber 1972). Glenn and Straight (1985b) reported that the dry venom from a *C. tortugensis* exhibited protease (109) and esterase (132) activities, and McCue (2005) noted phosphodiesterase activity.

Dry venom from this snake has a relative toxicity of less than 1.00 on a $\log_{10}$ scale (McCue 2005). The mean intraperitoneal LD_{50} for a standard mouse is 1.80 mg/kg (Glenn and Straight 1985b). No case history of envenomation by this species is available, and the human LD_{50} is unknown.

PREDATORS AND DEFENSE: No observation of predation on *C. tortugensis* has been reported, but the kingsnake, *Lampropeltis getula*, a known predator of rattlesnakes, occurs on Tortuga Island (Ruth 1974).

We have not observed defensive behavior in this snake, but Klauber (1972) thought it less excitable and nervous than its nearest relative *C. atrox*. However, Van Denburgh (1922) related that it rattles vigorously when approached; and Armstrong and Murphy (1979) noted that, when touched, it uses vertical body-bridges or lateral flexures to strike an annoying object, frequently with such force as to dislodge freshly killed rodents from a feeding forceps. It is a dangerous animal and should be treated with respect.

PARASITES AND PATHOGENS: The parasitic lingatulid tongue worms (Pentastomida) *Porocephalus crotali*, *P. tortugensis* (possibly a synonym of *P. crotali*), and *Raillietiella furcocerca* have also been found in the mouth of this snake (Klauber 1972; Murphy and Armstrong 1978; Riley and Self 1979).

POPULATIONS: Lindsay (1962), McCoy (1984), and Ruth (1974) reported *C. tortugensis* abundant on Tortuga Island, but no serious study of its population numbers or dynamics has been undertaken.

The Tortuga Island rattlesnake has been thought vulnerable to extinction because it is endemic on that island, but, although we need more information on its life requirements, at the present it seems to be holding its own.

REMARKS: *Crotalus tortugensis* is a member of the *C. atrox*-group, which, in addition to these two species, includes *C. adamanteus*, *C. catalinensis*, and *C. ruber* (Murphy et al. 2002). Of these, *C. tortugensis* seems to be molecularly closest to *C. atrox* (Castoe et al. 2007; Murphy et al. 2002). Beaman and Spencer (2004) and do Amaral (1929b) thought the color and pattern differences of *C. tortugensis* insufficient to warrant its separation as a separate species from *C. atrox*, and Castoe et al. (2007) suggested on the basis of molecular similarities that it be synonymized with *C. atrox*, as was done by Beaman and Hayes (2008). However, other reviewers of the genus *Crotalus* (Campbell and Lamar 2004; Gloyd 1940; Grismer 2002; Klauber 1972) have maintained it as a full species. We have taken a conservative stance on the status of *C. tortugensis*, and have kept it as a separate species.

The taxonomy and literature of the species have been reviewed by Beaman and Spencer (2004), Campbell and Lamar (2004), Gloyd (1940), and Klauber (1972).

Crotalus totonacus Gloyd and Kauffeld, 1940
Totonacan Rattlesnake
Tepocolcoatl

RECOGNITION: *Crotalus totonicus* is another poorly known Mexican rattlesnake. It is a stout-bodied snake with a TBL_{max} of 166.5 cm (Gloyd 1940), and many adults are longer than 150 cm. Anteriorly, a low, inconspicuous, vertebral keel is present on large adults. The body ranges from yellowish-brown or straw-colored to yellowish-orange, light olive-brown, to pale gray-brown, and some individuals living at higher elevations may be dark brown. The ground color lacks darker punctuations. Dorsally is a series of 30 (27–35) dark brown or black, white-, cream-, or yellow-bordered, dorsal blotches. The blotches are normally subhexagonal anteriorly, become more diamond-shaped or rhomboidal at midbody, and form dark cross bars posteriorly. The centers of the more posterior blotches usually are lighter. The blotches are separated by 1–2 scales anteriorly, but 3–4 scales more posteriorly on the body. On scale rows 1–4, a series of 2–4 scales wide, sometimes light-bordered, lateral blotches may also be present. These may fuse with the posterior dorsal blotches. Adults lack any other rows of lateral blotches, although small individuals may have an additional row. Six to nine dark bands cross the dark brown, gray, or black tail, particularly in juveniles; some adults lack the bands; TL averages 6% of TBL. Poorly defined paravertebral stripes may be present on the neck. The pitted body scales are keeled and occur in 25 (29) anterior and midbody scale rows, and 21 (19–23) rows near the tail. The cream to yellow venter contains a series of gray to grayish-brown irregular blotches on each side of the midline along most of the body, but not the chin and throat; these become darker posteriorly, and may expand to the lateral edges of the ventrals there. Present are 184–195 ventrals, 22–29 subcaudals, and an undivided anal plate. The head pattern is complicated. A black bar is normally present in front of the orbits. A darkly pigmented area

Crotalus totonacus
(Soto la Marina,
Tamaulipas;
Gladys Porter Zoo;
John H. Tashjian)

Crotalus totonacus
(Soto la Marina,
Tamaulipas;
Gladys Porter Zoo;
John H. Tashjian)

separated from the postocular stripe may occur behind each orbit. On each side a dark stripe extends diagonally backward from behind the supraocular to above the corner of the mouth. A pair of parietal stripes run backward behind the supraoculars to the back of the head, where they may form the paravertebral stripes, and eventually co-alesce with the first dark body blotch. Anteriorly, two parietal stripes often fuse to form a pointed projection or are only separated by 1–2 scales. Most dorsal head scales are small, but there is 1 large rostral, 2 large internasals, 2 large quadrangular canthals in contact along the midline (4 scales occur in the internasal-prefrontal region, and the upper loreal prevents the canthal from contacting the upper preocular), and 2 large supraoculars. Normally 2 (1–3) intersupraoculars are present, and the anterior parietal region is covered with relatively large unkeeled but striated scales. Laterally are 2 na-sals (prenasals widest, postnasals contact the loreals), 2 (3) loreals (the upper elongated and sometimes subdivided), 2 preoculars, 3–5 postoculars, 3–4 foveals surround the pit, 2 (3) interoculabials (separate the anterior suboculars from the supralabials), 2–3 suboculars, 13–14 (12–15) supralabials, and 14–15 (12–17) infralabials. Ventrally are a mental scale and a pair of wider than long chin shields.

The hemipenis is bifurcated with the forks of the sulcus spermaticus extending to the ends of the lobes, no spines in the crotch, and about 54 spines adorning each lobe (Klauber 1972).

The dentition has not been described, but is probably similar to that of *C. durissus*: 1 (0–2) palatine, 7–8 (9) pterygoids, and 10 (8–11) dentaries (Klauber 1972).

Males (TBL$_{max}$, 160 cm; Klauber 1972) have 184–192 (mean, 187) ventrals, 26–29 (mean, 27) subcaudals, and a TL/TBL of 7–8%. Females (TBL$_{max}$, 166.5 cm) have 193–195 (mean, 194) ventrals, 22–26 (mean, 23–24) subcaudals, and a TL/TBL of 6.0–6.5%.

GEOGRAPHIC VARIATION: None reported.

CONFUSING SPECIES: *C. totonacus* may be distinguished from other species of *Crotalus* in North America by using the above key. The allopatric *Crotalus durissus*, of which *C. totonacus* was formerly considered a subspecies, has a well-developed vertebral keel, lateral body blotches that do not contact the ventrals, distinct secondary lateral body blotches, a paravertebral stripe 2–4 dorsal scales wide, a mean of only 26 dorsal body blotches, usually 27 midbody scale rows, 155–170 ventrals in males and 168–190 in females, and 25–32 subcaudals in males and 26–29 in females (Campbell and Lamar 2004).

KARYOTYPE AND FOSSIL RECORD: Not reported.

DISTRIBUTION: This is the northernmost species of the *C. durissus*-complex. It occurs in northeastern Mexico from central Nuevo León and southern Tamaulipas (Armstrong and Murphy 1979; Lazcano et al. 2009c, Martin 1958), southward through eastern San Luis Potosí (Taylor 1950), and northern Veracruz (Gloyd and Kauffeld 1940) to Querétaro (Dixon et al. 1972).

HABITAT: *C. totonacus* is a denizen of lowland tropical thorn forests, semievergreen and tropical deciduous forests, and open oak-pine woodlands, at elevations from near sea level to at least 1,585–1,680 m (Dixon et al. 1972; Lazcano et al. 2009a; Martin 1958). Armstrong and Murphy (1979) thought that the snake preferred microhabitats around watercourses in dry areas in Tamaulipas, and it has been found in dry habitats in Querétaro (Dixon et al. 1972); but according to Campbell and Lamar (2004), it is found more often around rock outcrops and on hillsides with scattered rocks in the highlands than along waterways. Mammal burrows in termite mounds are often used as retreats. Typical plants within its habitats include shrubs such as ebony (*Pithecellobium flexicaule*), huisache (*Acacia fornesiana*), mesquite (*Prosopis* sp.), and cottonwood (*Populus* sp.) and bald cypress (*Taxodium mucronatum*) trees along streams.

Distribution of *Crotalus totonacus*.

BEHAVIOR AND ECOLOGY: Data for this species are practically nonexistent. Active individuals have been observed or collected from February to early August (Armstrong and Murphy 1979; Taylor 1950), but peak activity seems to occur in the wet season when it is often active during rain events; so the snake is probably annually active from February–March to October. However, the above dates may reflect the tendency for researchers to visit the area during the rainy season. Nothing is known of its over-wintering biology or thermal requirements. Its diel cycle has been reported as both crepuscular or nocturnal (Armstrong and Murphy 1979), and diurnal (at higher elevations) (Campbell and Lamar 2004).

REPRODUCTION: Like other rattlesnakes, *C. totonacus* is ovoviviparous. Armstrong and Murphy (1979) reported a 170 cm adult male and a large female were found within 12 m of each other in February, but it is not known if they would have engaged in breeding activity.

Klauber (1972) reported a litter of 37 young. Based on other Mexican rattlesnakes, parturition probably occurs in August or September. Neonates average 31.5 cm in TBL (Klauber 1972), and are more brightly patterned than adults. Campbell and Lamar (2004) reported that some *C. molossus* from northeastern Querétaro appear to have characteristics of both that species and *C. totonacus* and may be hybrids of the two snakes.

GROWTH AND LONGEVITY: No specific growth data have been published, but Armstrong and Murphy (1979) stated that the growth rates of captive young are the most rapid of any rattlesnake maintained by them. Snider and Bowler (1992) reported a survivorship at the Houston Zoo of 8 years, 3 months, and 16 days by a wild-caught female.

DIET AND FEEDING BEHAVIOR: *C. totonacus* seems to specialize in warm-blooded prey. Martin (1958) reported a cave rat (*Neotoma* sp.) and a tree squirrel (*Sciurus alleni*) as prey in Tamaulipas, and another juvenile from that state contained rodent hairs (Armstrong and Murphy 1979). Klauber (1972) found mammal hairs in three individuals and bird feathers in one. Captives will readily take house mice (*Mus musculus*) and brown rats (*Rattus norvegicus*; C. Ernst, personal observation).

VENOM DELIVERY SYSTEM: A male *C. totonacus* with an approximate TBL of 142 cm measured by us had fangs of 10.8 mm. Klauber (1939, 1972) reported TBL/FL and HL/FL ratios of 102% and 4.0%, respectively.

VENOM AND BITES: Klauber (1972) reported the venom yield from a fresh adult *C. totonacus* was 514 mg, but only an average yield for 9 *C. durissus* of 277 mg.

The venom has both neurotoxic and hemorrhagic properties, and causes hydrolase, phospholipase, and proteinase effects. Electrophoresis of the venom shows at least 14 protein bands at alkaline pH (8.3–8.6), and approximately 12 components at acidic pH (4.2–4.8) (Possani et al. 1980). It probably contains several strong toxins found in the venom of its close relative *C. durissus*, including myotoxin α (> 2.0 mg/mL, Bober et al. 1988) and the hemorrhagic toxin crotacetin (a C-type lectin and a homolog of convulxin) (Rádis-Baptista et al. 2005). Venom from *C. totonacus* lacks any direct lytic effect upon human red blood cells, and also poor hyaluronidase activity using protein contents up to 2.0 mg per assay (Possani et al. 1980). Tu (1977) reported the metal content. The specific gravity of the venom is above 1.09 (Klauber 1972).

The venom causes hemorrhaging in laboratory animals (Friederich and Tu 1971). An intraperitoneal LD_{50} of 2.5–8.0 mg/kg for mice was reported for venoms of *C. totonacus* and 2 subspecies of *C. durissus* by Glenn and Straight (1985a); Possani et al. (1980) also calculated it to be 2.5 for *C. totonacus*. The human LD_{50} is unknown, and no case histories involving bites of humans are available; but envenomation by this species can probably cause death, and in all cases should be considered serious.

PREDATORS AND DEFENSE: The only report of predation on this rattlesnake is of an adult ocelot (*Leopardus pardalis*) attacking and wounding a wild 170 cm adult male so badly that it died 2 weeks later (Armstrong and Murphy 1979).

Captives can be testy. If alarmed, they coil, raise their heads and body almost one-third off the ground, and bend the neck to bring the head closer to the disturbance (reminiscent of the defensive display of *C. atrox* or *C. durissus*). Armstrong and Murphy (1979) reported that young captives inflated their trunk, convulsed, and, on rare occasions, turned over onto their backs when confronted with a human. Because of its irritability, size, and virulent venom, *C. totonacus* should be shown great respect.

PARASITES AND PATHOGENS: The snake mentioned above under "Predators and Defense" regurgitated a large number of ascarid nematodes (Armstrong and Murphy 1979; Lazcano et al. 2009c).

POPULATIONS: Armstrong and Murphy (1979) thought *C. totonacus* to be common. DOR individuals have been reported (Armstrong and Murphy 1979, Lazcano et al. 2009c). Other data are lacking.

Because of its relatively small and fragmented distribution, Greene and Campbell (1992) thought *C. totonacus* vulnerable to extinction; but, in reality, we know so little of the population status and biology of this species that a firm conclusion as to its survival status cannot be made at this time.

REMARKS: *C. totonacus* was elevated to full species status from that of a subspecies of *C. durissus* on the basis of morphological characters and its allopatry by Campbell and Lamar (2004). This separation has been upheld in mtDNA tests performed by Wüster et al. (2005). Nevertheless, *C. totonacus* remains a member of the *C. durissus*-complex of rattlesnakes along with *C. basiliscus*, *C. durissus*, *C. enyo*, *C. molossus*, *C. unicolor*, and *C. vegrandis* (Murphy et al. 2002). Although it had been thought that its closest relative was *C. durissus*, mtDNA results show it to be more closely related to *C. basiliscus* and *C. molossus* (Wüster et al. 2005).

Crotalus viridis (Rafinesque, 1818)
Western Rattlesnake
Cascabel Occidental

RECOGNITION: The western rattlesnake (TBL_{max}, 151.5 cm; Klauber 1937) has a body that is either gray, greenish-gray, greenish-brown, olive, pale brownish-gray, yellowish-brown, tan, brown, pinkish, or reddish. Surprisingly, individuals occurring in the dark Pedro Armendariz lava habitat in New Mexico are not melanistic, as are other rattlesnakes there (Best and James 1984). Present are 33–57 (mean, 42–44) grayish-green, brownish-gray, brown, or black; straight or irregularly light-bordered; broader than long or hexagonal dorsal botches. The anterior blotches are often indented along both the front and rear edges, and those posterior may narrow into cross bands near the tail. Two to 15 (mean, 7–10) faint dark bands the same color as the body blotches are present on the tail, the last 1–2 of which are black, as is also the first rattle segment. As many as 12–16 rattle segments may be present, but usually much fewer (Fitch 1985b, 1998, 2002; Schmidt 2002). About 4.7% of the snakes lack a rattle (Holycross 2000b; Rowe et al. 2002). Two or 3 rows of poorly developed, small, dark, lateral blotches are normally present, and these may coalesce posteriorly with the dorsal blotches to form the tail bands. However, there is much ontogenetic variation. The pattern is usually vivid in juveniles, but fades with age, so the series of dorsal blotches may almost disappear in adults. Gloyd (1935b) and Slowinski and Rasmussen (1985) described color and pattern variations in *C. v. viridis* that included one with anterior stripes instead of blotches. Ashton et al. (1999), Gloyd (1958), Klauber (1972), Nickerson and Mays (1968), Slowinski and Rasmussen (1985) and Werner et al. (2004) have described other pattern variants, including patternless individuals. Albinos and xanthic individuals have been reported by Chace and Smith (1968), Ditmars (1923), Dyrkacz (1981), Harris (2006b), Hensley (1959), Killebrew and James (1983), and Klauber (1972). The normal pattern is similar to that of the nonvenomous snake *Pituophis catenatus*, with which *C. viridis* forms a mimic complex (Sweet 1985). Dorsal body scales occur in 23–27 anterior rows, 25–27 (21–29) rows at midbody, and 19–21 (22–23) rows close to the tail; their microdermoglyphic pattern is one of both straight and swirling linear ridges (Harris 2005 [see photograph]; Stille 1987). The venter is gray, cream, or white with little mottling or dark markings. Present are 155–196 ventrals, 14–31 subcaudals, and an undivided anal plate. Two light diagonal stripes occur on the side of the face; one extends backward from in front of the orbit to the supralabials, and the second from the rear of the orbit to in front of and above the corner of the mouth from where it continues onto the neck. A light narrow transverse stripe may cross each supraocular scale and connect them. Large adults may lack any pale dorsal head markings. *C. viridis* is the only rattlesnake with more than two internasals touching the rostral, which is usually higher than wide. Behind the rostral, and touching it, are 3–4 (1–8) internasals, a pair of canthals on each side (prefrontals are absent), at least 8 scales before the 4–6 (1–9) intersupraoculars, and 2 supraoculars. Lateral head scales consists of 2 nasals (the prenasal may or may not touch the first supralabials; the postnasal seldom contacts the upper preocular), 1 loreal (rarely 2–3, and then often only on one side), 2 preoculars, several suboculars, 2 (1–3) postoculars, 14–16 (10–19) supralabials, and 12–15 (11–19) infralabials.

Each lobe of the bilobed hemipenis is adorned with about 67–92 spines and 26–37 fringes, but no spines are present in the U-shaped crotch between the lobes. The calyces on the sulcomedial surface usually do not bear mesial spines. The sulcus spermaticus divides to extend up each lobe (Campbell and Lamar 2004; Klauber 1972). The hemipenis is illustrated in Dowling and Savage (1960).

The dentition on each side of the jaw includes a maxillary fang, 2–5 (mean, 3) palatines, 6–10 (mean, 7–8) pterygoids, and 6–11 (mean, 9–10) dentary teeth (C. Ernst, personal observation; Klauber 1972).

Males (TBL$_{max}$, 151.5 cm) have 155–189 (mean, 159–172) ventrals, 21–31 (mean, 26) subcaudals, and 5–15 (mean, 9–10) dark tail bands. Females (TBL$_{max}$, approximately

120 cm) have 169–196 (mean, 176–185) ventrals, 14–26 (mean, 19–20) subcaudals, and 4–11 (mean, 7–8) dark tail bands. Sexual dimorphism is not as important a factor as the adaptation for gape variation correlated to different prey types, seen in regional variation in head form (Smith and Collyer 2008).

GEOGRAPHIC VARIATION: Two subspecies are currently recognized. *Crotalus viridis viridis* (Rafinesque, 1818), the prairie rattlesnake or cascabel de pradera, ranges east of the Rocky Mountains from southeastern Alberta, southwestern Saskatchewan, Montana, eastern Idaho, and North Dakota south barely to northeastern and southeastern Arizona, New Mexico, western Texas, northern Chihuahua, and northwestern Coahuila. It is greenish-gray to brownish-gray (Ludlow 1981) with 33–55 (mean, 41–43) less distinctly colored, oval to quadrangular-shaped, light bordered, dorsal body blotches; and has 25–27 (mean, 26) midbody scale rows; 164–189 (mean, 178) ventrals in males and 170–196 (mean, 180–185) in females; 21–31 (mean, 26–27) subcaudals in males and 14–26 (mean, 20–21) in females; normally a single loreal; 11–15 (mean, 13) supralabials, 12–16 (mean, 13–14) infralabials; and a TBL up to 151.5 cm. *Crotalus v. nuntius* Klauber, 1935, the Hopi rattlesnake, lives in northeastern and north-central Arizona, and adjacent extreme northwestern New Mexico. McCoy (1962) has also reported *nuntius* from Montezuma County in southwestern Colorado, but Douglas et al. (2002) determined that it cannot be distinguished there from *C. v. viridis*.

Some differences in head morphology occur between regional populations of *C. v. viridis* in the Dakotas (Smith and Collyer 2008).

Crotalus v. nuntius is tan, pale brown, pinkish-brown, pink, red, or reddish-brown with 33–53 (mean, 42–43) distinctly colored, irregularly oval to rectangular, narrowly light-bordered, peripherally dark, dorsal body blotches; and has 21–27 (mean, 25) midbody scale rows, 163–178 (mean, 171) ventrals in males and 169–184 (mean, 176–177) in females; 21–28 (mean, 25) subcaudals in males and 14–22 (mean, 18–19) in females; normally 2 loreals; 12–17 (mean, 15) supralabials; 12–19 (mean, 15) infralabials; and a TBL of 80–100 cm (Olivier 2008b).

Crotalus viridis viridis
(R. D. Bartlett)

Crotalus viridis viridis (North Dakota; R. D. Bartlett)

C. v. viridis and *C. v. nuntius* intergrade in the Mesa Verde National Park, Colorado (Douglas 1966), and northwestern New Mexico (Gehlbach 1965).

CONFUSING SPECIES: Other North American *Crotalus* have facial stripes and body blotches, but only *C. viridis* has more than two internasals touching the rostral scale and poorly defined tail bands (see the above key). *C. v. nuntius* is a smaller snake (TBL, generally < 70 cm) than sympatric *C. v. viridis* (TBL, > 80 cm), and the subspecies *abyssus* (TBL, > 80 cm), *cerberus* (TBL, > 80 cm), *concolor* (TBL$_{max}$, 75.6 cm), and *lutosus* (TBL, > 80 cm) of *C. oreganus*. *Sistrurus catenatus* has nine large plate-like scales on the top of its head. The nonvenomous colubrid snakes, *Heterodon platirhinos* and *Pituophis catenifer*, have round pupils, and lack both a rattle and loreal pits.

KARYOTYPE: *C. viridis* has a karyotype consisting of 36 chromosomes: 16 macrochromosomes (4 metacentric, 6 submetacentric, and 4 subtelocentric), and 20 microchromosomes (Baker et al. 1972; Zimmerman and Kilpatrick 1973). Porter et al. (1991) reported 16 macrochromosomes (2 acrocenteric, 14 biarmed), and 20 microchromosomes. Males are ZZ, females ZW; the Z is metacentric or submetacentric, and the W is subtelocentric or submetacentric. Baker et al. (1972) reported the Z chromosome is submetacentric and the W is subtelocentric, but Zimmerman and Kilpatrick (1973) described these chromosomes as being metacentric and submetacentric, respectively. Extensive listings of the allozymes and isozymes of *C. v. viridis* from Chouteau County,

Montana, have been published by Crabtree and Murphy (1984) and Murphy and Crabtree (1985b).

FOSSIL RECORD: Fossils of *C. viridis* are known from as early as Miocene (Hemphillian) deposits in Nebraska (Brattstrom 1967), the Pliocene (Hemphillian) of Oklahoma (Brattstrom 1967; Parmley and Holman 1995), and the Pliocene (Blancan) of Kansas (Brattstrom 1967; Holman and Schloeder 1991; Peters 1953; Rogers 1976). In addition, undesignated Miocene (Barstovian) crotaline fangs and vertebrae from Webster County, southeastern Nebraska, could possibly be from *C. viridis* (Holman 1977b). Most fossil remains, however, are Pleistocene in age: Irvingtonian—Colorado (Holman 1995; Rogers et al. 1985), Nebraska (Holman 1995), and Kansas (Brattstrom 1967; Holman 1971; Preston 1979); and Rancholabrean—Kansas (Holman 1971). Also existing are Rancholabrean vertebrae from Arizona that may be either from this species or *C. mitchellii* (Mead and Bell 1994; Mead and Phillips 1981; Van Devender et al. 1977), and other undesignated Rancholabrean *Crotalus* fossils from Iowa and Texas that may represent *C. viridis* (Brattstrom 1954a; Gilmore 1938; Parmley 1990).

Banta (1974), based on Martin del Campo (1936), reported a pre-Columbian Native American drawing that depicts a larger rattlesnake ingesting a smaller one. The snakes involved resemble *C. viridis*, and this may be the first mention of cannibalism in the species.

DISTRIBUTION: *C. viridis* ranges from southeastern Alberta and southwestern Saskatchewan (Gannon 1978; Pendlebury 1977), Montana (Maxell et al. 2003; Werner et al. 2004), southeastern Idaho, and western North Dakota (Wheeler and Wheeler

Distribution of
Crotalus viridis.

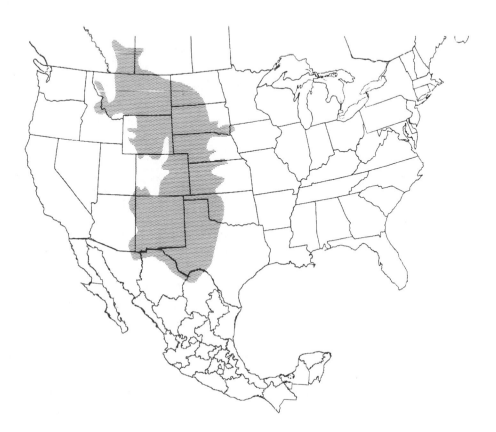

1966) south and southeastward, generally east of the Rocky Mountains, through South Dakota to extreme northwestern Iowa, eastern Wyoming, and Colorado, western Nebraska, Kansas and Oklahoma, most of New Mexico, and western Texas to extreme northern Chihuahua and northwestern Coahuila in Mexico. It is also found in northwestern and southwestern Colorado. Its distributional elevation covers 100–2,775 m (Banta 1968a; Campbell and Lamar 2004; Klauber 1972; Ludlow 1981), and it reaches its northern limit a little north of the western North American glacial border (Conant 1978).

HABITAT: Due to its extensive north-south and elevational distributions, *C. viridis* occupies a variety of vegetational habitats, but almost all have abundant prey and available retreats. Habitats such as both deciduous and coniferous woods, riparian zones and river bluffs, scrub areas, short and mixed prairie grasslands, mountain scrubland, shrub-steppes, lava fields, alluvial fans, buttes and mesas, sandstone outcrops, sand hills, and the margins of deserts are used. Usually south-facing rocky outcroppings with deep crevices or prairie dog towns (*Cynomys* sp., but mostly *C. ludovicianus*; C. Ernst, personal observation; Fogell 2005; Holycross and Fawcett 2002; Shipley and Reading 2006) are located within migratory distance. In Canada, the distribution coincides with the availability of suitable rock crevice hibernacula, and usually only involves river valleys (Gannon 1978; Pendlebury 1977). A detailed analysis of the species' habitats in Yellowstone National Park, Wyoming, is presented in Porter et al. (2002).

Some identified plants found within its habitats include the annual flowers *Helianthus annuus*, *H. cilaris*, *Pectis papposa*, cacti (*Coryphantha missouriensis*, *C. vivipara*; *Opuntia compressa*, *O. imbricata*, *O. polyacantha*), cedars and firs (*Juniperus monosperma*, *J. scolulorum*; *Pseudotsuga menziesii*), common cocklebur (*Xanthium* sp.), forbs and grasses (*Agripyron spicatum*, *Bouteloua gracilis*, *Bromus tectorum*, *Buchloe dactyoides*, *Elymus* sp., *Erysimum asperum*, *Grindelia squarrosa*, *Guiterrezia sarothrae*, *Hespeorostipa comata*, *Koeleria macrantha*, *Pascopyrum smithi*, *Plantago patagonica*, *Pleuraphis mutica*, *Poa secunda*, *Sporobolus cryptandrus*, *Stipa viridula*), greasewood (*Sarcobatus vermiculatus*), iodine weed (*Allenrolfea occidentalis*), juneberry (*Amelanchier* sp.), lead plant bush (*Amorpha canescens*), mesquite (*Prosopis* sp.), mountain mahogany (*Cercocarpus breviflorus*), mullein (*Verbascum thapsus*), oaks (*Quercus* sp.), pines (*Pinus ponderosa* and others), quaking aspen and cottonwood (*Populus deltoides*, *P. tremuloides*), rabbitbrush (*Chrysothamnus nauseosus*, *C. parryi*), Russian olive (*Elaeagnus angustifolia*), sagebrush (*Artemisia filifolia*, *A. tridentata*), saltbrush (*Atriplex canescens*), salt cedar (*Tamarix* sp.), and thistles (*Cirsium arvense*, *Salsola australis*).

BEHAVIOR AND ECOLOGY: This snake has been well studied, and much ecological and behavioral data are available for it, particularly the populations studied by Duvall and his associates.

C. viridis in south-central Wyoming may hibernate as long as 8.5 months each year (Duvall et al. 1985b). Typical hibernacula throughout the range are prairie dog towns and other deep animal burrows, the walls of cisterns or wells, caves, talus slopes, and crevices in south-facing rock outcrops. Emergence from hibernation is generally earlier in the south (late February to early April) and later in the north (late March to mid-April or early May); no sexual differences exist in the dates of spring emergence in Wyoming (Graves and Duvall 1990). The earliest recorded appearance of active individuals is 11 February (North Dakota; Wheeler 1947).

Crotalus viridis viridis, juvenile (Colorado, Carl H. Ernst)

Prairie rattlesnakes in Jefferson County, Colorado, are in hibernation from 1 January to late March (encounters are rare during this period). In late March to late April the snakes emerge from the dens and bask nearby, but re-enter the den each evening. The snakes gradually start to disperse from late April to late May, and are on the foraging range from late May to mid-September. In mid-September into early October they migrate back to the hibernaculum, and arrive there from early October into early November (some bask during at this time). Most snakes retreat for good into the dens from early November through December, and encounters become rare as this period advances (Ludlow 1981).

The normal annual activity period over most of range is probably April–September; but in New Mexico the snakes are active in March–December, with 40% of the observations in July–August (Degenhardt et al. 1996; Howell and Wood 1957). In Saskatchewan, gravid females were significantly more active in all seasons (Gannon and Secoy 1985). The pivotal BT for both arousal and dormancy is 10°C (Jacob and Painter 1980), but increases in ST are probably more responsible for the emergence than is AT.

After spring emergence these snakes stay near the hibernaculum and bask for several days to weeks before migrating to their summer feeding range. Northern populations begin their migration to the summer feeding range in early May (Gannon and Secoy 1985). Wyoming adults of both sexes forage for prey in the first half of the 3.5-month summer activity period, but males, not females, search for mates in the second half of the season; thus, males must search for and locate two stations at a time and space (food and mates) each season. Females continue to forage for the duration of the summer, but males are significantly more efficient in spatial searching during the vernal foraging times than are females, allowing them to concentrate their foraging activities into the first half of the season, so the second can be essentially devoted to finding a mate (Duvall et al. 1985b, 1990a; King and Duval 1990).

The summer home range is occupied until September, when the return migration to the hibernaculum begins (Gannon and Secoy 1985). From late September to

November, the snakes remain near the den, frequently basking, possibly to help digestion of a last meal or the ovaries to mature, but they feed less often as the AT drops. Eventually the hibernaculum is entered for the winter, often in December, and, although some may move about within it, most snakes lie there individually coiled or in bunches until the next spring (Duvall et al. 1985b; Gannon and Secoy 1985; Ludlow 1981). Some hibernacula have been used for centuries, and theoretically the *C. viridis* now inhabiting them are descendents of the first to overwinter there. Den fidelity is strong, with individuals returning year after year (Duvall et al. 1985b). The den may be shared with other reptiles, and once the rattlesnakes are torpid, homeotherms such as the possible predators, foxes (*Vulpes vulpes*; Starck 1987) and badgers (*Taxidea taxus*), may move into the den as well as potential prey species such as packrats (*Neotoma* sp.), prairie dogs (*Cynomys* sp.) and burrowing owls (*Athene cunicularia*).

The snakes probably seek the warmer portions of a hibernaculum, and move deeper into the den as winter progresses (Marion and Sexton 1984; Sexton and Marion 1981). The BTs of hibernating *C. viridis* may drop to 2–7°C (Jacob and Painter 1980). Cooling of the BT during hibernation is probably necessary for proper reproduction the following year (Tryon 1985). BT drops slower when the snakes are lying in groups than when alone, but physical contact with several other snakes does not help raise the BT (Graves and Duvall 1987; White and Lasiewski 1971). The longer a *C. viridis* is chilled, the longer arousal takes once its BT again rises in the early spring.

C. viridis forages in the morning and late afternoon or evening during the spring and fall, but in the summer it becomes more crepuscular or nocturnal (but 75% to full moonlight retards activity; Clarke et al. 1996). When not foraging, it either basks, lies coiled in the shade, or shelters underground, depending on the AT. Most are active at 20–35°C BTs (Brattstrom 1965; Graves and Duvall 1993; Stebbins 1954), but in Saskatchewan, apparently no significant difference in mean BT occurs between wild males and females (Gannon and Secoy 1985). The CT_{min} and CT_{max} for *C. viridis* are probably 0°C (C. Ernst, personal observation) and 38–42°C (Brattstrom 1965).

Ruben (1979) conducted laboratory tests comparing the blood physiology of *C. viridis* and the whipsnake, *Masticophis flagellum*. Although both snakes had a similar resting blood pH (rattlesnake, 7.38; whipsnake, 7.42), the normally more active *Masticophis* had a more efficient bicarbonate-carbonic acid blood buffering system. During maximal activity at 35°C, the pH of the rattlesnake's blood (6.69) declined 0.28 pH units more than that of the whipsnake (7.09). The blood buffering system of the latter snake was approximately 1.5 times as active as that of *C. viridis*. In contrast to the *Crotalus*, Ruben thought the *Masticophis* used its complex lung to facilitate function of the blood buffering system to speed the rate of CO_2 elimination from the body.

In another study of metabolism in *C. viridis*, Ruben (1976) determined that the maximal oxygen consumption at a BT of 35°C was 0.50 mL of O_2 per gram per hour, the whole body lactate concentration was approximately 1.25 mg/g, and the mean quantities of aerobically and anaerobically derived energy during 5 minutes of maximal activity were 10.59 and 16.20 mMol ATP $\times 10^{-3}$/g body weight, respectively.

Sexual dimorphism is present in the lung morphology of *C. viridis* (Keogh and Wallach 1999). Males have generally longer lungs than females at all body sizes, but total lung length increases isometrically with body size. Virtually all lung components of males are located more posteriorly than in females of the same body length. Males possess a longer vascular component than females, but no sexual dimorphism exists on the size of the avascular component. With increasing body length, the lung components are found more anteriorly, relative length of the vascular lung decreases, and relative length of the avascular lung increases in both sexes. Probably the greater size of the male lung is an adaptation aiding its generally greater annual activity.

C. viridis is capable of changing its body color in response to ETs, and becomes lighter with melanophores maximally contracted when at BTs of more than 34°C (Rahn 1942a). As BT drops to below 30°C, the melanophores disperse and the snakes gradually assume normal color when their BT is about 23°C. Snakes chilled to about 8°C exhibit maximal darkening. Such pigmental changes may aid in thermoregulation.

While mostly dependent on ingestion of prey for its moisture requirements during the annual activity period, if available, C. viridis will drink water to replace that lost through evaporation. Posthibernation individuals may seek out water to replace that lost during the winter dormant period (Starck 1984). The snake must maintain an extracellular fluid volume of about 41.9% of its mass (Smits and Lillywhite 1985).

In Alberta and Colorado, C. v. viridis may migrate as far as 11–25 km to and from the hibernaculum in the spring and fall (Duvall 1986; Jørgensen et al. 2008). In Alberta, 11 nongravid migrating female C. v. viridis exhibited 2 distinct movement patterns (Jørgensen et al. 2008). The first, shown by four females, was one of long-distance fixed-bearing migrations where the snakes left river breaks where dens occur to forage in upland prairie. Displacement from dens was 3–10 km, and the females maintained fixed-bearing routes despite encountering grazing, cultivation, roads, and pipelines, although they tended to use available edge habitats (fence lines and the margins of cultivated fields) in disturbed areas. The second movement pattern consisted of much shorter, 0.5–2.6 km, displacements from den sites. In spite of not moving far from den sites, the cumulative distances traveled by the females were similar to those making long-distance migrations from the den. Females that made the long-distance migrations experienced greater improvement in body condition than those that engaged in the complicated movements closer to the den.

Jørgensen and Gates (in Jørgensen et al. 2008) followed the movements of 19 Alberta females that made a mean 39.1 movements during the 133 days they were tracked. The snakes moved an average of 0.3 times per day at a mean distance of 0.21 km. The mean distance from their hibernacula was 2.76 km, and their mean total distance moved was 8.17 km.

C. viridis display chemosensory behaviors, sometimes erratic, at birth, that become more refined with age. Graves et al. (1986) recorded the rates of tongue-flicking, mouth-gaping, head-shaking, and face-wiping performed by neonates, and 2- to 3-month-old juveniles. Neonates averaged 6.28 tongue-flicks, 163 mouth-gapes, 0.22 head-shakes, and 1.70 face-wipes per minute when tested, compared with 0.28 tongue-flicks and no mouth-gapes, head-shakes, or face-wipes performed by the 2- to 3-month-old snakes (the mean latencies for these four behaviors were 0.54, 0.61, 0.68, and 1.82 minutes, respectively).

C. viridis can differentiate its own odor and those of its conspecifics and also of heterospecifics (Chiszar et al. 1991c; Scudder et al. 1988). However, such odors are probably not the major orientating mechanism in homing to the den (King et al. 1983). Instead, the snakes use fixed angle, sun compass orientation to find the hibernaculum

(Duvall et al. 1985b). In addition to a compass sense, *C. viridis* possesses a map sense that provides knowledge of its location in relation to a destination. It possibly also uses magnetic cues to orient (Jørgensen and Gates 2007).

Movements to and from hibernacula are not usually direct or nonstop, and may be related to prey availability (Duvall et al. 1985b; King and Duvall 1990). Mean distances covered per movement recorded by King and Duvall (1990) were 237 m for males and 137 m for females. Males average 25.3 days between movements; females only 6.7 days (Duvall et al. 1990b; King and Duvall 1990).

Olfaction plays the major role in finding prey-rich areas, and odor trails are detected either by tongue-flicking or, possibly, mouth-gaping (Chiszar et al. 1990; Duvall et al. 1985b; Graves and Duvall 1983). By traveling along paths of very high angular fixity or straightness, males more often than females minimize the likelihood of covering the same area twice. As the males search for either food or mates, detection of prey odor brings about a slowing of activity.

Once the summer feeding area is reached, a more or less permanent home range is established, and, unless prey become scarce, maintained for the rest of the summer.

A basking South Dakota *C. v. viridis* observed by C. Ernst had climbed almost 2 m up a slanting rock surface. *C. viridis* is also a good swimmer (Klauber 1972).

A possible case of "anting" has been reported. A Wyoming juvenile was observed motionless in a curious posture while red ants (*Formica* sp.) crawled over its body. The snake appeared to have recently shed and was lying on the ground with three points of its body just barely touching the soil, with its head, tail, cloaca, and other body portions held as high off the ground as possible. The tail was raised high, resulting in the exposure of just a small part of the cloacal cavity. The snake remained in this posture for about five minutes (Duvall et al. 1985b). Possibly, the small snake was allowing the ants to clean its body, but this has not been observed since. If that is the true behavior, as has been reported for some small North American colubrid snakes (Ernst and Ernst 2003) and supported by further study, it is a unique behavior in North American pitvipers.

Male *C. viridis* participate in dominance combat bouts during the mating seasons with behaviors similar to those of other species of *Crotalus* (Aldridge and Duvall 2002; Bennett 2004 [see photographic sequence]; Cage 2004; Duvall and Schuett 1997; Gloyd 1947; Klauber 1972; Taggart and Schmidt 2004; Thorne 1977).

REPRODUCTION: Because of the disparity in body size, the two subspecies mature at considerably different lengths. The shortest gravid female *C. v. nuntius* and *C. v. viridis* examined by Klauber (1972) had 39.5 cm and 89.0 cm TBLs, respectively, and a 55 cm captive female *C. v. nuntius* produced a litter (Sievert 2002b). In addition, as with *C. oreganus* (Macartney et al. 1990), males grow more rapidly than females, and mature at a smaller size and earlier age. The shorter subspecies probably mature at TBLs of 40–50 cm, and the larger subspecies at 55–65 cm (Duvall et al. 1992; Klauber 1972). Such lengths are usually attained in 3–4 years in males and 4–7 years in females. The length at which maturity is reached may even vary between populations of the same subspecies. Because males mature earlier, their potential lifetime reproductive output is greater than that of females. Duvall and Beaupre (1998) developed a model to explain the sexual strategy and size dimorphism in rattlesnakes, and we refer the reader to their paper for details too complex to discuss herein.

Aldridge (1979a, 1993) studied the gametic cycle and annual morphological changes in male *C. v. viridis*. Sertoli syncytium and spermatogonia are present in the seminiferous tubules during hibernation, and from mid-April to June the spermatogonia divide. Spermatids and spermatozoa are present from June to October. Spermatogenesis ceases in late August–November, but mature sperm are still present in the tubules. Peak sperm production occurs from mid-June through September, and spermiation apparently takes place from mid-June to mid-October. Sperm may be present in the vas deferens in every month. Seminiferous tubule diameter is smallest in the spring, is about twice the diameter in mid-July–September, and shrinks in October as sperm pass out of the tubules. ET plays a major role in initiating spermatogenesis.

The female cycle seems biennial, with about 50–70% of the females in any given population breeding in a designated year; but females may reproduce annually (Trans-Pecos, Texas; Werler and Dixon 2000), and others triennially (Fitch 1985a; King and Duvall 1990). A triennial cycle probably consists of two successive seasons of feeding and follicle development with a peak of sexual receptivity and mating occurring in the second year associated with maximum follicle development, and a third season in which spring ovulation takes place; fertilization is accomplished with stored sperm (Schuett 1992), summer pregnancy, and autumn birth of live young (King and Duvall 1990).

Vitellogenesis occurs twice in the female ovary during the year. Primary, microscopic to 4–6 mm follicles (present throughout the year) grow in summer after parturition to 15–20 mm by hibernation. Yolking ceases during the winter; but secondary vitellogenesis occurs in the spring, and some uterine muscular twisting may occur at that time (Almeida-Santos and Salomão 2002). Ovulation takes place in May or early June, when ripe females have follicles with mean lengths and widths of 25.4 mm and 12.7 mm, respectively, and their uterus is 4–8 mm wide (Aldridge 1979b; Gannon and Secoy 1984; Rahn 1942b). Reproductive females produce more body fat than those not reproducing that year.

The primary breeding season for *C. v. viridis* is the summer (Aldridge and Duvall 2002), when breeding activity occurs from mid-July into early September (Allen 1874; Duvall et al. 1985b; Graves and Duvall 1990; Hayes et al. 1992a; Holycross 1995; King and Duvall 1990; Wright and Wright 1957), and sperm can usually be found in the female's uterus at this time (Rahn 1942b). February courtship and mating have occurred in captive *C. v. nuntius* after a cooling period (Sievert 2002b), but spring matings in the wild seem to be rare (Collins 1993; Holycross 1995; Schuett and Buttenhoff 1993).

Males search for mates in the second half of the summer season, while females continue to forage for the duration of the season (Jørgensen et al. 2008; King and Duvall 1990). Males must locate two goals in time and space (food and mates; females search only for food), and exhibit significantly greater spatial searching efficiency during spring foraging than do females. This allows males to concentrate foraging activities into the first half of the annual activity season, so the second half can be almost exclusively devoted to mate searching. The mating system is best described as a prolonged male mate searching polygyny.

Males find and identify females by pheromones released from their skin. These are enhanced by ecdysis, and courtship usually occurs within 48 hours after shedding (Klauber 1972). Males perform straight-line, fixed-bearing movements while searching for females; and those searching along a straight path find more females than those moving in less straight paths (Duvall and Schuett 1997).

Once the female or her scent trail is found, the male follows her, and, once she is reached, maintains contact with her body, and searches with his tail for her vent. During courtship and coitus, the female usually remains quietly coiled; only occasionally does she flee, forcing the male to pursue her. After 3–4 tongue-flicks, the male performs forward-jerks or flexions of his body, and the female may move her tail back and forth laterally in response to the male's tactile movements. While this is occurring, he increases the rate of tongue-flicking toward the female's dorsum, often making contact, and may rub her with his chin. This flexion cycle ceases when the male initiates tail-search behavior. During this behavior, he generally withdraws his head and anterior trunk from the female and ceases both tongue-flicking and flexing. Outstretched positions may be necessary for copulation, as a coiled female hinders courtship. A period of no movement by the male follows until flexions begin again and the entire courtship sequence is repeated (Duvall et al. 1985a, 1992; Gloyd 1947; Hayes 1986; Hayes et al. 1992a, Wood 1933).

Although female *C. v. viridis* have the potential of mating with several males during a single breeding season, an analysis of allozymes from three gravid Montana females and their offspring by Crabtree and Murphy (1984) was inconclusive.

Copulation may last 90 minutes to several hours. As most mating occurs in the late summer, sexual encounters may normally occur at night. Because most matings occur in late summer when ovarian follicles are small and ovulation does not occur until the following spring, the female must store viable sperm over winter in her oviducts (King and Duvall 1990; Ludwig and Rahn 1943; Schuett 1992).

Females remaining in maternity colonies near the hibernaculum give birth sooner than those that migrate away from it. This may give their young a better chance of finding suitable overwintering sites (Duvall 1986).

The birthing period is from August to early October after a GP possibly lasting 100–120 days in *C. v. viridis* and 138 days in *C. v. nuntius* (in captivity; Sievert 2002b). Collins (1993) reported that parturition occurs in spring, summer, and fall, but, except possibly in captivity, spring births rarely occur, if at all. Females give birth underground in mammal burrows or in some other shelter.

The number of young produced is correlated to female body length (Fitch 1998); shorter (younger?) females produce smaller litters, averaging only about 6–8 young, while older (larger) females have 15 or more. Litters in the species average 7.8 young (n = 491) and may contain 3–21 (n = 530 literature records). Litter size is positively correlated with female body length (Fitch 1998), so *C. v. nuntius* produces litters with 3–10 (mean, 6.2; n = 16) neonates, and *C. v. viridis* produces litters of 3–21 (mean, 7.8; n = 475) neonates.

Neonates are more vividly colored and patterned than adults, and some young may have lighter pigment at the base of the tail. Their TBL averages 22.9 (range, 17.9–30.5; n = 72) cm, and their initial BM is 6.5–14.2 (mean, 9.7; n = 51) g. Neonate size is correlated with female length. Female *C. v. nuntius* produce young with TBLs of 17.8–23.5 (mean, 20.8; n = 13) cm and BMs of 6.5–11.0 (mean, 9.3; n = 10) g; those of *C. v. viridis* have 18.0–30.5 (mean, 24.9; n = 62) cm TBLs, 21.0–26.0 (mean, 23.5; n = 13) cm SVLs, and 10.1–16.0 (mean, 11.5; n = 17) g BMs. A few are born without a rattle (Holycross 2000b). Other developmental anomalies occasionally occur; a neonate from Saskatchewan was born with two heads (Reid 2005), and another had a cleft palate (Dean et al. 1980).

The observations of female *C. v. viridis* with accompanying newborns indicate that at least some remain with their young for a short period of time after giving birth, probably until after the young snakes' first ecdysis. Cunningham et al. (1996) reported 2 cases of South Dakota females with 4–6 young of newborn length, and Holycross and Fawcett (2002) found 3 postpartum females with a total of 31 neonates in Nebraska. All observations occurred in August.

Natural hybrids between *C. viridis* and *C. oreganus* are known (Douglas 1966; Klauber 1972; Murphy and Crabtree 1988). Porter et al. (1991) determined the ribosomal DNA sequences in *C. viridis*, as well as those from 3 other snakes and 16 species of lizards, to determine the location of chromosomal hybridization. They found that two pairs of microchromosomes are involved, where hybridization is restricted to one end of the chromosome. A natural hybridization with *C. atrox* has also occurred (see photograph in that account).

GROWTH AND LONGEVITY: Individual growth is quite variable and dependent on foraging success and climatic conditions, making it difficult to present an overall growth rate. However, growth slows with age and length in both sexes. Individuals are larger in cooler and more seasonal areas, the opposite of the trend in the *C. oreganus* clade (Ashton 2001).

Actual growth data related to a population are scarce. In Saskatchewan, *C. v. viridis* are 19.5–22.9 cm at birth, while those in their second summer are 35.0–59.5 cm; average growth for the first year is 33.7 cm, and 21.2 cm the second year (Gannon and Secoy 1984). Werner et al. (2004) stated that Montana *C. v. viridis* neonates grow about 30 cm in TBL their first year and 20 cm or more their second year.

Adult *C. viridis* shed 0–4 times per year (Gannon and Secoy 1984); the number is usually correlated with the snake's amount of prey intake and the weather conditions

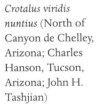

Crotalus viridis nuntius (North of Canyon de Chelley, Arizona; Charles Hanson, Tucson, Arizona; John H. Tashjian)

it experiences. Loomis (1951) reported that infestations with mites and certain skin injuries (loreal pit) may cause an increase in the rate of shedding.

The western rattlesnake has a potentially long lifespan. A captive *C. v. viridis* possibly lived 31 or 32 years at Tabor College, Kansas (M. Termin *in* Collins 1989). Another *C. v. viridis*, estimated to have been 2 years old when captured, survived 27 years, 9 months in captivity (Bailey et al. 1989), and still another wild-caught male of unknown age lived an additional 19 years, 3 months, and 10 days at the San Diego Zoo (Snider and Bowler 1992). A juvenile male *C. v. nuntius* survived an additional 12 years, 24 days at the Columbus Zoo (Snider and Bowler 1992).

DIET AND FEEDING BEHAVIOR: *C. viridis* is a generalist carnivore that occurs within the ranges of numerous species of vertebrate prey, and, consequently, has a quite variable diet. Newborn snakes and small juveniles prey on small mammals, whereas adults eat larger prey of a greater diversity, but prefer mammals. The annual feeding period usually lasts from mid-April to early October in most populations.

Recorded or suspected (based on the diet of *C. oreganus*), fresh-killed or carrion, prey of *C. v. viridis* are Mormon crickets (*Anabrus simplex*); fish—trout (*Salvelinus* sp. [?]); amphibians—spadefoots (*Spea bombifrons*), and frogs (*Rana* sp.); reptiles—lizards (*Cnemidophorus sexlineatus*; *Eumeces* sp.; *Holbrookia maculate*; *Phyrrnosoma douglassii*; *Sceloporus graciosus, S. undulatus*; *Uta stansburiana*), and snakes (*Crotalus viridis*, both natural and in captivity); birds (including eggs and nestlings)—juvenile pheasants (*Phasianus colchicus*), quail (*Callipepla gambelli*), domestic chickens (*Gallus gallus*), turkeys (*Meleagris gallopavo*), grouse (*Dendragapus obscurus*), burrowing owls (*Athene cunicularia*), mourning doves (*Zenaida macroura*), bushtit (*Psaltriparus minimus*), horned larks (*Eremophila alpestris*), and passeriforms (*Calamospiza melanocorys, Carpodacus mexicanus, Chondestes grammacus, Euphagus cyanocephalus, Junco hyemalis, Melospiza melodia, Mimus polyglottos, Pipilo erythrophthalmus, Pooecetes gramineus, Sialia mexicana, Sternella neglecta, Sturnus vulgaris, Turdus migratorius, Zonotrichia leucophrys*); mammals—pikas (*Ochotona princeps*), cottontails (*Sylvilagus audubonii, S. floridanus, S. nuttallii*), juvenile jackrabbits (*Lepus* sp.), pocket gophers (*Cratogeomys castanops*; *Geomys bursarius*; *Thomomys bottae, T. talpoides*), kangaroo rats (*Dipodomys ordii*), pocket mice (*Perognathus fasciatus, P. flavus, P. flavescens*), chipmunks (*Tamias minimus, T. quadrivittatus, T. striatus*), ground squirrels (*Ammospermophilus leucurus*; *Spermophilus columbianus, S. richardsonii, S. townsendii, S. tridecemlineatus*), juvenile prairie dogs (*Cynomys gunnisoni, C. leucurus, C. ludovicianus*), tree squirrels (*Sciurus aberti, S. carolinensis, S. niger*; *Tamiasciurus hudsonicus*), voles (*Clethrionomys gapperi*; *Lemmiscus curtatus*; *Microtus longicaudatus, M. montanus, M. ochrogaster, M. pennsylvanicus*), murid mice and rats (*Mus musculus* [captivity]; *Neotoma* sp.; *Peromyscus boylii, P. leucopus, P. maniculatus, P. truei*; *Rattus norvegicus* [captivity]; *Reithrodontomys megalotis*), and jumping mice (*Zapus princeps*). One had a cocklebur (*Xanthium* sp.) in its throat, which was probably swallowed accidentally.

References for the above prey list are Banta (1974), Bateman (1918), Brons (1882), Brown (1990), Bullock (1971), Campbell and Lamar (2004), Coues and Yarrow (1878), Degenhardt et al. (1996), Duvall (1986), Duvall et al. (1990a), Duvall et al. (1985a, 1985b), Elmore (1954), C. Ernst (personal observation), Gannon and Secoy (1984), Gehlbach (1965), Genter (1984), Giovanni et al. (2005), Gloyd (1933), Graves (1991), Hamilton (1950), Hammerson (1986), Hayes (1991b), Hayes and Duvall (1991), Hill (1943), Hill et al. (2001), Klauber (1972), Knight and Collins (1977), Koch and Peterson (1995), Lillywhite (1982), Ludlow (1981), Marr (1944), McKinney and Ballinger (1966), Mitchell (1986), Mosimann and Rabb (1952), Rodgers and Jellison (1942), Smith (1956), Stabler (1948), Stebbins (1954), Tennant (1985), Thompson and Nichols (1982), Turner (1955), Werler and Dixon (2000), Werner et al. (2004), Wheeler and Wheeler (1966), and Williston (1878).

In southeastern Alberta, Canada, the diet of *C. v. viridis*, based on gut contents and the contents of scats, consists of 53% (68% scats) sagebrush voles (*Lemmiscus curtatus*), 38% (8% scats) meadow voles (*Microtus pennsylvanica*), 9% unidentified passerine birds; and 8% of scat contents each for the olive-backed pocket mouse (*Perognathus fasciatus*), western jumping mouse (*Zapus princeps*), and Richardson's ground squirrel (*Spermophilus richardsonii*). Usually multiple prey items in a single snake were of the same species, suggesting that individual *C. v. viridis* tend to exploit patches where colonial burrowing prey species are abundant (Hill et al. 2001).

Similar prey data are few for *C. v. nuntius*. Klauber (1972) found mammal hair in five he examined, and thought that juveniles feed largely on lizards, particularly the common lesser earless lizard, *Holbrookia maculata*. Captives readily take house mice, *Mus musculus*, and small brown rats, *Rattus norvegicus* (C. Ernst, personal observation; Sievert 2002b). Because of its smaller size, the subspecies may be more dependent on lizard prey than its larger relative *C. v. viridis*.

C. viridis may either actively forage or catch prey from ambush. The first approach is used when prey are scattered, or often during migration to or from the den (Duvall et al. 1985b). Ambushing is used more often, particularly by males, when colonies of rodents have been identified and the snake can lie near the openings of burrows or rodent runs. It is only capable of short bouts of maximal activity before exhaustion (Ruben 1983), so the latter strategy is more efficient from an energy standpoint.

Prey is found either visually, by infrared detection (particularly for birds and mammals), or by olfaction. Warm, 39°C, rodent prey are struck quickly; but those with 26°C BTs are struck quickly only if rodent integumentary cues are also present (Chiszar and Smith 2008). Movement of the prey seems to be the primary visual component, which brings on a rise in the rate of tongue-flicking to provide additional olfactory information (Chiszar et al. 1981b; Scudder and Chiszar 1977). The snake prefers envenomated prey (Chiszar et al. 1980a; Duvall et al. 1978), and can distinguish the odor of prey (*Mus musculus*, *Rattus norvegicus*) from artificial scents such as perfume and distilled water (Melcer et al. 1988). Odors from the anterior end of mice help the snake to find the head for easier swallowing (Duvall et al. 1980).

The visual and infrared data are combined in biomodal neurons of the optic tectum, which receive input from both the retina and the pit organ (Newman and Hartline 1981, 1982). This is extremely important during crepuscular or nocturnal hunting. However, the pit organ may not be the only means of detecting prey body heat. *C. viridis* anesthetized so that the trigeminal nerve could not mediate electrophysiological responses of the pit organs to thermal stimulation still exhibited behavioral responses to thermal cues (Chiszar et al. 1986b). Either an auxiliary infrared-sensitive system (nociceptors) or the common temperature sense could be responsible. Intraoral thermal stimulation elicits response from the superficial maxillary branch of the trigeminal nerve (Dickman et al. 1987). Such responses to oral heat stimulation are independent of any responses associated with thermal stimulation of the pit organs. Histological preparations of tissues from the upper lip, palate, and fang sheath reveal dense ramifying neurons in the epidermal layers of the fang sheaths that are morphologically similar to the infrared sensitive neurons in pit organ membranes.

Visual and thermal cues are sufficient to bring on a strike-response by *C. viridis*. Warm objects (35°C) are struck more often than colder ones (25°C; Breidenbach 1990). The snake readily strikes when presented with objects 1.5–4.5°C warmer than AT.

Olfaction also plays a major role in prey detection, as indicated by the spring migratory and habitat selection patterns (Duvall et al. 1990a, 1990b; Theodoratus and Chiszar 2000), and the modified behavior of *C. viridis* with vomeronasal organs sutured (fewer tongue-flicks after striking prey increasing the time to find wounded prey) (Graves and Duvall 1985a; Stark et al. 2006). Such snakes have lower tongue-flick rates when presented with prey odor stimuli; unaltered *C. viridis* may mouth-gape and actually shake their heads 2–3 rapid horizontal jerks to help bring odors to the vomernasal organ (Graves and Duvall 1983, 1985b). Apparently prey-chemical preferences must be learned; they are absent in neonates, but present in adults (Chiszar and Radcliffe 1977). Food deprivation causes increases in tongue-flick rates in response to mouse odor (Chiszar et al. 1981a), and often strikes are hurried and inaccurate (Hayes 1993).

Rodent hygienic behavior may play an important role in successful predations by *C. viridis*. During laboratory tests conducted by Ackerman et al. (2007), *C. viridis* preferred a less odorous laboratory strain of *Mus musculus* over a more odorous one. It took longer to find and grasp the head of the odorous mouse and to swallow it. The least favored strain may have confused the snake's olfaction by handling fecal material and smearing it on themselves.

When prey wanders into striking range, it is usually struck only once, after which the snake normally allows it to flee while the venom digests it. The strike is usually aimed at the head and body, the largest targets, of moving prey (Schmidt et al. 1993). Lizards and small rodents are normally held, but large rodents are released (Chiszar et al. 1986d; Radcliffe et al. 1980). The larger the snake, the more likely it is to hold, rather than quickly release, prey.

If presented with several mice in succession, the later mice are often struck repeatedly or held in the mouth rather than released. This behavior probably results from the depletion of and reduced access to venom reserves (Hayes et al. 1992b). Although Hayes (1992a) reported that the quantity of venom released in a strike is not related to prey size, in later studies, Hayes (1995) and Herbert and Hayes (2008) found that juvenile *C. v. viridis* can meter the amount of venom released during a strike, injecting more venom into larger than into smaller prey. Possibly this is a learned response. Also, juveniles make longer fang contact with prey than do adults. Adult fang contact averages 0.16 second during a predatory strike to an adult mouse, but averages only 0.07 second during a defensive strike to a rat (Hayes 1991a, 1991b). Adults inject about 12 mg of venom during a predatory strike (Hayes et al. 1992b).

Once prey has been struck and tasted, a chemical search image is created, resulting in an increase in the rate of tongue-flicking (Chiszar and Radcliffe 1976; Chiszar et al. 1991a, 1991b, 1996; Haverly and Kardong 1996; B. O'Connell et al. 1982; Scudder et al. 1992). Chiszar et al. (1993b) concluded that the taste of the prey's blood is a critical component that sets off the strike-induced chemosensory searching by *C. viridis*; however, Smith and Kardong (2005) found this not to be the case in *C. oreganus*. Prey-trailing only occurs if it has been envenomated (Chiszar et al. 1988; Diller 1990; Furry et al. 1991; Golan et al. 1982; Lee et al. 1988, 1992). Taste buds present in the palatal mucosa, although few in number, may help *C. viridis* compile a taste image of envenomated prey (Berkhoudt et al. 2001). Once the odor trail of the bitten prey is located, the snake usually searches only about 15 minutes before returning to its hunting mode. Venom odors, and those from the head and nasal-oral tissues of the mouse, are important cues for finally detecting the dead mouse and finding its head for easier swallowing (Duvall et al. 1980; Lavín-Murcio and Kardong 1995). Ashton (2002) compared the directional swallowing of hairless mice and those with a normal hair coat, and found that the presence of hair retards the snake from finding the head-end of the mammal prey. However, lack of hair possibly provides stronger chemical and visual cues to the snake.

Small prey may simply be swallowed alive without envenomation by adults, thereby conserving venom. Retention of prey following a bite increases the severity of the envenomation, and a poor strike is usually rapidly followed by a second or third.

VENOM DELIVERY SYSTEM: FL is ontogenetic and positively correlated with TBL and SVL in venomous snakes, so adults of the larger *C. v. viridis* possess the longest fangs, while adults of the shorter *C. v. nuntius* have smaller fangs. The mean fang dimensions of 590 *C. v. viridis* with TBLs of 70–100 (mean, 85) cm, calculated from regression lines, were FL, 7.4 (range, 6.3–8.4) mm; TBL/FL, 115%; and HL/FL, 5.05%; their mean angle of fang curvature was 62°. Five *C. v. nuntius* with 45.8–54.0 (mean, 49.7) cm TBLs had 3.8–4.8 (mean, 4.1) mm FLs, and 122% and 5.76% means of TBL/FL and HL/FL, respectively; their mean angle of fang curvature was 59° (Klauber 1939).

VENOM AND BITES: Venom of *C. viridis* may produce hemorrhagic, neurologic, and proteolytic activities during the development of a single envenomation. Some New Mexican *C. v. viridis* contain the Mojave toxin usually associated with *C. scutulatus* (Glenn and Straight 1990), and some neurotoxic peptide *concolor* toxin has been identified from *C. v. viridis* in all parts of its range (Aird et al. 1991; Bober et al. 1988; Glenn and Straight 1990; Griffin and Aird 1990; Russell 1983; Ziolkowski et al. 1992). Myotoxin α occurs in large amounts in northern populations of *C. v. viridis*, but none to very small amounts in New Mexico and Texas populations (Straight et al. 1991), and not at all in *C. v. nuntius* (Henderson and Bieber 1986). The overall toxicity of venom from *C. viridis* is weaker than that of *C. atrox*, *C. helleri*, and *C. oreganus*, but the species is rather bad-tempered when aroused and still very dangerous. Fortunately, human bites are relatively infrequent.

Potential venom yield increases with body length, so the larger subspecies *C. v. viridis* produces more venom than does the smaller subspecies *C. v. nuntius*. Klauber (1972) collected a total of 46,697 mg of dried venom during 1,676 extractions from *C. v. viriidis*, an average of 44 mg per fresh adult. Do Amaral (1928) reported that the average amount of dry venom from a bite by young *C. v. viridis* is 50 mg, that of adults is 90 mg. Other reported average individual venom yields (mg) for *C. v. viridis* were 41 (Githens 1933) and 60 (Brown 1973). Reported maximum dry venom yields (mg) for that subspecies were 100 (Russell and Brodie 1974), 135 (Githens 1933), and 162 (Klauber 1972). Russell (1983) and Russell and Puffer (1970) gave the range of venom yield as 35–110 mg. Recorded average adult yields of dried venom (mg) from *C. v. nuntius* were 38 (Klauber 1935) and 51 (Klauber 1972); and the maximum recorded yield (mg) was 72 (Klauber 1972).

The lethal subcutaneous dose of dried venom from *C. viridis* for a 70 kg human adult is 504 mg, and the number of lethal doses per bite is 0.09–0.12 (Brown 1973). The venom of 2-week-old to 3-month-old juvenile *C. v. viridis* is about twice as deadly as that of an adult.

Hayes (1992b) reported that while envenomating *Peromyscus maniculatus*, *C. v. viridis* expends an average of 16 (range, 5–26) mg of venom; during a single envenomation, 15 (range, 5–25) mg; but during multiple bites, 21 (range, 16–26) mg. The average venom expenditure per bite is about 32% of that available (Hayes et al. 1992b).

Adults have yellow-colored venom, while that of juveniles remains colorless for several months, but is qualitatively similar to that of adults (Fiero et al. 1972).

Chemistry of venom from *C. viridis* has been well studied. It contains at least 21 different protein components, with both purines and pyrimidines involved, and several toxins. The percentage of protein in the venom is at least 75–84% (Brown 1973; McCue 2005).

Identified components, or indications of their activities, are arginine ester hydrolase (activity), bradykinin-releasing peptide (activity), chymotrypsin (activity), crotovirin, deoxyribonuclease, dipeptidyl peptidase IV (with some geographic variation as to content in *C. v. viridis*), disintegrins (RGD, MVD), endopeptidase (activity), hemorrhagic toxin, hyaluronidase, kalligrin-like enzymes, L-amino acid oxidase, metalloproteases (molecular weights 69,000–115,000), both Mojave toxins α (18% of crude venom) and β, NAD-nucleosidase, 5′nucleoside, phosphodiesterase, phospholipase A_2, phosphomonoesterase, ribonuclease 1 (venom RN-ase), trypsin (activity), and viriditoxins (viridin, viridistatin, viristitarin) (Aird 2005, 2008; Aird et al. 1991; Anaya et al. 1992; Arce et al. 2003; Bober et al. 1988; Glenn and Straight 1990; Graham et al. 2005; Griffin and Aird 1990; Komori and Nikai 1998; Komori et al. 1988; Li and Ownby 1994; Li et al. 1993; Liu et al. 1995; McCue 2005; Meier and Stocker 1995; Oshima et al. 1969; Ownby et al. 1976, 1979, 1984; Powell et al. 2008; Russell 1983; Soto et al. 1989, 2007; Takahashi and Mashiko 1998; Tan and Ponnudurai 1991; Tsai et al. 2001; Tu 1977; Tubbs et al. 2000; Wermelinger et al. 2005; White 2005; Young et al. 1980; Ziolkowski et al. 1992).

Venom lethalities for *C. viridis* have been reported for several test animals. Githens and George (1931) determined intravenous minimum lethal doses of 0.04 and 0.08 mg for a 350 g pigeon (*Columba livia*).

The ranges of minimum lethal dose (mg) and LD_{50} (mg) for dry venom from *C. v. viridis* for the mouse *Mus musculus* are subcutaneous—5.5–7.1 (juveniles), 14.3–14.8 (adults) (Fiero et al. 1972); intraperitoneal—minimum lethal dose, 1.25–2.25 (Githens and Wolff 1939; Khole 1991; Macht 1937); LD_{50}, 2.00–2.72 (Arce et al. 2003; Glenn and Straight 1977; Kocholaty et al. 1971; Russell 1967a) and 2.25 (Russell and Brodie 1974; Russell and Puffer 1970); and intravenous—LD_{50}, 1.01–1.77 (Arce et al. 2003; Friederich and Tu 1971; Gingrich and Hohenadel 1956; Kocholaty et al. 1971; Russell 1967a, 1967b, 1983; Russell and Brodie 1974; Russell and Puffer 1970). The interperitoneal LD_{50} for mice is 2.2 mg for dry venom from *C. v. nuntius* (Glenn and Straight 1977). Mice injected with greater amounts of venom die more quickly.

Dried venom stored in the dark in sealed containers at 6–28°C for 26 and 27 years retained most of its lethality to mice and cats (Russell et al. 1960). Mice had LD_{50} values of 0.40 mg (26 years) and 0.45 mg (27 years), respectively. Cats that were injected intravenously with a sublethal dose of venom from *C. v. viridis* experienced an immediate drop in systemic arterial pressure concomitant with an increase in venous and cisternal pressures, irregular breathing (apnea periods of 10–45 seconds) after arterial pressure dropped, and irregular electrocardiogram waves during heart beats.

Symptoms recorded during human envenomations include discoloration, swelling, pain, tingling or numbness (at the bitten area, tongue, mouth, or scalp); ecchymosis; stiffness; weakness; giddiness; breathing difficulty; hemorrhage; lowered blood flow and pressure; heart failure; nausea and vomiting; paralysis; unconsciousness or stupor; nervousness; excitability; secondary gangrene infection; and possibly an increase in vascular permeability of protein and erythrocytes (Audi and Seifert 2006; Cameron and Tu 1977; Dart et al. 1992; Hutchison 1929; Keyler 2005; Norris 2004; Russell 1967a, 1983; Russell and Michaelis 1960; Russell and Picchioni 1983; Ryan and Caravati 1994).

Death has occurred, particularly from untreated or poorly treated envenomations (Hutchison 1929; Straight and Glenn 1993). See also Klauber (1972), Keyler (2005), Koch and Peterson (1995), Lintner et al. 2006, Over (1928), and Seifert (2006) for case histories of nonfatal envenomations.

PREDATORS AND DEFENSE: *C. viridis* has many predatory enemies: mammals—rodents (*Neotoma* sp. [during hibernation]), mustelids (*Mephitis mephitis, Taxidea taxus*), cats (*Felis catus, Lynx rufus*), canids (*Canis familiaris, C. latrans; Vulpes vulpes; Urocyon cinereoargenteus*), and domestic hogs (*Sus scrofa*); birds—golden eagles (*Aquila chrysaetos*), hawks (*Buteo jamaicensis, B. regalis, B. swainsoni*), owls (*Bubo virginianus*), turkeys (*Meleagris gallopavo*), and roadrunners (*Geococcyx californianus*); snakes (*Charina bottae* [captivity]; *Coluber constrictor; Crotalus ruber* and *C. viridis* [captivity]; *Lampropeltis getula, Masticophis flagellum; Pituophis catenifer*); and possibly red ants (*Formica* sp.) (Banta 1974; Bent 1938; Bullock 1971; Campbell and Lamar 2004; Duvall 1986; Duvall et al. 1985b; Hurter 1893; Jackley 1938; Klauber 1972; Knight and Erickson 1976; Lillywhite 1982; Macartney and Weichel 1993; Olendorff 1976; Ross 1989; Starck 1987; Steenhof and Kochert 1985).

Accidents involving other wild animals often prove fatal; deer (*Odocoileus hemionus, O. virginianus*), pronghorn (*Antilocapra americana*), bison (*Bison bison*), domestic cattle (*Bos taurus*), domestic goats (*Capra hircus*), and horses (*Equus caballus*) often trample *C. viridis* (Klauber 1972).

When discovered, *C. viridis* will usually lie still, and sometimes hide its head (Duval et al. 1985b) and increase its tongue-flicking (its vomeronasal organ plays an important role in predator detection; Miller and Gutzke 1999). It may also hear the approach of a potential enemy; its ear is moderately sensitive to airborne sounds between 80 and 400 Hz, but is insensitive to tones above 1,000 Hz (Pylka et al. 1971).

The snake at first relies on its cryptic coloration and pattern to escape detection, especially when sufficient cover is present. This is particularly true of gravid females, but also smaller individuals (Kissner et al. 1997), and has led some persons to believe the snake timid (Baxter and Rahn 1941), but the opposite is true (see below). Lillywhite (1974) has noted an interesting case of procrypsis in recently burned chaparral. The dark coloration of *C. viridis* renders it cryptically colored during postfire years, as the snakes are easily confused with charred stalks of woody plants scattered within the burned area.

When *C. viridis* realizes it has been discovered, it normally tries to crawl directly away from the threat—stretching out its body so that the head and upper third of the body are positioned, or cocked, for a potential bite while it backs away with the posterior two-thirds of its body. When this does not help, the snake may simply hide its head under the central (usually widest) coil of the body (we have only seen this once). If further disturbed it forms a defensive coil with its head and anterior portion of the body raised high, rotates its coils so that it always faces the intruder, and rattles continuously (see a detailed account of C. Ernst's first experience with an irate *C. v. viridis* described in Ernst 1992).

The SVL-adjusted loudness of its rattling is about 75 dB, although Pylka et al. (1971) reported it to be 34 dB (1 dyne/cm^2). The frequency ranges from mean lows of 3.28–3.34 kHz to mean highs of 10.40–10.62 kHz (Cook et al. 1994; Fenton and Licht 1990). Most of the rattle energy is between 2 and 50 kHz, with little energy produced below 1,600 kHz (Pylka et al. 1971).

Rattling vibratory speed is temperature dependent. It is faster at higher temperatures than at low ones, averaging 41.0 cycles/second at 11°C and 57.9 cycles/second at 23°C (Chadwick and Rahn 1954). Kissner et al. (1997), however, found no consistent relationship between the distance from the danger and the start of rattling and the snake's BT.

If approached too closely, the snake will strike viciously (a South Dakota male even advanced toward C. Ernst each time it struck). An alarm pheromone may also be released from the cloacal glands that possibly warns other nearby *C. viridis* (Graves and Duvall 1988). The snake's heart rate increases as danger multiplies (Graves and Duvall 1988).

For males and nongravid females, defense is not tightly related to variation in BT, but the defensive behavior of gravid females apparently is temperature-dependent, as indicated by a shift from static, retaliatory defense at lower BTs to rapid escape at higher BTs (Goode and Duvall 1989).

Potentially predatory snakes (*Lampropeltis calligaster*, *L. getula*) are recognized by sight or their odors, and *C. viridis* will flee if given a chance. If not it either freezes in place, inflates its body, body-bridges, or strikes when approached by these snakes (Bogert 1941; Carpenter and Gillingham 1975; Gutzke et al. 1993; Weldon et al. 1992). The odor of either the kingsnake (*Lampropeltis getula*) or of methyl mercaptan from a spotted skunk (*Spilogale putorius*) causes an increase in the heart rate: kingsnake, 16% increase; skunk, 15% increase (Cowles and Phelan 1958).

C. viridis serves as the model in several mimicry systems. Gopher snakes (*Pituophis catenifer*) share aspects of coloration, pattern, and defensive behavior with sympatric subspecies of *C. viridis* in what seems to be a case of Batesian mimicry (Kardong 1980; Sweet 1985). The western hog-nosed snake (*Heterodon nasicus*) is also similarly colored and patterned. In addition, the hiss of the burrowing owl (*Athene cunicularia*) and the sound of the snake's rattle are very similar (Rowe et al. 1986).

PARASITES AND PATHOGENS: The parasitic blood protozoan *Haemogregarina corti* and other unidentified hemogregarines have been isolated from *C. viridis* (Campbell and Lamar 2004; Hull and Camin 1960; Laveran 1902). The following parasitic helminths have also been found in it: trematodes—*Manodistomum* sp.; cestodes—*Mesocestoides corti*, *Mesocestoides* sp. (tetrathyridia), and *Oochoristica osheroffi*; acanthocephalans—two unidentified species of oligacanthorynhoids; and nematodes—*Hexametra boddaërtii*,

H. leidyi; Kalicephalus inermis; Phasaloptera obtussima; Physaloptera sp., *Rhabdias* sp., and unidentified nematodes (Bolette 1998; Bowman 1984; Campbell and Lamar 2004; Ernst and Ernst 2006; Hanson 1976; Hanson and Widmer 1985; Klauber 1972; Mankau and Widmer 1977; Morgan 1943; Pfaffenberger et al. 1989; Sprent 1978; Widmer 1967; Widmer and Hanson 1983; Widmer and Olsen 1967; Widmer and Sprecht 1991; Widmer et al. 1995). In addition, the linguatulid crustacean tongue worm, *Porocephalus crotali*, has been found in *C. v. viridis* (Hill *in* Klauber 1972). Unidentified skin mites have been found attached to *C. v. viridis* from Nebraska (Loomis 1951) and South Dakota (C. Ernst, personal observation).

Captives may be prone to respiratory infections (Murphy and Armstrong 1978), and the species is probably susceptible to paramyxovirus infections. It may also suffer from fibrosarcomas (Ball *in* Klauber 1972; Wadsworth 1956).

POPULATIONS: Many colonies of the western rattlesnake were quite populous in the past, but today the snake has been reduced in many areas and extirpated in others (the affecting factors are discussed later). It may still be common in some local out-of-the way or protected areas where disturbance by humans is uncommon, but these are becoming more rare each year. In the Dakotas, Wyoming, and Montana it is the most common snake at some localities (C. Ernst, personal observation; Ludlow 1981; Smith and Smith 1972), and it is still commonly found at certain sites in northwestern Iowa (Fogell 2005).

The structure of most populations is centered around a specific hibernaculum or several closely located ones, especially in the north, and this has contributed to much of the trouble for the species. In many areas, populations have decreased or disappeared due to direct human predation at hibernacula. Formerly large populations overwintered in specific dens, but many of these have been systematically eradicated with gas, bullets, and explosives (Klauber 1972) so that now *C. viridis* is in danger of totally disappearing from areas where it was once exceedingly common.

In 1977, Pendlebury reported *C. viridis* to range from rare to common at several sites in southeastern Alberta and adjacent southwestern Saskatchewan, but Macartney and Weichel (1993) estimated the total population of *C. viridis* in southern Saskatchewan to be only 2,000–4,500 snakes. The population at a den in Saskatchewan was estimated to be 149 (Gannon and Secoy 1984), and a den in Wyoming contained 42 adults (Duvall et al. 1985b).

C. viridis was also uncommon during other studies in the United States. At the southern end of the range, Repp (1999) recorded only 5 (2%; 3, 8.6%, DOR) *C. v. nuntius* in a sample of 247 (35 DOR) rattlesnakes of 6 species during the El Niño year of 1997 in Arizona. In New Mexico, of 454 snakes recorded during a 4-year road-collecting study by Price and LaPointe (1990), only 6 (1.3%) were *C. viridis*; and of 158 captures of 13 species of snakes by Bateman et al. (2009), only 9 (5.7%) were this species. In an approximately 6.1 ha plot at Gran Quivira National Monument, New Mexico, Howell and Wood (1957) recorded 144 *C. viridis* during 1954–1956, a human contact ratio of about 1 every 6 days. At the Kiowa-Rita Blanca National Grasslands in that state, the species was the most abundant snake during a survey conducted by Collins et al. (2006) in 1999; but only 37 were recorded. Currently, *C. viridis* is only locally common in Yellowstone National Park, Wyoming (Koch and Peterson 1995).

Klauber (1939) examined the fangs of 590 *C. v. viridis* from Platteville, Colorado, but populations in that state have decreased in numbers since. Shipley and Reading (2006) captured only 13 (30.2% of 43 snakes) during a study of the herpetofauna and small mammal diversity at a prairie dog site in that state; and Ludlow (1981) recorded only 135 prairie rattlesnakes in 3 years at the Ken-Caryl Ranch, Jefferson County.

Of 33,117 snakes collected statewide in Kansas, only 32 were *C. viridis* (Fitch 1993), but in 1985b Fitch reported that rattles were collected from 425 Kansas prairie rattlesnakes, and later in 2002 from 283 individuals of the species.

Rattlesnake roundups give some idea of the abundance of *C. viridis* in certain areas, but help deplete the populations sampled: western Kansas—Sharon Springs, 487 (Fitch 1998); and from 200 to 179 between 1995 and 2001 (Schmidt 2002).

The normal sex ratio is probably 1:1. A sample taken in the last week of May at a south-central Wyoming site yielded 22 males, 10 gravid females, and 10 nongravid ones (Duvall et al. 1985b). Schmidt (2002) reported a combined 2-year adult ratio of 180 males and 196 females (0.92:1.00) at the Sharon Springs Roundup in western Kansas. The adult sex ratio of the Saskatchewan *C. viridis* examined by Gannon and Secoy (1984) was 1.13:1.00.

However, aberrant sex ratios have been reported. In Kansas, Fitch (1985b) measured the rattles of 92 adult males and 61 females (1.50:1.00). Samples taken at rattlesnake roundups are often sex biased; Fitch (1998) recorded 301 males to 186 females, a ratio of 1.62:1.00, at Sharon Springs, Kansas, and in 2002 he counted 175 males and 108 females (1.62:1.00) there. Of course, rattlesnake roundups are normally conducted during the months that males are more active than females.

Some juvenile (based on the presence of five or fewer rattle segments) to adult (>5 segments) ratios that have been recorded are 1.78:1.00 (Fitch 1985b), 0.10:1.00 (Fitch 1998), 0.37:1.00 (Fitch 2002), and a combined 2-year ratio of 0.16:1.00 (Schmidt 2002). Based on an adult SVL of 55 cm, the juvenile to adult ratio at Gannon and Secoy's (1984) Saskatchewan site was 0.57:1.00.

Overwintering mortality is high, particularly among juveniles. Gannon and Secoy (1984) reported that the proportion of first-year young decreased from 39% of their Saskatchewan population to only 12.7% from the fall of 1976 to the spring of 1977. Fitch (1985b) calculated a mortality rate of 60% for the first year of life and 50% in each subsequent year for a population in Kansas. He thought few survive longer than 8 years in the wild. Most populations are composed predominantly of adults with SVLs greater than 50 cm.

Humans cause this animal the most harm. They have killed it on sight (Collins et al. 2006); secondarily poisoned it (Campbell 1953b); slaughtered it on our highways (Campbell 1953b, 1956; Repp 1999); destroyed hibernacula, converted its prairie habitat to agricultural, commercial, or domestic use (Corn and Peterson 1996); and supported a commercial trade for the skin, meat, gall bladder, and various curios (usually taken from snakes at roundups; Fitzgerald and Painter 2000). The invasion of woody and other invasive plants is slowly altering its grassland habitats, possibly making them unsuitable to prey species (Fogell 2005). Another potential problem for the rattlesnakes living near nuclear reactors is the uptake of radiation from their prey; the snake seems quite susceptible to this (Cummings and Kappel 1966).

C. viridis is now considered endangered in Iowa and protected in Colorado, but more states and Canadian provinces must soon realize the plight of this snake.

REMARKS: *Crotalus viridis* was previously thought to be composed of nine subspecies (*abyssus, caliginus, cerberus, concolor, helleri, lutosus, nuntius, oreganus,* and *viridis*) until, in his doctoral dissertation, Quinn (1987) suggested that the species represented a complex of several taxa. Unfortunately Quinn did not publish his results, and further study of the systematics of the *viridis*-complex was delayed for a decade. More recently, the use of modern molecular techniques to examine the systematics of the taxa in the complex has brought about new arrangements that show that the original taxon *C. viridis* is a complex of species. Some confusion, however, still exists as to the exact number of species and their arrangement within the complex.

Using electrophoresis of plasma proteins, Minton (1992) found no significant difference between *C. v. viridis* and *C. v. lutosus,* but did find a difference between *C. v. viridis* and *C. v. concolor* approaching the difference he found between the former subspecies and *C. scutulatus.* Distinct differences between *C. v. viridus* and both *C. horridus* and *C. molossus* were also evident.

A phylogenetic study using mtDNA by Pook et al. (2000) indicates that two clades exist within the original species *C. viridis*—an eastern and southern clade found east of the Rocky Mountains consisting of only *viridis* and *nuntius;* and a second clade from west of the Rocky Mountains composed of *abyssus, cerberus, concolor, helleri, lutosus,* and *oreganus* (they did not recognize *caliginus* as a taxon separate from *helleri*), but with *cerberus* as a sister taxon to the other subspecies. Venoms containing lethal toxins similar to Mojave toxin evolved within the western clade (*concolor* and *helleri*); later comparison of mtDNA sequences from dried venom by Pook and McEwing (2005) confirmed the separation of populations of *viridis* from those of *lutosus, helleri,* and *oreganus.* Small body size apparently evolved separately in each clade, once in the eastern clade (*nuntius*) and twice in the western one (*caliginus,* an island population, and *concolor*). In addition, some other island populations of *lutosus, helleri,* and *oreganus* have evolved dwarf populations (Ashton 2000); Boback (2003) calculated the reduction in length was approximately 38%.

In 2001, Ashton and de Queiroz published the results of another molecular study using the D-loop region and ND2 gene of mtDNA, and reported that the *viridis*-complex is monophyletic and consists of two strongly divergent clades. They further recommended that the two clades be recognized as the distinct evolutionary species, *C. viridis* in the east and *C. oreganus* in the west, with *C. viridis* restricted to the subspecies *nuntius* and *viridis,* and the remaining subspecies assigned to *C. oreganus,* including *caliginus,* but recognizing *cerberus* as a sister taxon to the other western subspecies. *Cerberus* was also elevated to specific status by Grismer (2002).

Douglas et al. (2002) conducted a further study of the mtDNA markers of all nine subspecies included in the *viridis*-complex with emphasis on the Colorado Plateau. They thought this area, where six taxa (*abyssus, cerberus, concolor, lutosus, nuntius,* and *viridis*) potentially exchange genes, critical to understanding the evolution within the complex. Their results supported the monophyly of the *viridis*-complex. The eastern clade consisting of *viridis* and *nuntius* was 100% diagnosable, but was thought to be composed of a single taxon, because *nuntius* was only 52% separable from *viridis* and should be placed in the synonymy of *C. viridis.* The western clade was also well defined (87%) and composed of (1) *cerberus* (72%); (2) *concolor* (92%); (3) a *lutosus* and *abyssus* clade (92%) that contained *abyssus* (88%) and a paraphyletic *lutosus*; (4) a paraphyletic *oreganus* clade; and (5) a *helleri* clade including *caliginis.* *Cerberus* was the basal-most taxon in the western clade and distinct from the other western clades. This led them to propose the elevation of *abyssus, cerberus, concolor, helleri, lutosus*, and *oreganus* to specific status, and that *caliginis* be placed in the synonymy of *helleri*. They further suggested that the two undescribed clades within the western group, *lutosus*-like and *oreganus*-like, require additional sampling and molecular, morphological, and natural history analyses to clarify their taxonomic positions.

Campbell and Lamar (2004) recognized an eastern clade consisting of the subspecies *C. v. viridis* and *C. v. nuntius* of the single species *C. viridis*, and a western clade consisting of the species *C. oreganus* including the other seven subspecies. They thought it "prudent to await further evidence and justification" before elevating *cerberus* to a full species. More recently, Beaman and Hayes (2008) included the subspecies *abyssus, caliginus, concolor, helleri, lutosus*, and *oreganus* as subspecies of *C. oreganus*.

Confusing? It certainly is to us. Consequently, we have taken an even different, somewhat conservative, approach to the taxonomy of the original nine subsecies of *C. viridis*. We recognize three species: (1) the eastern *C. viridis* with its two subspecies *C. v. viridis* and *C. v. nuntius*; (2) the Pacific Coast and Great Basin *C. oreganus* with the five subspecies *C. o. oreganus, C. o. abyssus, C. o. cerberus, C. o. concolor*, and *C. o. lutosus*; and (3) the southern California, Baja California Norte and adjacent island populations of *C. helleri* with the subspecies *C. h. helleri* and *C. h. caliginis*.

The position of *C. viridis* to the other species assigned to the genus *Crotalus* has also been researched. Gloyd (1940), based on body pattern, thought *C. viridis* most closely related to *C. scutulatus*. However, Brattstrom (1964a) proposed the *C. viridis*-complex (including *C. helleri* and *C. oreganus*) is morphologically close to *C. atrox, C. cerastes, C. mitchellii*, and *C. tigris*. Klauber (1972) agreed, for the most part, with both Gloyd and Brattstrom, but electrophoretic studies of venom proteins by Foote and MacMahon (1977) placed *C. viridis* with *C. atrox* and *C. ruber*. More recent molecular research by Murphy et al. (2002) using mtDNA cytochrome *b*, ND5, 12S, RNA, tRNA[Val], and 16S RNA genes places the *C. viridis* in a separate evolutionary *viridis*-group that also includes *C. helleri, C. horridus, C. oreganus*, and *C. scutulatus*.

C. v. nuntius plays an important role in the culture of the Native American Hopi tribe, from which it derives its common name (sometimes *C. v. viridis* is also used). During the tribe's famous approximately 30-minute snake dance, held annually at 2–3 of the Hopi pueblos in central Navajo County, Arizona, it is carried in the mouths and hands of some of the participating Native priests. See Klauber (1972) for a detailed description and discussion of the dance.

Crotalus willardi Meek, 1906
Ridge-nosed Rattlesnake
Cascabel de Nariz-surcada

RECOGNITION: *Crotalus willardi* is the third and probably least known of the three small montane rattlesnakes in the southwestern United States and northern Mexico.

The holotype of *Crotalus willardi obscurus*, a captive male, grew to a TBL_{max} of 67 cm before being sacrificed (Harris and Simmons 1976), and a wild-caught individual of the same subspecies from New Mexico measured 68.8 cm (Keegan et al. 1999);

but most individuals are shorter than 55 cm. The snake is gray, brown, yellowish-brown, or reddish-brown with a series of 18–45 dark body blotches that normally have dark anterior and posterior edges. Separating the blotches is a white- to cream-colored cross band. Only 1–3 light bands cross the anterior portion of the tail; the posterior portion is either plain-colored or light-striped. Three longitudinal rows of small, indistinct, dark blotches occur along the sides of the body. Dorsal body scales are keeled and pitted. A scanning electron microscope shows the small linear ridges on their dorsal scale surface are mostly swirled; this pattern is illustrated in both Harris (2005) and Stille (1987). The body scales occur in 28–29 (25–31) anterior rows, 25 or 27 (23–31) midbody rows, and 19 (16–21) rows near the tail. The basal rattle segment is gray, grayish-brown, orangish, or reddish-brown, and 8–10 rattle-fringe scales are present. About 11% of the snakes are rattleless (Rowe et al. 2002). The pink, cream, or tan venter is finely mottled with dark brown or black, and is darker posteriorly. Its scalation consists of 140–160 ventrals, 21–36 subcaudals, and an undivided anal plate. The dorsal surface of the head may be patterned with irregularly shaped dark brown or black marks of varying sizes. Individuals from Arizona have two white longitudinal stripes on the side of the face, one extending diagonally backward from the prenasal scale below the eye to the corner of the mouth, and a second extending backward along the supralabials and infralabials. A median white, vertical stripe extends downward from the rostral and mental scales. The light postocular stripe is poorly developed or missing in some southwestern New Mexico and northwestern Chihuahuan individuals. Snakes from New Mexico also lack the rostral-mental stripe, and, if present, have only faded longitudinal stripes on the side of the face. The rostral is higher than wide, anteriorly pointed, and slightly upturned. It combines with the recurved outer edges of the two internasals and a pair of relatively large canthals on each side (prefrontals are absent) to produce a distinctly raised snout. Normally 20–40 (but up to 60) small scales occupy the intercanthal to parietal areas. Also present are 6–9 intersupra-

Crotalus willardi willardi (Huachuca Mountains, Arizona; Arizona-Sonora Desert Museum. John H. Tashjian)

Crotalus willardi amabilis (Sierra del Nido, Chihuahua; Dallas Zoo; John H. Tashjian)

oculars and 2 reduced supraoculars. Lateral head scales consist of 2 nasals (the post-nasal rarely touches the upper preocular, but the prenasal normally contacts the first supralabial), 2 (1–4) loreals, 2–3 preoculars, 3–4 (5) postoculars (the upper is not sub-divided), 2–3 suboculars, 1 (rarely, 2) interculabials, 13–15 (10–17) supralabials, and 13–14 (12–17) infralabials. Beneath are a small mental scale and two pairs of chin shields that are in contact medially (the posterior pair is the largest).

As in other species of *Crotalus*, the hemipenis is bilobed with a divided sulcus sper-maticus. About 56 short, heavy spines and 16 fringes are present on each lobe (reduc-tion of the lobal spines to reticulations is sudden; Harris and Simmons 1976), and 1–2 spines lie in the crotch between the lobes (Klauber 1972).

The dental formula for each side consists of a maxillary fang, 1–2 palatine, 6 (5–7) pterygoid, and 8 dentary teeth (Klauber 1972).

Males (TBL$_{max}$, 67 cm) have 140–158 (mean, 151) ventrals, 23–36 (mean, 29) sub-caudals, and tails 9.1–11.5 (mean, 10.5) % of TBL. Females (TBL$_{max}$, 56 cm) have 144–160 (mean, 155) ventrals, 21–32 (mean, 24) subcaudals, and tails only 7.9–9.8 (mean, 8.5) % of TBL.

GEOGRAPHIC VARIATION: The populations of *C. willardi* are not continuously dis-tributed, but instead occur in isolated pockets on separate mountain ranges, which has led to evolution of much variation.

Five subspecies are currently recognized (Barker 1992; Campbell and Lamar 2004). *Crotalus willardi willardi* Meek, 1906, the Arizona ridge-nosed rattlesnake or cascabel de nariz-surcada Arizona, ranges from the Huachuca, Patagonia, and Santa Rita moun-tains in southeastern Arizona southward into the Sierra de los Ajos, Sierra Azul, and Sierra de Cananea of northern Sonora, Mexico (Villa et al. 2007). It has well-developed white stripes on the sides of the face and vertically on the rostral and mental scales, a brownish to reddish-brown back, and little dark spotting on the head. It has 23 (19–25) dorsal body blotches, a mean TL/TBL of 10.2% in males and 8.0% in females, 25

(25–27) midbody scale rows, 150–151 (147–154) ventrals in males and 154–155 (151–159) in females, 27–28 (25–30) subcaudals in males and 22–23 (20–25) in females, 13–14 (12–15) supralabials, and 13–14 (12–16) infralabials.

  C. w. amabilis Anderson, 1962, the Del Nido Ridge-nosed rattlesnake or cascabel de nariz-surcada Del Nido, occurs only in several arroyos in the Sierra del Nido of north-central Chihuahua. The body color and ventral surface of the tail are reddish or pinkish, and the facial striping is well pronounced. It has 39 (34–45) dorsal body blotches, mean TL/TBL of 11.5% in males and 9.3% in females, 27 (27–29) midbody scale rows, 150–151 (147–153) ventrals in males and 154 (153–156) in females, 32–33 (31–34)

Crotalus willardi meridionalis (Durango; Thomas Porter, Reseda, California; John H. Tashjian)

Crotalus willardi meridionalis (Durango; Thomas Porter, Reseda, California; John H. Tashjian)

subcaudals in males and 28 (27–29) in females, 14–15 (13–16) supralabials, and 14–15 (13–17) infralabials.

C. w. meridionalis Klauber, 1949b, the southern ridge-nosed rattlesnake or cascabel meridional de nariz-surcada, is found in Durango and Zacatecas. The body is orangish-brown to brown, and the facial pattern is well pronounced. It has 30 (28–33) dorsal body blotches, a mean TL/TBL of 11.8% in males and 9.8% in females, 27 (27–29) midbody scale rows, 148 (147–149) ventrals in males and 148–149 (147–152) in females, 32–33 (31–34) subcaudals in males and 30 (29–31) in females, 14–15 (14–17) supralabials, and 14–15 (14–15) infralabials.

C. w. obscurus Harris and Simmons, 1976, the New Mexico ridge-nosed rattlesnake or cascabel de nariz-surcada Nuevo Mexico, lives only in the Animas and Peloncillo mountains of New Mexico, and the Sierra de San Luis of extreme northeastern Sonora and western Chihuahua, Mexico. It lacks the vertical white stripe on the rostral and mental scales, and the lateral facial strips are faded or absent. Its back may be either gray or brownish, and the head is heavily marked with dark spots. It has 23 (20–26) dorsal body blotches, a mean TL/TBL of 10.6% in males and 8.5% in females, 25 (24–26) midbody scale rows, 30 (29–33) subcaudals in males and 25–26 (24–30) in females, 13–14 (13–16) supralabials, and 14 (13–16) infralabials. The New Mexico population when first discovered by Bogert and Degenhardt (1961) was thought to represent the Mexican subspecies *C. w. silus*, but re-evaluation by Harris and Simmons (1976) showed it to be an undescribed form.

C. w. silus Klauber, 1949b, the Chihuahuan ridge-nosed rattlesnake or cascabel de nariz-surcada Chihuahua, is a mountain-dweller in western Chihuahua and northeastern Sonora. The body color is orangish-brown to brownish, and the facial pattern is obscure except for a broad white stripe extending backward mostly above the supralabials from just below the anterior edge of the orbit to the level of the corner of the mouth. This subspecies has 23 (20–27) dorsal body blotches, a mean TL/TBL of 10.6% in males and 8.9% in females, 27 (25–27) midbody scale rows, 153 (149–158) ventrals in males and 155–157 (154–159) in females, 30–31 (29–35) subcaudals in males and 26–27 (25–30) in females, 14 (13–16) supralabials, and 14 (13–15) infralabials. Armstrong and Murphy (1979) have observed that the *C. w. silus* from the Sierra de la Purca from above the waterfall (2,070 m) tend to be light brown and resemble the dominant pine needle ground cover there, while those from below the waterfalls tend to be much darker and resemble the more dominant oak (*Quercus*) leaf litter.

A plethora of body coloration variation exists within the different subspecies; see Campbell and Lamar (2004) for a detailed analysis.

A cladistic study by Barker (1992) using 20 morphological characters and 22 biochemical ones upheld the validity of the currently recognized 5 subspecies, and showed that *C. w. willardi* is a sister group to the other 4 subspecies. Greene (1994) proposed weighted "sister taxa" showing *C. w. willardi* separate from the other four subspecies, and *C. w. obscurus* and the Mexican *C. w. silus* closely related.

Six unique microsatellite loci are present in the DNA genome of *C. w. obscurus* (Holycross et al. 2002b). Data from the population in the Animas Mountains of New Mexico show that these loci are highly variable with 5–24 alleles per locus having observable heterozygosities of 0.32–0.91 and expected heterozygosities of 0.35–0.92, are sufficiently variable for assigning parentage with total exclusionary power for the first

Distribution of
Crotalus willardi.

parent of 0.96 and 0.99 for the second parent, and amplify similar size fragments in *C. atrox, C. o. lutosus, C. scutulatus,* and *C. tigris.*

CONFUSING SPECIES: The raised rostrum and pattern of white facial stripes of *C. willardi* distinguish it from all other rattlesnakes within its range. The western hog-nosed snake (*Heterodon nasicus*) also has an upturned snout, but lacks a loreal pit, elliptical pupils, white face stripes, and tail rattle.

KARYOTYPE: Undescribed.

FOSSIL RECORD: Not reported.

DISTRIBUTION: The greatest portion of the geographical range of the species is in Mexico, as it is known only from a few localities in southeastern Arizona and extreme southwestern New Mexico. In Mexico, it occurs from north-central and northwestern Sonora and eastern Chihuahua southward through west-central Durango to southwestern Zacatecas.

HABITAT: The ridged-nosed rattlesnake is a montane (elevation 1,475–2,800 m) species that usually occurs in pine-oak, pine-fir, or scrub-oak forest, but has been taken in open grassy or mesquite-grassland microhabitats within pine-oak woodlands. In these wooded habitats, it is most often found in canyons along streams, around cool shaded rock outcrops with crevices, in rock piles, or in old stumps or downed logs. A thick mat of leaf litter or pine needles normally covers the ground (see Heald 1951 for a detailed description of the mountainous habitat in Arizona).

Some prominent plants reported in this snake's high-elevation habitats (mostly in the United States) are alders (*Alnus oblongifolia*), ashes (*Fraxinus velutina*), aspens (*Populus tremuloides*), buckthorn (*Rhamnus* sp.), buckwheats (*Eriogonum* sp.), creosote (*Larrea tridentata*), ferns (Durango), firs (*Abies concolor, Pseudotsuga menziesii*), grasses (*Astrida* sp., *Bouteloua gracilis, Eragrostis intermedia, Muhlenbergia trifida, Sporobolus* sp.), junipers (*Juniperus deppeana, J. monosperma*), locust (*Robinia neomexicana*), maples (*Acer grandidentatum*), mesquite (*Prosopis* sp.), oaks (*Quercus arizonica, Q. emoryi, Q. gambelii, Q. hypoleucoides, Q. oblongifolia, Q. reticulata, Q. rugosa*), pines (*Pinus cembroides, P. engelmanni, P. leiophylla, P. strobiformis*), poison oak (*Rhus toxicodendron*, Durango), and sycamores (*Platanus wrightii*).

BEHAVIOR AND ECOLOGY: Some limited additional life history data have appeared lately for *C. willardi*.

The annual activity period of *C. w. willardi* in Arizona extends from April or early May to mid-November (Johnson 1983; Lowe et al. 1986), but a *C. w. obscurus* was observed foraging on 14 December in New Mexico (Setser et al. 2005). Hibernacula include rotting stumps, logs, rock crevices, and talus slopes. Several have been found hibernating 40–46 cm deep in talus formations in New Mexico (Degenhardt et al. 1996). Probably animal burrows are also used as overwintering sites.

Daily activity is dependent upon the local AT, and is mostly diurnal, particularly in the spring and fall. *C. willardi* is a basker. Some crepuscular or nocturnal foraging possibly occurs in the summer, although the rapid drop in AT after sunset in its higher-elevation habitats probably limits this. Daily activity peaks on warm, humid mornings, and, especially after rains, some will emerge late in the afternoon. Toward fall, when daytime ATs start to drop, most activity shifts to the afternoon.

Active *C. willardi* have been found at ATs from 6.8–26.0°C (Bogert and Degenhardt 1961; Degenhardt et al. 1996; Kauffeld 1943a; Setser et al. 2005) to 30–33°C (Brattstrom 1965; Flood and Wickland *in* Wright and Wright 1957), but most foraging and basking takes place when the AT is 24–29°C (Armstrong and Murphy 1979; Degenhardt et al. 1996).

Crotalus willardi meridionalis (Durango; Thomas Porter, Resedo, California; John H. Tashjian)

C. willardi has been seen in trees or bushes at heights of 42–61 cm (Kauffeld 1943a; Rossi and Feldner 1993), and Parker and Stotz (1977) reported arboreal foraging for birds.

Male-male combat has been observed during the summer in this species (Armstrong and Murphy 1979; Holycross and Goldberg 2001).

REPRODUCTION: Stebbins (2003) reported the TBL at maturity to be 38 cm for each sex. Gravid females with 45.2–48.1 cm TBLs have been found by Klauber (1949b, 1972) and Quinn (1977), and a 54.8 cm male has successfully mated (Tryon 1978). Holycross and Goldberg (2001) reported the SVLs of the smallest reproductively active male and female specimens they examined were 40.6 cm and 40.2 cm, respectively.

The male and female gametic cycles from combined populations of *C. w. willardi*, *C. w. obscurus*, and *C. w. silus* were reported by Holycross and Goldberg (2001). Males experience spermiogenesis from mid-June into October, and contain mature sperm in their vas deferens from the beginning of June to the beginning of September. The sexual segment of the kidney is enlarged in June–October (courtship and mating behavior has been observed during this same period, discussed later).

Females contain enlarged ovarian follicles from early May through October, but do not undergo vitellogenesis from the beginning of September into early October. Vitellogenesis begins again in late spring or early summer. Postpartum females are sexually inactive prior to hibernation. Ovulation and fertilization probably occur in the early spring (females are probably capable of at least storing sperm over winter; Schuett 1992). Gravid females are present from mid-April to mid-August (female *C. w. silus* and *C. w. willardi* carrying young have been found from early May to early September; Delgadillo Espinosa et al. 1999; Johnson 1983; Martin 1975b), and some females have undergone parturition during the period from late July to early September. The females examined by Holycross and Goldberg (2001) indicated either a biennial or a triennial breeding cycle.

Courtship and mating activities by *C. willardi* have occurred in captivity on 29 January; 17–19 April; 16, 19 and 22 June; 15 and 28 July; in early August; and on 8–9 September (Armstrong and Murphy 1979; Martin 1975b, 1976; Tryon 1978). Wild *C. willardi* experience a mostly summer breeding period; Holycross and Goldberg (2001) witnessed courtship by wild *C. w. obscurus* on 10 June, and Lowe et al. (1986) reported that courtship and copulation occur in wild *C. w. willardi* in July–August and sometimes in April–June. Male breeding activity is stimulated by female ecdysis. Observed matings have lasted 11–24 hours (Armstrong and Murphy 1979; Bryson and Lazcano 2002; Tryon 1978), but may have been even longer since entire sequences have not been witnessed.

When a male discovers a female, he immediately begins rapid, longitudinal chin-rubbing and tongue-flicks along the female's back and upper sides (Tryon 1978). If unreceptive, she immediately rapidly slaps her tail from side to side, and crawls away. This stimulates the male to greater activity. Once he catches her and assumes loose coils on and parallel to the female's body, she raises her tail slightly and stops moving. The male then loops his tail under hers and attempts to align their cloacal vents with several anterior-posterior strokes. If intromission is not accomplished, the male behavior sequence begins again. Once intromission is accomplished, most movement by both snakes ceases, but the male may periodically chin-rub and tongue-flick, and the female may rapidly bob her head. A captive courting male *C. w. amabilis* directed head-bobbing and tongue-flicking (3–5 flicks per 5 seconds) across a coiled, resting female's back, then moved his uplifted tail in both a horizontal and a vertical plane with an undulating motion (Guese *in* Armstrong and Murphy 1979). A captive male *C. w. silus* twitched (1 twitch/second) after discovering a female introduced into his cage, and tongue-flicked (2 flicks/second) the entire dorsal surface of her body for 5 minutes by holding his head at a 30° angle and sliding his mental scale forward for 1 cm (Armstrong and Murphy 1979).

In captivity, the GP has lasted as long as 13 months (Tryon 1978), but it is probably more like 4–5 months in the wild (Holycross and Goldberg 2001). Captive and wild birth dates range from 30 June to 10 September, with most records in August. Litters contain 2 (*C. w. silus*; Klauber 1949b, 1972) to 9 (*C. w. obscurus*, Martin 1976) young, and average 5.2 (n = 30 clutches). There is a positive correlation between the number of young produced and female length (Holycross and Goldberg 2001). Martin (1975c) and Mociño-Deloya and Setser (2007a) reported RCMs of 35.8% and 31% for litters by *C. w. willardi*; Delgadillo Espinosa et al. (1999) reported a RCM of 40.2% for *C. w. silus*; and a litter of *C. w. obscurus* had a RCM of 54.5% of the female's postpartum weight (Holycross 2000c).

Eighty-one neonates averaged 18.1 (15.0–22.2) cm TBLs, and had 6.3 (4.0–9.0) g BMs. Young *C. w. willardi* are brownish at birth, and have either yellowish or gray-striped tails. Young of *C. w. obscurus* are dark brown with both blackish and yellowish tails (Holycross 2000c; Martin 1976). Females have been observed under rocks with their litters (Greene et al. 2002), so they may remain with their neonates for at least a few days after parturition.

An apparent natural hybridization between *C. willardi obscurus* and *C. lepidus klauberi* has occurred in the Peloncillo Mountains of southwestern New Mexico (Campbell et al. 1989a).

GROWTH AND LONGEVITY: No data on growth rates have been published. However, *C. willardi* may have a relatively long lifespan. Snider and Bowler (1992) reported that a captive female *C. w. willardi*, wild-caught as a juvenile, lived an additional 21 years, 3 months, and 24 days in the collection of Charles R. Hackenbrook of Dallas, Texas. They also reported captive longevities of 14 years, 8 months, and 12 days for a male *C. w. meridionalis* (Houston Zoo; still alive when paper was submitted); 12 years, 9 months, and 23 days for a male *C. w. silus* at the Houston Zoo; and 8 years, 11 months, and 25 days for a male *C. w. obscurus* (Los Angeles Zoo).

DIET AND FEEDING BEHAVIOR: Wild *C. willardi* have eaten scorpions and centipedes (*Scolopendra*), lizards (*Cnemidophorus* sp., *Elgaria kingii*, *Sceloporus jarrovii*), birds (*Aimophila ruficeps, Catharus guttatus, Melospiza lincolnii, Myadestes townsendi, Spizella passerina, Troglodytes aedon, Wilsonia pusilla*, and an unidentified bird), shrews (*Sorex* sp.), chipmunks (*Tamias dorsalis*), murid mice (*Peromyscus boylii, P. maniculatus; Reithrodontomys* sp.), pocket mice (*Perognathus* sp.), and pocket gophers (*Thomomys umbrinus*) (Barker 1992; Bryson and Holycross 2001; Fowlie 1965; Greene 1994; M. Haecker, personal communication; Holycross et al. 2001, 2002a; Johnson 1983; Klauber 1949b, 1972; Lemos-Espinal et al. 2007; Marshall 1957; Martin 1975b, Mociño-Deloya and Setser 2007b; Parker and Stotz 1977; Setser et al. 2005; Woodin 1953). In captivity, it has consumed white laboratory mice (*Mus musculus*), lizards (*Anolis carolinensis, Phrynosoma*

Crotalus willardi obscurus (New Mexico; San Antoinio Zoo; John H. Tashjian)

solare, Urosaurus ornatus), snakes (*Hypsiglena torquata, Trimorphodon biscutatus*), and centipedes (*Scolopendra*) (Armstrong and Murphy 1979; Bogert and Degenhardt 1961; Holycross et al. 2002a; Johnson 1983; Kauffeld 1943b; Klauber 1972; Lowe et al. 1986; Manion 1968; Martin 1976; Vorhies 1948).

Juveniles are probably more dependent on lizards as prey than are adults. Juvenile *C. w. obscurus* examined by Holycross et al. (2002a) most frequently contained lizards (57.1% of prey) or centipedes (33.3%); but adults had a broader diet, feeding mostly on small mammals (62.3%), lizards (26.4%), and small birds (9.4%). Among the lizards, *Sceloporus* (particularly *S. jarrovii*) were taken 68.4% of the time; and among the mammals, *Peromyscus boylii* was the most frequent prey (64.9%).

C. willardi captures its prey either by striking it from ambush, or by actively hunting for it. While waiting in ambush the snake assumes either an *S*-shaped posture (Setser et al. 2005) or a coil (C. Ernst, personal observation) among or under branches or a log, or between two rocks. When bitten, rodents are usually released at once, and later trailed by olfactory cues; lizards are usually struck in the body region and retained until dead (C. Ernst, personal observation). Birds are retained by using the mouth and a loop of the body (Parker and Stotz 1977). Young *C. willardi* may use their light-colored tails to lure potential prey (Tryon *in* Greene 1992).

VENOM DELIVERY SYSTEM: Klauber (1939) reported that 5 adult, 51.7–57.9 cm, *C. willardi* (subspecies not given) had a mean FL of 5.7 (5.3–6.0) mm, a mean TBL/FL of 97, a mean HL/FL of 5.16, and a mean angle of fang curvature of 68°.

VENOM AND BITES: The colorless venom seems weak compared to that of other *Crotalus*, but this may be due to the small amount produced by this short rattlesnake. The total volume of venom available for injection is small; Minton and Weinstein (1984) reported a venom yield of 3.1 mg from *C. w. willardi*, and Klauber (1972) could extract only a total of 3.7 mg of dried venom from another individual of the same subspecies.

The chemical components of the venom have hardly been reported. It has a relatively high protein content, mean 1.003 (range, 0.998–1.007) mg, and shows a moderate amount of protease activity (Minton and Weinstein 1984).

The minimum LD_{50} of venom from *C. w. willardi* for a 20 g mouse is 0.24 mg (Githens and Wolff 1939). The intravenous LD_{50} (mg/kg) for 20–25 g mice is 1.61, and a subcutaneous dose of 0.33 mg from *C. w. willardi* causes extensive subcutaneous hemorrhaging and is lethal (Minton and Weinstein 1984). The LD_{50} of venom from *C. w. willardi* for a 350 g pigeon is only 0.1 mg (Githens and George 1931).

Venom toxicity changes over time with a lengthening of the individual snake, and is apparently related to a shift from a predominantly lizard diet in juveniles to a more rodent diet in adults (previously discussed). Venom from adult *C. w. obscurus* (mouse LD_{50}, 6.4 μg/g) is approximately 33% less toxic than juvenile venom (LD_{50}, 4.3 μg/g). Preliminary toxicity assays with the lizard *Uta stansburiana* indicate that juvenile venom kills the lizards significantly faster than does that of the adult (Mackessy *in* Holycross et al. 2002c). The apparent greater toxicity of juvenile venom may result in functional equivalency of the venoms due to the significantly smaller quantities of venom produced by juveniles when compared to adults. Also, due to size constraints, the range of potential prey available to juveniles is smaller than that for adults; higher venom toxicity ensures that prey is rapidly immobilized and retained once bitten. Holycross et al. (2002c) predicted that the ontogenetic venom toxicity differential will be even greater for lizards than that observed for mice.

Human envenomations by this species, both in the wild and in captivity, are extremely rare. Of 31 reported envenomations during academic research in the United States in a 26-year period, only one (3.2%) involved this species (Ivanyi and Altimari 2004). Russell (*in* Minton and Weinstein 1984) treated a human bite by *C. willardi* that showed only minimal local signs of envenomation.

PREDATORS AND DEFENSE: Natural predation on *C. willardi* is almost unknown. Holycross et al. (2001) reported the killing of a *C. w. obscurus* by either a red-tailed hawk (*Buteo jamaicensis*) or a Mexican spotted owl (*Strix occidentalis lucida*), but could not determine which bird was involved. Other probable natural enemies include ophiophagous snakes and carnivorous mammals.

This is a relatively secretive snake. Its disposition has been described as being either calm or testy; individual snakes react differently to perceived danger, but most are mild mannered. Normally, the snake either lies still without rattling or tries to crawl away, occasionally shaking its tail as it goes. Seldom do they coil and strike at an intruder, but they do have the unpleasant habit of turning the head to bite the hand holding their neck, so one must be alert to avoid such an accident. Usually all rattling ceases once they are placed in a collecting bag or container. Weldon and Burghardt (1979) and Weldon et al. (1992) reported that *C. w. willardi* bridge their body when confronted with a coachwhip (*Masticophis taeniatus*).

PARASITES AND PATHOGENS: Both cestodes (*Mesocestoides* sp. tetrathyridia) and acanthocephalans (unidentified oligacanthorhynchid larval cystacanths) as well as parasitic protozoans (*Sarcocystis* sp.) have been found in *C. willardi* (Ernst and Ernst 2006; Goldberg and Bursey 2000; McAllister et al. 1995).

In addition, serotypes of *Salmonella* bacteria have been found in a captive colony of *C. w. silus* and *C. w. willardi* where individuals were suffering from progressively debilitating osteomyelitis; the bacterium *S. enterica SS arizonae* was the suspected cause. One individual with characteristic bone lesions died, and the bacterium *Providencia rettgeri* was cultured from it, but no *Salmonella* (Grupka et al. 2006; Ramsay et al. 2002). *Salmonella* bacteria have caused fatal cases of septicemia in other rattlesnakes.

POPULATIONS: Because of its isolated and limited distribution, the species is often plentiful in some restricted local colonies. At some Arizona sites, *C. w. willardi* may be the most common snake; Johnson (1983) has found as many as six in two hours in one canyon. It seems most common in the Huachuca Mountains, followed by the Santa Rita and Patagonia mountains, in that order (Johnson and Mills 1982). Armstrong and Murphy (1979) collected five *C. w. meridionalis* in one day on a partially open hillside in Durango after a rain event. They also observed approximately 40 *C. w. silus* between 1970 and 1974 in the Sierra de la Purica in northern Sonora; and on one early July morning saw 11 of the snakes in a 2-hour period. Between 15 and 23 July 1973, 11 more were observed in the area.

A male to female ratio of 1.15:1.00 has been reported by Klauber (1936) and Quinn (1977) reported a ratio of 1.50:1.00, but Tryon (1978) found twice as many females as males were born in captive litters.

Smith et al. (2001), studying the effects of a prescribed fire in the Coronado National Forest of the Peloncillo Mountains in extreme southern Arizona and New Mexico, found that three radio-tagged *C. w. obscurus*, including a gravid female, survived the fire by retreating underground. The female had moved since being located before the fire, but she and a non-radio-tagged male exhibited smaller daily activity areas after the fire, but before the onset of the summer rainy season, than before the fire; however, another male had a greater daily activity area. During the rainy season the female and 2 males moved greater distances during a 24-hour period than from before the rains came. She showed a much smaller mean movement area, 4.1 m² (all records combined), than the 2 males (22.3 m² and 82.4 m², respectively). Such data indicate that

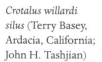

Crotalus willardi silus (Terry Basey, Ardacia, California; John H. Tashjian)

Crotalus willardi silus (R. D. Bartlett)

care should be taken with such fires in habitats containing endangered populations of *C. willardi*, but that the daily movements of individuals increase with recovery of the habitat after the summer monsoons commence.

The separated populations of *C. willardi* could lead to extinction of some of the smaller isolated colonies if human interference or natural conditions become intolerable. The major populations in Arizona are on federally protected lands, and the snake is considered threatened there. Overcollection of *C. w. obscurus* (Harris and Simmons 1974) has severely decimated the population at Animas, New Mexico, to the point that the species is now considered endangered by New Mexico and threatened under the federal Endangered Species Act. Fortunately, its microhabitat is not often visited by humans. H. M. Smith et al. (2005) found a DOR *C. w. silus* in Sonora. A recovery plan for the U.S. Fish and Wildlife Service was formulated by Baltosser and Hubbard in 1985 (U.S. Fish and Wildlife Service 1985).

An mtDNA study by Holycross and Douglas (2007) indicates that all populations of *C. w. obscurus* are genetically bottle-necked, and that the population in the Peloncillo Mountains of southwestern New Mexico, southeastern Arizona, and adjacent northern Chihuahua is sufficiently so to be declared an evolutionary significant unit. In addition, the populations in the Animas Mountain Range of New Mexico and the Sierra San Luis of northern Chihuahua comprise management subunits within a second evolutionarily significant unit. Both these units meet legal criteria for recognition as distinct population segments under the Endangered Species Act.

Someday soon we may have to manage all populations of *C. willardi* to ensure the species survival. In view of this, it is now imperative that more life history and ecological data be assembled for the three Mexican subspecies.

REMARKS: An electrophoretic study of venom proteins from *C. willardi* has shown that it is most closely related to the *Crotalus* species *cerastes*, *lepidus*, *pricei*, and *triseria-*

tus, and the Mexican *Crotalus ravus* (Foote and MacMahon 1977). Brattstrom (1964a) had previously proposed it close to *C. lepidus*, *C. pricei*, and *C. triseriatus* on morphological grounds. However, a more recent molecular study using mtDNA and RNA by Murphy et al. (2002) places *C. willardi* in the *C. polystictus*-group of *Crotalus* rattlesnakes, along with *C. cerastes*, *C. intermedius*, *C. pricei*, and *C. transversus*.

Glossary of Scientific Names

abyssus   bottomless, referring to the Grand Canyon

adamanteus   diamond-like, referring to the diamonded body pattern

amabilis   lovely

angelensis   belonging to Angel, referring to Angel Island, Mexico

atrox   savage, fierce, cruel; referring to its nasty disposition

basiliscus   royal, king-like; referring to its large size and potent venom

caliginis   bog-dweller

catalinensis   belonging to Catalina, referring to Santa Catalina Island, Mexico

cerastes   horned, referring to the horn-like supraocular scales

cerberus   black watchdog; the watchdog of Hades, referring to the dark body color

cercobombus   to buzz, referring to the rattling of the tail

cerralvensis   belonging to Cerralvo, referring to Cerralvo Island, Mexico

concolor   similarly colored, uniform; referring to the lack of blotches on the body

Crotalus   little bell, rattle; referring to the tail rattle

enyo   battle goddess; referring to the snake's disposition

estebanensis   belonging to Esteban, referring to San Esteban Island, Mexico

exsul   an exile; referring to the snake's distribution being limited to Cedros Island, Mexico

furvus   to rage, furious; referring to the snake's temperament

helleri   a patronym honoring Edmond Heller

klauberi   a patronym honoring rattlesnake expert Dr. Laurence M. Klauber

laterorepens   side-creeping; referring to sidewinding locomotion

lemosespinali   a patronym honoring Julio A. Lemos-Espinal, a Mexican herpetologist

lepidus   scaly, pretty, attractive; referring to the body coloration and pattern

lorenzoensis   belonging to Lorenzo; referring to San Lorenzo Island, Mexico

lucasensis   belonging to Lucas; referring to San Lucan Island, Mexico

lutosus   yellowish, muddy; referring to the body color

maculosus   spotted, stained, marked; referring to the body pattern

meridionalis   from the south

miquihuanus   belonging to Miquihuana; referring to the type-locality in the state of Tamaulipas, Mexico

mitchellii   a patronym honoring Dr. S. Weir Mitchell, an early researcher of rattlesnake venoms

molossus   for the Molossian wolf dog of antiquity

morulus   dark-colored, black; referring to body color

muertensis   belonging to Muerto; referring to El Muerto Island, Mexico

nigrescens   slightly black, dark; referring to body color

nuntius   messenger; referring to the rattling of the tail

oaxacus   belonging to Oaxaca; referring to the state of Oaxaca, Mexico

obscures   dusky, faded, hidden; referring to the less distinct body pattern

oliveri   a patronym honoring Bronx Zoo [New York] herpetologist Dr. James A. Oliver

oreganus   from Oregon; referring to the snake's northern Pacific range

pricei   a patronym honoring William Wightman Price, collector of the type specimen

pyrrhus   flame-colored, orangish, reddish; referring to body color

ruber   red; referring to body color

salvini   a patronym honoring herpetologist Osbert Salvin, an early researcher of Mexican reptiles *scutulatus*   small shield, diamond-like pattern; referring to the reduction in the number of scales in front of the suture separating the supraocular head scales, and the diamond body pattern

silus   snub-nosed

stephensi   a patronym honoring Frank Stephens, a member of the collecting team

taylori   a patronym honoring University of Kansas herpetologist Edward H. Taylor, who collected the type specimen

tigris   tiger; referring to the cross band body pattern

tortugensis   belonging to Tortuga; referring to Tortuga Island, Mexico

totonacus   named for the Totonac Native American tribe of northeastern Mexico

viridis   green; referring to the greenish-gray color of some individuals

willardi   a patronym honoring Frank C. Willard, who collected the type specimen

Bibliography

References are in alphabetical order by names of senior authors. Under a given senior author, single-author works are arranged by ascending year; works by two authors are ordered alphabetically by last name of second author; works by three or more authors are ordered by year only (with a, b, c extensions used to distinguish et al. citations within the same year).

Abramson, C. I., and A. J. Place. 2008. Learning in rattlesnakes: Issues and analysis, pp. 123–142. *In* W. K. Hayes, K. R. Beaman, M. D. Cardwell, and S. P. Bush (eds.), The biology of rattlesnakes. Loma Linda Univ. Press, Loma Linda, California.

Ackerman, G. L., D. Chiszar, and H. M. Smith. 2007. Rodent hygienic behavior and rattlesnake predation. Bull. Maryland Herpetol. Soc. 43: 30–34.

Adame, B. L., J. G. Soto, D. J. Secraw, J. C. Perez, J. L. Glenn, and R. C. Straight. 1990. Regional variation of biochemical characteristics and antigeneity in Great Basin rattlesnake (*Crotalus viridis lutosus*) venom. Comp. Biochem. Physiol. 97B: 95–101.

Agugliaro, J., and H. K. Reinert. 2005a. Comparative skin permeability of neonatal and adult timber rattlesnakes (*Crotalus horridus*). Comp. Biochem. Physiol. 141A: 70–75.

———. 2005b. A comparison of skin permeability within and between two populations of the timber rattlesnake (*Crotalus horridus*), p. 17. *In* Biology of the Rattlesnakes Symposium (program abstracts). Loma Linda Univ., Loma Linda, California.

Ahrenfeldt, R. H. 1955. Two British anatomical studies on American reptiles (1650–1750) II. Edward Tyson: Comparative anatomy of the timber rattlesnake. Herpetologica 11: 49–69.

Aird, S. D. 1985. A quantitative assessment of variation in venom constituents within and between three nominal rattlesnake subspecies. Toxicon 23: 1000–1004.

———. 2005. Taxonomic distribution and quantitative analysis of purine and pyrimidine nucleosides in snake venoms. Comp. Biochem. Physiol. 140B: 109–126.

———. 2008. Snake venom dipeptidyl peptidase IV: Taxonomic distribution and quantitative variation. Comp. Biochem. Physiol. 150B: 222–228.

Aird, S. D., and M. E. Aird. 1990. Rain-collecting behavior in a Great Basin rattlesnake (*Crotalus viridis lutosus*). Bull. Chicago Herpetol. Soc. 25: 217.

Aird, S. D., and I. I. Kaiser. 1985. Comparative studies on three rattlesnake toxins. Toxicon 23: 361–374.

Aird, S. D., C. S. Seebart, and I. I. Kaiser. 1988. Preliminary fractionation and characterization of the venom of the Great Basin rattlesnake (*Crotalus viridis lutosus*). Herpetologica 44: 71–85.

Aird, S. D., L. J. Thirkhill, C. S. Seebart, and I. I. Kaiser. 1989. Venoms and morphology of western diamondback/Mojave rattlesnake hybrids. J. Herpetol. 23: 131–141.

Aird, S. D., J. R. Yates III, P. A. Martino, J. Shabanowitz, D. F. Hunt, and I. I. Kaiser. 1990a. The amino acid sequence of the acidic subunit B-chain of crotoxin. Biochem. Biophys. Acta 1040: 217–224.

Aird, S. D., W. G. Kruggel, and I. I. Kaiser. 1990b. Amino acid sequence of the basic subunit of Mojave toxin from the venom of the Mojave rattlesnake (*Crotalus scutulatus*). Toxicon 28: 669–673.

Aird, S. D., W. G. Kruggel, and I. I. Kaiser. 1991. Multiple myotoxin sequences from the venom of a single prairie rattlesnake (*Crotalus viridis viridis*). Toxin 29: 265–268.

Aldridge, R. D. 1975. Environmental control of spermatogenesis in the rattlesnake, *Crotalus viridis*. Copeia 1975: 493–496.

————. 1979a. Seasonal spermatogenesis in sympatric *Crotalus viridis* and *Arizona elegans* (Reptilia, Serpentes) in New Mexico. J. Herpetol. 13: 187–192.

————. 1979b. Female reproductive cycles of the snakes *Arizona elegans* and *Crotalus viridis*. Herpetologica 35: 256–261.

————. 1993. Male reproductive anatomy and seasonal occurrence of mating and combat behavior of the rattlesnake *Crotalus v. viridis*. J. Herpetol. 27: 481–484.

————. 2002. The link between mating season and male reproductive anatomy in the rattlesnakes *Crotalus viridis oreganus* and *Crotalus viridis helleri*. J. Herpetol. 36: 295–300.

Aldridge, R. D., and W. S. Brown. 1995. Male reproductive cycle, age at maturity, and cost of reproduction in the timber rattlesnake (*Crotalus horridus*). J. Herpetol. 29: 399–407.

Aldridge, R. D., and D. Duvall. 2002. Evolution of the mating season in the pitvipers of North America. Herpetol. Monogr. 16: 1–25.

Alexander, C. G., and E. P. Alexander. 1957. *Oochoristica crotalicola*, a new anoplocephalid cestode from California rattlesnakes. J. Parasitol. 43: 365.

Alfonso, C., A. Alfonso, M. R. Vieytes, T. Yasumoto, and L. M. Botana. 2005. Quantification of yessotoxin using the fluorescence polarization technique and study of the adequate extraction procedure. Anal. Biochem. 344: 266–274.

Al-Joufi, A. M. H., and G. S. Bailey. 1994. A survey of kininase, tyrosine esterase, kininogenase and arginine esterase activities in some snake venoms. Comp. Biochem. Physiol. 108B: 221–224.

Allen, E. R., and W. T. Neill. 1950a. The eastern diamondback rattlesnake. Florida Wildl. 4(2): 10–11.

————. 1950b. The vertical position of the pupil in crocodilians and snakes. Herpetologica 6: 95–96.

Allen, E. R., and R. Slatten. 1945. A herpetological collection from the vicinity of Key West, Florida. Herpetologica 3: 25–26.

Allen, J. A. 1874. Notes on the natural history of portions of Dakota and Montana territories. Proc. Boston Soc. Nat. Hist. 17: 33–85.

Allen, M. J. 1933. Report on a collection of amphibians and reptiles from Sonora, Mexico, with the description of a new lizard. Occ. Pap. Mus. Zool. Univ. Michigan 259: 1–15.

Allsteadt, J., A. H. Savitzky, C. E. Petersen, and D. N. Naik. 2006. Geographic variation in the morphology of *Crotalus horridus* (Serpentes: Viperidae). Herpetol. Monogr. 20: 1–63.

Almeida-Santos, S. M., and M. G. Salomão. 2002. Reproduction in neotropical pitvipers, with emphasis on species of the genus *Bothrops*, pp. 445–462. *In* G. W. Schuett, M. Höggren, M. E. Douglas, and H. W. Greene (eds.), Biology of the vipers. Eagle Mountain Publ., Eagle Mountain, Utah.

Alt, N. 2002. *Crotalus molossus* in the El Pinacate Region, Sonora, Mexico. Bull. Chicago Herpetol. Soc. 37: 157.

Altimari, W. 1998. Venomous snakes: A safety guide for reptile keepers. Soc. Stud. Amphib. Rept. Herpetol. Circ. 26: i–iv, 1–24.

Alving, W. R., and K. V. Kardong. 1996. The role of the vomeronasal organ in rattlesnake (*Crotalus viridis oreganus*) predatory behavior. Brain Behav. Evol. 48: 165–172.

Amarello, M., and M. Goode. 2004. Variation in spatial ecology among populations of the tiger rattlesnake, p. 7. *In* Proceedings of the Snake Ecology Group 2004 Conference, Jackson County, Illinois.

Anaya, M., E. D. Rael, C. L. Lieb, J. Perez, and R. Salo. 1992. Antibody detection of venom proteins within a population of the rattlesnake *Crotalus v. viridis*. J. Herpetol. 26: 473–482.

Anderson, B. B., and F. H. Emmerson. 1970. The rattlesnake *Crotalus atrox* in southern Nevada. Great Basin Nat. 30: 107.

Anderson, C. D. 2006. Utility of a set of microsatellite primers developed for the massasauga rattlesnake (*Sistrurus catenatus*) for population genetic studies of the timber rattlesnake (*Crotalus horridus*). Mol. Ecol. Notes 6: 514–517.

————. 2010. Effects of movement and mating patterns on gene flow among overwintering hibernacula of the timber rattlesnake (*Crotalus horridus*). Copeia 2010: 54–61.

Anderson, C. D., and W. J. Drda. 2005. *Crotalus horridus* (timber rattlesnake): Behavior. Herpetol. Rev. 36: 456–457.

Anderson, J. D. 1962. A new subspecies of the ridged-nosed rattlesnake, *Crotalus willardi*, from Chihuahua, Mexico. Copeia 1962: 160–163.

Anderson, J. D., and W. Z. Lidicker, Jr. 1963. A contribution of our knowledge of the herpetofauna of the Mexican state of Aguascalientes. Herpetologica 19: 40–51.

Anderson, P. K. 1965. The reptiles of Missouri. Univ. Missouri Press, Columbia.

Andrews, K. M., and J. W. Gibbons. 2005. How do highways influence snake movement? Behavioral responses to roads and vehicles. Copeia 2005: 772–782.

Anon. 1929. The effect of snake venoms upon Protozoa. Bull. Antiv. Inst. Am. 3:89.

———. 1931. Relative resistance of Protozoa to *Crotalus atrox* and *Cobra* venoms. Bull. Antiv. Inst. Am. 5:28.

———. 1989. Venerable rattlesnake dies—Kansas snake held unofficial record for longevity. Kansas Herpetol. Soc. Newsl. 78: 8.

Anthony, A. W. 1893. Birds of San Pedro Matir, Lower California. Zoe 4: 228–247.

Antonio, F. B., and J. B. Barker. 1983. An inventory of phenotypic aberrancies in the eastern diamondback rattlesnake (*Crotalus adamanteus*). Herpetol. Rev. 14: 108–110.

Appleby, L. G. 1981. Rattlesnakes. Wildlife (London) 23(1): 42–44.

Arce, V., E. Rojas, C. L. Ownby, G. Rojas, and J. M. Gutiérrez. 2003. Preclinical assessment of the ability of polyvalent (Crotalinae) and anticoral (Elapidae) antivenoms produced in Costa Rica to neutralize the venoms of North American snakes. Toxicon 41: 851–860.

Armstrong, B. L., and J. B. Murphy. 1979. The natural history of Mexican rattlesnakes. Univ. Kansas Mus. Nat. Hist. Spec. Publ. 5: 1–88.

Arno, S. F. 1969. Interpreting the rattlesnake. Natl. Parks Mag. 43(267): 15–17.

Arthur, W. J., III, and D. H. Janke. 1986. Radionuclide concentrations in wildlife occurring at a solid radioactive waste disposal area. Northwest Sci. 60: 154–159.

Ashton, K. G. 1999. Shedding aggregations of *Crotalus viridis concolor*. Herpetol. Rev. 30: 211–213.

———. 2000. Notes on the island populations of the western rattlesnake, *Crotalus viridis*. Herpetol. Rev. 31: 214–217.

———. 2001. Body size variation among mainland populations of the western rattlesnake (*Crotalus viridis*). Evolution 55: 2523–2533.

———. 2002. Headfirst ingestion of prey by rattlesnakes: Are tactile cues used? J. Herpetol. 36: 500–502.

———. 2003. Movements and mating behavior of adult male midget faded rattlesnakes, *Crotalus oreganus concolor*, in Wyoming. Copeia 2003: 190–194.

Ashton, K. G., and A. de Queiroz. 2001. Molecular systematics of the western rattlesnake, *Crotalus viridis* (Viperidae), with comments on the utility of the D-loop in phylogenetic studies of snakes. Mol. Phyl. Evol. 21: 176–189.

Ashton, K. G., and J. Johnson. 1998. *Crotalus viridis concolor* (midget faded rattlesnake): Drinking from skin. Herpetol. Rev. 29: 170.

Ashton, K. G., and T. M. Patton. 2001. Movement and reproductive biology of female midget faded rattlesnakes, *Crotalus viridis concolor*, in Wyoming. Copeia 2001: 229–234.

Ashton, K. G., H. M. Smith, and D. Chiszar. 1999. A new pattern aberration in prairie rattlesnakes, *Crotalus viridis viridis*. Bull. Chicago Herpetol. Soc. 34: 153.

Ashton, R. E., Jr., and P. S. Ashton. 1981. Handbook of reptiles and amphibians of Florida. Part 1. The snakes. Windward Publ., Miami, Florida.

Atsatt, S. R. 1913. The reptiles of the San Jacinto area of southern California. Univ. California Publ. Zool. 12(3): 31–50.

Audi, J., and S. A. Seifert. 2006. Recurrence of coagulation abnormalities in an elderly patient after envenomation by an unseen crotaline snake in western Nebraska, p. 29. *In* S. A. Seifert (ed.), Snakebites in the new millennium: A state-of-the-art symposium, University of Nebraska Medical Center, 21–23 October 2005, Omaha, Nebraska. J. Med. Toxicol. 2: 29–45.

Auffenberg, W. 1963. The fossil snakes of Florida. Tulane Stud. Zool. 10: 131–216.

Avery, R. A. 1982. Field studies of body temperatures and thermoregulation, pp. 93–166. *In* C. Gans and F. H. Pough (eds.), Biology of the Reptilia, Vol. 12: Physiological ecology. Academic Press, New York.

Avila-Villegas, H. 2006. *Crotalus catalinensis* (Santa Catalina Island rattlesnake): Winter activity. Herpetol. Rev. 37: 476.

———. 2008. *Crotalus catalinensis* (Santa Catalina Island rattlesnake): Arboreality. Herpetol. Rev. 39: 468.

Avila-Villegas, H., and G. Arnaud. 2004. Diet of the Santa Catalina Island rattleless rattlesnake (*Crotalus catalinensis*), p. 8. *In* Proceedings of the Snake Ecology Group 2004 Conference, Jackson County, Illinois.

Avila-Villegas, H., C. S. Venegas-Barrera, and G. Arnaud. 2004. *Crotalus catalinensis* (Santa Catalina Island rattleless rattlesnake): Diet. Herpetol. Rev. 35: 60.

Avila-Villegas, H., A. Trejas, F. Torres, and G. Arnaud. 2005. *Crotalus catalinensis* (Santa Catalina Island rattlesnake): Diet and mortality. Herpetol. Rev. 36: 323.

Avila-Villegas, H., M. Martins, and G. Arnaud. 2007. Feeding ecology of the endemic rattleless rattlesnake, *Crotalus catalinensis*, of Santa Catalina Island, Gulf of California, Mexico. Copeia 2007: 80–84.

Axtell, R.W. 1959. Amphibians and reptiles of the Black Gap Wildlife Management Area, Brewster County, Texas. Southwest. Nat. 4: 88–109.

Axtell, R. W., and M. D. Sabath. 1963. *Crotalus pricei miquihuanus* from the Sierra Madre of Coahuila, Mexico. Copeia 1963: 161–164.

Baarslag, A. F. 1950. The pilot blacksnake and the timber rattlesnake in Vermont. Copeia 1950: 322–323.

Babcock, H. L. 1929. The snakes of New England. Nat. Hist. Guide, Boston Soc. Nat. Hist. (1): 1–30.

Babero, B. B., and F. H. Emmerson. 1974. *Thubunaea cnemidophorus* in Nevada rattlesnakes. J. Parasitol. 60: 595.

Backshall, S. 2008. Venomous animals of the world. Johns Hopkins Univ. Press, Baltimore.

Bailey, J. W. 1946. The mammals of Virginia. Privately published, Richmond, Virginia.

Bailey, V., M. R. Terman, and R. Wall. 1989. Noteworthy longevity in *Crotalus viridis viridis* (Rafinesque). Trans. Kansas Acad. Sci. 92: 116–117.

Baird, S. F. 1859. Reptiles of the boundary, pp. 1–35. *In* Report of the United States and Mexican Boundary Survey, U.S. 34th Congress, 1st Session, Exec. Doc. 108, vol. 2, part 2.

Baird, S. F., and C. Girard. 1853. Catalogue of North American reptiles in the Museum of the Smithsonian Institution. Part 1. Serpents. Smithsonian Misc. Coll. 2(5): 1–172.

Baker, M. R. 1987. Synopsis of the Nematoda parasitic in amphibians and reptiles. Memorial Univ. Newfoundland Occ. Pap. Biol. No. 11.

Baker, R. J., J. J. Bull, and G. A. Mengden. 1971. Chromosomes of *Elaphe subocularis* (Reptilia: Serpentes), with the description of an in vivo technique for preparation of snake chromosomes. Experientia 27: 1228–1229.

Baker, R. J., G. A. Mengden, and J. J. Bull. 1972. Karyotypic studies of thirty-eight species of North American snakes. Copeia 1972: 257–265.

Bakken, G. S., and A. R. Krochmal. 2007. The imaging properties and sensitivity of the facial pits of pitvipers as determined by optical and heat-transfer analysis. J. Exp. Biol. 210: 2801–2810.

Bakker, J. 2003. *Crotalus cerastes* (Hallowel, 1854): My experiences in keeping and breeding *Crotalus cerastes* Hallowell, 1854. Litt. Serpent. Engl. Ed. 23: 137–139, 141–144.

Balderas-Valdivia, C. J., and A. Ramírez-Bautista. 2005. Aversive behavior of beaded lizard, *Heloderma horridum*, to sympatric and allopatric predator snakes. Southwest. Nat. 50: 24–31.

Ball, J. C., R. H. Legere, D. S. Holland, and K. Bayless. 2003. *Crotalus pricei* (twin-spotted rattlesnake): Reproduction. Herpetol. Rev. 34: 373.

Ballard, S. 2004. Timber rattlesnake habitat protection guidelines in Illinois, p. 9. *In* Proceedings of the Snake Ecology Group 2004 Conference, Jackson County, Illinois.

Banta, B. H. 1965. A distributional check list of the Recent reptiles inhabiting the state of Nevada. Occs. Pap. Biol. Soc. Nevada (5): 1–8.

———. 1968a. The recent herpetofauna of the northern wet mountains, south-central Colorado. J. Herpetol. 1: 120.

———. 1968b. The Recent herpetofauna of the transect of prairie in El Paso County, Colorado. J. Herpetol. 2: 181–182.

———. 1974. A pre-Columbian record of cannibalism in the rattlesnake. Bull. Maryland Herpetol. Soc. 10: 56.

Baramova, E. N., J. D. Shannon, J. B. Bjarnason, and J. W. Fox. 1989. Degradation of extracellular matrix proteins by hemorrhagic metalloproteinases. Arch. Biochem. Biophys. 275: 63–71.

Barbour, R. W. 1950. The reptiles of Big Black Mountain, Harlan County, Kentucky. Copeia 1950: 100–107.

———. 1956. Poisonous snakes of Kentucky. Kentucky Happy Hunting Grounds 12(1): 18–19, 32.

———. 1971. Amphibians and reptiles of Kentucky. Univ. Press Kentucky, Lexington.

Barbour, T. 1920. Herpetological notes from Florida. Copeia 1920(84): 55–57.

———. 1922. Rattlesnakes and spitting snakes. Copeia 1922(105): 36–38.

———. 1926. Reptiles and amphibians: Their habits and adaptations. Houghton Mifflin, Boston.

Barker, D. G. 1991. An investigation of the natural history of the New Mexico ridgenose rattlesnake, *Crotalus willardi obscurus*. Report to Endangered Species Program, New Mexico Dept. Game and Fish, Santa Fe.

———. 1992. Variation, intraspecific relationships and biogeography of the ridgenose rattlesnake, *Crotalus willardi*, pp. 89–105. *In* J. A. Campbell and E. D. Brodie, Jr. (eds.), Biology of the pitvipers. Selva, Tyler, Texas.

Barron, J. N. 1997. Condition-adjusted estimator of reproductive output in snakes. Copeia 1997: 306–318.

Bartholomew, B. D., and R. D. Nohavec. 1995. Saltation in snakes with a note on escape saltation in a *Crotalus scutulatus*. Great Basin Nat. 55: 282–283.

Bartlett, R. D. 1988. In search of reptiles & amphibians. E. J. Brill, Leiden.

Bartlett, R. D., and P. Bartlett. 2003. Florida's snakes: A guide to their identification and habits. Univ. Press Florida, Gainesville.

Barton, A. J. 1950. Replacement fangs in newborn timber rattlesnakes. Copeia 1950: 235–236.

Bartz, A. D., and R. A. Sajdak. 2004. *Crotalus horridus* (timber rattlesnake): Arboreality, courtship. Herpetol. Rev. 35: 61.

Basey, H. E. 1988. Discovering Sierra reptiles and amphibians. 2nd ed. Yosemite Assoc., U.S. Dept. Interior.

Bateman, G. F. 1918. Are rattlesnakes beneficial? Outdoor Life 42(1): 501.

Bateman, H. L., A. Chung-MacCoubrey, H. L. Snell, and D. M. Finch. 2009. Abundance and species richness of snakes along the Middle Rio Grande riparian forest in New Mexico. Herpetol. Conserv. Biol. 4: 1–8.

Baxter, G. 1977. Rattlesnake fact and fiction. Wyoming Wildl. 41(6): 12–13, 15.

Baxter, G., and H. Rahn. 1941. Rattlesnakes of Wyoming. Wyoming Wildl. 6(4): 1–6.

Baxter, G. T., and M. D. Stone. 1980. Amphibians and reptiles of Wyoming. Wyoming Game Fish Dept. Bull. 16: 1–137.

Baxter, R. J., and G. R. Stewart. 1990 (1986). Excavation of winter burrows and relocation of desert tortoises (*Gopherus agassizii*) at the Luz Solar Generation Station, Kramer Junction, California. Proc. Symp. Gopher Tortoise Council 1986: 124–127.

Beaman, K. R., and E. A. Dugan. 2006. *Crotalus ruber*. Cat. Am. Amphib. Rept. 840: 1–17.

Beaman, K. R., and L. L. Grismer. 1994. *Crotalus enyo* (Cope): Baja California rattlesnake. Cat. Am. Amphib. Rept. 589: 1–6.

Beaman, K. R., and W. K. Hayes. 2008. Rattlesnakes: Research trends and annotated checklist, pp. 5–16. *In* W. K. Hayes, K. R. Beaman, M. D. Cardwell, and S. P. Bush (eds.), The biology of rattlesnakes. Loma Linda Univ. Press, Loma Linda, California.

Beaman, K. R., and C. L. Spencer. 2004. *Crotalus tortugensis* Van Denburgh and Slevin: Tortuga Island rattlesnake. Cat. Am. Amphib. Rept. 798: 1–5.

Beaman, K. R., and N. Wong. 2001. *Crotalus catalinensis*. Cat. Am. Amphib. Rept. 733: 1–4.

Beasley, R. J., E. L. Ross, P. M. Nave, D. H. Sifford, and B. D. Johnson. 1993. Phosphodiesterase activities of selected crotalid venoms. SAAS Bull. Biochem. Biotech. 6: 48–53.

Beaupre, S. J. 1993. An ecological study of oxygen consumption in the mottled rock rattlesnake, *Crotalus lepidus lepidus*, and the black-tailed rattlesnake, *Crotalus molossus molossus*, from two populations. Physiol. Zool. 66: 437–454.

————. 1995a. Comparative ecology of the mottled rock rattlesnake, *Crotalus lepidus*, in Big Bend National Park. Herpetologica 51: 45–56.

————. 1995b. Effects of geographically variable thermal environment on bioenergetics of mottled rock rattlesnakes. Ecology 76: 1655–1665.

————. 1996. Field metabolic rate, water flux, and energy budgets of mottled rock rattlesnakes, *Crotalus lepidus*, from two populations. Copeia 1996: 319–329.

————. 2002. Modeling time-energy allocation in vipers: Individual responses to environmental variation and implications for populations, pp. 463–481. *In* G. W. Schuett, M. Höggren, M. E. Douglas, and H. W. Greene (eds.), Biology of the vipers. Eagle Mountain Publ., Eagle Mountain, Utah.

————. 2005. Ratio representations of specific dynamic action (mass-specific SDA and SDA coefficient) do not standardize for body mass and meal size. Physiol. Biochem. Zool. 78: 126–131.

————. 2008. Annual variation in time-energy allocation by timber rattlesnakes (*Crotalus horridus*) in relation to food acquisition, pp. 111–122. *In* W. K. Hayes, K. R. Beaman, M. D. Cardwell, and S. P. Bush (eds.), The biology of rattlesnakes. Loma Linda Univ. Press, Loma Linda, California.

Beaupre, S. J., and D. Duvall. 1998a. Variation in oxygen consumption of the western diamondback rattlesnake (*Crotalus atrox*): Implications for sexual size dimorphism. J. Comp. Physiol. 168B: 497–506.

————. 1998b. Integrative biology of rattlesnakes: Contributions to biology and evolution. Bioscience 48: 531–538.

Beaupre, S. J., and F. Zaidan. 2001. Scaling of CO_2 production in the timber rattlesnake (*Crotalus horridus*) with comments on the cost of growth in neonates and comparative patterns. Physiol. Biochem. Zool. 74: 757–768.

Beaupre, S. J., D. J. Duvall, and J. O'Leile. 1998. Ontogenetic variation in growth and sexual size dimorphism in a central Arizona population of the western diamondback rattlesnake (*Crotalus atrox*). Copeia 1998: 40–47.

Beauvois, P. de. 1799. Memoir on amphibia. Serpents. Trans. Am. Phil. Soc. 4: 362–381.

Beavers, R. A. 1976. Food habits of the western diamondback rattlesnake, *Crotalus atrox*, in Texas. Southwest. Nat. 20: 503–515.

Bebarta, V. S., and R. C. Dart. 2005. Effectiveness of delayed use of Crotalidae polyvalent immune Fab (ovine) antivenom. *In* Snakebites in the new millennium: A state-of-the-art symposium (program abstracts). Univ. Nebraska Medical Center, Center for Continuing Educ., Omaha.

Bechtel, H. B., and E. Bechtel. 1991. Scaleless snakes and a breeding report of scaleless *Elaphe obsoleta lindheimeri*. Herpetol. Rev. 22: 12–14.

Beck, D. D. 1995. Ecology and energetics of three sympatric rattlesnake species in the Sonoran Desert. J. Herpetol. 29: 211–223.

————. 1996. Effects of feeding on body temperatures of rattlesnakes: A field experiment. Physiol. Zool.

————. 2005. Ambush-site selection by Sonoran desert rattlesnakes: A field experiment, p. 18. *In* Biology of the Rattlesnakes Symposium (program abstracts). Loma Linda Univ., Loma Linda, California.

Behler, J. L., and F. W. King. 1979. The Audubon Society field guide to North American reptiles and amphibians. Alfred A. Knopf, New York.

Bellairs, A. d'A., and G. Underwood. 1951. The origin of snakes. Biol. Rev. 26: 193–237.

Bennett, R. 2004. *Crotalus viridis* (prairie rattlesnake): Behavior. J. Kansas Herpetol. 12: 18.

Bent, A. C. 1937. Life histories of North American birds of prey. Part 1. U.S. Natl. Mus. Bull. 167.

————. 1938. Life histories of North American birds of prey. Part 2. U.S. Natl. Mus. Bull. 170.

Berg, C. S., and A. Bartz. 2005. Winter biology of timber rattlesnakes, *Crotalus horridus*, in Wisconsin, pp. 18–19. *In* Biology of the Rattlesnakes Symposium (program abstracts). Loma Linda Univ., Loma Linda, California.

Berg, C. S., R. A. Sajdak, and A. Bartz. 2005. Reproductive cycles of female timber rattlesnakes,

Crotalus horridus, in the upper Mississippi River Valley, p. 19. *In* Biology of the Rattlesnakes Symposium (program abstracts). Loma Linda Univ., California.

Berish, J. E. 1992. Annual size / sex class distribution of harvested rattlesnakes. Final report submitted to Bur. Wildl. Res., Florida Game Fresh Water Fish Comm., Tallahassee.

Berish, J. E. D. 1998. Characterization of rattlesnake harvest in Florida. J. Herpetol. 32: 551–557.

Berkhoudt, H., P. Wilson, and B. Young. 2001. Taste buds in the palatal mucosa of snakes. African Zool. 36: 185–188.

Bernstein, J. N., R. C. Dart, and D. Hardy. 1992. Natural history of envenomation by the twin spotted rattlesnake (*Crotalus p. pricei*). Vet. Human Toxicol. 34: 341.

Bertram, N., and K. W. Larsen. 2004. Putting the squeeze on venomous snakes: Accuracy and precision of length measurements taken with the "squeeze box." Herpetol. Rev. 35: 235–238.

Best, T. L., and H. C. James. 1984. Rattlesnakes (genus *Crotalus*) of the Pedro Armendariz Lava Field, New Mexico. Copeia 1984: 213–215.

Biardi, J. E. 2000. Adaptive variation and coevolution in California ground squirrel (*Spermophilus beecheyi*) and rock squirrel (*Spermophilus variegatus*) resistance to rattlesnake venom. Ph.D. diss., Univ. California, Davis.

———. 2006. Small mammals as a natural source of snake venom metalloprotease inhibitors, p. 42. *In* S. A. Seifert (ed.), Snakebites in the new millennium: A state-of-the-art symposium, University of Nebraska Medical Center, 21–23 October 2005, Omaha, Nebraska. J. Med. Toxicol. 2: 29–45.

———. 2008. The ecological and evolutionary context of mammalian resistance to rattlesnake venoms, pp. 557–568. *In* W. K. Hayes, K. R. Beaman, M. D. Cardwell, and S. P. Bush (eds.), The biology of rattlesnakes. Loma Linda Univ. Press, Loma Linda, California.

Biardi, J. E., D. C. Chien, and R. G. Coss. 2005. California ground squirrel (*Spermophilus beecheyi*) defenses against rattlesnake venom digestive and hemostatic toxins. J. Chem. Ecol. 31: 2501–2518.

Bieber, A. L., T. Tu, and A. T. Tu. 1975. Studies of an acidic cardiotoxin isolated from the venom of the Mojave rattlesnake (*Crotalus scutulatus*). Biochem. Biophys. Acta 400: 178–188.

Bieber, A. L., R. H. McParland, and R. R. Becker. 1987. Amino acid sequences of myotoxins from *Crotalus viridis concolor* venom. Toxicon 25: 677–680.

Bieber, A. L., R. R. Becker, R. McParland, D. F. Hunt, J. Shabanowitz, J. R. Yates III, P. A. Martino, and G. R. Johnson. 1990a. The complete sequence of the acidic subunit from Mojave toxin determined by Edman degradation and mass spectrometry. Biochem. Biophys. Acta 1037: 413–421.

Bieber, A. L., J. P. Mills, Jr., C. Ziolkowski, and J. Harris. 1990b. Rattlesnake neurotoxins: Biochemical and biological aspects. J. Toxicol. Toxin Rev. 9: 285–306.

Billing, W. M. 1930. The action of the toxin of *Crotalus adamanteus* on blood clotting. J. Pharm. Exp. Ther. 38: 173–196.

Bird, W., and P. Peak. 2007. A snake hunting guide: Methods, tools, and techniques for finding snakes. ECO Publications, Lansing, Michigan.

Bishopp, F. C., and H. L. Trembley. 1945. Distribution and hosts of certain North American ticks. J. Parasitol. 31: 1–54.

Bjarnason, J. B., and A. T. Tu. 1978. Hemorrhagic toxins from western diamondback rattlesnake (*Crotalus atrox*) venom: Isolation and characterization of five toxins and the role of zinc in hemorrhagic toxin e. Biochemistry 17: 3395–3404.

Bjarnason, J. B., A. Barish, G. S. Direnzo, R. Campbell, and J. W. Fox. 1983. Kallikrein-like enzymes from *Crotalus atrox* venom. J. Biol. Chem. 258: 12566–12573.

Blair, W. F. 1954. Mammals of the Mesquite Plains Biotic District in Texas and Oklahoma, and speciation in the central grasslands. Texas J. Sci. 6: 235–264.

Blanchard, F. N., and E. B. Finster. 1933. A method of marking living snakes for future recognition with discussion of some problems and results. Ecology 14: 334–347.

Blaney, R. M. 1971. An annotated check list and biogeographic analysis of the insular herpetofauna of the Apalachicola region, Florida. Herpetologica 27: 406–430.

Blum, H. F., and C. R. Spealman. 1933. Note on the killing of rattlesnakes by "sunlight." Copeia 1933: 150–151.

Boback, S. M. 2003. Body size evolution in snakes: Evidence from island populations. Copeia 2003: 81–94.

Bober, M. A., J. L. Glenn, R. C. Straight, and C. L. Ownby. 1988. Detection of myotoxin a-like proteins in various snake venoms. Toxicon 26: 665–673.

Bogert, C. M. 1941. Sensory cues used by rattlesnakes in their recognition of ophidian enemies. Ann. New York Acad. Sci. 41: 329–343.

———. 1942. Field note on the copulation of *Crotalus atrox* in California. Copeia 1942: 262.

———. 1943. Dentitional phenomena in cobras and other elapids with notes on adaptive modifications of fangs. Bull. Am. Mus. Nat. Hist. 81: 285–360.

———. 1947. Rectilinear locomotion in snakes. Copeia 1947: 253–254.

———. 1949. Thermoregulation in reptiles: A factor in evolution. Evolution 3: 195–211.

———. 1960. The influence of sound on the behavior of amphibians and reptiles, pp. 137–320. *In* W. E. Lanyon and W. N. Tavolga (eds.), Animal sounds and communication. Am. Inst. Biol. Sci. Publ. 7.

Bogert, C. M., and W. G. Degenhardt. 1961. An addition to the fauna of the United States, the Chihuahuan ridge-nosed rattlesnake in New Mexico. Am. Mus. Novit. 2064: 1–15.

Bogert, C. M., and J. A. Oliver. 1945. A preliminary analysis of the herpetofauna of Sonora. Bull. Am. Mus. Nat. Hist. 83: 297–426.

Bolaños, R. 1983 [1982]. Serpientes venenosos de Centro América: Distribución, caraterísticas, y patrones cardiológicos. Mem. Inst. Butantan 46: 275–291.

Bolette, D. P. 1997a. Oligacanthorhynchid cystacanths (Acanthocephala) in a long-nosed snake, *Rhinocheilus lecontei lecontei* (Colubridae) and a Mojave rattlesnake, *Crotalus scutulatus scutulatus* (Viperidae) from Maricopa County, Arizona. Southwest. Nat. 42: 232–236.

———. 1997b. First record of *Pachysentis canicola* (Acanthocephala: Oligacanthorhynchidae) and the occurrence of *Mesocestoides* sp. tetrathyridia (Cestoidea: Cyclophyllidea) in the western diamondback rattlesnake, *Crotalus atrox* (Serpentes, Viperidae). J. Parasitol. 83: 751–752.

———. 1998. Helminths of the prairie rattlesnake, *Crotalus viridis viridis* (Serpentes: Viperidae), from western South Dakota. Proc. Helminthol. Soc. Washington 65: 105–107.

Bond, G. R., and K. K. Burkhart. 1997. Thrombocytopenia following timber rattlesnake envenomation. Ann. Emerg. Med. 30: 40–44.

Bonilla, C. A. 1975. Defibrinating enzyme from timber rattlesnake (*Crotalus h. horridus*) venom: A potential agent for therapeutic defibrination. I. Purification and properties. Thromb. Res. 6: 151–169.

Bonilla, C. A., and M. K. Fiero. 1971. Comparative biochemistry and pharmacology of salivary gland secretions: II. Chromatographic separation of the basic proteins from some North American rattlesnake venoms. J. Chromatog. 56: 253–263.

Bonilla, C. A., and N. V. Horner. 1969. Comparative electrophoresis of *Crotalus* and *Agkistrodon* venoms from North American snakes. Toxicon 7: 327–329.

Bonilla, C. A., M. R. Faith, and S. A. Minton. 1973. L-amino acid oxidase, phosphodiesterase, total protein and other properties of juvenile timber rattlesnakes (*Crotalus h. horridus*) venom at different stages of growth. Toxicon 11: 301–303.

Bonine, K. E., E. W. Stitt, G. L. Bradley, and J. J. Smith. 2004. *Crotalus atrox* (western diamond-backed rattlesnake): Entrapment and opportunistic courtship. Herpetol. Rev. 35: 176–177.

Boone, A. R. 1937. Snake hunter catches snakes for fun. Pop. Sci. Monthly 131(4): 54–55.

Boquet, P. 1948. Venins de serpents et antivenins. Coll. Inst. Pasteur [Paris].

Borrell, B. J., T. J. La Duc, and R. Dudley. 2005. Respiratory cooling in rattlesnakes. Comp. Biochem. Physiol. 140A: 471–476.

Bostic, D. L. 1975. A natural history guide to the Pacific coast of north central Baja California and adjacent islands. Biological Educational Expeditions. Vista, California.

Boulenger, G. A. 1893–1896. Catalogue of the snakes in the British Museum (Natural History). 3 vols. The trustees (British Museum), London.

Boundy, J. 1997. Snakes of Louisiana. Louisiana Dept. Wildl. Fish., Baton Rouge.

Bouskila, A. 1995. Interactions between predation risk and competition: A field study of kangaroo rats and snakes. Ecology 76: 165–178.

———. 2001. A habitat selection game of interactions between rodents and their predators. Ann. Zool. Fennici 38: 55–70.

Bovee, E. C. 1962. The isoporan coccidian *Isopora dirumpens*, from the Florida diamondback rattlesnake, *Crotalus adamanteus*. J. Parasitol. 9(suppl.): 19.

Bowler, J. K. 1977. Longevity of reptiles and amphibians in North American collections. Soc. Stud. Amphib. Rept. Misc. Publ. Herpetol. Circ. 6: 1–32.

Bowman, D. D. 1984. *Hexametra leidyi* sp. n. (Nematoda: Ascarididae) from North American pit vipers (Reptilia: Viperidae). Proc. Helminthol. Soc. Washington 51: 54–61.

Boyer, D. R. 1957. Sexual dimorphism in a population of the western diamond-backed rattlesnake. Herpetologica 13: 213–217.

Brattstrom, B. H. 1952. Diurnal activities of a nocturnal animal. Herpetologica 8: 61–63.

———. 1953a. Records of Pleistocene reptiles and amphibians from Florida. Quart. J. Florida Acad. Sci. 16: 243–248.

———. 1953b. Records of Pleistocene reptiles from California. Copeia 1953: 174–179.

———. 1954a. The fossil pit-vipers (Reptilia: Crotalidae) of North America. Trans. San Diego Soc. Nat. Hist. 12: 31–46.

———. 1954b. Amphibians and reptiles from Gypsum Cave, Nevada. Bull. So. California Acad. Sci. 53: 8–12.

———. 1955a. Pliocene and Pleistocene amphibians and reptiles from southeastern Arizona. J. Paleontol. 29: 150–154.

———. 1955b. Records of some Pliocene and Pleistocene reptiles and amphibians from Mexico. Bull. So. California Acad. Sci. 54: 1–4.

———. 1958a. New records of Cenozoic amphibians and reptiles from California. Bull. So. California Acad. Sci. 57: 5–12.

———. 1958b. Additions to the Pleistocene herpetofauna of Nevada. Herpetologica 14: 36.

———. 1964a. Evolution of the pit vipers. Trans. San Diego Soc. Nat. Hist. 13: 185–268.

———. 1964b. Amphibians and reptiles from cave deposits in south-central New Mexico. Bull. So. California Acad. Sci. 63: 93–103.

———. 1965. Body temperatures of reptiles. Am. Midl. Nat. 73: 376–422.

———. 1967. A succession of Pliocene and Pleistocene snake faunas from the High Plains of the United States. Copeia 1967: 188–202.

———. 1996. Predators of the coast horned lizard, *Phrynosoma coronatum*. Phrynosomatics 1(2): 1–3.

Breckenridge, W. J. 1944. Reptiles and amphibians of Minnesota. Univ. Minnesota Press, Minneapolis.

Breen, A. R. 1984. Rhode Island's declining rattlers. Massachusetts Audubon 23: 866–868.

Breidenbach, C. H. 1990. Thermal cues influence strikes in pitless vipers. J. Herpetol. 24: 448–450.

Breithaupt, H., and E. Habermann. 1973. Biochemistry and pharmacology of phospholipase A from *Crotalus terrificus* venom as influenced by crotapotin, pp. 83–88. *In* E. Kaiser (ed.), Animal and plant toxins. Goldmann, München.

Brennan, T. C., and A. T. Holycross. 2004. *Crotalus oreganus concolor* (midget faded rattlesnake). Herpetol. Rev. 35: 190–191.

———. 2006. A field guide to amphibians and reptiles in Arizona. Arizona Game Fish Dept., Phoenix.

Bricker, J., L. M. Bushar, H. K. Reinert, and L. Gelbert. 1996. Purification of high quality DNA from shed skin. Herpetol. Rev. 27: 133–134.

Briggler, J. T., and J. W. Prather. 2002. *Crotalus horridus* (timber rattlesnake): Cave use. Herpetol. Rev. 33: 139.

Brimley, C. S. 1923. North Carolina herpetology. Copeia 1923(114): 3–4.

———. 1941–1942. The amphibians and reptiles of North Carolina: The snakes. Carolina Tips 4–5(19–26).

Bringsøe, H. 1998. Quo vadis? Three American CITES proposals for American reptiles. Herpetol. Rev. 29: 70–71.

Brock, O. G. 1981. Predatory behavior of eastern diamondback rattlesnakes (*Crotalus adamanteus*): Field enclosure and Y-maze laboratory studies, emphasizing prey trailing behaviors. Diss. Abst. Int. B41: 2510.

Brons, H. A. 1882. Notes on the habits of some western snakes. Am. Nat. 16: 564–567.

Brooking, W. J. 1934. Some reptiles and amphibians from Malheur County, in eastern Oregon. Copeia 1934: 93–94.

Brown, B. C. 1950. An annotated check list of the reptiles and amphibians of Texas. Baylor Univ. Press, Waco, Texas.

Brown, C. W., and C. H. Ernst. 1986. A study of variation in eastern timber rattlesnakes, *Crotalus horridus* Linnae (Serpentes: Viperidae). Brimleyana 12: 57–74.

Brown, D. E. (ed.). 1994. Biotic communities: Southwestern United States and northwestern Mexico. Univ. Utah Press, Salt Lake City.

Brown, D. G. 1990. Observation of a prairie rattlesnake (*Crotalus viridis viridis*) consuming neonatal cottontail rabbits (*Sylvilagus nuttalli*), with defense of the young cottontails by adult conspecifics. Bull. Chicago Herpetol. Soc. 25: 24–26.

Brown, D. G., and D. Duvall. 1993. Habitat associations of prairie rattlesnakes in Wyoming. Herpetol. Nat. Hist. 1: 5–12.

Brown, E. E. 1979a. Some snake food records from the Carolinas. Brimleyana 1: 113–124.

———. 1979b. Stray food records from New York and Michigan snakes. Am. Midl. Nat. 102: 200–203.

Brown, J. F. W., W. M. Marden, and D. L. Hardy, Sr. 2000. *Crotalus scutulatus scutulatus* (Mojave rattlesnake). Defensive behavior. Herpetol. Rev. 31: 45.

Brown, J. H. 1973. Toxicology and pharmacology of venoms from poisonous snakes. Charles C. Thomas, Springfield, Illinois.

Brown, J. R., C. A. Bishop, B. Baptiste, and R. J. Brooks. 2005. The effects of short-distance translocation on the northern Pacific rattlesnake (*Crotalus oreganus*) in southern British Columbia, Canada: Preliminary results and observations, p. 20. *In* Biology of the Rattlesnakes Symposium (program abstracts). Loma Linda Univ., Loma Linda, California.

Brown, N. L. 2003a. *Crotalus atrox* (western diamond-backed rattlesnake). Herpetol. Rev. 34: 168.

———. 2003b. *Crotalus scutulatus* (Mojave rattlesnake). Herpetol. Rev. 34: 168.

———. 2003c. *Crotalus mitchellii pyrrhus* (southwestern speckled rattlesnake). Herpetol. Rev. 34: 263.

———. 2003d. *Crotalus scutulatus* (Mojave rattlesnake). Herpetol. Rev. 34: 263.

Brown, P. R. 1997. A field guide to the snakes of California. Gulf Publishing Co., Houston, Texas.

Brown, T. K., J. M. Lemm, J.-P. Montagne, J. A. Tracey, and A. C. Alberts. 2008. Spatial ecology, habitat use, and survivorship of resident and translocated red diamond rattlesnakes (*Crotalus ruber*), pp. 377–394. *In* W. K. Hayes, K. R. Beaman, M. D. Cardwell, and S. P. Bush (eds.), The biology of rattlesnakes. Loma Linda Univ. Press, Loma Linda, California.

Brown, T. W. 1970. Autecology of the sidewinder (*Crotalus cerastes*) at Kelso Dunes, Mojave Desert California. Unpubl. Ph.D. diss., Univ. California, Los Angeles.

Brown, T. W., and H. B. Lillywhite. 1992. Autecology of the Mojave Desert sidewinder, *Crotalus cerastes cerastes*, at Kelso Dunes, Mojave Desert, California, U.S.A., pp. 279–308. *In* J. A. Campbell and E. D. Brodie, Jr. (eds.), Biology of the pitvipers. Selva, Tyler, Texas.

Brown, W. S. 1982. Overwintering body temperatures of timber rattlesnakes (*Crotalus horridus*) in northeastern New York. J. Herpetol. 16: 145–150.

———. 1984. Background information for the protection of the timber rattlesnake in New York state. Bull. Chicago Herpetol. Soc. 19: 94–97.

———. 1987. Hidden life of the timber rattler. Natl. Geogr. 172: 128–138.

———. 1988. Shedding rate and rattle growth in timber rattlesnakes. Am. Zoologist 28(4): 1015.

———. 1991. Female reproductive ecology in a northern population of the timber rattlesnake, *Crotalus horridus*. Herpetologica 47: 101–115.

———. 1992. Emergence, ingress, and seasonal captures at dens of northern timber rattlesnakes,

Crotalus horridus, pp. 251–258. *In* J. A. Campbell and E. D. Brodie, Jr. (eds.), Biology of the pitvipers. Selva, Tyler, Texas.

———. 1993. Biology, status, and management of the timber rattlesnake (*Crotalus horridus*): A guide for conservation. Soc. Stud. Amphib. Rept. Herpetol. Circ. 22: i–vi, 1–78.

———. 1995a. Heterosexual groups and the mating season in a northern population of timber rattlesnakes, *Crotalus horridus*. Herpetol. Nat. Hist. 3: 127–133.

———. 1995b. The female timber rattlesnake: A key to conservation. Rept. Amphib. Mag. September–October: 12–19.

———. 2008. Sampling timber rattlesnakes (*Crotalus horridus*): Phenology, growth, intimidation, survival, and a syndrome of undetermined origin in a northern population, pp. 235–256. *In* W. K. Hayes, K. R. Beaman, M. D. Cardwell, and S. P. Bush (eds.), The biology of rattlesnakes. Loma Linda Univ. Press, Loma Linda, California.

Brown, W. S., and D. B. Greenberg. 1992. Vertical-tree ambush posture in *Crotalus horridus*. Herpetol. Rev. 23: 67.

Brown, W. S., and F. M. MacLean. 1983. Conspecific scent-trailing by newborn timber rattlesnakes, *Crotalus horridus*. Herpetologica 39: 430–436.

Brown, W. S., and W. H. Martin. 1990. Geographic variation in female reproductive ecology of the timber rattlesnake, *Crotalus horridus*. Catesbeiana 10: 48.

Brown, W. S., and W. S. Parker. 1982. Niche dimensions and resource partitioning in a Great Basin Desert snake community, pp. 59–81. *In* N. J. Scott, Jr. (ed.), Herpetological communities. U.S. Fish Wildl. Serv. Wildl. Res. Rep. 13.

Brown, W. S., D. W. Pyle, K. R. Greene, and J. B. Friedlander. 1982. Movements and temperature relationships of timber rattlesnakes (*Crotalus horridus*) in northeastern New York. J. Herpetol. 16: 151–161.

Brown, W. S., M. Kéry, and J. E. Hines. 2007. Survival of timber rattlesnakes (*Crotalus horridus*) estimated by capture-recapture models in relation to age, sex, color morph, time, and birthplace. Copeia 2007: 656–671.

Browning, D. M., S. J. Beaupré, and L. Duncan. 2005. Using partitioned mahalanobis D^2 (*k*) to formulate a GIS-based model of timber rattlesnake hibernacula. J. Wildl. Manage. 69: 33–44.

Bryant, H. C. 1915. Rattlesnakes on Catalina Island. Copeia 1915(23): 48.

Bryson, R. W., Jr., and A. T. Holycross. 2001. *Crotalus willardi amabilis* (del Nido ridgenose rattlesnake): Diet. Herpetol. Rev. 32: 262.

Bryson, R. W., Jr., and D. Lazcano. 2002. Reproduction and mating behavior in the del Nido ridgenose rattlesnake, *Crotalus willardi amabilis*. Southwest. Nat. 47: 310–311.

———. 2003. *Crotalus pricei miquihuanus*: Notes on the natural history and captive husbandry of the Miquihuanan rattlesnake. Reptilia 28: 43–46.

Bryson, R. W., Jr., and J. M. Mueller. 2001. *Crotalus lepidus lepidus* (mottled rock rattlesnake). Herpetol. Rev. 32: 122.

Bryson, R. W., Jr., J. Banda, and D. Lazcano. 2002a. *Crotalus lepidus maculosus* (Durangan rock rattlesnake): Diet. Herpetol. Rev. 33: 139–140.

———. 2002b. *Crotalus pricei pricei* (twin-spotted rattlesnake): Habitat selection. Herpetol. Rev. 33: 140.

———. 2003. *Crotalus pricei miquihuanus* (Miquihuanan rattlesnake): Diet. Herpetol. Rev. 34: 65–66.

Bryson, R. W., Jr., J. M. Mueller, and D. Lazcano. 2008. Observations on the thermal ecology of montane Mexican rattlesnakes, pp. 287–290. *In* W. K. Hayes, K. R. Beaman, M. D. Cardwell, and S. P. Bush (eds.), The biology of rattlesnakes. Loma Linda Univ. Press, Loma Linda, California.

Buckle, A. C., J. Riley, and G. F. Hill. 1997. The in vitro development of the pentastomid *Porocephalus crotali* from the infective instar to the adult stage. Parasitology 115: 503–512.

Buijs, H. 1988. Ein Texaanse ratelslang (*Crotalus atrox*) bijt een soortgenoot. Lacerta 47: 72–73.

Bullard, K. M., and G. M. Fox. 2002. *Crotalus viridis oreganus* (northern Pacific rattlesnake): Diet. Herpetol. Rev. 33: 313.

Bullock, R. E. 1971. Cannibalism in captive rattlesnakes. Great Basin Nat. 31: 49–50.

Bullock, T. H., and R. B. Cowles. 1952. Physiology of an infrared receptor: The facial pit of pit vipers. Science (New York) 115: 541–543.

Bullock, T. H., and F. P. J. Diecke. 1956. Properties of an infra-red sense organ in the facial pit of pit vipers. J. Physiol. 134: 47–87.

Bullock, T. H., and W. Fox. 1957. The anatomy of the infra-red sense organ in the facial pit vipers. Quart. J. Micro. Sci. 98: 219–234.

Buntain, W. L. 1983. Successful venomous snakebite neutralization with massive antivenin infusion in a child. J. Trauma 23: 1012–1014.

Burger, J. W. 1934. The hibernation habits of the rattlesnake of the New Jersey Pine barrens. Copeia 1934: 142.

Burger, W. L., and M. M. Hensley. 1949. Notes on a collection of reptiles and amphibians from northwestern Sonora. Chicago Acad. Nat. Hist. Misc. 35: 1–6.

Burgess, J. L., and R. C. Dart. 1991. Snake venom coagulopathy: Use and abuse of blood products in the treatment of pit viper envenomation. Ann. Emerg. Med. 20: 795–801.

Burkett, R. D. 1966. Natural history of the cottonmouth moccasin, *Agkistrodon piscivorus* (Reptilia). Univ. Kansas Publ. Mus. Nat. Hist. 17: 435–491.

Bursey, C. R., S. R. Goldberg, and S. M. Secor. 1995. *Hexametra boddaërtii* (Nematoda: Ascaridae) in the sidewinder, *Crotalus cerastes* (Serpentes: Crotalidae), from California. Proc. Helminthol. Soc. Washington 62: 78–80.

Busby, W. H., J. T. Collins, and G. Suleiman. 2005. The snakes, lizards, turtles and amphibians of Fort Riley and vicinity. 2nd ed., revised. Lawrence, Kansas.

Busch, C., W. Lukas, H. M. Smith, D. Payne, and D. Chiszar. 1996. Strike-induced chemosensory searching (SICS) in northern Pacific rattlesnakes *Crotalus viridis oreganus* Holbrook, 1840, rescued from abusive husbandry conditions (Squamata: Serpentes: Viperidae). Herpetozoa 9(3/4): 99–104.

Bush, F. M. 1959. Foods of some Kentucky herptiles. Herpetologica 15: 73–77.

Bush, S. 2006. Snakebite and venom, pp. 271–275. *In* J. M. Lemm, Field guide to amphibians and reptiles of the San Diego region. Univ. California Press, Berkeley.

Bush, S. P., and M. D. Cardwell. 1999. Mojave rattlesnake (*Crotalus scutulatus scutulatus*) identification. Wild. Environ. Med. 10: 6–9.

Bush, S. P., and P. W. Jansen. 1995. Severe rattlesnake envenomation with anaphylaxis and rhabdomyolysis. Ann. Emerg. Med. 25: 845–848.

Bush, S. P., and E. Siedenburg. 1999. Neurotoxicity associated with suspected southern Pacific rattlesnake (*Crotalus viridis helleri*) envenomation. Wild. Environ. Med. 10: 247–249.

Bushar, L. M., H. K. Reinert, and L. Gelbert. 1998. Genetic variation and gene flow within and between local populations of the timber rattlesnake, *Crotalus horridus*. Copeia 1998: 411–422.

Bushar, L. M., M. Maliga, and H. K. Reinert. 2001. Cross-species amplification of *Crotalus horridus* microsatellites and their application in phylogenetic analysis. J. Herpetol. 35: 532–537.

Bushar, L. M., H. K. Reinert, and A. H. Savitzky. 2005. Isolation and reduced genetic variation in the timber rattlesnake, *Crotalus horridus,* of the New Jersey Pine Barrens, pp. 22–23. *In* Biology of the Rattlesnakes Symposium (program abstracts). Loma Linda Univ., Loma Linda, California.

Butler, J. A., T. W. Hull, and R. Franz. 1995. Neonate aggregations and maternal attendance of young in the eastern diamondback rattlesnake, *Crotalus adamanteus*. Copeia 1995: 196–198.

Butner, A. N. 1983. Rattlesnake bites in northern California. West. J. Med. 139: 179–183.

Cadle, J. E. 1992. Phylogenetic relationships among vipers: Immunological evidence, pp. 41–48. *In* J. A. Campbell, and E. D. Brodie, Jr. (eds.), Biology of the pitvipers. Selva, Tyler, Texas.

Cage, Y. 2004. Combat in the prairie rattlesnake. Sonoran Herpetol. 17: 107–108.

Cagle, F. R. 1968. Reptiles, pp. 213–268. *In* W. F. Blair, A. P. Blair, P. Brodkorb, F. R. Cagle, and G. A. Moore (eds.),Vertebrates of the United States. 2nd ed. McGraw-Hill, New York.

Cale, W. G., Jr., and J. W. Gibbons. 1972. Relationships between body size, size of the fat bodies, and total lipid content in the canebrake rattlesnake (*Crotalus horridus*) and the black racer (*Coluber constrictor*). Herpetologica 28: 51–53.

Calmonte, A. 1974. Die Felsenklapperschlange, *Crotalus lepidus lepidus* (Kennicott, 1861). Aquar. Aqua Terra 8: 460–462.

———. 1978. Die Schwarzschwanz klapperschlange in der Freiheit und im Terrarium beobachtet. Aquar. Aqua Terra 12: 221–223.

Cameron, D. L., and A. T. Tu. 1977. Characterization of myotoxin a from the venom of a prairie rattlesnake (*Crotalus viridis viridis*). Biochemistry 16: 2546–2552.

Camp, C. L. 1916. Notes on the local distribution and habitats of the amphibians and reptiles of southeastern California in the vicinity of the Turtle Mountains. Univ. California Publ. Zool. 12: 503–544.

Campbell, B. 1934. Report on a collection of reptiles and amphibians made in Arizona during the summer of 1933. Occ. Pap. Mus. Zool. Univ. Michigan 289: 1–10.

Campbell, H. 1950. Rattlesnakes tangled in wire. Herpetologica 6: 44.

———. 1953a. Probable strychnine poisoning in a rattlesnake. Herpetologica 8: 184.

———. 1953b. Observations on snakes DOR in New Mexico. Herpetologica 9: 157–160.

———. 1956. Snakes found dead on the roads of New Mexico. Copeia 1956: 124–125.

———. 1958 (1959). An unusually long rattle string in *Crotalus atrox* (Serpentes: Crotalidae). Southwest. Nat. 3: 233.

Campbell, J. A. 1979. *Crotalus scutulatus* (Viperidae) in Jalisco, Mexico. Southwest. Nat. 24: 693–694.

Campbell, J. A., and E. D. Brodie, Jr. (eds.). 1992. Biology of the pitvipers. Selva, Tyler, Texas.

Campbell, J. A., and W. W. Lamar. 1989. The venomous reptiles of Latin America. Comstock Publ. Assoc., Cornell Univ. Press, Ithaca, New York.

———. 2004. The venomous reptiles of the western hemisphere in two volumes, Comstock Publ. Assoc., Cornell Univ. Press, Ithaca, New York.

Campbell, J. A., E. D. Brodie, Jr., D. G. Barker, and A. H. Price. 1989a. An apparent natural hybrid rattlesnake and *Crotalus willardi* (Viperidae) from the Peloncillo Mountains of southwestern New Mexico. Herpetologica 45: 344–349.

Campbell, J. A., D. R. Formanowicz, Jr., and E. D. Brodie, Jr. 1989b. Potential impact of rattlesnake roundups on natural populations. Texas J. Sci. 41: 301–317.

Camper, J. D., and J. R. Dixon. 1988. Evaluation of a microchipmarking-system for amphibians and reptiles. Texas Parks Wildl. Dept. Res. Publ. 7: 100–159.

Caras, R. 1974. Venomous animals of the world. Prentice-Hall, Englewood Cliffs, New Jersey.

Cardwell, M. D. 2005a. Habitat utilization and behavior of the Mojave rattlesnake (*Crotalus scutulatus scutulatus*) in the western Mojave Desert, p. 23. *In* Biology of the Rattlesnakes Symposium (program abstracts). Loma Linda Univ., Loma Linda, California.

———. 2005b. *Crotalus scutulatus* (Mohave rattlesnake): Behavior. Herpetol. Rev. 36: 192.

———. 2006. Rain-harvesting in a wild population of *Crotalus s. scutulatus* (Serpentes: Viperidae). Herpetol. Rev. 37: 142–144.

———. 2008. The reproductive ecology of Mohave rattlesnakes. J. Zool. (London) 274: 65–76.

Cardwell, M. D., and B. Alexander. 2006. *Crotalus scutulatus scutulatus* (Mohave rattlesnake): Albinism. Herpetol. Rev. 37: 477.

Cardwell, M. D., and J. Banashek. 2006. *Crotalus scutulatus scutulatus* (Mohave rattlesnake): Morphology. Herpetol. Rev. 37: 477.

Cardwell, M. D., S. P. Bush, and R. T. Clark. 2005. Males biting males: Does testosterone shape both sides of the snakebite equation? p. 23. *In* Biology of the Rattlesnakes Symposium (program abstracts). Loma Linda Univ., Loma Linda, California.

Carl, G. C. 1960. The reptiles of British Columbia. British Columbia Prov. Mus. Handb. 3: 1–65.

Carmichael, E. B., and P. Petcher. 1945. Constituents of the blood of the normal and hibernating rattlesnake, *Crotalus horridus*. J. Biol. Chem. 161: 693–696.

Carnie, S. K. 1954. Food habits of nesting golden eagles in the coast ranges of California. Condor 56(1): 3–12.

Carpenter, C. C. 1986. An inventory of combat rituals in snakes. Smithsonian Herpetol. Inform. Serv. 69: 1–18.

Carpenter, C. C., and G. W. Ferguson. 1977. Variation and evolution of stereotyped behavior in reptiles, pp. 335–554. *In* C. Gans and D. W. Tinkle (eds.), Biology of the Reptilia, vol. 7. Academic Press, London.

Carpenter, C. C., and J. C. Gillingham. 1975. Postural responses to kingsnakes by crotaline snakes Herpetologica 31: 293–302.

Carpenter, C. C., J. C. Gillingham, and J. B. Murphy. 1976. The combat ritual of the rock rattle-snake (*Crotalus lepidus*). Copeia 1976: 764–780.

Carr, A. F., Jr. 1940. A contribution to the herpetology of Florida. Univ. Florida Biol. Sci. Ser. 3: 1–118.

Carr, A. F., Jr., and C. J. Goin. 1955. Guide to the reptiles, amphibians, and freshwater fishes of Florida. Univ. Florida Press, Gainesville.

Carroll, R. R., E. L. Hall, and C. S. Kitchens. 1997. Canebrake rattlesnake envenomation. Ann. Emerg. Med. 30: 45–48.

Cartwright, S. W. 1928. An unusual snake-bite accident. Bull. Antiv. Inst. Am. 2: 45.

Casas-Andreu, G., F. R. Méndez de la Cruz, and J. L. Camarillo. 1996. Anfibios y reptiles de Oaxaca: Lista, distribución y conservación. Acta Zool. Mexicana 69: 1–35.

Case, T. J. 1978. A general explanation for insular body size trends in terrestrial vertebrates. Ecology 59: 1–18.

———. 1983. Reptiles: Ecology, pp. 221–270. *In* T. J. Case and M. L. Cody (eds.). Island biogeography in the Sea of Cortez. Univ. California Press, Berkeley.

Castilonia, R. R., T. R. Pattabhiraman, and F. E. Russell. 1980. Neuromuscular blocking effects of Mojave rattlesnake (*Crotalus scutulatus scutulatus*) venom. Proc. West. Pharmacol. Soc. 23: 103–106.

Castilonia, R. R., T. R. Pattabhiraman, F. E. Russell, and H. Gonzalez. 1981. Electrophysiological studies on a protein fraction (K′) from Mojave rattlesnake (*Crotalus scutulatus scutulatus*) venom. Toxicon 19: 473–479.

Castoe, T. A., and C. L. Parkinson. 2006. Bayesian mixed models and the phylogeny of pitvipers (Viperidae: Serpentes). Mol. Phylogenet. Evol. 39: 91–110.

Castoe, T. A., C. L. Spencer, and C. L. Parkinson. 2007. Phylogeographic structure and historical demography of the western diamondback rattlesnake (*Crotalus atrox*): A perspective on North American desert biogeography. Mol. Phylogenet. Evol. 42: 193–212.

Cavanaugh, C. J. 1994. *Crotalus horridus* (timber rattlesnake). Longevity. Herpetol. Rev. 25: 70.

Chace, G. E., and H. M. Smith. 1968. Two additional examples of Gloyd's linked albinism in the prairie rattlesnake, *Crotalus viridis*. J. Herpetol. 2: 165–166.

Chadwick, L. E., and H. Rahn. 1954. Temperature dependence of rattling frequency in the rattlesnake, *Crotalus v. viridis*. Science (New York) 119: 442–443.

Charland, M. B. 1987. An examination of factors influencing first-year recruitment in the northern Pacific rattlesnake, *Crotalus viridis oreganus*, in British Columbia. Master's thesis, Univ. Victoria, British Columbia, Canada.

———. 1989. Size and winter survivorship in neonatal western rattlesnakes (*Crotalus viridis*). Can. J. Zool. 67: 1620–1625.

Charland, M. B., and P. T. Gregory. 1989. Feeding rate and weight gain in postpartum rattlesnakes: Do animals that eat more always grow more? Copeia 1989: 211–214.

———. 1990. The influence of female reproductive status on thermoregulation in a viviparous snake, *Crotalus viridis*. Copeia 1990: 1089–1098.

Chen, Y. H., Y. M. Wang, M. J. Hseu, and I. H. Tsai. 2004. Molecular evolution and structure-function relationships of crotoxin-like and asparagine-6-containing phospholipases A_2 in pit viper venoms. Biochem. J. 381: 25–34.

Chermock, R. L. 1952. A key to the amphibians and reptiles of Alabama. Univ. Alabama Mus. Pap. 33: 1–88.

Chiodini, R. J., J. P. Sundberg, and J. A. Czikowsky. 1982. Gross anatomy of snakes. Veterin. Med. Sm. Anim. Clin. 77: 413–419.

Chiszar, D., and C. W. Radcliffe. 1977. Absence of prey-chemical preferences in newborn rattlesnakes (*Crotalus cerastes*, *C. enyo*, and *C. viridis*). Behav. Biol. 21: 146–150.

Chiszar, D., and K. M. Scudder. 1980. Chemosensory searching by rattlesnakes during predatory episodes, pp. 125–129. *In* D. Muller-Schwarze and R. M. Silverstein (eds.), Chemical signals: Vertebrates and aquatic invertebrates. Plenum Press, New York.

Chiszar, D., and H. M. Smith. 2008. Effects of chemical and thermal cues on striking behavior and post-strike chemosensory searching in rattlesnakes, pp. 175–180. *In* W. K. Hayes, K. R.

Beaman, M. D. Cardwell, and S. P. Bush (eds.), The biology of rattlesnakes. Loma Linda Univ. Press, Loma Linda, California.

Chiszar, D., C. W. Radcliffe, and K. M. Scudder. 1977. Analysis of the behavioral sequence emitted by rattlesnakes during feeding episodes. I. Striking and chemosensory searching. Behav. Biol. 21: 418–425.

Chiszar, D., C. W. Radcliffe, and H. M. Smith. 1978a. Chemosensory searching for wounded prey by rattlesnakes is released by striking: A replication report. Herpetol. Rev. 9: 54–56.

Chiszar, D., K. Scudder, L. Knight, and H. M. Smith. 1978b. Exploratory behavior in prairie rattlesnakes (Crotalus viridis) and water moccasins (Agkistrodon piscivorus). Psychol. Rec. 28: 363–368.

Chiszar, D., D. Duvall, and K. Scudder. 1980a. Simultaneous and successive discriminations between envenomated and nonenvenomated mice by rattlesnakes (Crotalus durissus and C. viridis). Behav. Neurol. Biol. 29: 518–521.

Chiszar, D., C. W. Radcliffe, B. O'Connell, and H. M. Smith. 1980b. Strike-induced chemosensory searching in rattlesnakes (Crotalus enyo) as a function of disturbance prior to presentation of prey. Trans. Kansas Acad. Sci. 83: 230–234.

Chiszar, D., C. W. Radcliffe, H. M. Smith, and H. Bashinski. 1981a. Effect of prolonged food deprivation on response to prey odors by rattlesnakes. Herpetologica 37: 237–243.

Chiszar, D., S. W. Taylor, C. W. Radcliffe, H. M. Smith, and B. O'Connell. 1981b. Effects of chemical and visual stimuli upon chemosensory searching by garter snakes and rattlesnakes. J. Herpetol. 15: 415–423.

Chiszar, D., C. Andren, F. Nilson, B. O'Connell, J. S. Mestas, Jr., H. M. Smith, and C. W. Radcliffe. 1982a. Strike-induced chemosensory searching in Old World vipers and New World pit vipers. Anim. Learn. Behav. 10: 121–125.

Chiszar, D., C. W. Radcliffe, B. O'Connell, and H. M. Smith. 1982b. Analysis of the behavioral sequence emitted by rattlesnakes during feeding episodes. II. Duration of strike-induced chemosensory searching in rattlesnakes (Crotalus viridis, C. enyo). Behav. Neurol. Biol. 34: 261–270.

Chiszar, D., C. W. Radcliffe, K. M. Scudder, and D. Duvall. 1983a. Strike-induced chemosensory searching by rattlesnakes: The role of envenomation-related chemical cues in the post-strike environment, pp. 1–24. In D. Müller-Schwarze and R. M. Silverstein (eds.), Chemical signals in vertebrates, III. Plenum Press, New York.

Chiszar, D., K. Stimac, and T. Boyer. 1983b. Effect of mouse odor on visually-induced and strike-induced chemosensory searching in prairie rattlesnakes (Crotalus viridis). Chem. Senses 7: 301–308.

Chiszar, D., B. O'Connell, R. Greenlee, B. Demeter, T. Walsh, J. Chiszar, K. Moran, and H. M. Smith. 1985a. Duration of strike-induced chemosensory searching in long-term captive rattlesnakes at National Zoo, Audubon Zoo, and San Diego Zoo. Zoo Biol. 4: 291–294.

Chiszar, D., C. A. Castro, H. M. Smith, and C. Guyon. 1986a. A behavioral method for assessing utilization of thermal cues by snakes during feeding episodes, with a comparison of crotaline and viperine species. Ann. Zool., Acad. Zool. India 24: 123–131.

Chiszar, D., D. Dickman, and J. Colton. 1986b. Sensitivity to thermal stimulation in prairie rattlesnakes (Crotalus viridis) after bilateral anesthetization of the facial pits. Behav. Neurol. Biol. 45: 143–149.

Chiszar, D., C. W. Radcliffe, T. Byers, and R. Stoops. 1986c. Prey capture behavior in nine species of venomous snakes. Psychol. Rec. 36: 433–438.

Chiszar, D., C. W. Radcliffe, and F. Feiler. 1986d. Trailing behavior in banded rock rattlesnakes (Crotalus lepidus klauberi) and prairie rattlesnakes (C. viridis viridis). J. Comp. Psychol. 100: 368–371.

Chiszar, D., P. Nelson, and H. M. Smith. 1988. Analysis of the behavioral sequence emitted by rattlesnakes during feeding episodes. III. Strike-induced chemosensory searching and location of rodent carcasses. Bull. Maryland Herpetol. Soc. 24: 99–108.

Chiszar, D., T. Melcer, R. Lee, C. W. Radcliffe, and D. Duvall. 1990. Chemical cues used by prairie rattlesnakes (Crotalus viridis) to follow trails of rodent prey. J. Chem. Ecol. 16: 79–86.

Chiszar, D., G. Hobika, H. M. Smith, and J. Vidaurri. 1991a. Envenomation and acquisition of chemical information by prairie rattlesnakes. Prairie Nat. 23: 69–72.

Chiszar, D., C. W. Radcliffe, H. M. Smith, and P. Langer. 1991b. Strike-induced chemosensory searching: Do rattlesnakes make one decision or two? Bull. Maryland Herpetol. Soc. 27: 90–94.

Chiszar, D., H. M. Smith, C. M. Bogert, and J. Vidaurri. 1991c. A chemical sense of self in timber and prairie rattlesnakes. Bull. Psychon. Soc. 29: 153–154.

Chiszar, D., H. M. Smith, J. L. Glenn, and R. C. Straight. 1991d. Strike-induced chemosensory searching in venomoid pit vipers at Hogle Zoo. Zoo Biol. 10: 111–117.

Chiszar, D., R. K. K. Lee, H. M. Smith, and C. W. Radcliffe. 1992. Searching behaviors by rattlesnakes following predatory strikes, pp. 369–382. In J. A. Campbell and E. D. Brodie, Jr. (eds.), Biology of the pitvipers. Selva, Tyler, Texas.

Chiszar, D., H. M. Smith, and R. Defusco. 1993a. Crotalus viridis viridis (prairie rattlesnake). Diet. Herpetol. Rev. 24: 106.

Chiszar, D., G. Hobika, and H. M. Smith. 1993b. Prairie rattlesnakes (Crotalus viridis) respond to rodent blood with chemosensory searching. Brain Behav. Evol. 41: 229–233.

Chiszar, D., H. M. Smith, V. Veer, C. W. Radcliffe, and R. Smith. 1996. The new psychobiology: Assessing the competence of captive-raised herps. Proc. AZA Reg. Conf. 1996: 465–469.

Chiszar, D., K. DeWelde, M. Garcia, D. Payne, and H. M. Smith. 1999a. Strike-induced chemosensory searching (SICS) in northern Pacific rattlesnakes (Crotalus viridis oreganus, Holbrook, 1840) rescued from substandard husbandry conditions: II. Complete recovery of function after two years. Zoo Biol. 18: 141–146.

Chiszar, D., A. Walters, J. Urbaniak, H. M. Smith, and S. P. Mackessy. 1999b. Discrimination between envenomated and nonenvenomated prey by western diamondback rattlesnakes (Crotalus atrox): Chemosensory consequences of venom. Copeia 1999: 640–648.

Chiszar, D., A. Walters, and H. M. Smith. 2008. Rattlesnake preference for envenomated prey: Species specificity. J. Herpetol. 42: 764–767.

Chrapliwy, P. S., and C. M. Fugler. 1955. Amphibians and reptiles collected in Mexico in the summer of 1953. Herpetologica 11: 121–128.

Christie, T. 1994. Snake bite! Wyoming Wildl. 58(6): 22–29.

Christman, B. L., C. W. Painter, R. D. Jennings, and A. W. Lamb. 2000. Crotalus viridis cerberus (Arizona black rattlesnake). Herpetol. Rev. 31: 255.

Christman, B. L., A. T. Holycross, and C. W. Painter. 2004. Crotalus lepidus klauberi (banded rock rattlesnake): Morphology. Herpetol. Rev. 35: 62.

Christman, S. P. 1975. The status of the extinct rattlesnake Crotalus giganteus. Copeia 1975: 43–47.

———. 1980. Patterns of geographic variation in Florida snakes. Bull. Florida St. Mus. Biol. Sci. 25: 157–256.

Clamp, H. J. 1990. A snake story. Bull. South Carolina Herpetol. Soc. 1990 (September): 1–12.

Clark, A. M., P. E. Moler, E. E. Possardt, A. H. Savitzky, W. S. Brown, and B. W. Bowen. 2003. Phylogeography of the timber rattlesnake (Crotalus horridus) based on mtDNA sequences. J. Herpetol. 37: 145–154.

Clark, N., and F. Antonio. 2008. Reproduction in the eastern diamond-backed rattlesnake, Crotalus adamanteus, under optimal conditions in captivity, pp. 413–418. In W. K. Hayes, K. R. Beaman, M. D. Cardwell, and S. P. Bush (eds.), The biology of rattlesnakes. Loma Linda Univ. Press, Loma Linda, California.

Clark, R. F. 1949. Snakes of the hill parishes of Louisiana. J. Tennessee Acad. Sci. 24: 244–261.

Clark, R. F., B. S. Selden, and B. Furbee. 1993. The incidence of wound infection following crotalid envenomation. J. Emerg. Med. 11: 583–586.

Clark, R. F., S. R. Williams, S. P. Nordt, and L. V. Boyer-Hassen. 1997. Successful treatment of crotalid-induced neurotoxicity with a new polyspecific crotalid Fab antivenom. Ann. Emerg. Med. 30: 54–57.

Clark, R. W. 2002. Diet of the timber rattlesnake, Crotalus horridus. J. Herpetol. 36: 494–499.

———. 2004a. Feeding experience modifies the assessment of ambush sites by the timber rattlesnake, a sit-and-wait predator. Ethology 110: 471–483.

————. 2004b. Kin recognition in rattlesnakes. Proc. Royal Soc. London B (suppl.) 271: S243–S245.

————. 2004c. Timber rattlesnakes (*Crotalus horridus*) use chemical cues to select ambush sites. J. Chem. Ecol. 30: 607–617.

————. 2005a. Social lives of rattlesnakes: Because the snakes bask, breed, and hibernate together, recognizing their relatives is a key advantage, especially for females. Natur. Hist. 114: 36–42.

————. 2005b. Pursuit-deterrent communication between prey animals and timber rattlesnakes (*Crotalus horridus*): The response of snakes to harassment displays. Behav. Ecol. Sociobiol. 59: 258–261.

————. 2006a. Post-strike behavior of timber rattlesnakes (*Crotalus horridus*) during natural predation events. Ethology 112: 1089–1094.

————. 2006b. Fixed videography to study predation behavior of an ambush foraging snake, *Crotalus horridus*. Copeia 2006: 181–187.

————. 2007a. Timber rattlesnakes use public information to assess foraging locations. Herpetol. Rev. 38: 261.

————. 2007b. Public information for solitary foragers: Timber rattlesnakes use conspecific chemical cues to select ambush sites. Behav. Ecol. 18: 487–490.

Clark, R. W., W. S. Brown, R. Stechert, and K. R. Zamudio. 2008. Integrating individual behaviour and landscape genetics: The population structure of timber rattlesnake hibernacula. Mol. Ecol. 17: 719–730.

Clark, R. W., W. S. Brown, R. Stechert, and K. R. Zamudio. 2010. Roads, interrupted dispersal, and genetic diversity in timber rattlesnakes. Conserv. Biol. 24: 1059–1069.

Clark, W. W., and E. Schultz. 1980. Rattlesnake shaker muscle: 2. Fine structure. Tissue Cell 12: 335–351.

Clarke, G. K. 1961. Report on a bite by a red diamond rattlesnake, *Crotalus ruber ruber*. Copeia 1961: 418–422.

Clarke, G. K., and T. I. Marx. 1960. Heart rates of unanesthetized snakes by electrocardiography. Copeia 1960: 236–238.

Clarke, J. A., J. T. Chopko, and S. J. Mackessy. 1996. The effect of moonlight on activity patterns of adult and juvenile prairie rattlesnakes (*Crotalus viridis viridis*). J. Herpetol. 30: 192–197.

Clarke, R. F. 1958. An ecological study of reptiles and amphibians in Osage County, Kansas. Emporia State Res. Stud. 7: 1–52.

Clarkson, R. W., and J. C. deVos, Jr. 1986. The bullfrog, *Rana catesbeiana* Shaw, in the Lower Colorado River, Arizona-California. J. Herpetol. 20: 42–49.

Clench, W. J. 1925. A possible manner of snake distribution. Copeia 1925(142): 40.

Cliburn, J. W. 1979. Range revisions for some Mississippi reptiles. J. Mississippi Acad. Sci. 24: 31–37.

Cliff, F. S. 1954. Snakes of the islands in the Gulf of California, Mexico. Trans. San Diego Soc. Nat. Hist. 12: 67–98.

Cloudsley-Thompson, J. 2006. The colouration and displays of venomous reptiles: A review. Herpetol. Bull. 95: 25–30.

Clucas, B., M. P. Rowe, D. H. Owings, and P. C. Arrowood. 2008. Snake scent application in ground squirrels, *Spermophilus* spp.: A novel form of antipredator behaviour? Anim. Behav. 75: 299–307.

Cobb, V. A., and C. R. Peterson. 1999. *Crotalus viridus lutosus* (Great Basin rattlesnake): Mortality. Herpetol. Rev. 30: 45–46.

————. 2008. Thermal ecology of hibernation in a population of Great Basin rattlesnakes, pp. 291–302. *In* W. K. Hayes, K. R. Beaman, M. D. Cardwell, and S. P. Bush (eds.), The biology of rattlesnakes. Loma Linda Univ. Press, Loma Linda, California.

Cobb, V. A., J. J. Green, T. Worrall, J. Pruett, and B. Glorioso. 2005. Initial den location behavior in a litter of neonate *Crotalus horridus* (timber rattlesnakes). Southeast. Nat. 4: 723–730.

Cochran, D. A. 1954. Our snake friends and foes. Natl. Geogr. 106: 333–364.

Cochran, D. M. 1943. Poisonous reptiles of the World: A wartime handbook. Smithsonian Inst. Press, Washington, D.C.

————. 1944. Dangerous reptiles. Smithsonian Inst. Press, Washington, D.C.

Cochran, D. M., and C. J. Goin. 1970. The new field book of reptiles and amphibians. Putnam, New York.

Cochran, P. A. 2006. Timber rattlesnakes and residential development in fiction and in fact. Bull. Chicago Herpetol. Soc. 41: 197.

————. 2008. Phenology of timber rattlesnakes (*Crotalus horridus*) in southern Minnesota: Implications for conservation, pp. 441–446. *In* W. K. Hayes, K. R. Beaman, M. D. Cardwell, and S. P. Bush (eds.), The biology of rattlesnakes. Loma Linda Univ. Press, Loma Linda, California.

————. 2010. Rattlesnake eggs and the passing of a torch in Winona County, Minnesota. Arch. Natur. Hist. 37: 19–27.

Cochran, P. A., and S. J. Schmitt. 2009. Use of remote cameras to monitor rock ledge microhabitat of timber rattlesnakes (*Crotalus horridus*). Bull. Chicago Herpetol. Soc. 44: 161–165.

Cohen, A. C., and B. C. Myres. 1970. A function of the horns (supraocular scales) in the sidewinder rattlesnake, *Crotalus cerastes*, with comments on other horned snakes. Copeia 1970: 574–575.

Cole, C. J. 1990. Chromosomes of *Agkistrodon* and other viperid snakes, pp. 533–538. *In* H. K. Gloyd and R. Conant, Snakes of the *Agkistrodon* complex: A monographic review. Soc. Stud. Amphib. Rept. Contrib. Herpetol. 6.

Cole, T. 2005. Color/pattern morphs in rattlesnakes with emphasis on central Texas *atrox*, pp. 25–26. *In* Biology of the Rattlesnakes Symposium (program abstracts). Loma Linda Univ., Loma Linda, California.

Collins, J. T. 1964. A preliminary review of the snakes of Kentucky. J. Ohio Herpetol. Soc. 4: 69–77.

————. 1974. Amphibians and reptiles in Kansas. Univ. Kansas Mus. Nat. Hist. Publ. Ed. Ser. 1: 1–183.

————. 1990. Standard common and current scientific names for North American amphibians and reptiles. 3rd ed. Soc. Stud. Amphib. Rept. Herpetol. Circ. 19: 1–41.

————. 1991. Viewpoint: A new taxonomic arrangement for some North American amphibians and reptiles. Herpetol. Rev. 22: 42–43.

————. 1993. Amphibians and reptiles in Kansas. 3rd ed., revised. Univ. Kansas Mus. Nat. Hist. Publ. Ed. Ser. 13: 1–397.

————. 2006. A re-classification of snakes native to Canada and the United States. J. Kansas Herpetol. 19: 18–20.

Collins, J. T., and S. L. Collins. 2010. A pocket guide to Kansas snakes. 3rd ed. Great Plains Nature Center, Wichita, Kansas.

Collins, J. T., and J. L. Knight. 1980. *Crotalus horridus*. Cat. Am. Amphib. Rept. 253: 1–2.

Collins, J. T., S. L. Collins, and K. J. Irwin. 2006. A survey of amphibians, turtles, and reptiles of the eastern portion of the Kiowa National Grassland of New Mexico and the Rita Blanca National Grassland of adjacent Oklahoma and Texas. J. Kansas Herptol. 18: 10–20.

Collins, S. 1989. Venerable rattler dies: Kansas snake held unofficial record for longevity. Kansas Herpetol. Soc. Newsl. 78: 8.

Collins, S. L. 2003. An arboreal timber rattlesnake. J. Kansas Herpetol. 6: 7.

Conant, R. 1945. An annotated checklist of the amphibians and reptiles of the Del-Mar-Va Peninsula. Soc. Nat. Hist. Wilmington, Delaware.

————. 1951. The reptiles of Ohio. 2nd ed. Notre Dame Press, Notre Dame, Indiana.

————. 1955. Notes on three Texas reptiles, including an addition to the fauna of the state. Am. Mus. Novit. 1726: 1–6.

————. 1957. Reptiles and amphibians of the northeastern states. 3rd ed. Zoological Society, Philadelphia.

————. 1969. Some rambling notes on rattlesnakes. Arch. Environ. Health 19: 768–769.

————. 1975. A field guide to reptiles and amphibians of eastern and central North America. 2nd ed. Houghton Mifflin, Boston.

————. 1978. Distributional patterns of North American snakes: Some examples of the effects of Pleistocene glaciations and subsequent climatic changes. Bull. Maryland Herpetol. Soc. 14: 241–259.

Conant, R., and W. Bridges. 1939. What snake is that? Appleton Century, New York.

Conant, R., and J. T. Collins. 1998. A field guide to reptiles and amphibians: Eastern and central North America. 3rd ed., expanded. Houghton Mifflin, Boston.

Conceição, L. G., N. M. Argôlo Neto, A. P. Castro, L. B. A. Faria, and C. O. Fonterrada. 2007. Anaphylactic reaction after *Crotalus* envenomation treatment in a dog: Case report. J. Venom. Anim. Toxins incl. Trop. Dis. 13: 549–557.

Condon, P. T. 2005. Parasites on a tiger rattlesnake (*Crotalus tigris*). Sonoran Herpetol. 18: 55.

Conley, K. E., and S. L. Lindstedt. 1996. Minimal cost per twitch in rattlesnake tail muscle. Nature (London) 383: 71–72.

Conner, R. N., D. C. Rudolph, D. Saenz, R. R. Schaefer, and S. J. Burgdorf. 2003. Growth rates and post-release survival of captive neonate timber rattlesnakes, *Crotalus horridus*. Herpetol. Rev. 34: 314–317.

Contreras-Lozano, J. A., D. Lazcano, A. J. Contreras-Balderas, and P. A. Lavín-Murcio. 2010. Notes on Mexican herpetofauna 14: An update to the herpetofauna of Cerro El Potosí, Galeana, Nuevo León, México. Bull. Chicago Herpetol. Soc. 45: 41–46.

Cook, F. R. 1966. A guide to the amphibians and reptiles of Saskatchewan. Saskatchewan Mus. Nat. Hist. Pop. Ser. (13): 1–40.

———. 1984. Introduction to Canadian amphibians and reptiles. Natl. Mus. Canada, Ottawa.

Cook, P. M., M. P. Rowe, and R. W. Van Devender. 1994. Allometric scaling and interspecific differences in the rattling sounds of rattlesnakes. Herpetologica 50: 358–368.

Cook, S. F., Jr. 1955. Rattlesnake hybrids: *Crotalus viridis* × *Crotalus scutulatus*. Copeia 1955: 139–141.

Coombs, E. M. 1977. Wildlife observations of the hot desert region, Washington Co., Utah, with emphasis on reptilian species and their habitat in relation to livestock grazing. Utah Div. Wildl. Res. for USDI/BLM, Utah State Off. Contract No. YA-512-CT 6-102.

Cooper, J. E., and F. Groves. 1959. The rattlesnake, *Crotalus horridus*, in the Maryland Piedmont. Herpetologica 15: 33–34.

Cope, E. D. 1861a [1860]. Notes and descriptions of new and little known species of American reptiles. Proc. Acad. Nat. Sci. Philadelphia 12: 339–345.

———. 1861b [1862]. Contributions to the ophiology of Lower California, Mexico and Central America. Proc. Acad. Nat. Sci. Philadelphia 13: 292–306.

———. 1864. Contributions to the herpetology of tropical America. Proc. Acad. Nat. Sci. Philadelphia 16: 166–181.

———. 1867 [1866]. On the Reptilia and Batrachia of the Sonoran Province of the Nearctic Region. Proc. Acad. Nat. Sci. Philadelphia 18: 300–315.

———. 1892. A critical review of the characters and variations of snakes of North America. Proc. U.S. Natl. Mus. 14(882): 589–694.

Corn, P. S., and C. R. Peterson. 1996. Prairie legacies: Amphibians and reptiles, pp. 125–134. *In* F. B. Samson and F. L. Knopf (eds.), Prairie conservation: Preserving North America's most endangered ecosystem. Island Press, Covelo, California.

Corrigan, J. J., and M. A. Jeter. 1990. Mojave rattlesnake (*Crotalus scutulatus scutulatus*) venom: In vitro effects on platelets, fibrinolysis, and fibrinogen clotting. Vet. Human Toxicol. 32: 439–441.

Coss, R. G., K. L. Guse, N. S. Porin, and D. G. Smith. 1993. Development of antistrike defenses in California ground squirrels (*Spermophilus beechey*): II. Microevolutionary effects of relaxed selection from rattlesnakes. Behaviour 124: 137–164.

Cottam, C., W. C. Glazener, and G. G. Raun. 1959. Notes on food of moccasins and rattlesnakes from the Welder Wildlife Refuge, Sinton, Texas. Welder Wildl. Found. Contr. 45: 1–12.

Coues, E. 1875. Synopsis of the reptiles and batrachians of Arizona, pp. 585–663. *In* Report upon United States geographical surveys west of the one hundredth meridian (Wheeler Report), vol. 5: Zoology. U.S. Gov't. Print. Off., Washington, D.C.

Coues, E., and H. C. Yarrow. 1878. Notes on the herpetology of Dakota and Montana. Bull. U.S. Geol. Geogr. Surv. 4: 259–291.

Coupe, B. 2001. Arboreal behavior in timber rattlesnakes (*Crotalus horridus*). Herpetol. Rev. 32: 83–85.

————. 2002. Pheromones, search patterns, and old haunts: How do male timber rattlesnakes (*Crotalus horridus*) locate mates? pp. 139–148. *In* G. W. Schuett, M. Höggren, M. E. Douglas, and H. W. Greene (eds.), Biology of the vipers. Eagle Mountain Publ., Eagle Mountain, Utah.

Coupe, B., and J. E. Dawson. 2007. *Crotalus cerastes* (sidewinder): Attempted feeding and mortality. Herpetol. Rev. 38: 339–340.

Cowles, R. B. 1938. Unusual defense postures assumed by rattlesnakes. Copeia 1938: 13–16.

————. 1941. Observations on the winter activities of desert reptiles. Ecology 22: 125–140.

————. 1945. Some of the activities of the sidewinder. Copeia 1945: 220–222.

————. 1953. The sidewinder: Master of desert travel. Pacific Disc. 6(2): 12–15.

Cowles, R. B., and C. M. Bogert. 1936. The herpetology of the Boulder Dam region (Nev., Ariz., Utah). Herpetologica 1: 33–42.

————. 1944. A preliminary study of the thermal requirements of desert reptiles. Bull. Am. Mus. Nat. Hist. 83: 261–296.

Cowles, R. B., and R. L. Phelan. 1958. Olfaction in rattlesnakes. Copeia 1958: 77–83.

Crabtree, C. B., and R. W. Murphy. 1984. Analysis of maternal-offspring allozymes in *Crotalus viridis*. J. Herpetol. 18: 75–80.

Craig, L. E. W., J. C. Ramsay, and C. Edward. 2005. Spinal cord glioma in a ridge-nosed rattlesnake (*Crotalus willardi*). J. Zoo Wildl. Med. 36: 313–315.

Crimmins, M. L. 1927a. Facts about Texas snakes and their poisons. Texas St. J. Med. 23(3): 198–203.

————. 1927b. Notes on the Texas rattlesnakes. Bull. Antiv. Inst. Am. 1: 23–24.

————. 1931. Rattlesnakes and their enemies in the Southwest. Bull. Antiv. Inst. Am. 5: 46–47.

Cromwell, W. R. 1982. Underground desert toads. Pacific Disc. 35: 10–17.

Crother, B. I. (Chair.). 2000. Scientific and standard English names of amphibians and reptiles of North America north of Mexico, with comments regarding confidence in our understanding. Soc. Stud. Amphib. Rept. Herpetol. Circ.(29): 1–82.

Cruz, E., S. Gibson, K. Kandler, G. Sanchez, and D. Chiszar. 1987. Strike-induced chemosensory searching in rattlesnakes: A rodent specialist (*Crotalis viridis*) differs from a lizard specialist (*Crotalus pricei*). Bull Psychon. Soc. 25: 136–138.

Cruz, N. S., and R. G. Alvarez. 1994. Rattlesnake bite complications in 19 children. Ped. Emerg. Care 10: 30–33.

Cruz-Sáenz, D. S. Guerrero, D. Lazcano, and J. Téllez-López. 2009. Notes on the herpetofauna of western Mexico 1: An update on the herpetofauna of the state of Jalisco, Mexico. Bull. Chicago Herpetol. Soc. 44: 105–113.

Cummings, J. A., and J. Kappel. 1966. Radioautographic studies of *Crotalus viridis*. J. Colorado—Wyoming Acad. Sci. 5: 91–92.

Cundall, D. 2002. Envenomation strategies, head form, and feeding ecology in vipers, pp. 149–162. *In* G. W. Schuett, M. Höggren, M. E. Douglas, and H. W. Greene (eds.), Biology of the vipers. Eagle Mountain Publ., Eagle Mountain, Utah.

Cundall, D., and S. J. Beaupre. 2001. Field records of predatory strike kinematics in timber rattlesnakes, *Crotalus horridus*. Amphibia-Reptilia 22: 492–498.

Cunningham, G. R., S. M. Hickey, and C. M. Gowen. 1996. *Crotalus viridis viridis* (prairie rattlesnake): Behavior. Herpetol. Rev. 27: 24.

Cunningham, J. D. 1955. Arboreal habits of certain reptiles and amphibians in southern California. Herpetologica 11: 217–220.

————. 1959. Reproduction and food of some California snakes. Herpetologica 15: 17–19.

————. 1966. Field observations on the thermal relations of rattlesnakes. Southwest. Nat. 11: 140–141.

Curran, C. H. 1935. Rattlesnakes. Nat. Hist. 36: 331–340.

Curran, C. H., and C. Kauffeld. 1937. Snakes and their ways. Harper Bros., New York.

Curtis, L. 1949. The snakes of Dallas County, Texas. Field & Lab. (Southern Methodist Univ.) 17: 1–13.

Czaplewski, N. J., J. I. Mead, C. J. Bell, W. D. Peachey, and T.-L. Ku. 1999. Papago Springs Cave revisited, Part II: Vertebrate paleofauna. Occ. Pap. Oklahoma Mus. Nat. Hist. 5: 1–41.

DaLie, D. A. 1953. Poisonous snakes of America. J. Forest. 51: 243–248.

Dalrymple, G. H., F. S. Bernardino, Jr., T. M. Steiner, and R. J. Nodell. 1991. Patterns of species diversity of snake community assemblages, with data on two Everglades snake assemblages. Copeia 1991: 517–521.

Dammann, A. E. 1961. Some factors affecting the distribution of sympatric species of rattlesnakes (genus *Crotalus*) in Arizona. Ph.D. diss., Univ. Michigan, Ann Arbor.

Darlington, P. J., Jr. 1957. Zoogeography: The geographical distribution of animals. John Wiley and Sons, New York.

Dart, R. C., and R. A. Gustafson. 1991. Failure of electric shock treatment for rattlesnake envenomation. Ann. Emerg. Med. 37: 181–188.

Dart, R. C., J. T. McNally, D. W. Spaite, and R. Gustafson. 1992. The sequelae of pitviper poisoning in the United States, pp. 395–404. *In* J. A. Campbell and E. D. Brodie, Jr. (eds.), Biology of the pitvipers. Selva, Tyler, Texas.

Datta, G., A. Dong, J .Witt, and A. T. Tu. 1995. Biochemical characterization of basilase, a fibrinolytic enzyme from *Crotalus basiliscus basiliscus*. Arch. Biochem. Biophys. 317: 365–373.

Davenport, J. W. 1943. Field book of snakes of Bexar County, Texas and vicinity. White Mem. Mus., San Antonio, Texas.

David, P., and I. Ineich. 1999. Les serpents venimeux du monde: Systématique et répartition. Dumerilia 3: 3–499.

Davidson, T. M. 1988. Intravenous rattlesnake envenomation. West. J. Med. 148: 45–47.

Davis, D. D. 1936. Courtship and mating behavior in snakes. Field Mus. Nat. Hist. Zool. Ser. 20: 257–290.

Davis, R. A. 1980. Vipers among us. Cincinnati Mus. Nat. Hist. Quart. 17(2): 8–12.

Davis, R. W. 2010. *Crotalus horridus* (timber rattlesnake): Habitat use. Herpetol. Rev. 41: 235.

Davis, W. B., and J. R. Dixon. 1957. Notes on Mexican snakes (Ophidia). Southwest. Nat. 2: 19–27.

Dean, J. N., J. L. Glenn, and R. C. Straight. 1980. Bilateral cleft labial and palate in the progeny of a *Crotalus viridis viridis* Rafinesque. Herpetol. Rev. 11: 91–92.

Deckert, R. F. 1918. A list of reptiles from Jacksonville, Florida. Copeia 1918(54): 30–33.

de Cock Bunning, T. 1983. Thermal sensitivity as a specialization for prey capture and feeding in snakes. Am. Zool. 23: 363–375.

———. 1984. A theoretical approach to the heat sensitive pit organs of snakes. J. Theor. Biol. 111: 509–529.

———. 1985. Qualitative and quantitative explanation of the forms of heat sensitive organs in snakes. Acta Biotheor. 34: 193–205.

Degenhardt, W. G., C. W. Painter, and A. H. Price. 1996. Amphibians and reptiles of New Mexico. Univ. New Mexico Press, Albuquerque.

De Graaf, R. M., and D. D. Rudis. 1981. Forest habitat for reptiles & amphibians of the Northeast. U.S.D.A. For. Serv., Washington, D.C.

———. 1983. Amphibians and reptiles of New England: Habitats and natural history. Univ. Massachusetts Press, Amherst.

———. 1986. New England wildlife: Habitat, natural history, and distribution. U.S.D.A. For. Serv. Gen. Tech. Rep. NE-108.

Delgadillo Espinosa, J., E. Godinez Cano, F. Correa Sanchez, and A. Gonzalez Ruiz. 1999. *Crotalus willardi silus* (ridge-nosed rattlesnake): Reproduction. Herpetol. Rev. 30: 168–169.

De Mesquita, L. C. M., H. S. Selistre, and J. R. Giglio. 1991. The hypotensive activity of *Crotalus atrox* (western diamondback rattlesnake) venom: Identification of its origin. Am. J. Trop. Med. Hyg. 44: 90–105.

de Roodt, A. R., S. Litwin, and S. O. Angel. 2003. Hydrolysis of DNA by 17 snake venoms. Comp. Biochem. Physiol. 135C: 469–479.

Dessauer, H. C. 1961. Blood and other body fluids. *In* D. S. Dittmer (ed.), Biological handbooks. Fed. Am. Soc. Exp. Biol., Washington, D.C.

———. 1970. Blood chemistry of reptiles: Physiological and evolutionary aspects, pp. 1–72. *In* C. Gans (ed.), Biology of the Reptilia, vol. 3. Academic Press, New York.

Deutsch, H. F., and C. R. Diniz. 1955. Some proteolytic activities of snake venoms. J. Biol. Chem. 216: 17–26.

De Vault, T. L., and A. R. Krochmal. 2002. Scavenging by snakes: An examination of the literature. Herpetologica 58: 429–436.

Dickerman, R. W., and C. W. Painter. 2001. *Crotalus lepidus lepidus* (mottled rock rattlesnake):
Diet. Herpetol. Rev. 32: 46.

Dickinson, W. E. 1949. Field guide to the lizards and snakes of Wisconsin. Milwaukee Publ.
Mus., Pop. Sci. Handb. Ser. 2: 1–70.

Dickman, J. D., J. S. Colton, D. Chiszar, and C. A. Colton. 1987. Trigeminal responses to thermal stimulation of the oral cavity in rattlesnakes (*Crotalus viridis*) before and after bilateral
anesthetization of the facial pit organs. Brain Res. 400: 365–370.

Diller, L. V. 1981. Comparative ecology of Great Basin rattlesnakes (*Crotalus viridis lutosus*) and
Great Basin gopher snakes (*Pituophis melanoleucus deserticola*) and their impact on small
mammal populations in the Snake River Birds of Prey Natural Area. Ph.D. diss., Univ. Idaho,
Moscow.

———. 1990. A field observation on the feeding behavior of *Crotalus viridis lutosus*. J. Herpetol.
24: 95–97.

Diller, L. V., and D. R. Johnson. 1988. Food habits, consumption rates, and predation rates of
western rattlesnakes and gopher snakes in southwestern Idaho. Herpetologica 44: 228–233.

Diller, L. V., and R. L. Wallace. 1984. Reproductive biology of the Northern Pacific rattlesnake
(*Crotalus viridis oreganus*) in northern Idaho. Herpetologica 40: 182–193.

———. 1996. Comparative ecology of two snake species (*Crotalus viridis* and *Pituophis melanoleucus*) in southwestern Idaho. Herpetologica 52: 343–360.

———. 2002. Growth, reproduction, and survival in a population of *Crotalus viridis oreganus* in
north central Idaho. Herpetol. Monogr. 16: 26–45.

Ditmars, R. L. 1923. Reptiles of the Southwest. Bull. New York Zool. Soc. 26(2): 22–30.

———. 1927. Occurrence and habits of our poisonous snakes. Bull. Antiv. Inst. Am. 1: 3–5.

———. 1931a. The reptile book. Doubleday Doran and Co., Garden City, New York.

———. 1931b. Snakes of the world. Macmillan, New York.

———. 1936. The reptiles of North America. Doubleday, Doran and Co., Garden City, New
York.

———. 1939. A field book of North American snakes. Doubleday, Doran and Co., New York.

Dixon, J. R. 1956. The mottled rock rattlesnake, *Crotalus lepidus lepidus*, in Edwards County,
Texas. Copeia 1956: 126–127.

Dixon, J. R., M. Sabbath, and R. Worthington. 1962. Comments on snakes from central and
western Mexico. Herpetologica 18: 91–100.

Dixon, J. R., C. A. Ketchersid, and C. S. Lieb. 1972. The herpetofauna of Querétaro, Mexico,
with remarks on taxonomic problems. Southwest. Nat. 16: 225–237.

do Amaral, A.. 1927. Studies of Nearctic ophidia: II. *Crotalus pricei* Van Denburgh, 1896, a synonym of *C. triseriatus* (Wagler, 1830). Bull. Antiv. Inst. Am. 1: 48–54.

———. 1928. Studies on snake venoms. I. Amounts of venom secreted by Nearctic pit vipers.
Bull. Antiv. Inst. Am. 1: 103–104.

———. 1929a. Studies of Nearctic Ophidia. III. Notes on *Crotalus tigris* Kennicott, 1859. Bull.
Antiv. Inst. Am. 2: 82–85.

———. 1929b. Studies of Nearctic Ophidia. IV. On *Crotalus tortugensis* Vandenburgh and Slevin,
1921, *Crotalus atrox elegans* Schmidt, 1922, and *Crotalus atrox lucasensis* (Vandenburgh, 1920).
Bull. Antiv. Inst. Am. 2: 85–86.

———. 1929c. Studies of Nearctic ophidia: V. On *Crotalus confluentus* Say, 1823, and its allied
forms. Bull. Antiv. Inst. Am. 2: 86–97.

———. 1929d. Studies of Nearctic Ophidia. VI. Phylogeny of the rattlesnakes. Bull. Antiv. Inst.
Am. 3: 6–8.

Dobie, J. F. 1965. Rattlesnakes. Little, Brown and Co., Boston.

Dodd, C. K., and G. G. Charest. 1988. The herpetofaunal community of temporary ponds in
North Florida sandhills: Species composition, temporal use, and management implications,
pp. 87–97. *In* R. C. Szaro, K. E. Severson, and D. R. Patton (eds.), Management of amphibians, reptiles, and small mammals in North America. U.S.D.A. For. Serv. Gen. Tech. Rep.
RM-166.

Dodge, N. N. 1938. Amphibians and reptiles of Grand Canyon National Park. Grand Canyon Nat. Hist. Assn. Bull. 9: 1–55.

Domínguez-Laso, M., U. Padilla-García, E. Pérez-Ramos, and A. Quijada-Mascareñas. 2007. *Crotalus scutulatus salvini* (Humantlan rattlesnake). Herpetol. Rev. 38: 485.

Dorcas, M. E. 1992. Relationships among montane populations of *Crotalus lepidus* and *Crotalus triseriatus*, pp. 71–87. *In* J. A. Campbell and E. D. Brodie, Jr. (eds.), Biology of the pitvipers. Selva, Tyler, Texas.

Dorcas, M. E., W. A. Hopkins, and J. H. Roe. 2004. Effects of body mass and temperature on standard metabolic rate in the eastern diamondback rattlesnake (*Crotalus adamanteus*). Copeia 2004: 145–151.

Dornburg, A., and R. E. Weaver. 2009. First report of scavenging by the northern Pacific rattlesnake (*Crotalus oreganus oreganus*) in the wild. Northwest. Nat. 90: 55–57.

Douglas, C. L. 1966. Amphibians and reptiles of Mesa Verde National Park, Colorado. Univ. Kansas Publ. Mus. Nat. Hist. 15: 711–744.

Douglas, M. E., M. R. Douglas, G. W. Schuett, L. W. Porras, and A. T. Holycross. 2002. Phylogeography of the western rattlesnake (*Crotalus viridis*) complex, with emphasis on the Colorado Plateau, pp. 11–50. *In* G. W. Schuett, M. Höggren, M. E. Douglas, and H. W. Greene (eds.), Biology of the vipers. Eagle Mountain Publ., Eagle Mountain, Utah.

Douglas, M. E., M. R. Douglas, G. W. Schuett, and L. W. Porras. 2006. Evolution of rattlesnakes (Viperidae; *Crotalus*) in the warm deserts of western North America shaped by Neogene vicariance and Quaternary climate change. Mol. Ecol. 15: 3353–3374.

Douglas, M. E., M. R. Douglas, G. W. Schuett, L. W. Porras, and B. L. Thomason. 2007. Genealogical concordance between mitochondrial and nuclear DNAs supports species recognition of the panamint rattlesnake (*Crotalus mitchellii stephensi*). Copeia 2007: 920–932.

Dowling, H. G. 1957. A review of the amphibians and reptiles of Arkansas. Occ. Pap. Univ. Arkansas Mus. 3: 1–51.

———. 1958. Pleistocene snakes of the Ozark Plateau. Am. Mus. Novit. 1882: 1–9.

———. 1959. Classification of the Serpentes: A critical review. Copeia 1959: 38–52.

———. 1975. Yearbook of herpetology. HISS, New York.

Dowling, H. G., and J. M. Savage. 1960. A guide to the snake hemipenis: A survey of basic structure and systematic characteristics. Zoologica (New York) 45: 17–28.

Drake, F. G. 1921. The fishing habits of snakes. Outdoor Life 48: 50.

Du, X.-Y., and K. J. Clemetson. 2002. Snake venom L-amino acid oxidases. Toxicon 40: 659–665.

Dubnoff, J. W., and F. E. Russell. 1970a. Isolation of lethal protein and peptide from *Crotalus viridis helleri* venom. Proc. West. Pharm. Soc. 13: 98.

———. 1970b. Separation and purification of *Crotalus* venom fractions. Second International Symposium Animal Toxins, Tel Aviv, February 1970. Toxicon 8: 130.

———. 1971. Separation and purification of *Crotalus* venom fractions, pp. 361–367. *In* A. De-Vries and E. Kochva (eds.), Toxins of animal and plant origin, vol. 1. Gordon and Breach, London.

Duellman, W. E. 1954. The amphibians and reptiles of Jorullo Volcano, Michoacán, México. Occ. Pap. Mus. Zool. Univ. Michigan 560: 1–24.

———. 1958. A preliminary analysis of the herpetofauna of Colima, Mexico. Occ. Pap. Mus. Zoology, Univ. Michigan 589: 1–22.

———. 1961. The amphibians and reptiles of Michoacán, México. Univ. Kansas Publ. Mus. Nat. Hist. 15: 1–148.

———. 1965. A biogeographic account of the herpetofauna of Michoacán, México. Univ. Kansas Publ. Mus. Nat. Hist. 15: 627–709.

Dugan, E. A., and W. K. Hayes. 2005. Comparative ecology of red diamond (*Crotalus ruber*) and southern Pacific (*Crotalus helleri*) rattlesnakes in southern California, p. 27. *In* Biology of the Rattlesnakes Symposium (program abstracts). Loma Linda Univ., Loma Linda, California.

Dugan, E. A., and M. A. Melanson. 2005. *Crotalus atrox* (western diamond-backed rattlesnake): Diet. Herpetol. Rev. 36: 322–323.

Dugan, E., A. Figueroa, and W. K. Hayes. 2004. Specialists vs. generalists: Variation in habitat selection between red diamond (*Crotalus ruber*) and southern Pacific (*Crotalus helleri*) rattlesnakes, p. 15. *In* Proceedings of the Snake Ecology Group 2004 Conference, Jackson County, Illinois.

Dugan, E. A., B. Love, and M. G. Figueroa. 2006. *Crotalus basiliscus* (Mexican west coast rattlesnake): Predation. Herpetol. Rev. 37: 231.

Dugan, E. A., A. Figueroa, and W. K. Hayes. 2008. Home range size, movements, and mating phenology of sympatric red diamond (*Crotalus ruber*) and southern Pacific (*C. oreganus helleri*) rattlesnakes in southern California, pp. 353–364. *In* W. K. Hayes, K. R. Beaman, M. D. Cardwell, and S. P. Bush (eds.), The biology of rattlesnakes. Loma Linda Univ. Press, Loma Linda, California.

Dullemeijer, P. 1959. A comparative functional-anatomical study of the heads of some Viperidae. Morphol. Jarhb. 99: 881–985.

———. 1961. Some remarks on the feeding behaviour of rattlesnakes. Proc. Kon. Nederl. Akad. Wetenschap. Ser. C. 64: 383–396.

———. 1969. Growth and size of the eye in viperid snakes. Netherlands J. Zool. (London) 19: 249–276.

Dullemeijer, P., and G. D. E. Povel. 1972. The construction for feeding in rattlesnakes. Zool. Meded. 47: 561–578.

Dundee, H. A. 1994. *Crotalus horridus* (timber rattlesnake): Coloration. Herpetol. Rev. 25: 28.

Dundee, H. A., and D. A. Rossman. 1989. The amphibians and reptiles of Louisiana. Louisiana State Univ. Press, Baton Rouge.

Dunkle, D. H., and H. M. Smith. 1937. Notes on some Mexican ophidians. Occ. Pap. Mus. Zool. Univ. Michigan (363): 1–15.

Dunn, E. R. 1915. List of reptiles and amphibians from Clark County, Virginia. Copeia 1915(25): 62–63.

Durham, F. E. 1956. Amphibians and reptiles of the North Rim, Grand Canyon, Arizona. Herpetologica 12: 220–224.

Duvall, D. 1986. Shake, rattle, and roll. Nat. Hist. 95: 66–73.

Duvall, D., and S. J. Beaupre. 1998. Sexual strategy and size dimorphism in rattlesnakes: Integrating proximate and ultimate causation. Am. Zool. 38: 152–165.

Duvall, D., and D. Chiszar. 1990. Behavioural and chemical ecology of vernal migration and pre- and post-strike predatory activity in prairie rattlesnakes: Field and laboratory experiments, pp. 539–554. *In* D. W. MacDonald, D. Müller-Schwarze, and S. E. Natynczuk (eds.), Chemical signals in vertebrates 5. Oxford Univ. Press, New York.

Duvall, D., and G. W. Schuett. 1997. Straight-line movement and competitive mate searching in prairie rattlesnakes, *Crotalus viridis viridis*. Anim. Behav. 54: 329–334.

Duvall, D., D. Chiszar, and J. Trupiano. 1978. Preference for envenomated rodent prey by rattlesnakes. Bull. Psychon. Soc. 11: 7–8.

Duvall, D., K. M. Scudder, and D. Chiszar. 1980. Rattlesnake predatory behavior: Mediation of prey discrimination and release of swallowing cues arising from envenomated mice. Anim. Behav. 28: 674–683.

Duvall, D., M. B. King, and R. Miller. 1983. Rattler. Wyoming Wildl. 47(10): 26–30.

Duvall, D., K. Gutzwiller, and M. King. 1985a. Reconstructing the rattlesnake. BBC Wildl. 3: 80–82.

Duvall, D., M. B. King, and K. J. Gutzwiller. 1985b. Behavioral ecology and ethology of the prairie rattlesnake. Natl. Geogr. Res. 1: 80–111.

Duvall, D., D. Chiszar, W. K. Hayes, J. K. Leonhardt, and M. J. Goode. 1990a. Chemical and behavioral ecology of foraging in prairie rattlesnakes (*Crotalus viridis viridis*). J. Chem. Ecol. 16: 87–101.

Duvall, D., M. J. Goode, W. K. Hayes, J. K. Leonhardt, and D. G. Brown. 1990b. Prairie rattlesnake vernal migration: Field experimental analyses and survival value. Natl. Geogr. Res. 6: 457–469.

Duvall, D., S. J. Arnold, and G. W. Schuett. 1992. Pitviper mating systems: Ecological potential, sexual selection and microevolution, pp. 321–336. *In* J. A. Campbell and E. D. Brodie, Jr. (eds.), Biology of the pitvipers. Selva, Tyler, Texas.

Dyrkacz, S. 1981. Recent instances of albinism in North American amphibians and reptiles. Soc. Stud. Amphib. Rept. Herpetol. Circ. 11: 1–31.

Eaton, T. H., Jr. 1935. Amphibians and reptiles of the Navaho country. Copeia 1935: 150–151.

Ebert, J., and G. Westhoff. 2006. Behavioural examination of the infrared sensitivity of rattlesnakes (*Crotalus atrox*). J. Comp. Physiol. 192A: 941–947.

Edgren, R. A., Jr. 1948. Notes on a litter of young timber rattlesnakes. Copeia 1948: 132.

Edmund, A. G. 1960. Tooth replacement phenomena in the lower vertebrates. Royal Ontario Mus. Life Sci. Contrib. 52: 1–190.

Edstrom, A. 1992. Venomous and poisonous animals. Krieger, Malabar, Florida.

Effron, M., L. Griner, and K. Bernirschke. 1977. Nature and rate of neoplasia in captive wild mammals, birds, and reptiles at necropsy. J. Natl. Canc. Inst. 59: 185–198.

Ehret, S., M. Goode, D. Prival, and M. Amarello. 2005. Ecology of banded rock rattlesnakes (*Crotalus lepidus klauberi*) in the Chiricahua mountains of southeastern Arizona, p. 27. *In* Biology of the Rattlesnakes Symposium (program abstracts). Loma Linda Univ., Loma Linda, California.

Ehrlich, S. P. 1928. A case report of severe snake-bite poisoning. Bull. Antiv. Inst. Am. 2: 65–66.

Elmore, F. H. 1954. Gopher snake vs. ground squirrel. Yellowstone Nature Notes 28: 35–36.

Emery, J. A., and F. E. Russell. 1963. Lethal and hemorrhagic properties of some North American snake venoms, pp. 409–413. *In* H. L. Keegan and W. V. MacFarlane (eds.), Venomous and poisonous animals and noxious plants of the Pacific region. Pergamon Press, New York.

Emmerson, F. H. 1982. Western diamondback rattlesnake in southern Nevada: A correction and comments. Great Basin Nat. 42: 350.

Enderson, E. F. 1999. Predation of *Crotalus molossus molossus* (black-tail rattlesnake) by *Masticophis bilineatus bilineatus* (Sonoran whipsnake), Whetstone Mountains, Arizona. Sonoran Herpetol. 12: 72–73.

Enge, K. M. 2005. Florida's commercial trade in rattlesnakes and possible conservation strategies, pp. 212–221. *In* W. E. Meshaka, Jr., and K. J. Babbitt (eds.), Amphibians and reptiles: Status and conservation in Florida. Krieger, Malabar, Florida.

Enge, K. M., and K. N. Wood. 2002. A pedestrian road survey of an upland snake community in Florida. Southeast. Nat. 1: 365–380.

Engelhardt, G. P. 1932. Notes on poisonous snakes in Texas. Copeia 1932: 37–38.

Engelman, W., and F. J. Obst. 1981. Snakes: Biology, behavior and relationships to man. Exeter Books, New York.

Enzeroth, R., B. Chobotar, and E. Schlotseck. 1985. *Sarcocystis crotali*—new species with the Mojave rattlesnake (*Crotalus scutulatus scutulatus*) and mouse (*Mus musculus*) cycle. Arch. Protist. 129: 19–23.

Ernst, C. H. 1962. The comparative fang lengths of Nearctic snakes of the genus *Agkistrodon*. Master's thesis, West Chester Univ., West Chester, Pennsylvania.

———. 1982a. A study of the fangs of Russell's viper (*Vipera russellii*). J. Herpetol. 16: 67–71.

———. 1982b. A study of fangs of snakes belonging to the *Agkistrodon*-complex. J. Herpetol. 16: 72–80.

———. 1992. Venomous reptiles of North America. Smithsonian Inst. Press, Washington, D.C.

Ernst, C. H., and R. W. Barbour. 1989. Snakes of eastern North America. George Mason Univ. Press, Fairfax, Virginia.

Ernst, C. H., and E. M. Ernst. 2003. Snakes of the United States and Canada. Smithsonian Inst. Press, Washington, D.C.

———. 2006. Synopsis of helminths endoparasitic in snakes of the United States and Canada. Soc. Stud. Amphib. Rept. Herpetol. Circ. 34: 1–86.

Ernst, C. H., and G. R. Zug. 1996. Snakes in question: The Smithsonian answer book. Smithsonian Inst. Press, Washington, D.C.

Ernst, C. H., S. C. Belfit, S. W. Sekscienski, and A. F. Laemmerzahl. 1997. The amphibians and reptiles of Ft. Belvoir and Northern Virginia. Bull. Maryland Herpetol. Soc. 33: 1–62.

Essex, H. E., and J. Markowitz. 1930. The physiological action of rattlesnake venom (crotalin),

I. Effect on blood pressure: Symptoms and post-mortem observations. Am. J. Physiol. 92: 317–328.

Estep, K., T. Poole, C. W. Radcliffe, B. O'Connell, and D. Chiszar. 1981. Distance traveled by mice after envenomation by a rattlesnake (*C. viridis*). Bull. Psychon. Soc. 18: 108–110.

Estes, E. T. 1958. Timber rattlesnake in Jefferson County, Illinois. Herpetologica 14: 68.

Evans, F. C., and R. Holdenried. 1943. A population study of the Beechey ground squirrel in central California. J. Mamm. 24(2): 231–260.

Evermann, B. W. 1915. Do snakes swallow small mammals heads or tails first? Copeia 1915(14): 1–2.

Ewan, J. 1932. Pacific rattlesnake at high altitude on San Jacinto Peak, California. Copeia 1932: 36.

Fairley, N. H. 1929. The present position of snake bite and the snake bitten in Australia. Bull. Antiv. Inst. Am. 3: 65–76.

Falck, H. S. 1940. Food of the eastern rock rattlesnake in captivity. Copeia 1940: 135.

Fantham, H. B., and A. Porter. 1954. The endoparasites of some North American snakes and their effects on the Ophidia. Proc. Zool. Soc. London 123(4): 867–898.

Farstad, D., T. Thomas, T. Chow, S. Bush, and P. Stiegler. 1997. Mojave rattlesnake envenomation in southern California: A review of suspected cases. Wild. Environ. Med. 8: 89–93.

Faure, G. 1999. Les phospholipases A2 des venins de serpents. Bull. Soc. Zool. France 124: 149–168.

Fautin, R. W. 1946. Biotic communities of the northern desert shrub biome in western Utah. Ecol. Monogr. 16(4): 251–310.

Fenton, M. B., and L. E. Licht. 1990. Why rattle snake? J. Herpetol. 24: 274–279.

Ferguson, J. H., and R. M. Thornton. 1984. Oxygen storage capacity and tolerance of submergence of a non-aquatic reptile and an aquatic reptile. Comp. Biochem. Physiol. 77A: 183–187.

Ferrel, C. M., H. R. Leach, and D. F. Tillotson. 1953. Food habits of the coyote in California. California Fish Game 39: 301–341.

Fidler, H. K., R. D. Glasgow, and E. B. Carmichael. 1938. Pathologic changes produced by subcutaneous injection of rattlesnake (*Crotalus*) venom into *Macaca mulatta* monkeys. Proc. Soc. Exp. Biol. Med. 38: 892–894.

Fiero, M. K., M. W. Seifert, T. J. Weaver, and C. A. Bonilla. 1972. Comparative study of juvenile and adult prairie rattlesnake (*Crotalus viridis viridis*) venoms. Toxicon 10: 81–82.

Figueroa, A., E. A. Dugan, and W. K. Hayes. 2008. Behavioral ecology of neonate southern Pacific rattlesnakes (*Crotalus oreganus helleri*) tracked with externally-attached transmitters, pp. 365–376. *In* W. K. Hayes, K. R. Beaman, M. D. Cardwell, and S. P. Bush (eds.), The biology of rattlesnakes. Loma Linda Univ. Press, Loma Linda, California.

Fishbeck, D. W., and J. C. Underhill. 1959. A check list of the amphibians and reptiles of South Dakota. Proc. South Dakota Acad. Sci. 38: 107–113.

Fitch, H. S. 1949. Study of snake populations in central California. Am. Midl. Nat. 41: 513–579.

———. 1960. Criteria for determining sex and breeding maturity in snakes. Herpetologica 16: 49–51.

———. 1970. Reproductive cycles in lizards and snakes. Univ. Kansas Mus. Nat. Hist. Misc. Publ. 52: 1–247.

———. 1981. Sexual size differences in reptiles. Univ. Kansas Mus. Nat. Hist. Misc. Publ. 70: 1–72.

———. 1982. Resources of a snake community in prairie-woodland habitat of northeastern Kansas, pp. 93–97. *In* N. J. Scott, Jr. (ed.), Herpetological communities. U.S. Fish Wildl. Serv. Wildl. Res. Rep. 13.

———. 1985a. Variation in clutch and litter size in New World reptiles. Univ. Kansas Mus. Nat. Hist. Misc. Publ. 76: 1–76.

———. 1985b. Observation on rattle size and demography of prairie rattlesnakes (*Crotalus viridis*) and timber rattlesnakes (*Crotalus horridus*) in Kansas. Occ. Pap. Mus. Nat. Hist. Univ. Kansas 118: 1–11.

———. 1992. Methods of sampling snake populations and their relative success. Herpetol. Rev. 23: 17–19.

————. 1993. Relative abundance of snakes in Kansas. Trans. Kansas Acad. Sci. 96: 213–224.

————. 1998. The Sharon Springs Roundup and prairie rattlesnake demography. Trans. Kansas Acad. Sci. 101: 101–113.

————. 1999. A Kansas snake community: Composition and changes over 50 years. Krieger, Malabar, Florida.

————. 2000. Population structure and biomass of some common snakes in central North America. Univ. Kansas Mus. Nat. Hist. Sci. Pap. 17: 1–7.

————. 2002. A comparison of growth and rattle strings in three species of rattlesnakes. Univ. Kansas Mus. Nat. Hist. Sci. Pap. 24: 1–6.

Fitch, H. S., and A. F. Echelle. 2006. Abundance and biomass of twelve species of snakes native to northeastern Kansas. Herpetol. Rev. 37: 161–165.

Fitch, H. S., and B. Glading. 1947. A field study of a rattlesnake population. California Fish Game 33: 103–123.

Fitch, H. S., and G. R. Pisani. 1993. Life history traits of the western diamondback rattlesnake (*Crotalus atrox*) studied from roundup samples in Oklahoma. Occ. Pap. Mus. Nat. Hist. Univ. Kansas 156: 1–24.

————. 2002. Longtime recapture of a timber rattlesnake (*Crotalus horridus*) in Kansas. J. Kansas Herpetol. 2: 15–16.

————. 2006. The timber rattlesnake in northeastern Kansas. J. Kansas Herpetol. 19: 11–15.

Fitch, H. S., and H. Twining. 1946. Feeding habits of the Pacific rattlesnake. Copeia 1946: 64–71.

Fitch, H. S., G. R. Pisani, H. W. Greene, A. F. Echelle, and M. Zerwekh. 2004. A field study of the timber rattlesnake in Leavenworth County, Kansas. J. Kansas Herpetol. 11: 18–24.

Fitzgerald, L. A., and C. W. Painter. 2000. Rattlesnake commercialization: Long-term trends, issues, and implications for conservation. Wildl. Soc. Bull. 28: 235–253.

Flores-Barroeta, L., E. Hidalgo-Esculante, and F. Montero-Gei. 1961. Cestodes de vertebrados: vol. VIII. Rev. Biol. Trop. 9: 1887–1207.

Fogell, D. D. 2005. Status of the prairie rattlesnake (*Crotalus viridis*) in the Loess Hills of northwest Iowa, pp. 28–29. *In* Biology of the rattlesnakes symposium (program abstracts). Loma Linda Univ., Loma Linda, California.

Fogell, D. D., T. J. Leonard, and J. D. Fawcett. 2002a. *Crotalus horridus horridus* (timber rattlesnake): Habitat. Herpetol. Rev. 33: 211–212.

————. 2002b. *Crotalus horridus horridus* (timber rattlesnake): Climbing. Herpetol. Rev. 33: 212.

Follett, W. I. 1927. A California badger. California Fish Game 13: 220.

Foote, R., and J. A. MacMahon. 1977. Electrophoretic studies on rattlesnake (*Crotalus* and *Sistrurus*) venom: Taxonomic implications. J. Biochem. Physiol. 57B: 235–241.

Forbes, J. E. 1967. Respiratory activity of vibratory, tail epaxial and mid-body epaxial muscle in *Crotalus horridus*, *Agkistrodon contortrix*, and *Thamnophis sirtalis*. Master's thesis, Univ. Richmond, Richmond, Virginia.

Ford, N. B., and G. M. Burghardt. 1993. Perceptual mechanisms and the behavioral ecology of snakes, pp. 117–164. *In* R. A. Seigel and J. T. Collins (eds.), Snakes: Ecology and evolutionary biology. McGraw-Hill, New York.

Forks, J. E., and T. M. Hughes. 2007. *Crotalus molossus molossus* (northern black-tailed rattlesnake): Diet. Herpetol. Rev. 38: 205.

Forrester, D. J., R. M. Shealy, and S. H. Best. 1970. *Porocephalus crotali* (Pentastomida) in South Carolina. J. Parasitol. 56: 977.

Forstner, M. R. J., R. A. Hilsenbeck, and J. F. Scudday. 1997. Geographic variation in whole venom profiles from the mottled rock rattlesnake (*Crotalus lepidus lepidus*) in Texas. J. Herpetol. 31: 277–287.

Foster, C. D., S. Klueh, and S. J. Mullin. 2006. Extirpation of a relict timber rattlesnake (*Crotalus horridus*) population in Clark County, Illinois. Bull. Chicago Herpetol. Soc. 41: 147–148.

Foster, S., and R. A. Caras. 1994. A field guide to venomous animals and poisonous plants: North America north of Mexico. Houghton Mifflin, Boston.

Fouquette, M. J., Jr., and H. L. Lindsay, Jr. 1955. An ecological survey of reptiles in parts of northwestern Texas. Texas J. Sci. 7: 402–421.

Fouquette, M. J., Jr., and D. A. Rossman. 1963. Noteworthy records of Mexican amphibians and reptiles in the Florida State Museum and the Texas National History Collection. Herpetologica 19: 185–201.

Fowlie, J. A. 1965. The snakes of Arizona. Azul Quinta, Fallbrook, California.

Fox, J., and B. Hamilton. 2007. *Crotalus horridus* (timber rattlesnake): Behavior. Herpetol. Rev. 38: 86.

Fox, J. W., M. Elzinga, and A. T. Tu. 1979. Amino acid sequence and disulfide bond assignment of myotoxin-a isolated from the venom of prairie rattlesnake (*Crotalus virdis viridis*). Biochemistry 18: 678–684.

Freda, J. 1977. Fighting a losing battle. The story of a timber rattlesnake. HERP: Bull. New York Herpetol. Soc. 13: 35–38.

French, W. J., W. K. Hayes, S. P. Bush, M. D. Cardwell, J. O. Bader, and E. D. Rael. 2004. Mojave toxin in venom of *Crotalus helleri* (southern Pacific rattlesnake): Molecular and geographic characterization. Toxicon 44: 781–791.

Frey, J. K. 1996. *Crotalus lepidus* (rock rattlesnake). Aquatic behavior. Herpetol. Rev. 27: 145.

Friederich, C., and A. T. Tu. 1971. Role of metals in snake venoms for hemorrhagic, esterase and proteolytic activities. Biochem. Pharmacol. 20: 1549–1556.

Froom, B. 1972. The snakes of Canada. McClelland and Stewart, Toronto.

Fugler, C. M., and J. R. Dixon. 1961. Notes on the herpetofauna of the El Dorado area of Sinaloa, Mexico. Publ. Mus. Michigan St. Univ. Biol. Ser. 2: 1–24.

Fugler, C. M., and R. G. Webb. 1956. Distributional notes on some reptiles and amphibians from southern and central Coahuila. Herpetologica 12: 167–171.

Fujisawa, D., Y. Yamazaki, B. Lomonte, and T. Morita. 2008. Catalytically inactive phospholipase A(2) homologue binds to vascular endothelial growth factor receptor-2 via a C-terminal loop region. Biochem. J. 411: 515–522.

Funderburg, J. B. 1968. Eastern diamondback rattlesnake feeding on carrion. J. Herpetol. 2: 161–162.

Funderburg, J. B., and D. S. Lee. 1968. The amphibian and reptile fauna of pocket gopher (*Geomys*) mounds in central Florida. J. Herpetol. 1: 99–100.

Funk, R. S. 1964. On the food of *Crotalus m. molossus*. Herpetologica 20: 134.

———. 1965. Food of *Crotalus cerastes laterorepens* in Yuma County, Arizona. Herpetologica 21: 15–17.

Furlow, T. G., and L. V. Brennan. 1985. Purpura following timber rattlesnake (*Crotalus horridus horridus*) envenomation. Cutis 35: 234–236.

Furman, J. 2007. Timber rattlesnakes in Vermont & New York: Biology, history, and the fate of an endangered species. Univ. Press New England, Lebanon, New Hampshire.

Furry, K., T. Swain, and D. Chiszar. 1991. Strike-induced chemosensory searching and trail following by prairie rattlesnakes (*Crotalus viridis*) preying upon deer mice (*Peromyscus maniculatus*): Chemical discrimination among individual mice. Herpetologica 47: 69–78.

Gaffin, R. D., J. B. Scales, and R. L. Cate. 1995. Kallikrein-like enzyme from the venom of *Crotalus basiliscus basiliscus* (Serpentes: Crotalidae). Texas J. Sci. 47: 53–61.

Galán, J. A., E. E. Sánchez, A. Rodríguez-Acosta, and J. C. Pérez. 2004. Neutralization of venoms from two southern Pacific rattlesnakes (*Crotalus helleri*) with commercial antivenoms and endothermic animal sera. Toxicon 43: 791–799.

Galán, J. A., M. Guo, E. E. Sánchez, E. Cantu, A. Rodríguez-Acosta, J. C. Pérez, and W. A. Tao. 2008. Quantitative analysis of snake venoms using soluble polymer-based isotope labeling. Mol. Cell. Proteomics 7: 785–799.

Galligan, J. H., and W. A. Dunson. 1979. Biology and status of timber rattlesnake (*Crotalus horridus*) populations in Pennsylvania. Biol. Conserv. 15: 13–58.

Gannon, V. 1978. Factors limiting the distribution of the prairie rattlesnake. Blue Jay 36: 142–144.

Gannon, V., and D. M. Secoy. 1984. Growth and reproductive rates of a northern population of the prairie rattlesnake, *Crotalus v. viridis*. J. Herpetol. 18: 13–19.

———. 1985. Seasonal and daily activity patterns in a Canadian population of the prairie rattlesnake, *Crotalus viridis viridis*. Can. J. Zool. 63: 86–91.

Gardner, S. C., and E. Oberdörster. 2006. Toxicology of Reptiles. CRC Press, Boca Raton, Florida.

Gardner-Santana, L. C., and S. J. Beaupre. 2009. Timber rattlesnakes (*Crotalus horridus*) exhibit elevated and less variable body temperatures during pregnancy. Copeia 2009: 363–368.

Garman, S. W. 1883 [1884]. North American Reptilia. Part I: Ophidia. Mem. Mus. Comp. Zool. 8(3): 1–185.

Garrett, J. M., and D. G. Barker. 1987. A field guide to reptiles and amphibians of Texas. Texas Monthly Press, Austin.

Gartner, G. E. A., and R. S. Reiserer. 2003. *Crotalus mitchellii* (speckled rattlesnake): Mating. Herpetol. Rev. 34: 65.

Gates, G. O. 1957. A study of the herpetofauna in the vicinity of Wickenburg, Maricopa County, Arizona. Trans. Kansas Acad. Sci. 60: 403–418.

Gawade, S. P. 2004. Snake venom neurotoxins: Pharmacological classification. J. Toxicol. Toxin Rev. 23: 37–96.

Gehlbach, F. R. 1956. Annotated records of southwestern amphibians and reptiles. Trans. Kansas Acad. Sci. 59: 364–372.

———. 1965. Herpetology of the Zuni Mountains Region, northwestern New Mexico. Proc. U.S. Natl. Mus. 116: 243–332.

Geluso, K. 2007. *Crotalus lepidus lepidus* (mottled rock rattlesnake): Diet. Herpetol. Rev. 38: 86–87.

Genter, D. L. 1984. *Crotalus viridis* (prairie rattlesnake). Food. Herpetol. Rev. 15: 49–50.

Gerkin, R., K. C. Sergent, S. C. Curry, M. Vance, and D. R. Nielsen. 1987. Life-threatening airway obstruction from rattlesnake bite to the tongue. Ann. Emerg. Med. 16: 813–816.

Germano, D. J., and J. L. Brown. 2003. *Gambelia sila* (blunt-nosed leopard lizard): Predation. Herpetol. Rev. 32: 143–144.

Ghoniem, N., and M. Refai. 1969. Red diamond back rattlesnake as a reservoir of *Salmonella aqua* (30: Kil, 6). Vet Med. J. Giza 16: 179–181.

Gibbons, J. W. 1972. Reproduction, growth, and sexual dimorphism in the canebrake rattlesnake (*Crotalus horridus atricaudatus*). Copeia 1972: 222–226.

———. 1977. Snakes of the Savannah River Plant with information about snakebite prevention and treatment. ERDA's Savannah River Nat. Environ. Res. Park. SRO-NERP-1.

Gibbons, J. W., and M. Dorcas. 2005. Snakes of the Southeast. Univ. Georgia Press, Athens.

Gibbons, J. W., R. R. Haynes, and J. L. Thomas. 1990. Poisonous plants and venomous animals of Alabama and adjoining states. Univ. Alabama Press, Tuscaloosa.

Gibbs, H. L., and J. Diaz. 2010. Molecular diagnostics and DNA taxonomy: Identification of single copy nuclear DNA markers for North American pit vipers. Mol. Ecol. Resour. 10: 177–180.

Gibbs, J. P., A. R. Breisch, P. K. Ducey, G. Johnson, J. Behler, and R. Bothner. 2007. The amphibians and reptiles of New York state. Oxford Univ. Press, New York.

Gibson, J. D. 2002. Herpetofaunal survey of Sherando Lake Recreation Area, Loves Run Pond Complex, Green Pond, and Humpback Rocks. Catesbeiana 22: 3–13.

Gibson, S. E., Z. J. Walker, and B. A. Kingsbury. 2008. Microhabitat preferences of the timber rattlesnake (*Crotalus horridus*) in the hardwood forests of Indiana, pp. 275–286. *In* W. K. Hayes, K. R. Beaman, M. D. Cardwell, and S. P. Bush (eds.), The biology of rattlesnakes. Loma Linda Univ. Press, Loma Linda, California.

Gier, P. J., R. L. Wallace, and R. L. Ingermann. 1989. Influence of pregnancy on behavioral thermoregulation in the northern Pacific rattlesnake *Crotalus viridis oreganus*. J. Exp. Biol. 145: 465–469.

Gillingham, J. C., and R. E. Baker. 1981. Evidence for scavenging behavior in the western diamondback rattlesnake (*Crotalus atrox*). Z. Tierpsychol. 55: 217–227.

Gillingham, J. C., and D. L. Clark. 1981. An analysis of prey-searching behavior in the western diamondback rattlesnake, *Crotalus atrox*. Behav. Neurol. Biol. 32: 235–240.

Gillingham, J. C., C. C. Carpenter, and J. B. Murphy. 1983. Courtship, male combat and dominance in the western rattlesnake, *Crotalus atrox*. J. Herpetol. 17: 265–270.

Gilmore, C. W. 1938. Fossil snakes of North America. Geol. Soc. Am. Spec. Pap. 9: 1–93.

Gilmore, R. M. 1934. Rattlesnake and Cony. Yosemite Nat. Notes 13(9): 70.

Gingrich, W. C., and J. C. Hohenadel. 1956. Standardization of polyvalent antivenin, pp. 381–385. *In* E. E. Buckley and N. Porges (eds.), Venoms. Am. Assoc. Advan. Sci., Washington, D.C.

Giorgi, R., M. M. Bernardi, and Y. Cury. 1993. Analgesic effect evoked by low molecular weight substances extracted from *Crotalus durissus terrificus* venom. Toxicon 31: 1257–1266.

Giovanni, M. D., C. A. Taylor, and G. Perry. 2005. *Crotalus viridis viridis* (prairie rattlesnake): Diet. Herpetol. Rev. 36: 323.

Githens, T. S. 1931. Antivenin: Its preparation and standardization. Bull. Antiv. Inst. Am. 4: 81–85.

———. 1933. Unpublished report: Data of Mulford Biological Laboratories, no pagination.

Githens, T. S., and L. W. Butz. 1929. Venoms of North American snakes and their relationship. Bull. Antiv. Inst. Am. 2: 100–104.

Githens, T. S., and I. D. George. 1931. Comparative studies on the venoms of certain rattle-snakes. Bull. Antiv. Inst. Am. 5: 31–34.

Githens, T. S., and N. O. Wolff. 1939. The polyvalency of crotalidic antivenins. J. Immunol. 37: 33–51.

Glading, B. 1938. Studies of the nesting cycle of the California valley quail in 1937. California Fish Game 24: 318–340.

Glaser, H. S. R. 1948. Bactericidal activity of *Crotalus* venom *in vitro*. Copeia 1948: 245–247.

Glass, T. G., Jr. 1969. Cortisone and immediate fasciotomy in the treatment of severe rattle-snake bite. Texas Med. 69(7): 40–47.

Glaudas, X. 2009. Rain-harvesting by the southwestern speckled rattlesnake (*Crotalus mitchellii pyrrhus*). Southwest. Natur. 54: 518–521.

Glenn, J. L., and H. E. Lawler. 1987. *Crotalus scutulatus salvini* (Huamantlan rattlesnake): Behavior. Herpetol. Rev. 18: 15–16.

Glenn, J. L., and R. C. Straight. 1977. The midget faded rattlesnake (*Crotalus viridis concolor*) venom: Lethal toxicity and individual variability. Toxicon 15: 129–133.

———. 1978. Mojave rattlesnake *Crotalus scutulatus scutulatus* venom: Variation in toxicity with geographical origin. Toxicon 16: 81–84.

———. 1982. The rattlesnakes and their venom yield and lethal toxicity, pp. 3–119. *In* A. T. Tu (ed.), Rattlesnake venoms: Their actions and treatment. Marcel Dekker, New York.

———. 1985a. Distribution of proteins immunologically similar to Mojave toxin among species of *Crotalus* and *Sistrurus*. Toxicon 23: 28.

———. 1985b. Venom properties of the rattlesnakes (*Crotalus*) inhabiting the Baja California region of Mexico. Toxicon 23: 769–775.

———. 1987. Variations in the venom of *Crotalus lepidus klauberi*. Toxicon 25: 142.

———. 1989. Intergradation of two different venom populations of the Mojave rattlesnake (*Crotalus scutulatus scutulatus*) in Arizona. Toxicon 27: 411–418.

———. 1990. Venom characteristics as an indicator of hybridization between *Crotalus viridis viridis* and *Crotalus scutulatus scutulatus* in New Mexico. Toxicon 28: 857–862.

Glenn, J. L., R. C. Straight, and C. C. Snyder. 1972. Yield of venom obtained from *Crotalus atrox* by electrical stimulation. Toxicon 10: 575–579.

Glenn, J. L., R. C. Straight, M. C. Wolfe, and D. L. Hardy. 1983. Geographical variation in *Crotalus scutulatus scutulatus* (Mojave rattlesnake) venom properties. Toxicon 21: 119–130.

Glenn, J. L., R. C. Straight, and T. B. Wolt. 1994. Regional variation in the presence of cane-brake toxin in *Crotalus horridus* venom. Comp. Biochem. Physiol. 107C: 337–346.

Glenn, W. G., W. G. Malette, J. B. Fitzgerald, A. T. K. Cockett, and T. G. Glass, Jr. 1963. Some characteristics of rattlesnake (*Crotalus atrox*) venom and electrophoretically separated components. Tex. Rep. Biol. Med. 21: 188.

Glissmeyer, H. R. 1951. Egg production of the Great Basin rattlesnake. Herpetologica 7: 24–27.

Gloyd, H. K. 1928. The amphibians and reptiles of Franklin County, Kansas. Trans. Kansas Acad. Sci. 31: 115–141.

———. 1933. An unusual feeding record for the prairie rattlesnake. Copeia 1933: 98.

———. 1935a. The cane-brake rattlesnake. Copeia 1935: 175–178.

———. 1935b. Some aberrant color patterns in snakes. Pap. Michigan Acad. Sci. Arts Lett. 20: 661–668.

———. 1936a. A Mexican subspecies of *Crotalus molossus* Baird and Girard. Occ. Pap. Mus. Zool. Univ. Michigan 325: 1–5.

————. 1936b. The subspecies of *Crotalus lepidus*. Occ. Pap. Mus. Zool. Univ. Michigan 337: 1–5.

————. 1937. A herpetological consideration of faunal areas in southern Arizona. Bull. Chicago Acad. Sci. 5: 79–136.

————. 1940. The rattlesnakes, genera *Sistrurus* and *Crotalus*. Chicago Acad. Sci. Spec. Publ. 4: 1–266.

————. 1947. Notes on the courtship and mating behavior of certain snakes. Chicago Acad. Sci. Nat. Hist. Misc. 12: 1–4.

————. 1948. Description of a neglected subspecies of rattlesnake from Mexico. Chicago Acad. Sci. Nat. Hist. Misc. 17: 1–4.

————. 1958. Aberrations in the color patterns of some crotalid snakes. Bull. Chicago Acad. Sci. 10: 185–195.

Gloyd, H. K., and C. F. Kauffeld. 1940. A new rattlesnake from Mexico. Bull. Chicago Acad. Sci. 6: 11–14.

Gloyd, H. K., and H. M. Smith. 1942. Amphibians and reptiles from the Carmen Mountains, Coahuila. Bull. Chicago Acad. Sci. 6: 231–235.

Goin, C. J., O. B. Goin, and G. R. Zug. 1978. Introduction to herpetology. 3rd ed. Freeman, San Francisco.

Golan, L., C. Radcliffe, T. Miller, B. O'Connell, and D. Chiszar. 1982. Trailing behavior in prairie rattlesnakes (*Crotalus viridis*). J. Herpetol. 16: 287–293.

Gold, B. S., and W. A. Wingert. 1994. Snake venom poisoning in the United States: A review of therapeutic practice. South. Med. J. 87: 579–589.

Goldberg, C. S., T. Edwards, M. E. Kaplan, and M. Goode. 2003. PCR primers for microsatellite loci in the tiger rattlesnake (*Crotalus tigris*, Viperidae). Mol. Ecol. Notes 3: 539–541.

Goldberg, S. R. 1999a. Reproduction in the tiger rattlesnake, *Crotalus tigris* (Serpentes: Viperidae). Texas J. Sci. 51: 31–36.

————. 1999b. Reproduction in the red diamond rattlesnake in California. California Fish Game 85: 177–180.

————. 1999c. Reproduction in the blacktail rattlesnake, *Crotalus molossus* (Serpentes: Viperidae). Texas J. Sci. 51: 323–328.

————. 2000a. Reproduction in the rock rattlesnake, *Crotalus lepidus* (Serpentes: Viperidae). Herpetol. Nat. Hist. 7: 83–86.

————. 2000b. Reproduction in the twin-spotted rattlesnake, *Crotalus pricei* (Serpentes: Viperidae). West. N. Am. Nat. 60: 98–100.

————. 2000c. Reproduction in the speckled rattlesnake, *Crotalus mitchellii* (Serpentes: Viperidae). Bull. So. California Acad. Sci. 99: 101–104.

————. 2004. Reproductive cycle of the sidewinder, *Crotalus cerastes* (Serpentes: Viperidae), from California. Texas J. Sci. 56: 55–62.

————. 2007. Testicular cycle of the western diamondback rattlesnake, *Crotalus atrox* (Serpentes: Viperidae), from Arizona. Bull. Maryland Herpetol. Soc. 43: 103–107.

Goldberg, S. R., and K. R. Beaman. 2003a. *Crotalus catalinensis* (Santa Catalina Island rattleless rattlesnake): Reproduction. Herpetol. Rev. 34: 249–250.

————. 2003b. Reproduction in the Baja California rattlesnake, *Crotalus enyo* (Serpentes: Viperidae). Bull. So. California Acad. Sci. 102: 39–42.

Goldberg, S. R., and C. R. Bursey. 1999a. *Crotalus lepidus* (rock rattlesnake), *Crotalus molossus* (blacktail rattlesnake), *Crotalus pricei* (twin-spotted rattlesnake), *Crotalus tigris* (tiger rattlesnake): Endoparasites. Herpetol. Rev. 30: 44–45.

————. 1999b. First reported occurrence of *Ophidascaris labiatopapillosa* (Nematoda: Ascarididae) in the red diamond rattlesnake. California Fish Game 85: 181–182.

————. 2000. *Crotalus mitchellii* (speckled rattlesnake) and *Crotalus willardi* (ridgenose rattlesnake): Endoparasites. Herpetol. Rev. 31: 104.

————. 2002. *Crotalus cerastes* (sidewinder): Endoparasites. Herpetol. Rev. 33: 138.

————. 2004. *Crotalus cerberus* (Arizona black rattlesnake): Endoparasites. Herpetol. Rev. 35: 400.

Goldberg, S. R., and P. C. Rosen. 2000. Reproduction in the Mojave rattlesnake, *Crotalus scutulatus* (Serpentes: Viperidae). Texas J. Sci. 52: 101–109.

Goldberg, S. R., C. R. Bursey, and A. T. Holycross. 2002a. *Abbreviata terrapenis* (Nematoda: Physalopteridae): An accidental parasite of the banded rock rattlesnake (*Crotalus lepidus klauberi*). J. Wildl. Dis. 38: 453–456.

Goldberg, S. R., C. R. Bursey, and C. W. Painter. 2002b. Helminths of the western diamondback rattlesnake, *Crotalus atrox*, from southeast New Mexico rattlesnake roundups. Southwest. Nat. 47: 307–310.

Goldberg, S. R., C. R. Bursey, and K. R. Beaman. 2003a. First reported occurrence of *Porocephalus crotali* (Pentastomida: Porocephalida) in the Santa Catalina Island rattleless rattlesnake. California Fish Game 89: 51–53.

———. 2003b. *Crotalus enyo* (Baja California rattlesnake): Endoparasites. Herpetol. Rev. 34: 64–65.

Goldberg, S. R., K. R. Beaman, and E. A. Dugan. 2005. Notes on reproduction in the Mexican west coast rattlesnake, *Crotalus basiliscus* (Serpentes: Viperidae). Texas J. Sci. 57: 197–201.

Goldberg, S. R., C. R. Bursey, K. R. Beaman, and E. A. Dugan. 2006. *Crotalus basiliscus* (Mexican west coast rattlesnake): Endoparasites. Herpetol. Rev. 37: 94.

Gomis-Rüth, L. F., J. Kress, J. Kellermann, I. Mayr, X. Lee, R. Huber, and W. Brode. 1994. Refined 2·0 Å x-ray crystal structure of the snake venom zinc-endopeptidase adamalysin II: Primary and tertiary structure determination, refinement, molecular structure and comparison with astacin, collagenase and thermolysin. J. Mol. Biol. 239: 513–544.

González-Romero, A., and S. Alvarez-Cardenas. 1989. Herpetofauna de la region del Pinacate, Sonora, Mexico: Un inventario. Southwest. Nat. 34: 519–526.

Goode, M. J., and D. Duvall. 1989. Body temperature and defensive behaviour of free-ranging prairie rattlesnakes, *Crotalus viridis viridis*. Anim. Behav. 38: 360–362.

Goode, M., and J. F. Smith. 2005. Calm before the storm: Tiger rattlesnakes and urban development, pp. 29–30. *In* Biology of the Rattlesnakes Symposium (program abstracts). Loma Linda Univ., Loma Linda, California.

Goode, M., J. J. Smith, and M. Amarello. 2008. Seasonal and annual variation in home range and movements of tiger rattlesnakes (*Crotalus tigris*) in the Sonoran Desert of Arizona, pp. 327–334. *In* W. K. Hayes, K. R. Beaman, M. D. Cardwell, and S. P. Bush (eds.), The biology of rattlesnakes. Loma Linda Univ. Press, Loma Linda, California.

Goodman, R. H., Jr., G. L. Stewart, and T. J. Moisi. 1997. *Crotalus cerastes cerastes* (Mojave Desert sidewinder): Longevity. Herpetol. Rev. 28: 89.

Goodrich, R. L., R. W. Murphy, J. Nyhan, and M. J. Wong. 1978. Geographic distribution: *Crotalus ruber lucasensis*. Herpetol. Rev. 9: 108.

Grace, T. G., and G. E. Omer. 1980. The management of upper extremity pit viper wounds. J. Hand Surg. 5: 168–177.

Graf, W., S. G. Jewett, Jr., and K. L. Gordon. 1939. Records of amphibians and reptiles from Oregon. Copeia 1939: 101–104.

Graham, R. L. J., C. Graham, S. McClean, T. Chen, M. O'Rourke, D. Hirst, D. Theakston, and C. Shaw. 2005. Identification and functional analysis of a novel bradykinin inhibitory peptide in the venoms of New World *Crotalinae* pit vipers. Biochem. Biophys. Res. Commun. 338: 1587–1592.

Graham, R. L. J., C. Graham, D. Theakston, G. McMullan, and C. Shaw. 2008. Elucidation of trends within venom components from the snake families Elapidae and Viperidae using gel filtration chromatography. Toxicon 51: 121–129.

Grajales-Tam, K. M., R. Rodríguez-Estrella, and J. Cancino Hernández. 2003. Dieta estacional del coyote *Canis latrans* durante el periodo 1996–1997 en el desierto de Vizcaíno. Baja California Sur, México. Acta Zool. Mex. 89: 17–28.

Grams, F., R. Huber, L. F. Kress, L. Moroder, and W. Bode. 1993. Activation of snake venom metalloproteinases by a cysteine switch-like mechanism. FEBS Lett. 335: 76–80.

Grant, G. S. 1970. Rattlesnake predation on the clapper rail. Chat 34: 20–21.

Graves, B. M. 1989. Defensive behavior of female prairie rattlesnakes (*Crotalus viridis*) changes after parturition. Copeia 1989: 793–794.

———. 1991. Consumption of an adult mouse by a free-ranging neonate prairie rattlesnake. Southwest. Nat. 36: 143.

Graves, B. M., and D. Duvall. 1983. Occurrence and function of prairie rattlesnake mouth gaping in a non-feeding context. J. Exp. Zool. 227: 471–474.

————. 1985a. Avomic prairie rattlesnakes (*Crotalus viridis*) fail to attack rodent prey. Z. Tierpsychol. 67: 161–166.

————. 1985b. Mouth gaping and head shaking by prairie rattlesnakes are associated with vomeronasal organ olfaction. Copeia 1985: 496–497.

————. 1987. An experimental study of aggregation and thermoregulation in prairie rattlesnakes (*Crotalus viridis viridis*). Herpetologica 43: 259–264.

————. 1988. Evidence of an alarm pheromone from the cloacal sacs of prairie rattlesnakes. Southwest. Nat. 33: 339–345.

————. 1990. Spring emergence patterns of wandering garter snakes and prairie rattlesnakes in Wyoming. J. Herpetol. 24: 351–356.

————. 1993. Reproduction, rookery use, and thermoregulation in free-ranging, pregnant *Crotalus v. viridis*. J. Herpetol. 27: 33–41.

Graves, B. M., M. B. King, and D. Duvall. 1986. Natural history of prairie rattlesnake (*Crotalus viridis viridis*) in Wyoming. Herptile 11: 5–10.

Gray, B. S. 2006. The serpent's cast: A guide to identification of shed skins from snakes of the Northeast and Mid-Atlantic states. Cent. N. Am. Herptile Monogr. 1: 1–88.

Green, N. B., and T. K. Pauley. 1987. Amphibians & reptiles in West Virginia. Univ. Pittsburgh Press, Pittsburgh, Pennsylvania.

Greenberg, D. B. 2005. The effects of surface insolation on shelter site selection by *Crotalus mitchellii* and *C. ruber*, p. 30. *In* Biology of the rattlesnakes symposium (program abstracts). Loma Linda Univ., Loma Linda, California.

Greenberg, D. B., and W. J. McClintock. 2008. Remember the third dimension: Terrain modeling improves estimates of snake home range size. Copeia 2008: 801–806.

Greene, H. W. 1983. Dietary correlates of the origin and radiation of snakes. Am. Zool. 23: 431–441.

————. 1990. A sound defense of the rattlesnake. Pacific Disc. 43(4): 10–19.

————. 1992. The ecological and behavioral context of pitviper evolution, pp. 107–118. *In* J. A. Campbell and E. D. Brodie, Jr. (eds.), Biology of the pitvipers. Selva, Tyler, Texas.

————. 1994. Systematics and natural history, foundations for understanding and conserving biodiversity. Am. Zool. 34: 48–56.

————. 1997. Snakes: The evolution of mystery in nature. Univ. California Press, Berkeley.

Greene, H. W., and J. A. Campbell. 1992. The future of pitvipers, pp. 421–428. *In* J. A. Campbell and E. D. Brodie, Jr. (eds.), Biology of the pitvipers. Selva, Tyler, Texas.

Greene, H. W., P. G. May, D. L. Hardy, Sr., J. M. Sciturro, and T. M. Farrell. 2002. Parental behavior by vipers, pp. 179- 205. *In* G. W. Schuett, M. Höggren, M. E. Douglas, and H. W. Greene (eds.), Biology of the vipers. Eagle Mountain Publ., Eagle Mountain, Utah.

Gregory, P. T. 1984. Communal denning in snakes, pp. 57–76. *In* R. A. Seigel, L. E. Hunt, J. L. Knight, L. Malaret, and N. L. Zuschlag (eds.), Vertebrate ecology and systematics: A tribute to Henry S. Fitch. Univ. Kansas Mus. Nat. Hist. Spec. Publ. 10.

————. 2009. Northern lights and seasonal sex: The reproductive ecology of cool-climate snakes. Herpetologica 65:1–13.

Gregory, P. T., and R. W. Campbell. 1984. The reptiles of British Columbia. British Columbia Prov. Mus. Handb. 44: 1–103.

Gregory, P. T., J. M. Macartney, and K. W. Larsen. 1987. Spatial patterns and movements, pp. 366–395. *In* R. A. Seigel, J. J. Collins, and S. S. Nowak (eds.), Snakes: Ecology and evolutionary biology. McGraw-Hill, New York.

Gregory, V. M., F. E. Russell, J. R. Brewer, and L. R. Zawadski. 1984. Seasonal variations in rattlesnake venom proteins. Proc. West. Pharmacol. Soc. 27: 233–236.

Gregory-Dwyer, V. M., N. B. Egen, A. B. Bosisio, P. G. Righetti, and F. E. Russell. 1986. An isoelectric focusing study of seasonal variation in rattlesnake venom proteins. Toxicon 24: 995–1000.

Grenard, S. 2000. Is rattlesnake venom evolving? Nat. Hist. 109(6): 44–49.

Griffin, P. R., and S. D. Aird. 1990. A new small myotoxin from the venom of the prairie rattlesnake (*Crotalus viridis viridis*). FEBS Lett. 274: 43–47.

Grinnell, J. 1908. The biota of the San Bernardino mountains. Univ. California Pub. Zool. 5(1): 1–170.

Grinnell, J., and H. W. Grinnell. 1907. Reptiles of Los Angeles County, California. Throop Inst. Bull. 35: 1–64.

Grismer, L. L. 1994a. The evolutionary and ecological biogeography of the herpetofauna of Baja California and the Sea of Cortez, Mexico. Ph.D. diss., Loma Linda Univ., Loma Linda, California.

———. 1994b. Geographic origins for the reptiles on islands in the Gulf of California, México. Herpetol. Nat. Hist. 2: 17–40.

———. 1994c. Ecogeography of the peninsular herpetofauna of Baja California, México and its utility in historical biogeography, pp. 89–126. In P. R. Brown and J. W. Wright (eds.), Herpetology of North American deserts. Southwest. Herpetol. Soc. Spec. Pub. 5.

———. 1994d. The origin and evolution of the peninsular herpetofauna of Baja California, Mexico. Herpetol. Nat. Hist. 2: 51–106.

———. 1999a. An evolutionary classification of reptiles on islands in the Gulf of California, México. Herpetologica 55: 446–469.

———. 1999b. Checklist of amphibians and reptiles on islands in the Gulf of California, Mexico. Bull. So. California Acad. Sci. 98: 45–56.

———. 2002. Amphibians and reptiles of Baja California, including its Pacific islands and the islands in the Sea of Cortés, México: Natural history, distribution and identification. Univ. California Press, Berkeley.

Groombridge, B. 1986. Comments on the M. pterygoideus glandulae of crotaline snakes (Reptilia: Viperidae). Herpetologica 42: 449–457.

Grube, G. E. 1963. Albino timber rattlesnake in Pennsylvania. Turtox News 41(2): 70.

Grupka, L. M., E. C. Ramsay, and D. A. Bemis. 2006. Salmonella surveillance in a collection of rattlesnakes (Crotalus spp.). J. Zoo Wildl. Med. 37: 306–312.

Guidry, E. V. 1953. Herpetological notes from southeastern Texas. Herpetologica 9: 49–56.

Guilday, J. E. 1962. The Pleistocene local fauna of the Natural Chimneys, Augusta County, Virginia. Ann. Carnegie Mus. 36: 87–122.

Guilday, J. E., P. S. Martin, and A. D. McCrady. 1964. New Paris 4: A Pleistocene cave deposit in Bedford County, Pennsylvania. Natl. Speleol. Soc. Bull. 26: 121–194.

Guilday, J. E., H. W. Hamilton, and A. D. McCrady. 1966. The bone breccia of Bootlegger Sink, York County, Pennsylvania. Ann. Carnegie Mus. 8: 145–163.

Guisto, J. A. 1995. Severe toxicity from crotalid envenomation after early resolution of symptoms. Ann. Emerg. Med. 26: 387–389.

Gumbart, T. C., and K. A. Sullivan. 1990. Predation on yellow-eyed junco nestlings by twin-spotted rattlesnakes. Southwest. Nat. 35: 367–368.

Günther, A. C. L. G. 1885–1902. Reptilia and Batrachia, pp. xx–366. In F. D. Goodman and O. Slavin (eds.), Biologia Centrali-Americana: Zoology. Dulau, London.

Gut, H. J., and C. E. Ray. 1963. The Pleistocene vertebrate fauna of Reddick, Florida. Quart. J. Florida Acad. Sci. 26: 315–328.

Guthrie, J. E. 1926. The snakes of Iowa. Iowa St. Coll. Agric. Mech. Arts, Agric. Exp. Stn. Bull. 239: 146–192.

Gutzke, W. H. N., C. Tucker, and R. T. Mason. 1993. Chemical recognition of kingsnakes by crotalines: Effects of size on the ophiophage defensive response. Brain Behav. Evol. 41: 234–238.

Haast, W. E., and R. Anderson. 1981. Complete guide to snakes of Florida. Phoenix Publ. Co., Miami, Florida.

Hachimori, Y., M. A. Wells, and D. J. Hanahan. 1971. Observations on the phospholipase A_2 of Crotalus atrox: Molecular weight and other properties. Biochemistry 10: 4084–4089.

Halama, K. J., A. J. Malisch, M. Aspell, J. T. Rotenberry, and M. F. Allen. 2008. Modeling the landscape niche characteristics of red diamond rattlesnakes (Crotalus ruber): Implications for biology and conservation, pp. 463–472. In W. K. Hayes, K. R. Beaman, M. D. Cardwell, and S. P. Bush (eds.), The biology of rattlesnakes. Loma Linda Univ. Press, Loma Linda, California.

Hall, E. R. 1929. A "den" of rattlesnakes in eastern Nevada. Bull. Antiv. Inst. Am. 3: 79–80.

———. 1953. A westward extension of known geographic range for the timber rattlesnake in southern Kansas. Trans. Kansas Acad. Sci. 56: 89.

Hall, H. H., and H. M. Smith. 1947. Selected records of reptiles and amphibians from southeastern Kansas. Trans. Kansas Acad. Sci. 49: 447–454.

Haller, R. 1971. The diamondback rattlesnakes. Herpetology 3(3): 1–34.

Halliday, T. R., and P. A. Verrell. 1988. Body size and age in amphibians and reptiles. J. Herpetol. 22: 253–265.

Hallowell, E. 1854. Descriptions of reptiles from California. Proc. Acad. Nat. Sci. Philadelphia 7: 91–97.

Hamako, J., Y. Suzuki, N. Hayashi, M. Kimura, Y. Ozeki, K. Hashimoto, and T. Matsui. 2007. Amino acid sequence and characterization of C-type lectin purified from the snake venom of *Crotalus ruber*. Comp. Biochem. Physiol. 146B: 299–306.

Hamilton, B. 2005. Use of solar models and GIS to evaluate potential hibernacula in the Great Basin rattlesnake (*Crotalus lutosus*), pp. 30–31. *In* Biology of the Rattlesnakes Symposium (program abstracts). Loma Linda Univ., Loma Linda, California.

Hamilton, B., and D. Richard. 2010. *Crotalus oreganus lutosus* (great basin rattlesnake): Elevation. Herpetol. Rev. 41: 90.

Hamilton, B. T., and E. M. Nowak. 2009. Relationships between insolation and rattlesnake hibernacula. West. No. Am. Natur. 69: 319–328.

Hamilton, P. S., and J. Wrieden. 2004. *Crotalus molossus molossus* (black-tailed rattlesnake): Male-male fighting. Herpetol. Rev. 35: 63.

Hamilton, W. J., Jr. 1950. Food of the prairie rattlesnake (*Crotalus v. viridis* Rafinesque). Herpetologica 6: 34.

Hamilton, W. J., Jr., and J. A. Pollack. 1955. The food of some crotalid snakes from Fort Benning, Georgia. Chicago Acad. Sci. Nat. Hist. Misc. 140: 1–4.

Hammerson, G. A. 1981. Opportunistic scavenging by *Crotalus ruber* not field-proven. J. Herpetol. 15: 125.

———. 1986. Amphibians and reptiles in Colorado. Colorado Div. Wildl., Denver.

Hampton, P. M., and N. E. Haertle. 2009. A new view from a novel squeeze box design. Herpetol. Rev. 40: 44.

Hansen, A. 1931. Attempts at purification of certain snake venoms. Bull. Antiv. Inst. Am. 5: 48–49.

Hanson, G. B. 1976. Immunological responses in the prairie rattlesnake, *Crotalus viridis viridis* Rafinesque 1818, to laboratory infection with tetrathyridia of *Mesocestoides corti* Hoepple 1925 (Eucestoda: Mesocestoididea). Diss. Abstr. Int. (B) 37: 2019.

Hanson, G. B., and E. A. Widmer. 1985. Asexual multiplication of tetrathyridia of *Mesocestoides corti* in *Crotalus viridis viridis*. J. Wildl. Dis. 21: 20–24.

Hanson, J., and R. B. Hanson. 1997. 50 common reptiles and amphibians of the Southwest. Southwest Parks Monuments Assoc., Tucson, Arizona.

Happ, W. M. 1951. My snake bite experience. Bull. San Diego Co. Med. Soc. 37(3): 20, 22, 26.

Harding, J. H. 1997. Amphibians and reptiles of the Great Lakes Region. Univ. Michigan Press, Ann Arbor.

Harding, K. A., and K. R. G. Welch. 1980. Venomous snakes of the World: A checklist. Pergamon Press, New York.

Hardy, D. L. 1983. Envenomation by the Mojave rattlesnake (*Crotalus scutulatus scutulatus*) in southern Arizona, U.S.A. Toxicon 21: 111–118.

———. 1986. Fatal rattlesnake envenomation in Arizona: 1969–1984. Clinical Toxicol. 24: 1–10.

———. 1992. A review of first aid measures for pitviper bite in North America with an appraisal of Extractor™ suction and stun gun electroshock, pp. 405–414. *In* J. A. Campbell and E. D. Brodie, Jr. (eds.), Biology of the pitvipers. Selva, Tyler, Texas.

———. 1997. Fatal bite by a captive rattlesnake in Tucson, Arizona. Sonoran Herpetol. 10: 38–39.

———. 1998. Male-male copulation in captive Mojave rattlesnakes (*Crotalus s. scutulatus*): Its possible significance in understanding the behavior and physiology of crotaline copulation. Bull. Chicago Herpetol. Soc. 33: 258–262.

Hardy, D. L., Sr., and H. W. Greene. 1995. *Crotalus molossus molossus* (blacktail rattlesnake): Maximum length. Herpetol. Rev. 26: 101.

Hardy, D. L., M. Jeter, and J. J. Corrigan, Jr. 1982. Envenomation by the northern blacktail rattlesnake (*Crotalus molossus molossus*): Report of two cases and the *in vitro* effects of the venom
on fibrinolysis and platelet aggregation. Toxicon 20: 487–492.

Hardy, L. M., and R. W. McDiarmid. 1969. The amphibians and reptiles of Sinaloa, México.
Univ. Kansas Publ. Mus. Nat. Hist. 18: 39–252.

Harlan, R. 1827. Genera of North American Reptilia and a synopsis of the species. J. Acad. Nat.
Sci. Philadelphia 5: 317–372.

Harris, H. S., Jr. 1975. Distributional survey (Amphibia/Reptilia): Maryland and the District of
Columbia. Bull. Maryland Herpetol. Soc. 11: 73–167.

———. 2005. Scanning electron microscopy: Scale topography in *Crotalus* and *Sistrurus*. Bull.
Maryland Herptol. Soc. 41: 101–115.

———. 2006a. Serum electrophoretic patterns in a select group of Mexican Montane rattlesnakes (*Crotalus*). Bull. Maryland Herpetol. Soc. 42: 30–34.

———. 2006b. Some notes on color phases in serpents. Bull. Maryland Herpetol. Soc. 42:
181–183.

———. 2007. The distribution in Maryland, both recent and historic, of the timber rattlesnake,
Crotalus horridus. Bull. Maryland Herpetol. Soc. 43: 8–13.

Harris, H. S., Jr., and R. S. Simmons. 1972. An April birth record for *Crotalus lepidus* with a summary of annual broods of rattlesnakes. Bull. Maryland Herpetol. Soc. 8: 54–56.

———. 1974. The New Mexican ridge-nosed rattlesnake. Natl. Parks Conserv. Mag. 48(3): 22–24.

———. 1975. An endangered species, the New Mexican ridge-nosed rattlesnake. Bull. Maryland Herpetol. Soc. 11: 1–7.

———. 1976. The paleogeography and evolution of *Crotalus willardi*, with a formal description
of a new subspecies from New Mexico, United States. Bull. Maryland Herpetol. Soc. 12:
1–22.

———. 1977a. A preliminary account of insular rattlesnake populations, with special reference
to those occurring in the Gulf of California and off the Pacific coast. Bull. Maryland Herpetol. Soc. 13: 92–110.

———. 1977b. Additional notes concerning cannibalism in pit vipers. Bull. Maryland Herpetol.
Soc. 13: 121–122.

———. 1978. A preliminary account of the rattlesnakes with the descriptions of four new subspecies. Bull. Maryland Herpetol. Soc. 14: 105–211.

Harrison, H. H. 1949–1950. Pennsylvania reptiles and amphibians. Pennsylvania Fish. Comm.,
Harrisburg.

———. 1971. The world of the snake. Lippincott, New York.

Harshbarger, J. C. 1974. Activities report (of the) Registry of Tumors in lower animals, 1965–
1973. Smithsonian Inst., Washington, D.C.

Hartline, P. H. 1974. Thermoreception in snakes, pp. 297–312. *In* A. Fessard (ed.), Handbook of
sensory physiology, vol. 3. Springer-Verlag, New York.

Hartline, P. H., and H. W. Campbell. 1969. Auditory and vibratory responses in the midbrains
of snakes. Science (New York) 163: 1221–1223.

Hartline, P. H., L. Kass, and M. S. Loop. 1978. Merging of modalities in the optic tectum: Infrared and visual integration in rattlesnakes. Science (New York) 199: 1225–1229.

Hartman, F. A. 1911. Description of a little-known rattlesnake, *Crotalus willardi*, from Arizona.
Proc. U.S. Natl. Mus. 39(1800): 569–570.

Hartnett, W. G. 1931. Poisonous snake bite: Etiology and treatment with case report. Ohio St.
Med. J. 27: 636–639.

Hartweg, N., and J. A. Oliver. 1940. A contribution to the herpetology of the Isthmus of Tehu-
antepec. IV. Misc. Publ. Mus. Zool. Univ. Michigan 47: 1–31.

Harwig, S. H. 1966. Rattlesnakes are where and when you find them. J. Ohio Herpetol. Soc. 5:
163.

Haupt, H., Jr. 1915. Hibernation of reptiles. Copeia 1915(20): 18–19.

Haverly, J. E., and K. V. Kardong. 1996. Sensory deprivation effects on the predatory behavior of
the rattlesnake, *Crotalus viridis oreganus*. Copeia 1996: 419–428.

Hawgood, B. J. 1982. Physiology and pharmacological effects of rattlesnake venoms, pp. 121–

162. *In* A. T. Tu (ed.), Rattlesnake venoms: Their actions and treatment. Marcel Dekker, New York.

Hay, O. P. 1917. Vertebrates mostly from Stratum No. 3 at Vero, Florida; together with descriptions of a new species. Ann. Rep. Florida Geol. Surv. 9: 43–68.

Hayes, C. E., and A. L. Bieber. 1986. The effects of myotoxin from midget faded rattlesnake (*Crotalus viridis concolor*) venom on neonatal rat myotubes in cell culture. Toxicon 24: 169–173.

Hayes, W. K. 1986. Observations of courtship in the rattlesnake, *Crotalus viridis oreganus*. J. Herpetol. 20: 246–249.

Hayes, W. K. 1991a. Envenomation strategies of prairie rattlesnakes. Ph.D. diss., Univ. Wyoming, Laramie.

———. 1991b. Ontogeny of striking, prey-handling and envenomation behavior of prairie rattlesnakes (*Crotalus v. viridis*). Toxicon 29: 867–875.

———. 1992a. Factors associated with the mass of venom expended by prairie rattlesnakes (*Crotalus v. viridis*) feeding on mice. Toxicon 30: 449–460.

———. 1992b. Prey-handling and envenomation strategies of prairie rattlesnakes (*Crotalus v. viridis*) feeding on mice and sparrows. J. Herpetol. 26: 496–499.

———. 1993. Effects of hunger on striking, prey-handling, and venom expenditure of prairie rattlesnakes (*Crotalus v. viridis*). Herpetologica 49: 305–310.

———. 1995. Venom metering by juvenile prairie rattlesnakes, *Crotalus v. viridis*: Effects of prey size and experience. Anim. Behav. 50: 33–40.

———. 2008. The snake venom-metering controversy: Levels of analysis, assumptions, and evidence, pp. 191–220. *In* W. K. Hayes, K. R. Beaman, M. D. Cardwell, and S. P. Bush (eds.), The biology of rattlesnakes. Loma Linda Univ. Press, Loma Linda, California.

Hayes, W. K., and D. Duvall. 1991. A field study of prairie rattlesnake predatory strikes. Herpetologica 47: 78–81.

Hayes, W. K., and J. G. Galusha. 1984. Effects of rattlesnake (*Crotalus viridis oreganus*) envenomation upon mobility of male wild and laboratory mice (*Mus musculus*). Bull. Maryland Herpetol. Soc. 20: 135–144.

Hayes, W. K., and D. M. Hayes. 1993. Stimuli influencing the release and aim of predatory strikes of the northern Pacific rattlesnake (*Crotalus viridis oreganus*). Northwest. Natur. 74: 1–9.

Hayes, W. K., and S. P. Mackessy. 2010. Sensationalistic journalism and tales of snakebite: Are rattlesnakes rapidly evolving more toxic venom? Wilderness & Environmental Med. 21: 35–45.

Hayes, W. K., D. Duvall, and G. W. Schuett. 1992a. A preliminary report on the courtship behavior of free-ranging prairie rattlesnakes, *Crotalus viridis viridis* (Rafinesque), in south-central Wyoming, pp. 45–48. *In* P. Strimple and J. Strimple (eds.), Contributions in herpetology. Greater Cincinnati Herpetol. Soc., Ohio.

Hayes, W. K., I. I. Kaiser, and D. Duvall. 1992b. The mass of venom expended by prairie rattlesnakes when feeding on rodent prey, pp. 383–388. *In* J. A. Campbell and E. D. Brodie, Jr. (eds.), Biology of the pitvipers. Selva, Tyler, Texas.

Hayes, W. K., E. A. Verde, and F. E. Hayes. 1994. Cardiac responses during courtship, male-male fighting, and other activities in rattlesnakes. J. Tennessee Acad. Sci. 69(1): 7–9.

Hayes, W. K., S. S. Herbert, G. C. Rehling, and J. F. Gennaro. 2002. Factors that influence venom expenditure in viperids and other snake species during predatory and defensive contexts, pp. 207–233. *In* G. W. Schuett, M. Höggren, M. E. Douglas, and H. W. Greene (eds.), Biology of the vipers. Eagle Mountain Publ., Eagle Mountain, Utah.

Hayes, W. K., K. R. Beaman, M. D. Cardwell, and S. P. Bush (eds.). 2008. The biology of rattlesnakes. Loma Linda Univ. Press, Loma Linda, California.

Heald, W. F. 1951. Sky islands of Arizona. Nat. Hist. 60: 53–63, 95–96.

Henderson, J. T., and A. L. Bieber. 1986. Antigenic relationships between Mojave toxin subunits, Mojave toxin and some crotalid venoms. Toxicon 24: 473–479.

Hennessy, D. F., and D. H. Owings. 1988. Rattlesnakes create a context for localizing their search for potential prey. Ethology 77: 317–329.

Hensley, M. M. 1959. Albinism in North American amphibians and reptiles. Publ. Mus. Michigan St. Univ. Biol. Ser. 1: 133–159.

Herbert, S. S. 1998. Factors influencing venom expenditure during defensive bites by cotton-mouths (*Agkistrodon piscivorus*) and rattlesnakes (*Crotalus viridis, Crotalus atrox*). Master's thesis, Loma Linda Univ., Loma Linda, California.

Herbert, S. S., and W. K. Hayes. 2008. Venom expenditure by rattlesnakes and killing effectiveness in rodent prey: Do rattlesnakes expend optimal amounts of venom? pp. 221–228. *In* W. K. Hayes, K. R. Beaman, M. D. Cardwell, and S. P. Bush (eds.), The biology of rattlesnakes. Loma Linda Univ. Press, Loma Linda, California.

Hermann, J. A. 1950. Mammals of the Stockton Plateau of northeastern Terrell County, Texas. Texas J. Sci. 2: 368–393.

Herrington, R. E. 1988. Talus use by amphibians and reptiles in the Pacific Northwest, pp. 216–221. *In* R. C. Szaro, K. E. Severson, and D. R. Patton (eds.), Management of amphibians, reptiles, and small mammals in North America. U.S.D.A. For. Serv. Gen Tech. Rep. RM-166.

Hersek, M. J., D. H. Owings, and D. F. Hennessy. 1992. Combat between rattlesnakes (*Crotalus viridis oreganus*) in the field. J. Herpetol. 26: 105–107.

Heyrend, F. L., and A. Call. 1951. Growth and age in western striped racer and Great Basin rattlesnake. Herpetologica 7: 28–40.

Hibbard, C. W. 1936. The amphibians and reptiles of Mammoth Cave National Park proposed. Trans. Kansas Acad. Sci. 39: 277–281.

Hill, H. R. 1935. New host records of the linguatulid, *Kiricephalus coarctatus* (Diesing) in the United States. Bull. So. California Acad. Sci. 34: 226–267.

Hill, J. G., III, I. Hanning, S. J. Beaupre, S. C. Ricke, and M. M. Slavik. 2008. Denaturing gradient gel electrophoresis for the determination of bacterial species diversity in the gastrointestinal tracts of two crotaline snakes. Herpetol. Rev. 39: 433–438.

Hill, M. M. A., G. L. Powell, and A. P. Russell. 2001. Diet of the prairie rattlesnake, *Crotalus viridis viridis,* in southeastern Alberta. Can. Field-Nat. 115: 241–246.

Hill, S. 1943. Rattlesnakes of northwestern Colorado and southern Wyoming. Trail and Timberline 293: 59–60.

Hinderliter, M. G., and J. R. Lee. 2006. *Masticophis flagellum* (coachwhip): Ophiophagy. Herpetol. Rev. 37: 232–233.

Hinshaw, W. R., and E. McNeil. 1946. Paracolon type 10 from captive rattlesnakes. J. Bacteriol. 51: 397–398.

Hinze, J. D., J. A. Barker, T. R. Jones, and R. E. Winn. 2001. Life threatening upper airway edema caused by a distal rattlesnake bite. Am. Emerg. Med. 37: 79–82.

Hirth, H. F. 1966a. Weight changes and mortality of three species of snakes during hibernation. Herpetologica 22: 8–12.

———. 1966b. The ability of two species of snakes to return to a hibernaculum after displacement. Southwest. Nat. 11: 49–53.

Hirth, H. F., and A. C. King. 1968. Biomass densities of snakes in the cold desert of Utah. Herpetologica 24: 333–335.

———. 1969. Body temperatures of snakes in different seasons. J. Herpetol. 3: 99–100.

Hirth, H. F., R. C. Pendleton, A. C. King, and T. R. Downard. 1969. Dispersal of snakes from a hibernaculum in northwestern Utah. Ecology 50: 332–339.

Hite, L. A., J. D. Shannon, J. B. Bjarnason, and J. W. Fox. 1992. Sequence of a cDNA clone encoding the zinc metalloproteinase hemorrhagic toxin e from *Crotalus atrox*: Evidence for signal, zymogen, and disintegrin-like structures. Biochemistry 31: 6203–6211.

Hoessle, C. 1963. A breeding pair of western diamondback rattlesnakes, *Crotalus atrox.* Bull. Philadelphia Herpetol. Soc. 11: 65–66.

Hoff, G., and D. O. Trainer. 1973. Arboviruses in reptiles: Isolation of a Bunyamwera group virus from a naturally infected turtle. J. Herpetol. 7: 55–62.

Hoge, A. R., and S. A. R. W. D. L. Romano. 1971. Neotropical pit vipers, sea snakes, and coral snakes, pp. 211–293. *In* W. Bücherl and E. E. Buckley (eds.), Venomous animals and their venoms, vol. II: Venomous vertebrates. Academic Press, New York.

Holbrook, J. E. 1836–1842. North American herpetology; or, a description of the reptiles inhabiting the United States. Vols. 1–33. J. Dobson, Philadelphia.

Hollingsworth, B. D., and E. Mellink. 1996. *Crotalus exsul lorenzoensis* (San Lorenzo Island rattle-snake): Arboreal behavior. Herpetol. Rev. 27: 143–144.

Holman, J. A. 1958. The Pleistocene herpetofauna of Saber-tooth Cave, Citrus County, Florida. Copeia 1958: 276–280.

———. 1959a. Amphibians and reptiles from the Pleistocene (Illinoian) of Williston, Florida. Copeia 1959: 96–102.

———. 1959b. A Pleistocene herpetofauna near Orange Lake, Florida. Herpetologica 15: 121–124.

———. 1965. A late Pleistocene herpetofauna from Missouri. Trans. Illinois St. Acad. Sci. 58: 190–194.

———. 1966. The Pleistocene herpetofauna of Miller's Cave, Texas. Texas J. Sci. 18: 372–377.

———. 1967. A Pleistocene herpetofauna from Ladd's Georgia. Bull. Georgia Acad. Sci. 25: 154–166.

———. 1969. Herpetofauna of the Slaton local fauna of Texas. Southwest. Nat. 14: 203–212.

———. 1970. A Pleistocene herpetofauna from Eddy County, New Mexico. Texas J. Sci. 22: 29–39.

———. 1971. Herpetofauna of the Sundahl local fauna (Pleistocene: Illinoian) of Kansas. Contr. Mus. Paleontol. Univ. Michigan 23: 349–355.

———. 1974. Late Pleistocene herpetofauna from southwestern Missouri. J. Herpetol. 8: 343–346.

———. 1977a. The Pleistocene (Kansan) herpetofauna of Cumberland Cave, Maryland. Ann. Carnegie Mus. 46: 157–172.

———. 1977b. Upper Miocene snakes (Reptilia, Serpentes) from southeastern Nebraska. J. Herpetol. 11: 323–335.

———. 1978. The late Pleistocene herpetofauna of Devil's Den Sinkhole, Levy County, Florida. Herpetologica 34: 228–237.

———. 1979. A review of North American tertiary snakes. Publ. Mus. Michigan St. Univ. Paleontol. Ser. 1: 201–260.

———. 1980. Paleoclimatic implications of Pleistocene herpetofauanas of eastern and central North America. Trans. Nebraska Acad. Sci. 8: 131–140.

———. 1981. A review of North American Pleistocene snakes. Publ. Mus. Michigan St. Univ. Paleontol. Ser. 1: 261–306.

———. 1982. The Pleistocene (Kansas) herpetofauna of Trout Cave, West Virginia. Ann. Carnegie Mus. 51: 391–404.

———. 1986. The known herpetofauna of the Late Quaternary of Virginia poses a dilemma, pp. 36–42. *In* J. N. McDonald and S. O. Bird (eds.), The Quarternary of Virginia—a symposium volume. Commonwealth of Virginia, Dept. of Mines, Minerals, and Energy, Division of Mineral Resources, Charlottesville.

———. 1991. North American Pleistocene herpetological stability and its impact on the interpretation of modern herpetofaunas: An overview. Illinois St. Mus. Sci. Pap. 23: 227–235.

———. 1995. Pleistocene amphibians and reptiles in North America. Oxford Univ. Press, New York.

———. 1996. The large Pleistocene (Sangamonian) herpetofauna of the Williston IIIA Site, North-central Florida. Herpetol. Nat. Hist. 4: 35–47.

———. 2000a. Fossil snakes of North America: Origin, evolution, distribution, paleoecology. Indiana Univ. Press, Bloomington.

———. 2000b. Snake fauna associated with the "earliest recent" mammalian fauna in northeastern North America. Ann. Carnegie Mus. 69: 5–9.

———. 2009. The amphibians and reptiles of Michigan: A Quaternary and Recent adventure. Wayne St. Univ. Press, Detroit, Michigan.

Holman, J. A., and C. J. Clausen. 1984. Fossil vertebrates associated with paleo-Indian artifact at Little Salt Spring, Florida. J. Vert. Paleontol. 4: 146–154.

Holman, J. A., and F. Grady. 1987. Herpetofauna of New Trout Cave. Natl. Geogr. Res. 3: 305–317.

———. 1989. The fossil herpetofauna (Pleistocene: Irvingtonian) of Hamilton Cave, Pendelton County, West Virginia. Natl. Speleol. Soc. Bull. 51: 34–41.

Holman, J. A., and R. L. Richards. 1981. Late Pleistocene occurrence in southern Indiana of the smooth green snake, *Opheodrys vernalis*. J. Herpetol. 15: 123–124.

Holman, J. A., and M. E. Schloeder. 1991. Fossil herpetofauna of the Lisco C Quarries (Pliocene: early Blancan) of Nebraska. Trans. Nebraska Acad. Sci. 18: 19–29.

Holman, J. A., and A. J. Winkler. 1987. A mid-Pleistocene (Irvingtonian) herpetofauna from a cave in southcentral Texas. Texas Mem. Mus. Univ. Texas Pearce-Sellards Ser. 44: 1–17.

Holman, J. A., J. H. Harding, M. M. Hensley, and G. R. Dudderar. 1989. Michigan snakes: A field guide and pocket reference. Michigan St. Univ. Coop. Ext. Serv., East Lansing.

Holman, J. A., G. Bell, and J. Lamb. 1990. A Late Pleistocene herpetofauna from Bell Cave, Alabama. Herpetol. J. 1: 521–529.

Holstege, C. P., M. B. Miller, M. Wermuth, B. Furbee, and S. C. Curry. 1997. Crotalid snake envenomation. Crit. Care Clin. 13: 889–921.

Holycross, A. T. 1995. *Crotalus viridis* (western rattlesnake): Phenology. Herpetol. Rev. 26: 37–38.

———. 2000a. *Crotalus atrox* (western diamondback rattlesnake): Morphology. Herpetol. Rev. 31: 177–178.

———. 2000b. *Crotalus viridis viridis* (prairie rattlesnake): Morphology. Herpetol. Rev. 31: 178.

———. 2000c. *Crotalus willardi obscurus* (New Mexico ridgenose rattlesnake): Caudal dichromatism. Herpetol. Rev. 31: 246.

———. 2001. *Crotalus molossus* (black-tailed rattlesnake). Herpetol. Rev. 32: 194.

———. 2002. Conservation biology of two rattlesnakes, *Crotalus willardi obscurus* and *Sistrurus catenatus edwardsii*. Ph.D. diss., Arizona St. Univ., Tempe.

Holycross, A. T., and M. E. Douglas. 2007. Geographic isolation, genetic divergence, and ecological non-exchangeability define ESUs in a threatened sky-island rattlesnake. Biol. Conserv. 134: 142–154.

Holycross, A. T., and J. D. Fawcett. 2002. Observations on neonatal aggregations and associated behaviors in the prairie rattlesnake, *Crotalus viridis viridis*. Am. Midl. Nat. 148: 181–184.

Holycross, A. T., and S. R. Goldberg. 2001. Reproduction in northern populations of the ridgenose rattlesnake, *Crotalus willardi* (Serpentes: Viperidae). Copeia 2001: 473–481.

Holycross, A. T., L. K. Kamees, and C. W. Painter. 2001. Observations of predation on *Crotalus willardi obscurus* in the Animas Mountains, New Mexico. Southwest. Nat. 46: 363–364.

Holycross, A. T., C. W. Painter, D. B. Prival, D. E. Swann, M. J. Schroff, T. Edwards, and C. R. Schwalbe. 2002a. Diet of *Crotalus lepidus klauberi* (banded rock rattlesnake). J. Herpetol. 36: 589–597.

Holycross, A. T., M. E. Douglas, J. R. Higbee, and R. H. Bogden. 2002b. Isolation and characterization of microsatellite loci from a threatened rattlesnake (New Mexico ridge-nosed rattlesnake, *Crotalus willardi obscurus*). Mol. Ecol. Notes 2: 537–539.

Holycross, A. T., C. W. Painter, D. G. Barker, and M. E. Douglas. 2002c. Foraging ecology of the threatened New Mexico ridge-nosed rattlesnake (*Crotalus willardi obscurus*), pp. 243–251. *In* G. W. Schuett, M. Höggren, M. E. Douglas, and H. W. Greene (eds.), Biology of the vipers. Eagle Mountain Publ., Eagle Mountain, Utah.

Holzer, M., and S. P. Mackessy. 1996. An aqueous endpoint assay of snake venom phospholipase A_2. Toxicon 34: 1149–1155.

Horn, E. E., and H. S. Fitch. 1942. Interrelations of rodents and other wildlife of the range (San Joaquin Experimental Range). Univ. California Agric. Exp. Sta. Bull. 663: 96–129.

Hornocker, M. G., R. C. Pendleton, J. P. Messick, J. S Whitman, and J. Copeland. 1978. Dynamics of predation upon a raptor prey base in the Snake River Birds of Prey Natural Area, Idaho. Final Rept., U.S.D.I. Bur. Land Manage., Boise, Idaho.

Horton, C. W. 1951. The near ultraviolet absorption spectrum of *Crotalus* venom. Herpetologica 7: 173–174.

Hoss, S. K., C. Guyer, L. L. Smith, and G. W. Schuett. 2010. Multiscale influences of landscape composition and configuration on the spatial ecology of eastern diamond-backed rattlesnakes (*Crotalus adamanteus*). J. Herpetol. 44: 110–123.

Houston, T. C. 2006. *Crotalus concolor*: The midget faded rattlesnake, natural history notes and captive maintenance. Reptilia 45: 33–37.

Howard, P., and K. Kopf. 2003. *Crotalus adamanteus* (eastern diamondback rattlesnake): Diet. Herpetol. Rev. 34: 373.

Howell, C. T., and S. F. Wood. 1957. The prairie rattlesnake at Gran Quivara National Monument, New Mexico. Bull. So. California Acad. Sci. 56: 97–98.

Huang, S. Y., and J. C. Perez. 1980. Comparative study on hemorrhagic and proteolytic activities of snake venoms. Toxicon 18: 421–426.

Huang, S. Y., J. C. Perez, E. D. Rael, C. Lieb, M. Martinez, and S. A. Smith. 1992. Variation in the antigenic characteristics of venom from the Mojave rattlesnake (*Crotalus scutulatus scutulatus*). Toxicon 30: 387–396.

Hubbard, D. H. 1941. The vertebrate animals of Friant Reservoir Basin with special reference to the possible effects upon them of the Friant Dam. California Fish Game 27: 198–215.

Hubbard, W. E. 1939. *Entonyssus ewingi* n. sp., an ophidian lung mite. Am. Midl. Nat. 21: 657–662.

Hubbs, B., and B. O'Connor. 2009. A guide to the rattlesnakes of the United States. Tricolor Books, Tempe, Arizona.

Hudson, D. M., and B. H. Brattstrom. 1977. A small herpetofauna from the Late Pleistocene of Newport Beach Mesa, Orange County, California. Bull. So. California Acad. Sci. 76: 16–20.

Hudson, G. E. 1942. The amphibians and reptiles of Nebraska. Univ. Nebraska, Nebraska Conserv. Bull. 24: 1–146.

Hudson, R., and G. Carl. 1985. *Crotalus horridus* (timber rattlesnake): Coloration. Herpetol. Rev. 16: 28–29.

Huey, L. M. 1942. A vertebrate faunal survey of the Organ Pipe Cactus National Monument, Arizona. Trans. San Diego Soc. Nat. Hist. 9: 353–376.

Huheey, J. E., and A. Stupka. 1967. Amphibians and reptiles of Great Smoky Mountains National Park. Univ. Tennessee Press, Knoxville.

Hulbert, R. C., Jr., and G. S. Morgan. 1989. Stratigraphy, paleoecology, and vertebrate fauna of the Leisley Shell Pit local fauna, early Pleistocene (Irvingtonian) of southwestern Florida. Pap. Florida Paleontol. 2: 1–19.

Hull, R. W., and J. H. Camin. 1960. Haemogregarines in snakes: The incidence and identity of the erythrocytic stages. J. Parasitol. 46: 515–523.

Hulse, A. C. 1973. Herpetofauna of the Fort Apache Indian Reservation, east central Arizona. J. Herpetol. 7: 275–282.

Hulse, A. C., C. J. McCoy, and E. J. Censky. 2001. Amphibians and reptiles of Pennsylvania and the Northeast. Cornell Univ. Press, Ithaca, New York.

Hung, C.-C., and S.-H. Chiou. 1994. Isolation of multiple isoforms of α-fibrinogenase from the western diamondback rattlesnake, *Crotalus atrox*: N-terminal sequence homology with ancrod, an antithrombotic agent from Malayan viper. Biochem. Biophys. Res. Comm. 201: 1414–1423.

Hunter, J. S. 1898. Hawk killed by rattlesnake. Osprey 3: 46.

Hunter, M. L., Jr., J. Albright, and J. Arbuckle (eds.). 1992. The amphibians and reptiles of Maine. Maine Agric. Exp. St. Bull. 838.

Hunter, M. L., Jr., A. J. K. Calhoun, and M. McCollough. 1999. Maine amphibians and reptiles. Univ. Maine Press, Orono.

Hurter, J. 1893. Catalogue of reptiles and batrachians found in the vicinity of St. Louis, Mo. Trans. St. Louis Acad. Sci. 6: 251–261.

———. 1911. Herpetology of Missouri. Trans. St. Louis Acad. Sci. 20: 59–274.

Hutchison, R. H. 1929. On the incidence of snake-bite poisoning in the United States and the results of the new methods of treatment. Bull. Antiv. Inst. Am. 3: 43–57.

———. 1930. Further notes on the incidence of snakebite poisoning in the United States. Bull Antiv. Inst. Am. 4: 40–43.

Ineich, I., X. Bonnet, R. Shine, T. Shine, F. Brischoux, M. Lebreton, and L. Chirio. 2006. What, if anything, is a "typical" viper? Biological attributes of basal viperid snakes (genus *Causus* Wagler, 1830). Biol. J. Linnean Soc. 89: 575–588.

Infante, J. P., R. C. Kirwan, and J. T. Brenna. 2001. High levels of docosahexaenoic acid (22: 6n-3)-containing phospholipids in high-frequency contraction muscles of hummingbirds and rattlesnakes. Comp. Biochem. Physiol. 130B: 291–298.

International Commission on Zoological Nomenclature. 2000. Opinion 1960. *Crotalus ruber* Cope, 1892 (Reptilia, Serpentes): Specific name given precedence over that of *Crotalus exsul* Garman, 1884. Bull. Zool. Nomencl. 57: 189–190.

Irvine, F. R. 1954. Snakes as food for man. British J. Herpetol. 1: 183–189.

Ivanyi, C., and W. Altimari. 2004. Venomous reptile bites in academic research. Herpetol. Rev. 35: 49–50.

Ivanyi, C. S., and H. K. McCrystal. 2005. Translocation of venomous reptiles in Pima County, Arizona: Advice and consent or dissent? p. 32. *In* Biology of the rattlesnakes symposium (program abstracts). Loma Linda Univ., Loma Linda, California.

Ivanyi, C., S. Poulin, C. Wicker, K. Jacobs, and R. Ivanyi. 2003. Can a snake walk? Sonoran Herpetol. 16: 76.

Iverson, J. B. 1975. Notes on Nebraska reptiles. Trans. Kansas Acad. Sci. 78: 51–62.

Jackley, A. M. 1938. Badgers feed on rattlesnakes. J. Mammal. 19: 374–375.

Jackson, D. G., and T. S. Githens. 1931. Treatment of *Crotalus atrox* venom poisoning in dogs. Bull. Antiv. Inst. Am. 5: 1–6.

Jackson, J. J. 1983. Snakes of the southeastern United States. Georgia Ext. Serv., Athens.

Jackson, K. 2003. The evolution of venom-delivery systems in snakes. Zoo. J. Linnean Soc. 137: 337–354.

Jacob, J. S. 1977. An evaluation of the possibility of hybridization between the rattlesnakes *Crotalus atrox* and *C. scutulatus* in the southwestern United States. Southwest. Nat. 22: 469–485.

———. 1980. Heart rate-ventilatory response of seven terrestrial species of North American snakes. Herpetologica 36: 326–335.

Jacob, J. S., and J. S. Altenbach. 1977. Sexual color dimorphism in *Crotalus lepidus klauberi* Gloyd (Reptilia, Serpentes, Viperidae). J. Herpetol. 11: 81–84.

Jacob, J. S., and C. W. Painter. 1980. Overwinter thermal ecology of *Crotalus viridis* in the north-central plains of New Mexico. Copeia 1980: 799–805.

Jacob, J. S., S. R. Williams, and R. P. Reynolds. 1987. Reproductive activity of male *Crotalus atrox* and *C. scutulatus* (Reptilia: Viperidae) in northeastern Chihuahua, Mexico. Southwest Nat. 32: 273–276.

Jacobsen, N. 1977. The prairie rattler and its bite. North Dakota Outdoors 40(3): 15–17.

Jacobson, E. R., and J. M. Gaskin. 1992. Paramyxoviral infection of viperid snakes, pp. 415–420. *In* J. A. Campbell and E. D. Brodie, Jr. (eds.), Biology of the pitvipers. Selva, Tyler, Texas.

Jacobson, E., J. M. Gaskin, C. F. Simpson, and T. G. Terrell. 1980. Paramyxo-like virus infection in a rock rattlesnake. J. Am. Vet. Med. Assoc. 177(9): 796–799.

Jaksic, F. M., and H. W. Greene. 1984. Empirical evidence of non-correlation between tail loss frequency and predation intensity on lizards. Oikos (Copenhagen) 42: 407–411.

Jameson, D. L., and A. G. Flury. 1949. The reptiles and amphibians of the Sierra Vieja range of southwestern Texas. Texas J. Sci. 1: 54–77.

Jansen, P. W., R. M. Perkin, and D. Van Stralen. 1992. Mojave rattlesnake envenomation: Prolonged neurotoxicity and rhabdomyolysis. Ann. Emerg. Med. 21: 322–323.

Jayne, B. C. 1986. Kinematics of terrestrial snake locomotion. Copeia 1986: 915–927.

Jenkins, C. L., and C. R. Peterson. 2005. Linking landscape disturbance to the population ecology of Great Basin Rattlesnakes (*Crotalus oreganus lutosus*) in the Upper Snake River Plain. Idaho BLM Technical Bulletin 2005-07. Idaho State Office, U.S. Department of the Interior, Bureau of Land Management.

———. 2008. A trophic-based approach to the conservation biology of rattlesnakes: Linking landscape disturbance to rattlesnake populations, pp. 265–274. *In* W. K. Hayes, K. R. Beaman, M. D. Cardwell, and S. P. Bush (eds.), The biology of rattlesnakes. Loma Linda Univ. Press, Loma Linda, California.

Jenkins, C. L., C. R. Peterson, S. C. Doering, and V. A. Cobb. 2009. Microgeographic variation in reproductive characteristics among western rattlesnake (*Crotalus oreganus*) populations. Copeia 2009: 774–780.

Jennings, M. R.. 1987. Annotated check list of the amphibians and reptiles of California. South-west. Herpetol. Soc. Spec. Pub. 3: 1–48.

Jensen, J. B., C. D. Camp, W. Gibbons, and M. J. Elliott (eds.). 2008. Amphibians and reptiles of Georgia. Univ. Georgia Press, Athens.

Jiménez-Porras, J. M. 1961. Biochemical studies on venom of the rattlesnake, *Crotalus atrox atrox*. J. Exp. Zool. 148: 251–258.

———. 1971. Biochemistry of snake venoms, pp. 43–85. *In* S. A. Minton (ed.), Snake venoms and envenomation. Marcel Dekker, New York.

Jochimsen, D., and C. R. Peterson. 2004. Factors influencing the road mortality of snakes on the eastern Snake River Plain, p. 18. *In* Proceedings of the Snake Ecology Group 2004 Conference, Jackson County, Illinois.

John, T. R., L. A. Smith, and I. I. Kaiser. 1994. Genomic sequences encoding the acidic and basic subunits of Mojave toxin: Unusually high sequence identity of non-coding regions. Gene 139: 229–234.

Johnson, B. D. 1967. Some interrelationships of selected *Crotalus* and *Agkistrodon* venom properties and their relative lethalities. Diss. Abstr. 12B: 4606.

Johnson, B. D., J. C. Tullar, and H. L. Stahnke. 1966. A quantitative protozoan bio-assay method for determining venom potencies. Toxicon 3: 297–300.

Johnson, B. D., H. L. Stahnke, and R. Koonce. 1967. A method for estimating *Crotalus atrox* venom concentrations. Toxicon 5: 35–38.

Johnson, B. D., J. Hoppe, R. Rogers, and H. L. Stahnke. 1968a. Characteristics of venom from the rattlesnake *Crotalus horridus atricaudatus*. J. Herpetol. 2: 107–112.

Johnson, B. D., H. L. Stahnke, and J. A. Hoppe. 1968b. Variations of *Crotalus scutulatus* raw venom concentrations. J. Arizona Acad. Sci. 5: 41–42.

Johnson, D. H., M. D. Bryant, and A. H. Miller. 1948. Vertebrate animals of the Providence Mountains area of California. Univ. California Publ. Zool. 48: 221–376.

Johnson, E. K. 1987. Stability of venoms from the northern Pacific rattlesnake (*Crotalus viridis oreganus*). Northwest. Sci. 61: 110–113.

Johnson, E. K., K. V. Kardong, and C. L. Ownby. 1987. Observations on white and yellow venoms from an individual southern Pacific rattlesnake (*Crotalus viridis helleri*). Toxicon 25: 1169–1180.

Johnson, E. M. 1975. Faunal and floral material from a Kansas City Hopewell site: Analysis and interpretation. Occ. Pap. Mus. Texas Tech. Univ. 36: 1–37.

Johnson, G. 1995. Spatial ecology, habitat preference, and habitat management of the eastern massasauga, *Sistrurus c. catenatus* in a New York weakly-minerotrophic peatland. Ph.D. diss., State Univ. New York, Coll. Env. Sci. Forest., Syracuse, New York.

Johnson, G. R., and A. L. Bieber. 1988. Mojave toxin: Rapid purification, heterogeneity and resistance to denaturation by urea. Toxicon 26: 337–351.

Johnson, M. L. 1942. A distributional check-list of the reptiles of Washington. Copeia 1942: 15–18.

———. 1995. Reptiles of the state of Washington. Northwest Fauna 3: 5–80.

Johnson, R. G. 1955. The adaptive and phylogenetic significance of vertebral form in snakes. Evolution 9: 367–388.

———. 1956. The origin and evolution of the venomous snakes. Evolution 10: 56–65.

Johnson, T. B. 1983. Status Report: *Crotalus willardi willardi* (Meek, 1905). U.S. Fish Wildl. Serv. Contr. 14-16-0002-81-224.

———. 1987. Banded rock rattlesnake. Wildl. Views 30(5): 4.

Johnson, T. B., and G. S. Mills. 1982. A preliminary report on the status of *Crotalus lepidus*, *C. pricei* and *C. willardi* in southeastern Arizona. U.S. Fish Wildl. Serv. Contr. 14-16-0002-81-224.

Johnson, T. R. 1987. The amphibians and reptiles of Missouri. Missouri Department of Conservation, Jefferson City.

Jones, A. 1997. Big reptiles, big lies. Rept. Amphib. Mag. 51: 22–27.

Jones, M., J. Troy, and M. R. J. Forstner. 2008. *Crotalus atrox* (western diamondback rattlesnake). Herpetol. Rev. 39: 240.

Jørgensen, D., and C. C. Gates. 2007. *Crotalus viridis viridis* (prairic rattlesnake): Behavior. Herpetol. Rev. 38: 87–88.

Jørgensen, D., C. C. Gates, and D. P. Whiteside. 2008. Movements, migrations, and mechanisms: A review of radiotelemetry studies of prairie (*Crotalus viridis viridis*) and western (*Crotalus oreganus*) rattlesnakes, pp. 303–316. *In* W. K. Hayes, K. R. Beaman, M. D. Cardwell, and S. P. Bush (eds.), The biology of rattlesnakes. Loma Linda Univ. Press, Loma Linda, California.

Joseph, J. S., M. C. M. Chung, P. J. Mirtschin, and R. M. Kini. 2002. Effect of snake venom procoagulants on snake plasma: Implications for the coagulation cascade of snakes. Toxicon 40: 175–183.

Juckett, G., and J. G. Hancox. 2002. Venomous snakebites in the United States: Management review and update. Am. Fam. Physician 65(7): 1367–1374, 1377.

Julian, G. 1951. Sex ratios of the winter populations. Herpetologica 7: 21–24.

Jurado, J. D., E. D. Rael, C. S. Lieb, E. Nakayasu, W. K. Hayes, S. P. Bush, and J. A. Ross. 2007. Complement inactivating proteins and intraspecies venom variation in *Crotalus oreganus helleri*. Toxicon 49: 339–350.

Kain, P. 1995. Home range, seasonal movements, and behavior of the eastern diamondback rattlesnake (*C. adamanteus*). Master's thesis, Southeastern Louisiana Univ., Hammond.

Kaiser, I. I., J. L. Middlebrook, M. H. Crumrine, and W. W. Stevenson. 1986. Cross-reactivity and neutralization by rabbit antisera raised against crotoxin, its subunits, and two related toxins. Toxicon 24: 669–678.

Kallech-Ziri, O., J. Luis, M. El Ayeb, and N. Marrakchi. 2007. Snake venom disintegrins: Classification and therapeutic potential. Arch. Inst. Pasteur Tunis 84(1–4): 29–37.

Kardong, K. V. 1975. Prey capture in the cottonmouth snake (*Agkistrodon piscivorus*). J. Herpetol. 9: 169–175.

———. 1980. Gopher snakes and rattlesnakes: Presumptive Batesian mimicry. Northwest. Sci. 54: 1–4.

———. 1982a. The evolution of the venom apparatus in snakes from colubrids to viperids & elapids. Mem. Inst. Butantan 46: 105–118.

———. 1986a. Predatory strike behavior of the rattlesnake, *Crotalus viridis oreganus*. J. Comp. Psychol. 100: 304–314.

———. 1986b. The predatory strike of the rattlesnake: When things go amiss. Copeia 1986: 816–820.

———. 1990. General skull, bone, and muscle variation in *Agkistrodon* and related genera, pp. 573–581. *In* H. K. Gloyd and R. Conant, Snakes of the *Agkistrodon* complex: A monographic review. Soc. Stud. Amphib. Rept. Contrib. Herpetol. 6.

———. 1993. The predator behavior of the northern Pacific rattlesnake (*Crotalus viridis oreganus*): Laboratory versus wild mice as prey. Herpetologica 49: 457–463.

———. 1996. Mechanical damage inflicted by fangs on prey during predatory strikes by rattlesnakes, *Crotalus viridis oreganus*. Bull. Maryland Herpetol. Soc. 32: 113–118.

Kardong, K. V., and V. L. Bels. 1998. Rattlesnake strike behavior: Kinematics. J. Exp. Biol. 201: 837–850.

Kardong, K. V., and S. P. Mackessy. 1991. The strike behavior of a congenitally blind rattlesnake. J. Herpetol. 25: 208–211.

Kardong, K. V., and T. L. Smith. 2002. Proximate factors involved in rattlesnake predatory behavior: A review, pp. 253–266. *In* G. W. Schuett, M. Höggren, M. E. Douglas, and H. W. Greene (eds.), Biology of the vipers. Eagle Mountain Publ., Eagle Mountain, Utah.

Karnes, M., and R. Tumlison. 2003. *Crotalus atrox* (western diamond-backed rattlesnake). Herpetol. Rev. 34: 387.

Kauffeld, C. F. 1939. If you like danger—there are snakes. Outdoor Life 83(3): 32–33, 67–68.

———. 1943a. Field notes on some Arizona reptiles and amphibians. Am. Midl. Nat. 29: 342–359.

———. 1943b. Growth and feeding of new-born Price's and green rock rattlesnakes. Am. Midl. Nat. 29: 607–614.

———. 1955. Off with their skins. Animaland 12(5): 1–4.

———. 1957. Snakes and snake hunting. Hanover House, Garden City, New York.

———. 1961. Massasauga land. Bull. Philadelphia Herpetol. Soc. 9(3): 7–13.

————. 1969. Snakes: The keeper and the kept. Doubleday, Garden City, New York.

Kawata, K. 2004. Carl and his rattlesnakes: Herpetology at the Staten Island Zoo. Herpetol. Rev. 35: 316–320.

Keegan, H. L. 1944. Indigo snakes feeding upon poisonous snakes. Copeia 1944: 59.

Keegan, H. L., and T. F. Andrews. 1942. Effects of crotalid venom of North American snakes. Copeia 1942: 251–254.

Keegan, K. A., R. N. Reed, A. T. Holycross, and C. W. Painter. 1999. *Crotalus willardi* (ridgenose rattlesnake): Maximum length. Herpetol. Rev. 30: 100.

Keenlyne, K. D. 1972. Sexual differences in feeding habits of *Crotalus horridus horridus*. J. Herpetol. 6: 234–237.

————. 1978. Reproductive cycles in two rattlesnakes. Am. Midl. Nat. 100: 368–375.

Keiser, E. D., Jr. 1971. The poisonous snakes of Louisiana and the emergency treatment of their bites. Louisiana Wildl. Fish. Comm., Baton Rouge.

Keiser, E. D., Jr., and L. D. Wilson. 1979. Checklist and key to the herpetofauna of Louisiana. 2nd ed. Lafayette Natur. Hist. Mus. Tech. Bull. 1: 1–49.

Kelly, H. A., A. W. Davis, and H. C. Robertson. 1936. Snakes of Maryland. Nat. Hist. Soc. Maryland, Baltimore.

Kemnitzer, J. W., Jr. (ed.). 2006. Rattlesnake adventures: Hunting with the oldtimers. Krieger, Malabar, Florida.

Kemper, W. F., S. L. Lindstedt, L. K. Hartzler, J. W. Hicks, and K. E. Conley. 2001. Shaking up glycolysis: Sustained, high lactate flux during aerobic rattling. Proc. Nat. Acad. Sci. 98: 723–728.

Kennedy, J. P. 1964. Natural history notes on some snakes of eastern Texas. Texas J. Sci. 16: 210–215.

Kennicott, R. 1861. On three new forms of rattlesnakes. Proc. Acad. Nat. Sci. Philadelphia 13: 206–208.

Keogh, J. S., and V. Wallach. 1999. Allometry and sexual dimorphism in the lung morphology of prairie rattlesnakes, *Crotalus viridis viridis*. Amphibia-Reptilia 20: 377–389.

Kerchove, C. M., M. S. A. Luna, M. B. Zablith, M. F. M. Lazari, S. S. Smaili, and N. Yamanouye. 2008. α_1-adrenoceptors trigger the snake venom production cycle in secretory cells by activating phosphatidylinositol 4,5-bisphosphate hydrolysis and ERK signaling pathway. Comp. Biochem. Physiol. 150A: 431–437.

Keyler, D. E. 2005. Venomous snakebites: Minnesota & Upper Mississippi River Valley 1982–2002. Minnesota Herpetol. Soc. Occ. Pap. 7: 1–28.

————. 2008. Timber rattlesnake (*Crotalus horridus*) envenomations in the upper Mississippi River Valley, pp. 569–580. *In* W. K. Hayes, K. R. Beaman, M. D. Cardwell, and S. P. Bush (eds.), The biology of rattlesnakes. Loma Linda Univ. Press, Loma Linda, California.

Keyler, D., and B. Oldfield. 1992. Velvet tails in the Blufflands. Minnesota Volunteer May–June: 32–43.

Khole, V. 1991. Toxicities of snake venoms and their components, pp. 405–470. *In* A. T. Tu (ed.), Reptile venoms and toxins: Handbook of natural toxins, vol. 5. Marcel Dekker, New York.

Killebrew, F. C., and T. L. James. 1983. *Crotalus viridis viridis* (prairie rattlesnake): Coloration. Herpetol. Rev. 14: 74.

Kilmon, J., and H. Shelton (eds.). 1981. Rattlesnakes in America. Shelton Press, Sweetwater, Texas.

Kimon, J. A. 1976. High tolerance to snake venom by the Virginia opossum, *Didelphis virginiana*. Toxicon 14: 337–340.

King, K. A. 1975. Unusual food item of the western diamondback rattlesnake (*Crotalus atrox*). Southwest. Nat. 20: 416–417.

King, M., and D. Duvall. 1990. Prairie rattlesnake seasonal migrations: Episodes of movement, vernal foraging and sex differences. Anim. Behav. 39: 924–935.

King, M., D. McCarron, D. Duvall, G. Baxter, and W. Gern. 1983. Group avoidance of conspecific but not interspecific chemical cues by prairie rattlesnakes (*Crotalus viridis*). J. Herpetol. 17: 196–198.

King, W. 1939. A survey of the herpetology of Great Smoky Mountains National Park. Am. Midl. Nat. 21: 531–582.

Kinney, C., G. Abishahin, and B. A. Young. 1998. Hissing in rattlesnakes: Redundant signaling or inflationary epiphenomenon? J. Exp. Zool. 280: 107–113.

Kissner, K. J., M. R. Forbes, and D. M. Secoy. 1997. Rattling behavior of prairie rattlesnakes (*Crotalus viridis viridis*, Viperidae) in relation to sex, reproductive status, body size, and body temperature. Ethology 103: 1042–1050.

Kitchens, C. S. 1992. Hemostatic aspects of envenomation by North American snakes. Hematology/Oncology Clinics of North America 6: 1188–1195.

Kitchens, C. S., and L. H. S. Van Mierop. 1983. Mechanism of defibrination in humans after envenomation by the eastern diamondback rattlesnake. Am. J. Hematol. 14: 345–354.

Kitchens, C. S., S. Hunter, and L. H. S. Van Mierop. 1987. Severe myonecrosis in a fatal case of envenomation by the canebrake rattlesnake (*Crotalus horridus atricaudatus*). Toxicon 25: 455–458.

Klauber, L. M. 1926. The snakes of San Diego County, California. Copeia 1926(155): 144.

———. 1927. Some observations on the rattlesnakes of the extreme Southwest. Bull. Antiv. Inst. Am. 1: 7–21.

———. 1928. The collection of rattlesnake venom. Bull. Antiv. Inst. Am. 2: 11–18.

———. 1929. Range extensions in California. Copeia 1929(170): 15–22.

———. 1930a. Differential characteristics of southwestern rattlesnakes allied to *Crotalus atrox*. Bull. Zool. Soc. San Diego 6: 1–73.

———. 1930b. Some new and renamed subspecies of *Crotalus confluentis* Say, with remarks on related species. Trans. San Diego Soc. Nat. Hist. 6: 95–144.

———. 1931. A statistical survey of the snakes of the southern border of California. Bull. Zool. Soc. San Diego 8: 1–93.

———. 1932. Amphibians and reptiles observed enroute to Hoover Dam. Copeia 1932: 118–128.

———. 1934. An addition to the fauna of New Mexico and a delegation. Copeia 1934: 52.

———. 1935. A new subspecies of *Crotalus confluentus*, the prairie rattlesnake. Trans. San Diego Soc. Nat. Hist. 8: 75–90.

———. 1936. A statistical study of the rattlesnakes: I. Introduction. II. Sex ratio. III. Birth rate. Occ. Pap. San Diego Soc. Nat. Hist. 1: 1–24.

———. 1936–40. A statistical study of the rattlesnakes. Parts I–VII. Trans. San Diego Soc. Nat. Hist. 1–6.

———. 1937. A statistical study of the rattlesnakes: IV. The growth of the rattlesnake. Occ. Pap. San Diego Soc. Nat. Hist. 3: 1–56.

———. 1938a. A statistical study of the rattlesnakes. Occ. Pap. San Diego Soc. Nat. Hist. 4: 1–53.

———. 1938b. Notes from a herpetological diary, I. Copeia 1938: 191–197.

———. 1939. A statistical study of the rattlesnakes. VI. Fangs. Occ. Pap. San Diego Soc. Nat. Hist. 5: 1–61.

———. 1941. Four papers on the application of statistical methods to herpetological problems. Bull. Zool. Soc. San Diego 17: 1–95.

———. 1943a. Tail-length differences in snakes with notes on sexual dimorphism and the coefficient of divergence. Bull. Zool. Soc. San Diego 18: 1–60.

———. 1943b. The correlation of variability within and between rattlesnake populations. Copeia 1943: 115–118.

———. 1944. The sidewinder, *Crotalus cerastes*, with description of a new subspecies. Trans. San Diego Soc. Nat. Hist. 10: 91–126.

———. 1949a. The relationship of *Crotalus ruber* and *Crotalus lucasensis*. Trans. San Diego Soc. Nat. Hist. 11: 57–60.

———. 1949b. Some new and revived subspecies of rattlesnakes. Trans. San Diego Soc. Nat. Hist. 11: 61–116.

———. 1949c. The subspecies of the ridge-nosed rattlesnake, *Crotalus willardi*. Trans. San Diego Soc. Nat. Hist. 11: 121–140.

———. 1952. Taxonomic studies of the rattlesnakes of mainland Mexico. Bull. Zool. Soc. San Diego 26: 1–143.

———. 1963. A new insular subspecies of the speckled rattlesnake. Trans. San Diego Soc. Nat. Hist. 13: 73–80.

————. 1972. Rattlesnakes: Their habits, life histories, and influence on mankind. 2nd ed. Univ. California Press, Berkeley.

Klemens, M. W. 1993. Amphibians and reptiles of Connecticut and adjacent regions. Bull. Connecticut St. Geol. Nat. Hist. Surv. 112: 1–318.

Klenzendorf, S. A., D. J. Lee, M. R. Vaughan, and R. B. Duncan, Jr. 2004. *Crotalus horridus* (timber rattlesnake): Defense and black bear death. Herpetol. Rev. 35: 61–62.

Klingenberg, R. J. 1993. Understanding reptile parasites: A basic manual for herpetoculturists & veterinarians. Spl. Ed. Adv. Vivar. Syst. Herpetol. Libr., Lakeside, California.

Knight, A., D. Styer, S. Pelikan, J. A. Campbell, L. D. Densmore III, and D. P. Mindell. 1993. Choosing among hypotheses of rattlesnake phylogeny: A best-fit rate test for DNA sequence data. Syst. Biol. 42: 356–367.

Knight, J. L., and J. T. Collins. 1977. The amphibians and reptiles of Cheyenne County, Kansas. Rep. St. Biol. Surv. Kansas 15: 1–18.

Knight, R. L., and A. W. Erickson. 1976. High incidence of snakes in the diet of nesting redtailed hawks. Raptor Res. 10: 108–111.

Koch, E. D., and C. R. Peterson. 1995. Amphibians & reptiles of Yellowstone and Grand Teton National Parks. Univ. Utah Press, Salt Lake City.

Kocholaty, W. F., E. B. Ledford, J. G. Daly, and T. A. Billings. 1971. Toxicity and some enzymatic properties and activities of venoms of Crotalidae, Elapidae and Viperidae. Toxicon 9: 131–138.

Kochva, E., and C. Gans. 1966. Histology and histochemistry of venom glands of some crotaline snakes. Copeia 1966: 506–515.

Koenig, H. F., and J. L. La Grone. 2000. *Crotalus molossus* (blacktail rattlesnake). Herpetol. Rev. 31: 254–255.

Komori, Y., and T. Nikai. 1998. Chemistry and biochemistry of kallikrein-like enzyme from snake venoms. J. Toxicol. Toxin Rev. 17: 261–277.

Komori, Y., T. Nikai, and H. Sugihara. 1988. Biochemical and physiological studies on a kallikrin-like enzyme from the venom of *Crotalus viridis viridis* (prairie rattlesnake). Biochem. Biophys. Acta 967: 92–102.

Krebs, J., T. G. Curro, and L. G. Simmons. 2006. The use of a venomous reptile restraining box at Omaha's Henry Doorly Zoo, p. 40. *In* S. A. Seifert (ed.), Snakebites in the new millennium: A state-of-the-art symposium, University of Nebraska Medical Center, 21–23 October 2005, Omaha, Nebraska. J. Med. Toxicol. 2: 29–45.

Krochmal, A. R., and G. S. Bakken. 2003. Thermoregulation is the pits: Use of thermal radiation for retreat site selection by rattlesnakes. J. Exp. Biol. 206: 2539–2545.

————. 2005. Einstein and Klauber road cruise in heaven: How understanding physics can unlock the secrets of rattlesnake biology, p. 35. *In* Biology of the Rattlesnakes Symposium (program abstracts). Loma Linda Univ., Loma Linda, California.

Krochmal, A. R., G. S. Bakken, and T. J. LaDuc. 2004. Heat in evolution's kitchen: Evolutionary perspectives on the functions and origin of the facial pit of pitvipers (Viperidae: Crotalinae). J. Exp. Biol. 207: 4231–4238.

Kruilikowski, L. 2004. Snakes of New England: Photographic and natural history study. Luv Life Publ., Old Lyme, Connecticut.

Kuhn, B. F., M. J. Rochelle, and K. V. Kardong. 1991. Effects of rattlesnake (*Crotalus viridis oreganus*) envenomation upon the mobility and death rate of laboratory mice (*Mus musculus*) and wild mice (*Peromyscus maniculatus*). Bull. Maryland Herpetol. Soc. 27: 189–194.

Kurecki, T., and L. F. Kress. 1985a. Purification and partial characterization of the hemorrhagic factor from the venom of *Crotalus adamanteus* (eastern diamondback rattlesnake). Toxicon 23: 657–668.

————. 1985b. Purification and partial characterization of a high molecular weight metalloproteinase from the venom of *Crotalus adamanteus* (eastern diamondback rattlesnake). Toxicon 23: 855–863.

La Bonte, J. P. 2008. Ontogeny of prey preference in the southern Pacific rattlesnake, *Crotalus oreganus helleri*, pp. 169–174. *In* W. K. Hayes, K. R. Beaman, M. D. Cardwell, and S. P. Bush (eds.), The biology of rattlesnakes. Loma Linda Univ. Press, Loma Linda, California.

La Duc, T. J. 2002. Does a quick offense equal a quick defense? Kinematic comparisons of predatory and defensive strikes in the western diamond-backed rattlesnake (*Crotalus atrox*), pp. 267–278. *In* G. W. Schuett, M. Höggren, M. E. Douglas, and H. W. Greene (eds.), Biology of the vipers. Eagle Mountain Publ., Eagle Mountain, Utah.

La Duke, T. C. 1991. The fossil snakes of Pit 91, Rancho La Brea, California. Los Angeles Co. Nat. Hist. Mus. Contrib. Sci. 424: 1–28.

La Grange, R. G., and F. E. Russell. 1970. Blood platelet studies in man and rabbits following *Crotalus* envenomation. Proc. West. Pharmacol. Soc. 13: 99–105.

Lamberski, N., K. Seurynck, C. Gregory, T. A. Moore, C. S. Pfass, and R. Hines. 2002. Recovery of a pathogenic *Salmonella arizona* from four snakes at the Riverbanks Zoo. J. Herpetol. Med. Surg. 12(4): 17–22.

Lamson, G. H. 1935. The reptiles of Connecticut. Connecticut St. Geol. Nat. Hist. Surv. Bull. 54: 1–35.

Landers, J. L. 1987. Prescribed burning for managing wildlife in southeastern pine forests. U.S. For. Serv. Gen. Tech. Rep. 50(65): 19–27.

Landreth, H. F. 1973. Orientation and behavior of the rattlesnake, *Crotalus atrox*. Copeia 1973: 26–31.

Langebartel, D. A., and H. M. Smith. 1954. Summary of the Norris Collection of reptiles and amphibians from Sonora, Mexico. Herpetologica 10: 125–136.

La Rivers, I., III. 1973. Effects of rattlesnake venom on rattlesnakes *Crotalus viridis oreganus/ Crotalus viridis lutosus*. HISS News-J. 1(5): 161–162.

———. 1976. I—Some comments on snakebite treatment in the United States, and II—An account of a human envenomation by an adult western diamondback rattlesnake (*Crotalus atrox* Baird & Girard). Occ. Pap. Biol. Soc. Nevada 42: 1–6.

Larson, P., and L. Larson. 1990. The deserts of the Southwest: A Sierra Club naturalist's guide. 2nd ed. Sierra Club Books, San Francisco.

Lasher, D. N. 1980. A bicephalic *Crotalus horridus* from Alabama. Herpetol. Rev. 11: 89.

Lasky, W. R. 1980. A case of spitting rattlesnake. Herpetology 10(3): 14.

Latreille, P. A. 1801. *In* C. S. Sonnini and P. A. Latreille, Histoire naturelle des reptiles . . . Vol. 3. Deterville, Paris.

Laughlin, H. E., and B. J. Wilks. 1962. The use of sodium pentobarbital in population studies of poisonous snakes. Texas J. Sci. 14: 188–191.

Laveran, A. 1902. Sur quelques Hemogregarines des Ophidiens. Comp. Rend. Acad. Sci. 135: 1036–1040.

Lavín-Murcio, P. A., and K. V. Kardong. 1995. Scents related to venom and prey as cues in the poststrike trailing behavior of rattlesnakes, *Crotalus viridis oreganus*. Herpetologica 51: 39–44.

Lavín-Murcio, P. A., B. G. Robinson, and K. V. Kardong. 1993. Cues involved in relocation of struck prey by rattlesnakes, *Crotalus viridis oreganus*. Herpetologica 49: 463–469.

Lawson, R., J. B. Slowinski, B. I. Crother, and F. T. Burbrink. 2005. Phylogeny of the Colubroidea (Serpentes): New evidence from mitochrondrial and nuclear genes. Mol. Phylogenet. Evol. 37: 581–601.

Lazcano, D., J. B. Leal, and R. W. Bryson, Jr. 2004. *Crotalus lepidus* (rock rattlesnake): Diet. Herpetol. Rev. 35: 62–63.

Lazcano, D., Jr., A. Kardon, and R. W. Bryson. 2005. Reproduction in montane rattlesnakes: Endoherpetology of *Crotalus lepidus morulus*, p. 36. *In* Biology of the rattlesnakes symposium (program abstracts). Loma Linda Univ., Loma Linda, California.

Lazcano, D., D. J. Galvan, C. Garcia de la Pena, and G. Castaneda. 2007a. *Crotalus lepidus* captive maintenance of the rock rattlesnake. Reptilia (Great Britain) 53: 27–31.

———. 2007b. Terrarienhaltung von Gebirgsklapperschlangen. Reptilia-D (German) 12(4): 38–41.

Lazcano, D., A. Sánchez-Almazán, C. García-de la Peña, G. Castañeda, and A. J. Contreras-Balderas. 2007c. Notes on Mexican herpetofauna 9: Herpetofauna of a fragmented *Juniperus* forest in the state natural protected area of San Juan y Puentes, Aramberri, Nuevo León, Mexico. Bull. Chicago Herpetol. Soc. 42: 1–6.

Lazcano, D., J. A. Contreras-Lozano, J. Gallardo-Valdez, C. García del Peña, and G. Castañeda.

2009a. Notes on Mexican herpetofauna 11: Herpetological diversity in Sierra "Cerro de La Silla" (Saddleback Mountain), Nuevo León, Mexico. Bull. Chicago Herpetol. Soc. 44: 21–27.

Lazcano, D., M. A. Salinas-Camarena, and J. A. Contreras-Lozano. 2009b. Notes on Mexican herpetofauna 12: Are roads in Nuevo León, Mexico, taking their toll on snake populations? Bull. Chicago Herp. Soc. 44: 69–75.

Lazcano, D., W. L. Farr, P. A. Lavín-Murcio, J. A. Contreras-Lozano, A. Kardon, S. Narváez-Torres, and J. A. Chávez-Cisneros. 2009c. Notes on Mexican herpetofauna 13: DORs in the municipality of Aldama, Tamaulipas, Mexico. Bull. Chicago Herpetol. Soc. 44: 181–195.

Le Clere, J. 1996. Snakes of Minnesota. Minnesota Herpetol. Soc. Newsl. 16(1): 9–10.

Lee, C.-Y. 1979. Snake venoms. Springer-Verlag, New York.

Lee, J. R. 2009. The herpetofauna of the Camp Shelby Joint Forces Training Center in the Gulf Coastal Plain of Mississippi. Southeast. Natur. 8: 639–652.

Lee, R. K. K., D. A. Chiszar, and H. M. Smith. 1988. Post-strike orientation of the prairie rattlesnake facilitates location of envenomated prey. Ethology 6: 129–134.

Lee, R. K. K., D. A. Chiszar, H. M. Smith, and K. Kandler. 1992. Chemical and orientational cues mediate selection of prey trails by prairie rattlesnakes (*Crotalus viridis*). J. Herpetol. 26: 95–98.

Lee, W.-H., and Y. Zhang. 2003. Research development of snake venom C-type lectins. Zool. Res. 24(2): 151–160.

Lemm, J. M. 2006. Field guide to amphibians and reptiles of the San Diego region. Univ. California Press, Berkeley.

Lemos-Espinal, J. A., and H. M. Smith. 2008a. Amphibians & reptiles of the state of Colima, Mexico. Bibliomania, Salt Lake City, Utah.

———. 2008b. Amphibians & reptiles of the state of Chihuahua, Mexico. Bibliomania, Salt Lake City, Utah.

Lemos-Espinal, J. A., D. Chiszar, and H. M. Smith. 2000. *Crotalus lepidus lepidus* (mottled rock rattlesnake). Herpetol. Rev. 31: 113.

———. 2004. *Crotalus basiliscus* (Mexican west coast rattlesnake). Herpetol. Rev. 35: 83.

Lemos-Espinal, J. A., P. Heimes, and H. M. Smith. 2007. *Crotalus willardi amabilis* (del Nido ridge-nosed rattlesnake): Diet. Herpetol. Rev. 38: 205.

Levine, N. D. 1980. Some corrections of coccidian (Apicomplexa: Protozoa) nomenclature. J. Parasitol. 66: 830–834.

Leviton, A. E. 1972. Reptiles and amphibians of North America. Doubleday, Garden City, New York.

Lewis, T. H. 1949. Dark coloration in the reptiles of the Tularosa Malpais, New Mexico. Copeia 1949: 181–184.

———. 1950. The herpetofauna of the Tularosa Basin and Organ Mountains of New Mexico with notes on some ecological features of the Chihuahuan Desert. Herpetologica 6: 1–10.

———. 1951. Dark coloration in the reptiles of the Malpais of the Mexican Border. Copeia 1951: 311–312.

Li, Q., and C. L. Ownby. 1994. Cross reactivities of monoclonal antibodies against hemorrhagic toxins of prairie rattlesnake (*Crotalus viridis viridis*) venom. Comp. Biochem. Physiol. 107B: 51–59.

Li, Q., T. R. Colberg, and C. L. Ownby. 1993. Purification and characterization of two high molecular weight hemorrhagic toxins from *Crotalus viridis viridis* venom using monoclonal antibodies. Toxicon 31: 711–722.

Lillywhite, H. B. 1974. Activity of snakes in recently burned chaparral. Bull. Ecol. Soc. Am. 55: 43.

———. 1982. Cannibalistic carrion ingestion by the rattlesnake, *Crotalus viridis*. J. Herpetol. 16: 95.

———. 1993. Subcutaneous compliance and gravitational adaptation in snakes. J. Exp. Zool. 267: 557–562.

———. 2008. Dictionary of herpetology. Krieger, Melbourne, Florida.

Lillywhite, H. B., P. de Delva, and B. P. Noonan. 2002. Patterns of gut passage time and the chronic retention of fecal mass in viperid snakes, pp. 497–506. In G. W. Schuett, M. Höggren, M. E. Douglas, and H. W. Greene (eds.), Biology of the vipers. Eagle Mountain Publ., Eagle Mountain, Utah.

Linder, A. D., and E. Fichter. 1977. The amphibians and reptiles of Idaho. Idaho St. Univ. Press, Pocatello.

Lindsay, G. E. 1962. The Belvedere expedition to the Gulf of California. Trans. San Diego Soc. Nat. Hist. 13: 1–44.

Liner, E. A. 1994. Scientific and common names for the amphibians and reptiles of Mexico in English and Spanish. Soc. Stud. Amphib. Rept. Herpetol. Circ. 23.

Liner, E. A., and A. H. Chaney. 1986. *Crotalus lepidus lepidus* (mottled rock rattlesnake): Reproduction. Herpetol. Rev. 17: 89.

Linnaeus, C. 1758. Systema naturae per regna tria naturae, secundum classes, ordines, genera, species, cum characteribus, differentiis, synonymis, locis. 10th ed. Vol. 1. Stockholm, Sweden.

Linsdale, J. M. 1927. Amphibians and reptiles of Doniphan County, Kansas. Copeia 1927(164): 75–81.

———. 1940. Amphibians and reptiles in Nevada. Proc. Am. Acad. Arts Sci. 73: 197–257.

Linsdale, J. M., and L. P. Tevis, Jr. 1951. The dusky-footed wood rat. Univ. California Press, Berkeley.

Lintner, C. P., D. E. Keyler, and E. F. Bilden. 2006. Prairie rattlesnake (*Crotalus viridis viridis*) envenomation: Recurrent coagulopathy in a child treated with Immune Fab, pp. 33–34. *In* S. A. Seifert (ed.), Snakebites in the new millennium: A state-of-the-art symposium, University of Nebraska Medical Center, 21–23 October 2005, Omaha, Nebraska. J. Med. Toxicol. 2: 29–45.

Linzey, D. W. 1979. Snakes of Alabama. Strode, Huntsville, Alabama.

Linzey. D. W., and M. J. Clifford. 1981. Snakes of Virginia. Univ. Press Virginia, Charlottesville.

Linzey. D. W., and A. V. Linzey. 1968. Mammals of the Great Smoky Mountains National Park. J. Elisha Mitchell Sci. Soc. 84: 383–414.

Little, E. L., Jr. 1940. Amphibians and reptiles of the Roosevelt Reservoir area, Arizona. Copeia 1940: 260–265.

Liu, C.-Z., H.-C. Peng, and T.-F. Huang. 1995. Crotavirin, a potent platelet aggregation inhibitor purified from the venom of the snake *Crotalus viridis*. Toxicon 33: 1289–1298.

Loewen, S. L. 1940. On some reptilian cestodes of the genus *Oochoristica* (Anoplocephalidae). Trans. Am. Micro. Soc. 59: 511–518.

Logier, E. B. S. 1939. The reptiles of Ontario. Royal Ontario Mus. Handbk. (4): 1–63.

———. 1958. The snakes of Ontario. Univ. Toronto Press, Toronto, Canada.

Logier, E. B. S., and G. C. Toner. 1961. Check list of amphibians and reptiles of Canada and Alaska. Contrib. Royal Ontario Mus. Zool. Palaeontol. 53: 1–92.

Lohoefener, R., and R. Altig. 1983. Mississippi herpetology. Mississippi St. Univ. Res. Cent. Nat. Space Tech. Lab. Bull. 1: 1–66.

Lomonte, B., Y. Angulo, and C. Santamaría. 2003. Comparative study of synthetic peptides corresponding to region 115–129 in Lys[49] myotoxic phospholipases A_2 from snake venoms. Toxicon 42: 307–312.

Long, T., and A. Pires-Dasilva. 2009. The snake box: A novel approach for safely restraining venomous snakes. Herpetol. Bull. 108: 34–35.

Lönnberg, A. J. E. 1894. Notes on reptiles and batrachians collected in Florida in 1892 and 1893. Proc. U.S. Natl. Mus. 17(1003): 317–339. (Separates issued 15 November 1894 and completed volume in 1895.)

Loomis, R. B. 1951. Increased rate of ecdysis in *Crotalus,* caused by chiggers damaging a facial pit. Herpetologica 7: 83–84.

Loomis, R. B., and R. C. Stephens. 1967. Additional notes on snakes taken in and near Joshua Tree National Monument, California. Bull. So. California Acad. Sci. 66: 1–22.

Loprinzi, C. L., J. Hennessee, L. Tamsky, and T. E. Johnson. 1983. Snake antivenin administration in a patient allergic to horse serum. South. Med. J. 76: 501–502.

Lowe, C. H. 1942. Notes on mating of desert rattlesnakes. Copeia 1942: 261–262.

———. 1948. Territorial behavior in snakes and the so-called courtship dance. Herpetologica 4: 129–135.

———. 1964. The vertebrates of Arizona. Univ. Arizona Press, Tucson.

Lowe, C. H., and K. S. Norris, Jr. 1950. Aggressive behavior in male sidewinders, *Crotalus cerastes,* with a discussion of aggressive behavior and territoriality in snakes. Chicago Acad. Sci. Nat. Hist. Misc. 66: 1–13.

———. 1954. Analysis of the herpetofauna of Baja California, Mexico. Trans. San Diego Soc. Nat. Hist. 12: 47–64.

———. 1955. Analysis of the herpetofauna of Baja California, Mexico. III. New and revived reptilian subspecies of Isla de San Estéban, Gulf of California, Sonora, Mexico, with notes on other satellite islands of Isla Tiburón. Herpetologica 11: 89–96.

Lowe, C. H., D. S. Hinds, P. J. Lardner, and K. E. Justice. 1967. Natural free-running period in vertebrate animal populations. Science (New York) 156(3774): 531–534.

Lowe, C. H., C. R. Schwalbe, and T. B. Johnson. 1986. The venomous reptiles of Arizona. Arizona Game Fish Dept., Phoenix.

Lowell, J. A. 1957. A bite by a sidewinder rattlesnake. Herpetologica 13: 135–136.

Luck, J. M., and L. Keeler. 1929. The blood chemistry of two species of rattlesnakes, *Crotalus atrox* and *Crotalus oregonus*. J. Biol. Chem. 82: 703–707.

Ludlow, M. E. 1981. Observations on *Crotalus v. viridis* (Rafinesque) and the herpetofauna of the Ken-Caryl Ranch, Jefferson County, Colorado. Herpetol. Rev. 12: 50–52.

Ludwig, M., and H. Rahn. 1943. Sperm storage and copulatory adjustment in the prairie rattlesnake. Copeia 1943: 15–18.

Lundelius, E. L., Jr., R. W. Graham, E. Anderson, J. Guilday, J. A. Holman, D. W. Steadman, and S. D. Webb. 1983. Terrestrial vertebrate faunas, pp. 311–353. *In* S. Porter (ed.), The Late Pleistocene. Univ. Minnesota Press, Minneapolis.

Lutterschmidt, W. I., J. J. Lutterschmidt, and H. K. Reinert. 1996. An improved timing device for monitoring pulse frequency of temperature-sensing transmitters in free-ranging animals. Am. Midl. Nat. 136: 172–180.

Lutterschmidt, W. I., D. I. Lutterschmidt, R. T. Mason, and H. K. Reinert. 2009. Seasonal variation in hormonal responses of timber rattlesnakes (*Crotalus horridus*) to reproductive and environmental stressors. J. Comp. Physiol. 179B: 747–757.

Lyman, D., Jr. 2006. *Crotalus cerastes* (sidewinder): Diet. Herpetol. Rev. 37: 476–477.

Lynch, V. J. 2007. Inventing an arsenal: Adaptive evolution and neofunctionalization of snake venom phospholipase A_2 genes. BMC Evol. Biol. 7(2): 1–14.

Lynn, W. G. 1931. The structure and function of the facial pit of the pitvipers. Am. J. Anat. 49: 97–139.

Lyons, W. J. 1971. Profound thrombocytopenia associated with *Crotalus ruber* envenomation: A clinical case. Toxicon 9: 237–240.

Macartney, J. M. 1985. The ecology of the northern Pacific rattlesnake, *Crotalus viridis oreganus*, in British Columbia. Master's thesis, Univ. Victoria, British Columbia, Canada.

———. 1989. Diet of the northern Pacific rattlesnake, *Crotalus viridis oreganus*, in British Columbia. Herpetologica 45: 299–304.

Macartney, J. M., and P. T. Gregory. 1988. Reproductive biology of female rattlesnakes (*Crotalus viridis*) in British Columbia. Copeia 1988: 47–57.

Macartney, J. M., and B. Weichel. 1993. Status of the prairie rattlesnake and the eastern yellow-bellied racer in Saskatchewan. Prov. Mus. Alberta Nat. Hist. Occ. Pap. 19: 291–299.

Macartney, J. M., P. T. Gregory, and K. W. Larson. 1988. A tabular survey of data on movements and home ranges of snakes. J. Herpetol. 22: 61–73.

Macartney, J. M., K. W. Larson, and P. T. Gregory. 1989. Body temperatures and movements of hibernating snakes (*Crotalus* and *Thamnophis*) and thermal gradients of natural hibernacula. Can. J. Zool. 67: 108–114.

Macartney, J. M., P. T. Gregory, and M. B. Charland. 1990. Growth and sexual maturity of the western rattlesnake, *Crotalus viridis*, in British Columbia. Copeia 1990: 528–542.

Mace, G. M., and R. Lande. 1991. Assessing extinction threats: Towards a reevaluation of IUCN threatened species categories. Conserv. Biol. 5: 148–147.

MacGregor, G. A., and H. K. Reinert. 2001. The use of passive integrated transponders (PIT tags) in snake foraging studies. Herpetol. Rev. 32: 170–172.

Macht, D. I. 1937. Comparative toxicity of sixteen specimens of *Crotalus* venom. Proc. Soc. Exp. Biol. Med. 36: 499–501.

Mackessy, S. P. 1985. Fractionation of red diamond rattlesnake (*Crotalus ruber ruber*) venom: Protease, phosphodiesterase, L-amino acid oxidase activities and effects of metal ions and inhibitors on protease activity. Toxicon 23: 337–340.

———. 1988. Venom ontogeny in the Pacific rattlesnake *Crotalus viridis helleri* and *C. v. oreganus*. Copeia 1988: 92–101.

———. 1991. Morphology and ultrastructure of the venom glands of the northern Pacific rattlesnake *Crotalus viridis oreganus*. J. Morphol. 208: 109–128.

———. 1993a. Fibrinogenolytic proteases from the venoms of juvenile and adult northern Pacific rattlesnakes (*Crotalus viridis oreganus*). Comp. Biochem. Physiol. 106B: 181–189.

———. 1993b. Kallikrein-like and thrombin-like proteases from the venom of juvenile northern Pacific rattlesnakes (*Crotalus viridis oreganus*). J. Natural Toxins 2: 223–239.

———. 1996. Characterization of the major metalloprotease isolated from the venom of the northern Pacific rattlesnake, *Crotalus viridis oreganus*. Toxicon 34: 1277–1285.

———. 1998. Phosphodiesterases, ribonucleases and deoxyribonucleases, pp. 361–404. *In* G. S. Bailey (ed.), Enzymes from snake venom. Alaken, Fort Collins, Colorado.

Mackessy, S. P., and L. M. Baxter. 2006. Bioweapons synthesis and storage: The venom gland of front-fanged snakes. Zool. Anz. 245: 147–159.

Mackessy, S. P., K. Williams, and K. G. Ashton. 2003. Ontogenetic variation in venom composition and diet of *Crotalus oreganus concolor*: A case of venom paedomorphosis? Copeia 2003: 769–782.

MacMahon, J. A., and A. H. Hammer. 1975a. Hematology of the sidewinder (*Crotalus cerastes*). Comp. Biochem. Physiol. 51A: 53–58.

———. 1975b. Effects of temperature and photoperiod on oxygenation and other blood parameters of the sidewinder (*Crotalus cerastes*): Adaptive significance. Comp. Biochem. Physiol. 51A: 59–69.

Madrid-Sotelo, C. A., and C. J. Balderas-Valdivia. 2008. *Crotalus molossus nigrescens* (Mexican black-tailed rattlesnake): Behavior. Herpetol. Rev. 39: 468–469.

Maeda, N., N. Tamiya, T. R. Pattabhiraman, and F. E. Russell. 1978. Some chemical properties of the venom of the rattlesnake, *Crotalus viridis helleri*. Toxicon 16: 431–441.

Mahaney, P. A. 1997a. *Crotalus pricei* (twin-spotted rattlesnake): Reproduction. Herpetol. Rev. 28: 205.

———. 1997b. Taxon management account: Twin-spotted rattlesnake *Crotalus pricei* ssp. Am. Zoo Aquar. Assoc. Snake Adv. Group: 1–9.

Mahrdt, C. R., and B. H. Banta. 1997. *Aneides lugubris* (arboreal salamander). Predation. Herpetol. Rev. 28: 81.

Maity, G., and D. Bhattacharyya. 2005. Assay of snake venom phospholipase A_2 using scattering mode of a spectrofluorimeter. Curr. Sci. 89: 1004–1008.

Malawy, M. A. 2005. Thermal biology of *Crotalus atrox*, p. 38. *In* Biology of the rattlesnakes symposium (program abstracts). Loma Linda Univ., Loma Linda, California.

Manion, S. 1968. *Crotalus willardi*—The Arizona ridge-nosed rattlesnake. Herpetology 2(3): 27–30.

Mankau, S. K., and E. A. Widmer. 1977. Prevalence of *Mesocestoides* (Eucestoda: Mesocestoididea) tetrathyridia in southern California reptiles with notes on the pathology in the Crotalidae. Japan. J. Parasitol. 26: 256–259.

Mara, W. P. 1993. Venomous snakes of the World. T. F. H. Publ., Neptune City, New Jersey.

Marchisin, A. 1980. Predator-prey interactions between snake-eating snakes. Ph.D. diss., Rutgers Univ., Newark, New Jersey.

Marcy, D. 1945. Birth of a brood of *Crotalus basiliscus*. Copeia 1945: 169–170.

Marion, K. R., and O. J. Sexton. 1984. Body temperatures and behavioral activities of hibernating prairie rattlesnakes, *Crotalus viridis*, in artificial dens. Prairie Nat. 16: 111–116.

Markland, F. S., Jr. 1991. Inventory of α- and β-fibrinogenases from snake venoms: For the Subcommittee on Nomenclature of Exogenous Hemostatic Factors of the Scientific and Standardization Committee of the International Society on Thrombosis and Haemostasis. Thromb. Haemostas. 65: 438–443.

Markland, F. S., and P. S. Damus. 1971. Purification and properties of a thrombin-like enzyme from the venom of *Crotalus adamanteus* (eastern diamondback rattlesnake). J. Biol. Chem. 246: 6460–6473.

Markland, F. S., Jr., and A. Perdon. 1986. Comparison of two methods for proteolytic enzyme detection in snake venom. Toxicon 24: 385–393.

Markland, F. S., Jr., and H. Pirkle. 1977. Thrombin-like enzyme from the venom of *Crotalus adamanteus* (eastern diamondback rattlesnake). Thrombosis Res. 10: 487–494.

Marlow, R. W. 1988. Red diamond rattlesnake, pp. 236–237. *In* D. C. Zeiner, W. F. Laudenslayer, Jr., and K. E. Mayer (eds.), California wildlife, vol. 1: Amphibians and reptiles. California Dept. Fish Game, Sacramento.

Marmie, W., S. Kuhn, and D. Chiszar. 1990. Behavior of captive-raised rattlesnakes (*Crotalus enyo*) as a function of rearing conditions. Zoo Biol. 9: 241–246.

Marr, J. C. 1944. Notes on amphibians and reptiles from the central United States. Am. Midl. Nat. 32: 478–490.

Marshall, J. T. 1957. Birds of pine-oak woodland in southern Arizona and adjacent Mexico. Pacific Coast Avifauna 32: 1–125.

Martin, B. 1974. Distribution and habitat adaptations in rattlesnakes of Arizona. HERP: Bull. New York Herpetol. Soc. 10(3–4): 3–12.

Martin, B. E. 1975a. Notes on a brood of the Arizona ridge-nosed rattlesnake *Crotalus willardi willardi*. Bull. Maryland Herpetol. Soc. 11: 64–65.

———. 1975b. An occurrence of the Arizona ridge-nosed rattlesnake, *Crotalus willardi willardi*, observed feeding in nature. Bull. Maryland Herpetol. Soc. 11: 66–67.

———. 1975c. A brood of Arizona ridge-nosed rattlesnakes (*Crotalus willardi willardi*) bred and born in captivity. Bull. Maryland Herpetol. Soc. 12: 187–189.

———. 1976. A reproductive record for the New Mexican ridge-nosed rattlesnake (*Crotalus willardi obscurus*). Bull. Maryland Herpetol. Soc. 12: 126–128.

Martin, J. H., and R. M. Bagby. 1972. Temperature-frequency relationship of the rattlesnake rattle. Copeia 1972: 482–485.

———. 1973. Effects of fasting on the blood chemistry of the rattlesnake, *Crotalus atrox*. Comp. Biochem. Physiol. 44A: 813–820.

Martin, J. R., and J. T. Wood. 1955. Notes on the poisonous snakes of the Dismal Swamp area. Herpetologica 11: 237–238.

Martin, P. J. 1930. Snake hunt nets large catch. Bull. Antiv. Inst. Am. 4: 77–78.

Martin, P. S. 1958. A biogeography of reptiles and amphibians in the Gomez Farias region, Tamaulipas, Mexico. Univ. Michigan Mus. Zool. Misc. Publ. 101: 1–102.

Martin, W. H. 1976. Reptiles observed on the Skyline Drive and Blue Ridge Parkway, Va. Virginia Herpetol. Soc. Bull. 81:1–3.

———. 1979. The timber rattlesnake in Virginia: Its distribution and present status. Virginia Herpetol. Soc. Bull. 89: 1–4.

———. 1981. The timber rattlesnake in the Northeast: Its range, past and present. HERP: Bull. New York Herpetol. Soc. 17: 15–20.

———. 1988. Life history of the timber rattlesnake. Catesbeiana 8: 9–12.

———. 1990. The timber rattlesnake, *Crotalus horridus*, in the Appalachian Mountains of eastern North America. Catesbeiana 10: 49.

———. 1992. Phenology of the timber rattlesnake (*Crotalus horridus*) in an unglaciated section of the Appalachian Mountains, pp. 259–278. *In* J. A. Campbell and E. D. Brodie, Jr. (eds.), Biology of the pitvipers. Selva, Tyler, Texas.

———. 1993. Reproduction of the timber rattlesnake (*Crotalus horridus*) in the Appalachian Mountains. J. Herpetol. 27: 133–143.

———. 1996a. Timber rattlesnakes and Pleistocene climates. Virginia Herpetol. Soc. Newsl. 6(2): 1–3, 10.

———. 1996b. *Crotalus horridus* (timber rattlesnake): Reproductive phenology. Herpetol. Rev. 27: 144–145.

———. 2002. Life history constraints on the timber rattlesnake (*Crotalus horridus*) at its climatic limits, pp. 285–306. *In* G. W. Schuett, M. Höggren, M. E. Douglas, and H. W. Greene (eds.), Biology of the vipers. Eagle Mountain Publ., Eagle Mountain, Utah.

Martin, W. H., and D. B. Means. 2000. Distribution and habitat relationships of the eastern diamondback rattlesnake (*Crotalus adamanteus*). Herpetol. Nat. Hist. 7: 9–34.

Martin, W. H., W. H. Smith, S. H. Harwig, R. O. Magram, and R. Stechert. 1990. Distribution
and status of the timber rattlesnake (*Crotalus horridus*) in Pennsylvania. Unpubl. Rep. Carnegie Mus. Nat. Hist. and Pennsylvania Fish Comm.

Martin, W. H., J. C. Mitchell, and R. Hoggard. 1992. *Crotalus horridus* (timber rattlesnake). Herpetol. Rev. 23: 91.

Martin, W. H., W. S. Brown, E. Possardt, and J. B. Sealy. 2008. Biological variation, management
units, and a conservation action plan for the timber rattlesnake (*Crotalus horridus*), pp. 447–
462. *In* W. K. Hayes, K. R. Beaman, M. D. Cardwell, and S. P. Bush (eds.), The biology of
rattlesnakes. Loma Linda Univ. Press, Loma Linda, California.

Martin del Campo, R. 1936. Los batracios y reptiles segun los codices y relatos de los antiguos
mexicanos. Anales del Instituto de Biologia 7(4): 489–512.

Martínez, M., E. D. Rael, and N. L. Maddux. 1990. Isolation of a hemorrhagic toxin from Mojave rattlesnake (*Crotalus scutulatus scutulatus*) venom. Toxicon 28: 685–694.

Martins, M., G. Arnaud, and R. Murillo-Quero. 2008. Exploring hypotheses about the loss of
the rattle in rattlesnakes: How arboreal is the Isla Santa Catalina rattleless rattlesnake, *Crotalus catalinensis?* So. Am. J. Herpetol. 3: 162–167.

Martof, B. S. 1956. Amphibians and reptiles of Georgia, a guide. Univ. Georgia Press, Athens.

Martof, B. S., W. M. Palmer, J. R. Bailey, J. R. Harrison III, and J. Dermid. 1980. Amphibians and
reptiles of the Carolinas and Virginia. Univ. North Carolina Press, Chapel Hill.

Marx, H., J. S. Ashe, and L. E. Watrous. 1988. Phylogeny of the viperine snakes (Viperinae):
Part I. Character analysis. Fieldiana: Zool. 51: i–iv, 1–16.

Marx, N., and G. B. Rabb. 1972. Phyletic analysis of fifty characters of advanced snakes. Fieldiana: Zool. 63: 1–321.

Mashiko, H., and H. Takahashi. 1998. Haemorrhagic factors from snake venoms: II. Structures
of haemorrhagic factors and types and mechanisms of haemorrhage. J. Toxicol. Toxin Rev.
17: 493–512.

Mata-Silva, V., S. Dilks, and J. D. Johnson. 2010. *Crotalus lepidus* (rock rattlesnake): Diet. Herpetol. Rev. 41: 235–236.

Matias-Ferrer, N., and S. Murillo. 2004. *Crotalus atrox* (western diamond-backed rattlesnake).
Herpetol. Rev. 35: 190.

Matlack, R. S., and R. L. Rehmeier. 2002. Status of the western diamondback rattlesnake (*Crotalus atrox*) in Kansas. Southwest. Nat. 47: 312–314.

Matsuda, B. M., D. M. Green, and P. T. Gregory. 2006. Amphibians and reptiles of British Columbia. Royal British Columbia Mus., Victoria, British Columbia.

Mata-Silva, V., S. Dilks, and J. D. Johnson. 2010. *Crotalus lepidus* (rock rattlesnake): Diet. Herpetol. Rev. 41: 235–236.

Mattison, C. 1986. Snakes of the World. Facts on File Publ., New York.

———. 1988. Keeping and breeding snakes. Blandford Press, London.

———. 1996. Rattler! A natural history of rattlesnakes. Blandford Press, London.

Mauldin, H. K. 1968. History of Clear Lake, Mt. Konocti and the Lake County cattle industry.
Anderson Printing, Kelseyville, California.

Maxell, B. A., J. K. Werner, P. Hendricks, and D. L. Flath. 2003. Herpetology in Montana: A history, status summary, checklists, dichotomous keys, accounts for native, potentially native,
and exotic species, and indexed bibliography. Northwest Fauna 5.

Mayhew, W. W. 1963. Biology of the granite spiny lizard, *Sceloporus orcutti*. Am. Midl. Nat. 69:
310–327.

McAllister, C. T., S. R. Goldberg, H. J. Holshuh, and S. E. Trauth. 1993. Disseminated mycotic
dermatitis in a wild-caught timber rattlesnake, *Crotalus horridus* (Serpentes: Viperidae), from
Arkansas. Texas J. Sci. 45: 279–281.

McAllister, C. T., S. J. Upton, S. E. Trauth, and J. R. Dixon. 1995. Coccidian parasites (Apicomplexa) from snakes in the southcentral and southwestern United States: New host and geographic records. J. Parasitol. 81: 63–68.

McAllister, C. T., C. R. Bursey, and J. F. Roberts. 2004. *Physocephalus sexalatus* (Nematoda: Spirurida: Spirocercidae) in three species of rattlesnakes, *Crotalus atrox*, *Crotalus lepidus*, and *Crotalus scutulatus*, from southwestern Texas. J. Herpetol. Med. Surg. 4(3): 10–12.

McAllister, C. T., H. W. Robison, and D. Arbour. 2008. *Crotalus atrox* (western diamond-backed rattlesnake). Herpetol. Rev. 39: 370–371.

McCallion, J. 1945. Notes on Texas reptiles. Herpetologica 2: 197–198.

McCauley, R. H., Jr. 1945. The reptiles of Maryland and the District of Columbia. Privately published, Hagerstown, Maryland.

McCoy, C. J. 1961. Birth season and young of *Crotalus scutulatus* and *Agkistrodon contortrix lacticinctus*. Herpetologica 17: 140.

———. 1962. Noteworthy amphibians and reptiles from Colorado. Herpetologica 18: 60–62.

———. 1980. Identification guide to Pennsylvania snakes. Carnegie Mus. Nat. Hist. Educ. Bull. 1: 1–12.

———. 1982. Amphibians and reptiles in Pennsylvania: Checklist, bibliography, and atlas of distribution. Carnegie Mus. Nat. Hist. Spec. Publ. 6: 1–91.

———. 1984. Rattlesnake island. Notes from NOAH 11(4): 43–44.

McCrady, W. B., J. B. Murphy, C. M. Garrett, and D. T. Roberts. 1994. Scale variation in a laboratory colony of amelanistic diamondback rattlesnakes (*Crotalus atrox*). Zoo Biol. 13: 95–106.

McCranie, J. R. 1980a. *Crotalus adamanteus*. Cat. Am. Amphib. Rept. 252: 1–2.

———. 1980b. *Crotalus pricei*. Cat. Am. Amphib. Rept. 266: 1–2.

———. 1981. *Crotalus basiliscus*. Cat. Am. Amphib. Rept. 283: 1–2.

McCranie, J. R., and L. D. Wilson. 1979. A preliminary account of the rattlesnakes with the description of four new subspecies, by Herbert S. Harris, Jr. and Robert S. Simmons [A review]. Herpetol. Rev. 10: 18–21.

McCrystal, H. K., and M. J. McCord. 1986. *Crotalus mitchellii* (Cope): Speckled rattlesnake. Cat. Am. Amphib. Rept. 388: 1–4.

McCue, M. D. 2005. Enzyme activities and biological functions of snake venoms. Appl. Herpetol. 2: 109–123.

———. 2006. Cost of producing venom in three North American pitviper species. Copeia 2006: 818–825.

———. 2007a. Western diamondback rattlesnakes demonstrate physiological and biochemical strategies for tolerating prolonged starvation. Physiol. Biochem. Zool. 80(1): 25–34.

———. 2007b. Evaluating the digestive advantages conferred by prey envenomation in rattlesnakes. Comp. Biochem. Physiol. 148A: S85.

———. 2007c. Snakes survive starvation by employing supply- and demand-side economic strategies. Zoology 110: 318–327.

———. 2007d. Prey envenomation does not improve digestive performance in western diamondback rattlesnakes (*Crotalus atrox*). J. Exp. Zool. 307A: 568–577.

McDiarmid, R. W. (ed.). 1978. Rare and endangered biota of Florida, vol. 3: Amphibians and reptiles. Univ. Press Florida, Gainesville.

McDiarmid, R. W., J. A. Campbell, and T'S. A. Touré. 1999. Snake species of the world: A taxonomic and geographic reference, vol. 1. Herpetologists' League, Washington, D.C.

McDonald, H. C., and E. Anderson. 1975. A Late Pleistocene vertebrate fauna from southeastern Idaho. Tebiwa 18: 20–37.

McDowell, S. B. 1986. The architecture of the corner of the mouth of colubroid snakes. J. Herpetol. 20: 353–407.

McGowan, E. M. 2004. Importance of male body size for male mating success in *Crotalus horridus*: Effects of female habitat selection and behavior, p. 21. *In* Proceedings of the Snake Ecology Group 2004 Conference, Jackson County, Illinois.

McGowan, E. M., and D. M. Madison. 2008. Timber rattlesnake (*Crotalus horridus*) mating behavior in southeastern New York: Female defense in a search-based mating system, pp. 419–430. *In* W. K. Hayes, K. R. Beaman, M. D. Cardwell, and S. P. Bush (eds.), The biology of rattlesnakes. Loma Linda Univ. Press, Loma Linda, California.

McGuire, J. A. 1991. *Crotalus enyo cerralvensis* (Cerralvo Island rattlesnake): Behavior. Herpetol. Rev. 22: 100.

McKee, E. D., and C. M. Bogert. 1934. The amphibians and reptiles of Grand Canyon National Park. Copeia 1934: 178–180.

McKeller, M. R., and J. C. Pérez. 2002. The effects of western diamondback rattlesnake (*Crotalus atrox*) venom on the production of antihemorrhagins and/or antibodies in the Virginia opossum (*Didelphis virginiana*). Toxicon 40: 427–439.

McKenna, M. G., and G. Allard. 1976. Rattlesnake research. North Dakota Outdoors 38(9): 11–13.

McKinney, C. O., and R. E. Ballinger. 1966. Snake predators of lizards in western Texas. Southwest. Nat. 11: 410–412.

McMillan, M., and C. S. Lieb. 2005. Morphological variation in *Crotalus basiliscus* and *Crotalus molossus* (Serpentes: Viperidae) with an evaluation of hybridization, p. 38. *In* Biology of the rattlesnakes symposium (program abstracts). Loma Linda Univ., Loma Linda, California.

McPartland, J. M., and R. Foster. 1988. Stun-guns and snakebites. Lancet 2: 1141.

McPeak, R. H. 2000. Amphibians and reptiles of Baja California. Sea Challengers, Monterey, California.

Mead, J. I., and C. J. Bell. 1994. Late Pleistocene and Holocene herpetofaunas of the Great Basin and Colorado Plateau, pp. 255–275. *In* K. T. Harper, L. L. St. Clair, K. H. Thorne, and W. M. Hess (eds.), Natural history of the Colorado Plateau and Great Basin. Univ. Press Colorado, Niwot.

Mead, J. I., and A. M. Phillips III. 1981. The late Pleistocene and Holocene fauna and flora of Vulture Cave, Grand Canyon, Arizona. Southwest. Nat. 26: 257–288.

Mead, J. I., and T. R. Van Devender. 1981. Late Holocene diet of *Bassariscus astutus* in the Grand Canyon, Arizona. J. Mammal. 62: 439–442.

Mead, J. I., R. S. Thompson, and T. R. Van Devender. 1982. Late Wisconsinan and Holocene fauna from Smith Creek Canyon, Snake Range, Nevada. Trans. San Diego Soc. Nat. Hist. 20: 1–16.

Mead, J. I., E. L. Roth, T. R. Van Devender, and D. W. Steadman. 1984. The late Wisconsinan vertebrate fauna from Deadman Cave, southern Arizona. Trans. San Diego Soc. Nat. Hist. 20: 247–276.

Mead, J. I., T. H. Heaton, and E. M. Mead. 1989. Late Quaternary reptiles from two caves in the east-central Great Basin. J. Herpetol. 23: 186–189.

Meade, G. P. 1940. Observations on Louisiana captive snakes. Copeia 1940: 165–168.

Meade, L. E. 2005. Kentucky snakes: Their identification, variation and distribution. Kentucky St. Nat. Pres. Comm., Frankfort.

Means, D. B. 1985. Radio-tracking the eastern diamondback rattlesnake. Natl. Geogr. Soc. Res. Rep. 18: 529–536.

———. 1986. Life history and ecology of the eastern diamondback rattlesnake (*Crotalus adamanteus*). Final Project Rept. Tallahassee: Florida Game Fresh Water Fish Commission.

———. 2009. Effects of rattlesnake roundups on the eastern diamondback rattlesnake (*Crotalus adamanteus*). Herpetol. Conserv. Biol. 4: 132–141.

Mebs, D., and F. Kornalik. 1984. Intraspecific variation in content of a basic toxin in eastern diamondback rattlesnake (*Crotalus adamanteus*) venom. Toxicon 22: 831–833.

Mecham, J. S. 1959. Some Pleistocene amphibians and reptiles from Friesenhahn Cave, Texas. Southwest. Nat. 3: 17–27.

Medica, P. A. 2009. *Crotalus scutulatus scutulatus* (Mojave rattlesnake): Defensive behavior/saltation and head hiding. Herpetol. Rev. 40: 95–97.

Meek, S. E. 1906 [1905]. An annotated list of a collection of reptiles from southern California and northern Lower California. Field Columbian Mus. Publ. Zool. Ser. 7: 3–19.

Mehrtens, J. M. 1987. Living snakes of the world in color. Sterling Publ. Co., New York.

Meier, J., and K. F. Stocker. 1995. Biology and distribution of venomous snakes of medical importance and the composition of snake venoms, pp. 367–412. *In* J. Meier and J. White (eds.), Handbook of clinical toxicology of animal venoms and poisons. CRC Press, New York.

Meik, J. M. 2005. *Crotalus mitchellii stephensi* (Panamint rattlesnake): Diet. Herpetol. Rev. 36: 68.

———. 2008. Morphological analysis of the contact zone between the rattlesnakes *Crotalus mitchellii stephensi* and *Crotalus m. pyrrhus*, pp. 39–46. *In* W. K. Hayes, K. R. Beaman, M. D.

Cardwell, and S. P. Bush (eds.), The biology of rattlesnakes. Loma Linda Univ. Press, Loma Linda, California.

Meik, J. M., B. E. Fontenot, C. J. Franklin, and C. King. 2008. Apparent natural hybridization between the rattlesnakes *Crotalus atrox* and *C. horridus*. Southwest. Nat. 53: 196–200.

Melcer, T., and D. Chiszar. 1989a. Strike-induced chemical preferences in prairie rattlesnakes (*Crotalus viridis*). Anim. Learn. Behav. 17: 368–372.

———. 1989b. Striking prey creates a specific chemical search image in rattlesnakes. Anim. Behav. 37: 477–486.

Melcer, T., K. Kandler, and D. Chiszar. 1988. Effects of novel chemical cues on predatory responses of rodent-specializing rattlesnakes. Bull. Psychon. Soc. 26: 580–582.

Melcer, T., D. Chiszar, and H. M. Smith. 1990. Strike-induced chemical preferences in rattlesnakes: Role of chemical cues arising from the diet of prey. Bull. Maryland Herpetol. Soc. 26: 1–4.

Mellink, E. 1990. *Crotalus scutulatus* (Mojave rattlesnake). Reproduction. Herpetol. Rev. 21: 93.

Mello, K. 1978. *Crotalus viridis cerberus* (Arizona black rattlesnake). Herpetol. Rev. 9: 22.

Mendelson, J. R., and W. B. Jennings. 1992. Shifts in the relative abundance of snakes in a desert grassland. J. Herpetol. 26: 38–45.

Merrow, J. S., and T. Aubertin. 2005. *Crotalus horridus* (timber rattlesnake): Reproduction. Herpetol. Rev. 36: 192.

Metsch, R. B., A. Dray, and F. E. Russell. 1984. Effects of the venom of the southern Pacific rattlesnake, *Crotalus viridis helleri*, and its fractions on striated and smooth muscle. Proc. West. Pharmacol. Soc. 27: 395–398.

Metter, D. E. 1963. A rattlesnake that encountered a porcupine. Copeia 1963: 161.

Meylan, P. A. 1982. The squamate reptiles of the Ingles IA Fauna (Irvingtonian: Citrus County, Florida). Bull. Florida St. Mus. Biol. Sci. 27: 1–85.

———. 1995. Pleistocene amphibians and reptiles from the Leisey Shell Pit, Hillsborough County, Florida. Bull. Florida St. Mus. Nat. Hist. 37: 273–297.

Miller, A. H., and R. C. Stebbins. 1964. The lives of desert animals in Joshua Tree National Monument. Univ. California Press, Berkeley.

Miller, D. M., R. A. Young, T. W. Gatlin, and J. A. Richardson. 1982. Amphibians and reptiles of the Grand Canyon National Park. Grand Canyon Natur. Hist. Assoc. Monogr. (4): 1–144.

Miller, L. 1942. A Pleistocene tortoise from the McKittrick Asphalt. Trans. San Diego Soc. Nat. Hist. 9: 439–442.

Miller, L. R., and W. H. N. Gutzke. 1999. The role of the vomeronasal organ of crotalines (Reptilia: Serpentes: Viperidae) in predator detection. Anim. Behav. 58: 53–57.

Miller, W. E. 1971. Pleistocene vertebrates of the Los Angeles Basin and vicinity (exclusive of Rancho La Brea). Nat. Hist. Mus. Los Angeles Co., Sci. Bull. 10: 1–24.

———. 1980. The late Pliocene Las Tunas local fauna from southernmost Baja California, Mexico. J. Paleontol. 54: 762–805.

Milstead, W. W., J. S. Mecham, and H. McClintock. 1950. The amphibians and reptiles of the Stockton Plateau in northern Terrell County, Texas. Texas J. Sci. 2: 543–562.

Minton, S. A., Jr. 1953. Variation in venom samples from copperheads (*Agkistrodon contortrix mokeson*) and timber rattlesnakes (*Crotalus horridus horridus*). Copeia 1953: 212–215.

———. 1956. Some properties of North American pit viper venoms and their correlation with phylogeny, pp. 145–151. *In* E. Buckley and N. Porges (eds.), Venoms. American Association for the Advancement of Science, Washington, D.C.

———. 1959 [1958]. Observations on amphibians and reptiles of the Big Bend Region of Texas. Southwest. Nat. 3: 28–54.

———. 1969. The feeding strike of the timber rattlesnake. J. Herpetol. 3: 121–124.

———. 1971. Snake venoms and envenomation. Marcel Dekker, New York.

———. 1972. Amphibians and reptiles of Indiana. Indiana Acad. Sci., Monogr. 3: 1–346.

———. 1974. Venom diseases. Charles C. Thomas, Springfield, Illinois.

———. 1975. A note on the venom of an aged rattlesnake. Toxicon 13: 73–74.

———. 1977. Toxicity of venoms from some little known Mexican rattlesnakes. Toxicon 15: 580–581.

————. 1980. Snakebites in the U.S.A. The Snake 12: 141.

————. 1982. Snake bite, pp. 283–301. *In* G. V. Hillyer and C. E. Hopla (eds.), Parasitic zoonoses. CRC Press, Boca Raton, Florida.

————. 1992. Serologic relationships among pitvipers: Evidence from plasma albumins and immunodiffusion, pp. 155–161. *In* J. A. Campbell and E. D. Brodie, Jr. (eds.), Biology of the pitvipers. Selva, Tyler, Texas.

————. 2001. Amphibians and reptiles of Indiana. 2nd ed., revised. Indiana Acad. Sci., Indianapolis.

Minton, S. A., Jr., and M. R. Minton. 1969. Venomous reptiles. Charles Scribner's, New York.

————. 1991. Rattlesnakes and Mexican folk medicine. Herpetol. Rev. 22: 116.

Minton, S. A., Jr., and S. A. Weinstein. 1984. Protease activity and lethal toxicity of venoms from some little known rattlesnakes. Toxicon 22: 828–830.

————. 1986. Geographic and ontogenetic variation in venom of the western diamondback rattlesnake (*Crotalus atrox*). Toxicon 24: 71–80.

Miranda, L., Jr., V. Mata-Silva, S. Dilks, H. Riveroll, Jr., and J. D. Johnson. 2008. *Crotalus molossus* (blacktail rattlesnake): Morphology. Herpetol. Rev. 39: 97.

Mitchell, J. C. 1974. The snakes of Virginia, parts 1 and 2. Virginia Wildl. 35(February): 16–18, 28; 35(April): 12–15.

————. 1980. Viper's brood. A guide to identifying some of Virginia's juvenile snakes. Virginia Wildl. 41(9): 8–10.

————. 1986. Cannibalism in reptiles: A worldwide review. Soc. Stud. Amphib. Rept. Herpetol. Circ. 15: i–iii, 1–37.

————. 1994a. The reptiles of Virginia. Smithsonian Inst. Press, Washington, D.C.

————. 1994b. Timber rattlesnakes (*Crotalus horridus*) in Prince William Forest Park: Released captives or native population? Banisteria 3: 21–24.

Mitchell, J. C., and R. A. Beck. 1992. Free-ranging domestic cats' predation on native vertebrates in rural and urban Virginia. Virginia J. Sci. 43: 197–207.

Mitchell, J. C., and W. H. Martin III. 1981. Where the snakes are. Virginia Wildl. 42(6): 8–9.

Mitchell, J. C., and C. Ruckdeschel. 2008. *Crotalus adamanteus* (eastern diamond-backed rattlesnake): Prey. Herpetol. Rev. 39: 467–468.

Mitchell, J. C., and D. Schwab. 1991. Canebrake rattlesnake, *Crotalus horridus atricaudatus* Latreille, pp. 462–464. *In* K. Terwilliger (ed.), Virginia's endangered species: Proceedings of a symposium. McDonald Woodward Publ. Co., Blacksburg, Virginia.

Mitchell, J. C., and J. R. Webb. 2000. *Crotalus horridus* (timber rattlesnake). Catesbeiana 20(1): 40.

Mitchell, J. D. 1903. The poisonous snakes of Texas, with notes on their habits. Texas Med. News 12: 411–437.

Mitchell, S. W. 1860. Researches upon the venom of the rattlesnake: With an investigation of the anatomy and physiology of the organs concerned. Smithsonian Contrib. Knowl. 12(6).

Mociño-Deloya, E., and K. Setser. 2007a. *Crotalus willardi* (ridge-nosed rattlesnake): Reproduction. Herpetol. Rev. 38: 205–206.

————. 2007b. *Crotalus willardi* (ridge-nosed rattlesnake): Diet. Herpetol. Rev. 38: 206.

Moesel, J. 1918. The prairie rattler in western and central New York. Copeia 1918(58): 67–68.

Molenaar, G. J. 1992. Anatomy and physiology of infrared sensitivity of snakes, pp. 367–435. *In* C. Gans and P. S. Ulinski (eds.), Biology of the Reptilia, vol. 17. Univ. Chicago Press, Chicago.

Moler, P. E. (ed.). 1992. Rare and endangered biota of Florida, vol. III: Amphibians and reptiles. Univ. Press Florida, Gainesville.

Molina, O., R. K. Seriel, M. Martinez, M. L. Sierra, A. Varela-Ramirez, and E. D. Rael. 1990. Isolation of two hemorrhagic toxins from *Crotalus basiliscus basiliscus* (Mexican west coast rattlesnake) venom and their effect on blood clotting and complement. Intl. J. Biochem. 22: 253–261.

Moll, E. O. 2003. Patronyms of the pioneer west. V. *Crotalus pricei*, Van Denburgh, 1895—Twin-spotted rattlesnake. Sonoran Herpetol. 16: 110–112.

Monroe, J. E. 1962. Chromosomes of rattlesnakes. Herpetologica 17: 217–220.

Moon, B. R. 2001. Muscle physiology and the evolution of the rattling system in rattlesnakes. J. Herpetol. 35: 497–500.

———. 2005. From physiology to fitness: The cost of self defense in rattlesnakes, p. 39. *In* Biology of the rattlesnakes symposium (program abstracts). Loma Linda Univ., Loma Linda, California.

———. 2006. From physiology to fitness: The costs of a defensive adaptation in rattlesnakes. Physiol. Biochem. Zool. 79(1): 133–139.

Moon, B. R., and A. M. Rabatsky. 2008. Is there an optimal length for the rattlesnake rattle? pp. 101–110. *In* W. K. Hayes, K. R. Beaman, M. D. Cardwell, and S. P. Bush (eds.), The biology of rattlesnakes. Loma Linda Univ. Press, Loma Linda, California.

Moon, B. R., K. E. Conley, S. L. Lindstedt, and M. R. Urquhart. 2003. Minimal shortening in a high-frequency muscle. J. Exp. Biol. 206: 1291–1297.

Moon, B. R., C. S. Ivanyi, and J. Johnson. 2004a. Identifying individual rattlesnakes using tail pattern variation. Herpetol. Rev. 35: 154–156.

Moore, R. G. 1978. Seasonal and daily activity patterns and thermoregulation in the southwestern speckled rattlesnake (*Crotalus mitchelli pyrrhus*) and the Colorado Desert sidewinder (*Crotalus cerastes laterorepens*). Copeia 1978: 439–442.

Moran, J. B., and C. R. Geren. 1979. A comparison of biological and chemical properties of three North American (Crotalidae) snake venoms. Toxicon 17: 237–244.

Morgan, B. B. 1943. The *Physaloptera* (Nematoda) of reptiles. Le Naturaliste Canadien 70: 179–185.

Mori, N., and H. Sugihara. 1988. Kallikrein-like enzyme from *Crotalus ruber ruber* (red rattlesnake) venom. Intl. J. Biochem. 20: 1425–1433.

———. 1989a. Comparative study of two arginine ester hydrolases, E-I and E-II from the venom of *Crotalus ruber ruber* (red rattlesnake). Comp. Biochem. Physiol. 92B: 537–547.

———. 1989b. Characterization of kallikrein-like enzyme from *Crotalus ruber ruber* (red rattlesnake) venom. Intl. J. Biochem. 21: 83–90.

Mori, N., T. Nikai, and H. Sugihara. 1987a. Phosphodiesterase from the venom of *Crotalus ruber ruber*. Intl. J. Biochem. 19: 115–119.

Mori, N., T. Nikai, H. Sugihara, and A. Tu. 1987b. Biochemical characterization of hemorrhagic toxins with fibrinogenase activity isolated from *Crotalus ruber ruber* venom. Arch. Biochem. Biophys. 253: 108–121.

Mosauer, W. 1932a. The amphibians and reptiles of the Guadalupe Mountains of New Mexico and Texas. Occ. Pap. Mus. Zool. Univ. Michigan 246: 1–19.

———. 1932b. Adaptive convergence in the sand reptiles of the Sahara and of California: A study of structure and behavior. Copeia 1932: 72–78.

———. 1932c. On the locomotion of snakes. Science (New York) 76: 583–585.

———. 1933. Locomotion and diurnal range of *Sonora occipitalis*, *Crotalus cerastes*, and *Crotalus atrox* as seen from their tracks. Copeia 1933: 14–16.

———. 1935a. How fast can snakes travel? Copeia 1935: 6–9.

———. 1935b. The reptiles of a sand dune area and its surroundings in the Colorado Desert, California: A study in habitat preference. Ecology 16: 13–27.

Mosauer, W., and E. L. Lazier. 1933. Death from insolation in desert snakes. Copeia 1933: 149.

Mosimann, J. E., and G. B. Rabb. 1952. The herpetology of Tiber Reservoir area, Montana. Copeia 1952: 23–27.

Mount, R. H. 1975. Reptiles and amphibians of Alabama. Auburn Univ. Agric. Exp. Stat., Auburn, Alabama.

——— (ed.). 1984. Vertebrate wildlife of Alabama. Alabama Agric. Exp. St., Auburn.

Muir, J. H. 1982. Notes on the climbing ability of a captive timber rattlesnake, *Crotalus horridus*. Bull. Chicago Herpetol. Soc. 17: 22–23.

———. 1990. Three anatomically aberrant albino *Crotalus atrox* neonates. Bull. Chicago Herpetol. Soc. 25: 41–42.

Mulcahy, D. G., J. R. Mendelson III, K. W. Setser, and E. Hollenbeck. 2003. *Crotalus cerastes* (sidewinder): Prey/predator weight-ratio. Herpetol. Rev. 34: 64.

Munekiyo, S. M., and S. P. Mackessy. 2005. Presence of peptide inhibitors in rattlesnake venoms and their effects on endogenous metalloproteases. Toxicon 45: 255–263.

Munguia-Vega, A., K. Pelz-Serrano, M. Goode, and M. Culver. 2009. Eleven new microsatellite loci for the tiger rattlesnake (*Crotalus tigris*). Mol. Ecol. Res. 9: 1267–1270.

Munro, D. F. 1947. Effect of a bite by *Sistrurus* on *Crotalus*. Herpetologica 4: 57.

Murphy, J. B. 1973. A review of diseases and treatment of captive chelonians: Bacterial and viral infections, part two of a series. HISS News-J. 1(3): 77–81.

Murphy, J. B., and B. L. Armstrong. 1978. Maintenance of rattlesnakes in captivity. Univ. Kansas Mus. Nat. Hist. Spec. Publ. 3: 1–40.

Murphy, J. B., and J. A. Shadduck. 1978. Reproduction in the eastern diamondback rattlesnake, *Crotalus adamanteus*, in captivity, with comments regarding taratoid birth anomaly. British J. Herpetol. 5: 727–733.

Murphy, J. B., J. E. Joy, Jr., and J. A. Shadduck. 1975. Colonic torsion in a western diamondback rattlesnake, *Crotalus atrox*. J. Herpetol. 9: 248.

Murphy, J. B., J. E. Rehg, P. F. A. Maderson, and W. B. McCrady. 1987. Scutellation and pigmentation defects in a laboratory colony of western diamondback rattlesnakes (*Crotalus atrox*): Mode of inheritance. Herpetologica 43: 292–300.

Murphy, J. C. 1990. A model for regional herpetological society field studies: The snakes of southern California, a proposal, and some thoughts on collecting. Bull. Chicago Herpetol. Soc. 25: 42–45.

Murphy, R. W. 1983a. Paleobiogeography and genetic differentiation of the Baja California herpetofauna. Occ. Pap. California Acad. Sci. 137: 1–48.

———. 1983b. The reptiles: Origin and evolution, pp. 130–158. *In* T. J. Case and M. L. Cody (eds.), Island biogeography in the Sea of Cortéz. Univ. California Press, Berkeley.

———. 1983c. A distributional checklist of the reptiles and amphibians on the islands in the Sea of Cortez, pp. 429–437. *In* T. J. Case and M. L. Cody (eds.), Island biogeography in the Sea of Cortez. Univ. California Press, Berkeley.

Murphy, R. W., and G. Aguirre L. 2002a. The evolution of reptiles, pp. 181–220. *In* T. J. Case, M. L. Cody, and E. Ezcurra (eds.), Island biogeography in the Sea of Cortéz. Oxford Univ. Press, Oxford.

———. 2002b. A distributional checklist of the amphibians and reptiles on the islands in the Sea of Cortéz. Appendixes 6.2 and 6.3, pp. 580–591. *In* T. J. Case, M. L. Cody, and E. Ezcurra (eds.), Island biogeography in the Sea of Cortéz. Oxford Univ. Press, Oxford.

Murphy, R. W., and C. B. Crabtree. 1985a. Genetic relationships of the Santa Catalina Island rattleless rattlesnake, *Crotalus catalinensis* (Serpentes: Viperidae). Acta Zool. Mex. 9: 1–16.

———. 1985b. Evolutionary aspects of isozyme patterns, number of loci, and tissue-specific gene expression in the prairie rattlesnake, *Crotalus viridis viridis*. Herpetologica 41: 451–470.

———. 1988. Genetic identification of a natural hybrid rattlesnake: *Crotalus scutulatus scutulatus* × *C. viridis viridis*. Herpetologica 44: 119–123.

Murphy, R. W., and J. R. Ottley. 1984. Distribution of amphibians and reptiles on islands in the Gulf of California. Ann. Carnegie Mus. 53: 207–230.

Murphy, R. W., D. J. Morafka, and R. D. MacCulloch. 1989. Phylogenetic relationships of rattlesnakes as revealed by protein electrophoresis (abstract). Symposium of the Texas Herpetol. Soc., p. 10.

Murphy, R. W., V. Kovac, O. Haddrath, G. S. Allen, A. Fishbein, and N. E. Mandrak. 1995. mtDNA gene sequence, allozyme, and morphological uniformity among red diamond rattlesnakes, *Crotalus ruber* and *Crotalus exsul*. Can. J. Zool. 73: 270–281.

Murphy, R. W., J. Fu, A. Lathrop, J. V. Feltham, and V. Kovac. 2002. Phylogeny of the rattlesnakes (*Crotalus* and *Sistrurus*) inferred from sequences of five mitochondrial DNA genes, pp. 69–92. *In* G. W. Schuett, M. Höggren, M. E. Douglas, and H. W. Greene (eds.), Biology of the vipers. Eagle Mountain Publ., Eagle Mountain, Utah.

Murray, K. F. 1955. Herpetological collections from Baja California. Herpetologica 11: 33–48.

Myers, C. W. 1954. Subspecific identity of *Crotalus horridus* in Washington County, Missouri. Copeia 1954: 300–301.

———. 1956. An unrecorded food item of the timber rattlesnake. Herpetologica 12: 326.

Nair, B. C., C. Nair, and W. B. Elliott. 1979. Isolation and partial characterization of a phospholipase A_2 from the venom of *Crotalus scutulatus salvini*. Toxicon 17: 557–569.

Neill, W. T. 1948. Hibernation of amphibians and reptiles in Richmond County, Georgia. Herpetologica 4: 107–114.

———. 1951. Notes on the natural history of certain North American snakes. Publ. Res. Div. Ross Allen's Rept. Inst. 1: 47–60.

———. 1957. Some misconceptions regarding the eastern coral snake, *Micrurus fulvius*. Herpetologica 13: 111–118.

———. 1958. The occurrence of amphibians and reptiles in saltwater areas, and a bibliography. Bull. Mar. Sci. Gulf Caribbean 8: 1–97.

———. 1960. The caudal lure of various juvenile snakes. Quart. J. Florida Acad. Sci. 23: 173–200.

———. 1961. River frog swallows eastern diamondback rattlesnake. Bull. Philadelphia Herpetol. Soc. 9(1): 19.

———. 1963. Polychromatism in snakes. Quart. J. Florida Acad. Sci. 26: 194–216.

Nellis, D. W. 1997. Poisonous plants and animals of Florida and the Caribbean. Pineapple Press, Sarasota, Florida.

Newman, E. A., and P. N. Hartline. 1981. Integration of visual and infrared information in bimodal neurons of the rattlesnake optic tectum. Science (New York) 213: 789–791.

———. 1982. The infrared "vision" of snakes. Sci. Am. 246: 116–125.

Nichol, A. A., V. Douglas, and L. Peck. 1933. On the immunity of rattlesnakes to their venom. Copeia 1933: 211–213.

Nickerson, M. A., and C. E. Mays. 1968. More aberrations in the color patterns of rattlesnakes (genus *Crotalus*). Wasmann J. Biol. 26: 125–131.

———. 1969. A preliminary herpetofaunal analysis of Graham (Pinaleno) mountain region, Graham County, Arizona, with ecological comments. Trans. Kansas Acad. Sci. 72: 492–505.

Nicoletto, P. 1985. Some reptiles from Sinking Creek and Gap Mountains, Montgomery County, Virginia, April–June 1983. Catesbeiana 5(1): 13–15.

Nieto, N. C., J. E. Foley, J. Bettaso, and R. S. Lane. 2009. Reptile infection with *Anaplasma phagocytophilum*, the causative agent of granulocytic anaplasmosis. J. Parasitol. 95: 1165–1170.

Nikai, T., R. Kito, N. Mori, H. Sugihara, and A. T. Tu. 1983. Isolation and characterization of fibrinogenase from western diamondback rattlesnake venom and its comparison to the thrombin-like enzyme, crotalase. Comp. Biochem. Physiol. 76B: 679–686.

Noble, G. K., and A. Schmidt. 1937. The structure and function of the facial and labial pits of snakes. Proc. Am. Philos. Soc. 77: 263–288.

Nohavec, R. 1995. Update on Hogle Zoo's *Crotalus viridis concolor*. Intermontanus 4(2): 12.

Norris, R. 2004. Venom poisoning by North American reptiles, pp. 683–708. *In* J. A Campbell and W. W. Lamar (eds.), The venomous reptiles of the western hemisphere. Comstock Publ. Assoc., Cornell Univ. Press, Ithaca, New York.

Norris, R. L. 2005. First report of a bite by the mottled rock rattlesnake (*Crotalus lepidus lepidus*). Toxicon 46: 414–417.

Norris, R. L., Jr., and S. P. Bush. 2001. North American venomous reptile bites, pp. 896–926. *In* P. S. Auerbach (ed.), Wilderness medicine. 4th ed. Mosby, St. Louis, Missouri.

Norris, R. L., and S. P. Bush. 2007. Bites by venomous reptiles in the Americas, pp. 1051–1085. *In* P. S. Auerbach (ed.), Wilderness medicine. 5th ed. Mosby Elsevier, Philadelphia.

Nowak, E. M. 1998. Implications of nuisance rattlesnake relocation at Montezuma Castle National Monument. Sonoran Herpetol. 11: 2–5.

———. 2005. Nuisance rattlesnake movements and ecology in Arizona (USA) national parks, pp. 40–41. *In* Biology of the rattlesnakes symposium (program abstracts). Loma Linda Univ., Loma Linda, California.

Nowak, E. M., T. Hare, and J. McNally. 2002. Management of "nuisance" vipers: Effects of translocation on western diamond-back rattlesnakes (*Crotalus atrox*), pp. 533–560. *In* G. W. Schuett, M. Höggren, M. E. Douglas, and H. W. Greene (eds.), Biology of the vipers. Eagle Mountain Publ., Eagle Mountain, Utah.

Nussbaum, R. A., E. D. Brodie, Jr., and R. M. Storm. 1983. Amphibians and reptiles of the Pacific Northwest. Univ. Press Idaho, Moscow.

Obrecht, C. B. 1946. Notes on South Carolina reptiles and amphibians. Copeia 1946: 71–74.

O'Connell, B., D. Chiszar, and H. M. Smith. 1981. Effect of poststrike disturbance on strike-induced chemosensory searching in the prairie rattlesnake (*Crotalus v. viridis*). Behav. Neurol. Biol. 32: 343–349.

O'Connell, B., T. Poole, P. Nelson, H. M. Smith, and D. Chiszar. 1982. Strike-induced searching by prairie rattlesnakes (*Crotalus v. viridis*) after predatory and defensive strikes which made contact with mice. Bull. Maryland Herpetol. Soc. 18: 152–160.

O'Connell, B., D. Chiszar, and H. M. Smith. 1983. Strike-induced chemosensory searching in prairie rattlesnakes (*Crotalus viridis*) during daytime and at night. J. Herpetol. 17: 193–196.

O'Connell, R. R., J. Greenlee, J. Bacon, and D. Chiszar. 1982. Strike-induced chemosensory searching in old world vipers and new world pit vipers at San Diego Zoo. Zoo Biol. 1: 287–294.

Odum, R. A. 1979. The distribution and status of the New Jersey timber rattlesnake including an analysis of Pine Barren populations. HERP: Bull. New York Herpetol. Soc. 15: 27–35.

Offerman, S. R., S. P. Bush, J. A. Moynihan, and R. F. Clark. 2002. Crotaline Fab antivenom for the treatment of children with rattlesnake envenomation. Pediatrics 110: 968–971.

Ogawa, T., T. Chijiwa, N. Oda-Ueda, and M. Ohno. 2005. Molecular diversity and accelerated evolution of C-type lectin-like proteins from snake venom. Toxicon 45: 1–14.

Oldfield, B. L., and D. E. Keyler. 1989. Survey of timber rattlesnake (*Crotalus horridus*) distribution along the Mississippi River in western Wisconsin. Trans. Wisconsin Acad. Sci., Arts, Lett. 77: 27–34.

———. 1997. Timber rattlesnakes: Velvet tails of Minnesota's blufflands, pp. 22–26. *In* J. J. Moriarty and D. Jones (eds.), Minnesota's amphibians and reptiles: Their conservation and status; Proceedings of a symposium. Serpent's Tale Natural History Book Distributors, Lanesboro, Minnesota.

Olendorff, R. R. 1976. The food habits of North American golden eagles. Am. Midl. Nat. 95: 231–236.

Oliver, J. A. 1955. The natural history of North American amphibians and reptiles. Van Nostrand, Princeton, New Jersey.

———. 1958. Snakes in fact and fiction. Macmillan, New York.

Oliver, J. A., and J. R. Bailey. 1939. Amphibians and reptiles of New Hampshire, pp. 195–217. *In* Biol. Surv. Connecticut Watershed, Concord, New Hampshire.

Olivier, R. 2008a. De Baja California ratelslang, *Crotalus enyo* (Cope, 1861) en een kweekverslag van de Rosario ratelslang, *Crotalus enyo furvus* (Lowe & Norris, 1954). Lacerta 66(1–3): 30–39.

———. 2008b. De Hopiratelslang, *Crotalus viridis nuntius* (Klauber 1936): Grote verwarring rond een "kleine" ratelslang. Lacerta 66(1–3): 40–46.

Omori-Satoh, T., J. Lang, H. Breithaupt, and E. Habermann. 1975. Partial amino acid sequence of the basic *Crotalus* phospholipase A. Toxicon 13: 69–71.

Orr, H. C., L. E. Harris, Jr., A. V. Bader, R. I. Kirschstein, and P. G. Probst. 1972. Cultivation of cells from a fibroma in a rattlesnake, *Crotalus horridus*. J. Natl. Cancer Inst. 48: 259–264.

Ortenburger, A. I. 1922. Some cases of albinism in snakes. Copeia 1922(113): 90.

Ortenburger, A. I., and B. Freeman. 1930. Notes on some reptiles from western Oklahoma. Publ. Univ. Oklahoma Biol. Surv. 2: 209–239.

Ortenburger, A. I., and R. D. Ortenburger. 1927 [1926]. Field observations on some amphibians and reptiles of Pima County, Arizona. Proc. Oklahoma Acad. Sci. 6: 101–121.

Oshima, G., T. Sato-Ohmori, and T. Suzuki. 1969. Proteinase, arginineester hydrolase and a kinin releasing enzyme in snake venoms. Toxicon 7: 229–233.

Otten, E. J., and D. McKimm. 1983. Venomous snakebite in a patient allergic to horse serum. Ann. Emerg. Med. 12: 624–627.

Ouyang, C., C. M. Teng, and T. F. Huang. 1992. Characterization of snake venom components acting on blood coagulation and platelet function. Toxicon 30: 945–966.

Over, W. H. 1923. Amphibians and reptiles of South Dakota. South Dakota Geol. Nat. Hist. Surv. Bull. 12: 1–31.

———. 1928. A personal experience with rattlesnake bite. Bull. Antiv. Inst. Am. 2: 8–10.

Owings, D. H., and R. G. Coss. 1977. Snake mobbing by California ground squirrels: Adaptive variation and ontogeny. Behaviour 138: 575–595.

————. 2008. Hunting California ground squirrels: Constraints and opportunities for northern Pacific rattlesnakes, pp. 155–168. *In* W. K. Hayes, K. R. Beaman, M. D. Cardwell, and S. P. Bush (eds.), The biology of rattlesnakes. Loma Linda Univ. Press, Loma Linda, California.

Owings, D. H., R. G. Coss, D. McKernon, M. R. Rowe, and P. C. Arrowood. 2001. Snake-directed antipredator behavior of rock squirrels (*Spermophilus variegatus*): Population differences and snake-species discrimination. Behaviour 138: 575–595.

Ownby, C. L. 1982. Pathology of rattlesnake envenomation, pp. 163–209. *In* A. T. Tu (ed.), Rattlesnake venoms: Their actions and treatment. Marcel Dekker, New York.

Ownby, C. L., A. T. Tu, and R. A. Kainer. 1975. Effect of diethylenetriaminepentaacetic acid and procaine on hemorrhage induced by rattlesnake venom. J. Clin. Pharmacol. 15: 419–426.

Ownby, C. L., D. Cameron, and A. T. Tu. 1976. Isolation of myotoxic component from rattlesnake (*Crotalus viridis viridis*) venom. Am. J. Pathol. 85: 149–166.

Ownby, C. L., J. Bjarnason, and A. T. Tu. 1978. Hemorrhagic toxins from rattlesnake (*Crotalus atrox*) venom: Pathogenesis of hemorrhage induced by three purified toxins. Am. J. Pathol. 93: 201–218.

Ownby, C. L., W. M. Woods, and G. V. Odell. 1979. Antiserum to myotoxin from prairie rattlesnake (*Crotalus viridis viridis*) venom. Toxicon 17: 373–380.

Ownby, C. L., T. R. Colberg, P. L. Claypool, and G. V. Odell. 1984. In vivo test of the ability of antiserum to myotoxin α from prairie rattlesnake (*Crotalus virdis viridis*) venom to neutralize local myonecrosis induced by myotoxin α and homologous crude venom. Toxicon 22: 99–105.

Oyler-McCance, S. J., J. St. John, J. M. Parker, and S. H. Anderson. 2005. Characterization of microsatellite loci isolated in midget faded rattlesnake (*Crotalus viridis concolor*). Mol. Ecol. Notes 5: 452–453.

Pack, H. J. 1930. Snakes of Utah. Utah Agric. Exp. Sta. Bull. 221: 1–32.

Pain, S. 1999. Sssss is for danger. A ground squirrel knows exactly how to rattle a rattlesnake. But why does it go out of its way to provoke a deadly enemy? Stephanie Pain investigates. New Scientist 2216: 34–37.

Painter, C. W., and L. Fitzgerald. 1998. Rattlesnake roundups and commercial trade of the western diamondback rattlesnake *Crotalus atrox*. Sonoran Herpetol. 11: 16–18.

Painter, C. W., L. A. Fitzgerald, and M. L. Heinrich. 1999. *Crotalus atrox* (western diamondback rattlesnake). Morphology. Herpetol. Rev. 30: 44.

Palisot de Beauvois, A. M. F. J. 1799. Memoir on Amphibia. Serpents. Trans. Am. Philos. Soc. 4: 362–381.

Palmer, R. S. 1946. The rattlesnake in Maine. Chicago Acad. Sci. Natural History Miscellanea 2:1–3.

————. 1974. Poisonous snakes of North Carolina. St. Mus. Nat. Hist. North Carolina, Raleigh.

Palmer, W. M., and A. L. Braswell. 1995. Reptiles of North Carolina. Univ. North Carolina Press, Chapel Hill.

Pandya, B. V., and A. Z. Budzynski. 1984. Anticoagulant proteases from western diamondback rattlesnake (*Crotalus atrox*) venom. Biochemistry 23: 460–470.

Panfoli, I., S. Ravera, D. Calzia, E. Dazzi, S. Gandolfo, I. M. Pepe, and A. Morelli. 2007. Inactivation of phospholipase A_2 and metalloproteinase from *Crotalus atrox* venom by direct current. J. Biochem. Mol. Toxicol. 21: 7–12.

Parker, H. W. 1977. Snakes of the World. Dover Publ., New York.

Parker, J. M., and S. H. Anderson. 2002. *Crotalus viridis concolor* (midget faded rattlesnake): Maximum length. Herpetol. Rev. 33: 140.

————. 2007. Ecology and behavior of the midget faded rattlesnake (*Crotalus oreganus concolor*) in Wyoming. J. Herpetol. 41: 41–51.

Parker, M. R. 2005. Airborne cues and their role in rattlesnake predatory behavior, p. 41. *In* Biology of the rattlesnakes symposium (program abstracts). Loma Linda Univ., Loma Linda, California.

Parker, M. R., and K. V. Kardong. 2005. Rattlesnakes can use airborne cues during post-strike prey relocation, pp. 397–402. *In* R. T. Mason, M. P. LeMaster, and D. Müller-Schwarze (eds.), Chemical signals in vertebrates 10. Springer, New York.

Parker, M. R., B. A. Young, and K. V. Kardong. 2008. The forked tongue and edge detection in snakes (*Crotalus oreganus*): An experimental test. J. Comp. Psychol. 122: 35–40.

Parker, S. A., and D. Stotz. 1977. An observation on the foraging behavior of the Arizona ridge-nosed rattlesnake, *Crotalus willardi willardi* (Serpentes: Crotalidae). Bull. Maryland Herpetol. Soc. 13: 123.

Parker, W. S. 1974. Home range, growth, and population density of *Uta stansburiana* in Arizona. J. Herpetol. 8: 135–139.

Parker, W. S., and W. S. Brown. 1973. Species composition and population changes in two complexes of snake hibernacula in northern Utah. Herpetologica 28: 319–326.

———. 1974. Mortality and weight changes of Great Basin rattlesnakes (*Crotalus viridis*) at a hibernaculum in northern Utah. Herpetologica 30: 234–239.

Parkinson, C. L., J. A. Campbell, and P. T. Chippindale. 2002. Multigene phylogenetic analysis of pitvipers, with comments on their biogeography, pp. 93–110. *In* G. W. Schuett, M. Höggren, M. E. Douglas, and H. W. Greene (eds.), Biology of the vipers. Eagle Mountain Publ., Eagle Mountain, Utah.

Parmalee, P. W., R. D. Oesch, and J. E. Guilday. 1969. Pleistocene and Recent vertebrate faunas from Crankshaft Cave, Missouri. Illinois St. Mus. Rep. Invest. 14: 1–37.

Parmley, D. 1990. Late Pleistocene snakes from Fowlkes Cave, Culberson County, Texas. J. Herpetol. 24: 266–274.

Parmley, D., and J. A. Holman. 1995. Hemphillian (late Miocene) snakes from Nebraska, with comments on Arikareean through Blancan snakes of midcontinental North America. J. Vert. Paleontol. 15: 79–95.

Parmley, D., and A. M. Parmley. 2001. Food habits of the canebrake rattlesnake (*Crotalus horridus atricaudatus*) in central Georgia. Georgia J. Sci. 59: 172–178.

Parrish, H. M. 1963. Analysis of 460 fatalities from venomous animals in the United States. Am. J. Med. Sci. 245(2): 12–141.

———. 1964a. Snakebite injuries in Louisiana. J. Louisiana St. Med. Soc. 116: 249–257.

———. 1964b. Texas snakebite statistics. Texas St. J. Med. 60: 592–598.

———. 1966. Incidence of treated snakebites in the United States. Public Health Reprints 81: 269–276.

Parrish, H. M., and L. P. Donovan. 1964. Facts about snakebites in Alabama. J. Med. Assoc. St. Alabama 33: 297–305.

Parrish, H. M., and R. E. Thompson. 1958. Human envenomation from bites of recently milked rattlesnakes: A report of three cases. Copeia 1958: 83–86.

Parrish, H. M., A. W. MacLaurin, and R. L. Tuttle. 1956. North American pit vipers: Bacterial flora of the mouths and venom glands. Virginia Med. Monthly 83: 383–385.

Parsons, H., and M. Sarell. 2005. Managing a landscape with rattlesnakes, p. 42. *In* Biology of the rattlesnakes symposium (program abstracts). Loma Linda Univ., Loma Linda, California.

Pattabhiraman, T. R., and F. E. Russell. 1973. A lethal protein from the venom of the southern pacific rattlesnake *Crotalus viridis helleri*. Proc. West. Pharmacol. Soc. 16: 107–110.

Pattabhiraman, T. R., D. C. Buffkin, and F. E. Russell. 1974. Some chemical and pharmacological properties of toxic fractions from the venom of the southern Pacific rattlesnake, *Crotalus viridis helleri*—II, Proc. West. Pharmacol. Soc. 17: 223–227.

Pattabhiraman, T. R., H. Whigman, A. H. F. Lui, and F. E. Russell. 1976. Effects of the venom of the rattlesnake *Crotalus viridis helleri* on vascular smooth muscle. Proc. West. Pharmacol. Soc. 19: 385–391.

Patten, R. B. 1981. Author's reply. J. Herpetol. 15: 126.

Patten, R. B., and B. H. Banta. 1980. A rattlesnake, *Crotalus ruber*, feeds on a road-killed animal. J. Herpetol. 14: 111–112.

Pauly, G. B., and M. F. Benard. 2002. *Crotalus viridis oreganus* (northern Pacific rattlesnake): Costs of feeding. Herpetol. Rev. 33: 56–57.

Pavlik, S. 2007. Arboreal behavior in the tiger rattlesnake (*Crotalus tigris*). Sonoran Herpetol. 20: 56.

Pendlebury, G. B. 1977. Distribution and abundance of the prairie rattlesnake, *Crotalus viridis viridis*, in Canada. Can. Field-Nat. 91: 122–129.

Penn, G. H. 1942. The life history of *Porocephalus crotali*, a parasite of the Louisiana muskrat. J. Parasitol. 28: 277–283.

Perales, J., A. G. C. Neves-Ferreira, R. H. Valente, and G. B. Domont. 2005. Natural inhibitors of snake venom hemorrhagic metalloproteinases. Toxicon 45: 1013–1020.

Pérez, J. C., and E. E. Sánchez. 1999. Natural protease inhibitors to hemorrhagins in snake venoms and their potential use in medicine. Toxicon 37: 703–728.

Pérez, J. C., S. Pichyangkul, and V. E. Garcia. 1979. The resistance of three species of warm-blooded animals to rattlesnake (*Crotalus atrox*) venom. Toxicon 17: 601–607.

Pérez-Higareda, G., and H. M. Smith. 1991. Ofidiofauna de Veracruz. Análisis taxonómico y zoogeográfico. Universidad Nacional Autónoma de México. Publicación Especial 7.

Perkins, C. B. 1943. Notes on captive-bred snakes. Copeia 1943: 108–112.

———. 1951. Hybrid rattlesnakes. Herpetologica 7: 146.

Perkins, R. M., and M. J. R. Lentz. 1934. Contribution to the herpetology of Arkansas. Copeia 1934: 139–140.

Perry, A. 1920. *Crotalus horridus* rattles for half hour. Copeia 1920(86): 133–134.

Peters, J. 1953. A fossil snake of the genus *Heterodon* from the Pliocene of Kansas. J. Paleontol. 27: 328–331.

Peters, J. A. 1954. The amphibians and reptiles of the coast and coastal Sierra of Michoacán, Mexico. Occ. Pap. Mus. Zool. Univ. Michigan 554: 1–37.

———. 1964. Dictionary of herpetology. Hafner Publ. Co., New York.

Peters, J. A., and B. Orejas-Miranda. 1970. Catalogue of the Neotropical Squamata Part I. Snakes. Smithsonian Inst. Press, Washington, D.C.

Petersen, R. C. 1970. Connecticut's venomous snakes: Timber rattlesnake and northern copperhead. Connecticut St. Geol. Nat. Hist. Surv. Bull. 103: 1–39.

Peterson, A. 1990. Ecology and management of a timber rattlesnake (*Crotalus horridus* L.) population in south-central New York state, pp. 255–261. *In* R. S. Mitchell, C. J. Sheviak, and D. J. Leopold (eds.), Ecosystem management: Rare species and significant habitats. Proc. 15th Ann. Nat. Areas Conf., New York St. Mus. Bull. 471.

Peterson, K. H. 1983 [1982]. Reproduction of captive *Crotalus mitchelli mitchelli* and *Crotalus durissus* at the Houston Zoological Gardens. 6th Ann. Rept. Symp. on Captive Propagation and Husbandry [National Zoological Park, Washington, D.C., 28–31 July 1982], pp. 323–327.

Pfaffenberger, G. S., N. M. Jorgensen, and D. D. Woody. 1989. Parasites of prairie rattlesnakes (*Crotalus viridis viridis*) and gopher snakes (*Pituophis melanoleucus sayi*) from the eastern high plains of New Mexico. J. Wildl. Dis. 25: 305–306.

Phelps, T. 1981. Poisonous snakes. Blandford Press, Poole, Dorset, United Kingdom.

Philpott, C. H. 1931. Relative resistance of protozoa to *Crotalus atrox* and cobra venom. Bull. Antiv. Inst. Am. 5: 28.

Picado, T. C. 1931. Epidermal microornaments of the Crotalinae. Bull. Antiv. Inst. Am. 4: 104–105.

Pickwell, G. V. 1972. Amphibians and reptiles of the Pacific states. Dover Publ., New York.

Picolo, G., R. Giorgi, M. M. Bernardi, and Y. Cury. 1998. The antinociceptive effect of *Crotalus durissus terrificus* snake venom is mainly due to a supraspinally integrated response. Toxicon 36: 223–227.

Pinney, R. 1981. The snake book. Doubleday, Garden City, New York.

Pisani, G. R., and H. S. Fitch. 1993. A survey of Oklahoma's rattlesnake roundups. Kansas Herpetol. Soc. Newsl. 92: 7–15.

Pisani, G. R., and B. R. Stephenson. 1991. Food habits in Oklahoma *Crotalus atrox* in fall and early spring. Trans. Kansas Acad. Sci. 94: 137–141.

Pisani, G. R., J. T. Collins, and S. R. Edwards. 1973. A re-evaluation of the subspecies of *Crotalus horridus*. Trans. Kansas Acad. Sci. 75: 255–263.

Place, A. J., and C. I. Abramson. 2004. A quantitative analysis of the ancestral area of rattlesnakes. J. Herpetol. 38: 152–156.

Place, A. J., and C. I. Abramson. 2005. Learning in snakes, II: Empirical studies of learning in rattlesnakes, p. 43. *In* Biology of the Rattlesnakes Symposium (program abstracts). Loma Linda Univ., Loma Linda, California.

———. 2008. Habituation of the rattle response in western diamondback rattlesnakes, *Crotalus atrox*. Copeia 2008: 835–843.

Platt, S. G., and T. R. Rainwater. 2008. A new maximum size record for *Crotalus molossus* (Baird and Girard, 1853). J. Kansas Herpetol. 28: 14–15.

———. 2009. An elevation record for *Crotalus lepidus lepidus* (Kennicott, 1861) in the Davis Mountains of West Texas. J. Kansas Herpetol. 30: 12.

Platt, S. G., K. R. Russell, W. E. Snyder, L. W. Fontenot, and S. Miller. 1999. Distribution and conservation status of selected amphibians and reptiles in the piedmont of South Carolina. J. Elisha Mitchell Sci. Soc. 115(1): 8–19.

Platt, S. G., C. G. Brantley, and T. R. Rainwater. 2001a. Canebrake fauna: Wildlife diversity in a critically endangered ecosystem. J. Elisha Mitchell Sci. Soc. 117: 1–19.

Platt, S. G., A. W. Hawkes, and T. R. Rainwater. 2001b. Diet of the canebrake rattlesnake (*Crotalus horridus atricaudatus*): An additional record and review. Texas J. Sci. 53: 115–120.

Plowman, D. M., T. L. Reynolds, and S. M. Joyce. 1995. Poisonous snakebite in Utah. West. J. Med. 163: 547–551.

Plummer, M. V. 2000. *Crotalus scutulatus* (Mojave rattlesnake). Thermal stress. Herpetol. Rev. 31: 104–105.

Pokriefka, R. A., C. A. Weatherby, A. F. Ognjan, F. A. Paul, and R. E. Amenta. 1993. Handbook of antimicrobial therapy for reptiles and amphibians. Herpetol. Ichthyol. Infect. Dis. Assoc., Detroit, Michigan.

Ponce-Campos, P., R. Romero-Contreras, and S. M. Huerta-Ortega. 2000. *Crotalus lepidus maculosus* (Durangan rock rattlesnake). Herpetol. Rev. 31: 113.

Pook, C. E., and R. McEwing. 2005. Mitochondrial DNA sequences from dried snake venom: A DNA barcoding approach to the identification of venom samples. Toxicon 46: 711–715.

Pook, C. E., W. Wüster, and R. S. Thorpe. 2000. Historical biogeography of the western rattlesnake (Serpentes: Viperidae: *Crotalus viridis*), inferred from mitochondrial DNA sequence information. Mol. Phylogenet. Evol. 15: 269–282.

Pool, W. R., and A. L. Bieber. 1981. Fractionation of midget faded rattlesnake (*Crotalus viridis concolor*) venom: Lethal fractions and enzymatic activities. Toxicon 19: 517–527.

Pope, C. H. 1937. Snakes alive and how they live. Viking Press, New York.

———. 1944a. Amphibians and reptiles of the Chicago area. Chicago Natur. Hist. Mus.

———. 1944b. The poisonous snakes of the New World. New York Zool. Soc., New York.

———. 1946. Snakes of the northeastern United States. New York Zool. Soc., New York.

———. 1950. The snake bite problem. Bull. Chicago Nat. Hist. Mus. 21: 6–7.

———. 1955. The reptile world. Alfred A. Knopf, New York.

Poran, N. S., R. G. Coss, and E. Benjamini. 1987. Resistance of California ground squirrels (*Spermophilus beecheyi*) to the venom of the northern Pacific rattlesnake (*Crotalus viridis oreganus*): A study of adaptive variation. Toxicon 25: 767–777.

Porras, L. W. 2000. In a world of its own: The midget-faded rattlesnake (*Crotalus viridis concolor*). Reptiles 8(9): 28–36.

Porter, C. A. 1994. Organization and chromosomal location of repetitive DNA sequences in three species of squamate reptiles. Chromosome Res. 2: 263–273.

Porter, C. A., M. J. Hamilton, J. W. Sites, Jr., and R. J. Baker. 1991. Location of ribosomal DNA in chromosomes of squamate reptiles: Systematic and evolutionary implications. Herpetologica 47: 271–280.

Porter, K. R. 1972. Herpetology. Saunders, Philadelphia.

Porter, T. 1983. Induced cannibalism in *Crotalus mitchelli*. Bull. Chicago Herpetol. Soc. 18: 48.

Porter, W. P., J. L. Sabo, C. R. Tracy, O. J. Reichman, and N. Ramankutty. 2002. Physiology on a landscape scale: Plant-animal interactions. Integ. and Comp. Biol. 42: 431–453.

Possani, L. D., B. P. Sosa, A. C. Alagón, and P. M. Burchfield. 1980. The venom from the snakes *Agkistrodon bilineatus taylori* and *Crotalus durissus totonacus*: Lethality, biochemical and immunological properties. Toxicon 18: 356–360.

Possardt, E. E., W. H. Martin, W. S. Brown, and J. Sealy. 2005. A range wide action plan for the timber rattlesnake (*Crotalus horridus*): Hope for actual conservation progress or more paper?

p. 43. *In* Biology of the rattlesnakes symposium (program abstracts). Loma Linda Univ., Loma Linda, California.

Pough, F. H. 1966. Ecological relationships in southeastern Arizona with notes on other species. Copeia 1966: 676–683.

Pough, F. H., R. M. Andrews, J. E. Cadle, M. L. Crump, A. H. Savitzky, and K. D. Wells. 2004. Herpetology. 3rd ed. Pearson/Prentice Hall, Upper Saddle River, New Jersey.

Powell, R., M. Inboden, and D. D. Smith. 1990. Erstnachweis von Hybriden zwischen den Klapperschlagen *Crotalus cerastes laterorepens* Klauber, 1944 und *Crotalus scutulatus scutulatus* (Kennicott, 1861). Salamandra 26: 319–329.

Powell, R., J. T. Collins, and E. D. Hooper, Jr. 1998. A key to amphibians & reptiles of the continental United States and Canada. Univ. Press Kansas, Lawrence.

Powell, R. L., and C. S. Lieb. 2008. Perspective on venom evolution in *Crotalus*, pp. 551–556. *In* W. K. Hayes, K. R. Beaman, M. D. Cardwell, and S. P. Bush (eds.), The biology of rattlesnakes. Loma Linda Univ. Press, Loma Linda, California.

Powell, R. L., C. S. Lieb, and E. D. Rael. 2004a. Identification of a neurotoxic venom component in the tiger rattlesnake, *Crotalus tigris*. J. Herpetol. 38: 149–152.

Powell, R. L., M. McMillan, J. D. Jurado, and D. I. Lannutti. 2004b. *Crotalus molossus* (black-tailed rattlesnake): Fatal ingestion. Herpetol. Rev. 35: 400.

Powell, R. L., E. E. Sánchez, and J. C. Pérez. 2006. Farming for venom: Survey of snake venom extraction facilities worldwide. Appl. Herpetol. 3: 1–10.

Powell, R. L., C. S. Lieb, and E. D. Rael. 2008. Geographic distribution of Mojave toxin and Mojave toxin subunits among selected *Crotalus* species, pp. 537–550. *In* W. K. Hayes, K. R. Beaman, M. D. Cardwell, and S. P. Bush (eds.), The biology of rattlesnakes. Loma Linda Univ. Press, Loma Linda, California.

Powers, A. 1972. An instance of cannibalism in captive *Crotalus viridis helleri* with a brief review of cannibalism in rattlesnakes. Bull. Maryland Herpetol. Soc. 8: 60–61.

———. 1973. A review of the purpose of the rattle in crotalids as a defensive diversionary mechanism. Bull. Maryland Herpetol. Soc. 9: 30–32.

Prado Vera, I. 1971. Estudio taxonomico de algunos nemátodos parasitos de reptiles de México. Thesis, Fac. Cienc., Univ. Nac. Auton. México.

Prange, H. D., and S. P. Christman. 1976. The allometrics of rattlesnake skeletons. Copeia 1976: 542–545.

Preston, R. E. 1979. Late Pleistocene cold-blooded vertebrate fauna from the mid-continental United States. I. Reptilia; Testudines, Crocodilia. Univ. Michigan Mus. Paleontol. Pap. Paleontol. 19: 1–53.

Price, A. H. 1980. *Crotalus molossus*. Cat. Am. Amphib. Rept. 242: 1–2.

———. 1982. *Crotalus scutulatus*. Cat. Am. Amphib. Rept. 291: 1–2.

———. 1988. Observations on maternal behavior and neonate aggregation in the western diamondback rattlesnake, *Crotalus atrox* (Crotalidae). Southwest. Nat. 33: 370–374.

———. 1998. Poisonous snakes of Texas. Texas Parks Wildl. Press, Austin.

Price, A. H., and J. L. LaPointe. 1990. Activity patterns of a Chihuahuan Desert snake community. Ann. Carnegie Mus. 59: 15–23.

Prieto, A. A., and E. R. Jacobson. 1968. A new locality for melanistic *Crotalus molossus molossus* in southern New Mexico. Herpetologica 24: 339–340.

Prival, D. 2000a. No *pricei* is too high: Ecology and conservation of the twin-spotted rattlesnake. Sonoran Herpetol. 13: 14–18.

———. 2000b. It's not easy being green: Selected natural history aspects and a new hypothesis regarding the evolution of sexual dichromatism in *Crotalus lepidus klauberi*, p. 43. *In* Biology of the rattlesnakes symposium (program abstracts). Loma Linda Univ., Loma Linda, California.

———. 2008. Morphology, reproduction, and habitat use of a northern population of banded rock rattlesnakes, *Crotalus lepidus klauberi*, pp. 431–440. *In* W. K. Hayes, K. R. Beaman, M. D. Cardwell, and S. P. Bush (eds.), The biology of rattlesnakes. Loma Linda Univ. Press, Loma Linda, California.

Prival, D. B., M. J. Goode, D. E. Swann, C. R. Schwalbe, and M. J. Schroff. 2002. Natural history of a northern population of twin-spotted rattlesnakes, *Crotalus pricei*. J. Herpetol. 36: 598–607.

Prival, D. B., M. J. Goode, and C. R. Schwalbe. 2003. *Crotalus pricei* (twin-spotted rattlesnake): Winter activity. Herpetol. Rev. 34: 250.

Punzo, F. 1976. Analysis of the pH and electrolyte components found in the blood plasma of several species of west Texas reptiles. J. Herpetol. 10: 49–52.

Puskar, A. M. 1999. Captive breeding of the timber rattlesnake (*Crotalus horridus*). Bull. Chicago Herpetol. Soc. 34: 156–158.

Pylka, J. M., J. A. Simmons, and E. G. Wever. 1971. Sound production and hearing in the rattlesnake. Herpetol. Rev. 3: 107.

Quinn, H. 1981. *Crotalus lepidus lepidus* (mottled rock rattlesnake): Coloration. Herpetol. Rev. 12: 79–80.

Quinn, H. R. 1977. Further notes on reproduction in *Crotalus willardi* (Reptilia, Serpentes, Crotalidae). Bull. Maryland Herpetol. Soc. 13: 111.

———. 1987. Morphology, isozymes, and mitochondrial DNA as systematic indicators in *Crotalus viridis*. Ph.D. diss., Univ. Houston, Texas.

Quinn, J. S. 1985. Caspian terns respond to rattlesnake predation on colony. Wilson Bull. 97: 233–234.

Rabatsky, A. M. 2005. Rattle loss in insular rattlesnake species, p. 44. *In* Biology of the rattlesnakes symposium (program abstracts). Loma Linda Univ., Loma Linda, California.

———. 2006. Rattle reduction and loss in rattlesnakes endemic to islands in the Sea of Cortés. Sonoran Herpetol. 19(7): 80–81.

Rach, M., and P. Delis. 2005. Thermoregulation of gravid timber rattlesnakes (*Crotalus horridus*) at basking sites in north central Pennsylvania. J. Pennsylvania Acad. Sci. 78(Abstract and index issue): 128.

Radcliffe, C. W., and T. P. Maslin. 1975. A new subspecies of the red rattlesnake, *Crotalus ruber*, from San Lorenzo Sur Island, Baja California Norte, Texico. Copeia 1975: 490–493.

Radcliffe, C. W., D. Chiszar, and B. O'Connell. 1980. Effects of prey size on poststrike behavior in rattlesnakes (*Crotalus durissus*, *C. enyo*, and *C. viridis*). Bull. Psychon. Soc. 16: 449–450.

Radcliffe, C. W., K. Estep, T. Boyer, and D. Chiszar. 1986. Stimulus control of predatory behavior in red spitting cobras (*Naja mossambica pallida*) and prairie rattlesnakes (*Crotalus viridis*). Anim. Behav. 34: 804–814.

Rader, W. L. 1995. The Mojave rattlesnake (*Crotalus scutulatus*) of the eastern Mojave Desert of California. Herpetology 25(1): 1–7.

Rádis-Baptista, G., F. B. M. B. Moreno, L. L. Nogueira, A. M. C. Martins, D. O. Toyama, M .H. Toyama, W. F. Azevedo, Jr., B. S. Cavada, and T. Yamane. 2005. Crotacetin, a novel snake venom c-type lectin, is homolog of convulxin. J. Venom. Anim. Toxins Incl. Trop. Dis. 11(4): 557–578.

Rado, T. A., and P. G. Rowlands. 1981. A range extension and low elevational record for Arizona ridgenose rattlesnake (*Crotalus w. willardi*). Herpetol. Rev. 12: 15.

Rael, E. D., R. A. Knight, and H. Zepeda. 1984. Electrophoretic variants of Mojave rattlesnake (*Crotalus scutulatus scutulatus*) venoms and migration differences of Mojave toxin. Toxicon 22: 980–985.

Rael, E. D., R. J. Salo, and H. Zepeda. 1986. Monoclonal antibodies to Mojave toxin and use for isolation of cross-reacting proteins in *Crotalus* venoms. Toxicon 24: 661–668.

Rael, E. D., J. D. Johnson, O. Molina, and H. K. McCrystal. 1992. Distribution of Mojave toxin-like protein in rock rattlesnake (*Crotalus lepidus*) venom, pp. 163–168. *In* J. A. Campbell and E. D. Brodie, Jr. (eds.), Biology of the pitvipers. Selva, Tyler, Texas.

Rael, E. D., C. S. Lieb, N. Maddux, A. Varela-Ramirez, and J. Perez. 1993. Hemorrhagic and Mojave toxins in the venoms of the offspring of two Mojave rattlesnakes (*Crotalus scutulatus scutulatus*). Comp. Biochem. Physiol. 106B: 595–600.

Rael, E. D., J. Z. Rivas, T. Chen, N. Maddux, E. Huizar, and C. S. Lieb. 1997. Differences in fibrinolysis and complement inactivation by venom from different northern blacktailed rattlesnakes (*Crotalus molossus molossus*). Toxicon 35: 505–513.

Rafinesque, C. S. 1818. Further account of discoveries in natural history in the western states. Am. Month. Mag. Crit. Rev. 4: 39–42.

Rage, J. C. 1984. Serpentes. Handbuch der Palaeoherpetologie. Part II. Gustav Fischer, Stuttgart.

Ragsdale, F. R., and R. L. Ingermann. 1991. Influence of pregnancy on the oxygen affinity of red cells from the northern Pacific rattlesnake *Crotalus viridis oreganus*. J. Exp. Biol. 159: 501–505.

Ragsdale, F. R., J. K. Herman, and R. L. Ingermann. 1995. Nucleoside triphosphate levels versus oxygen affinity of rattlesnake red cells. Resp. Physiol. 102: 63–69.

Rahn, H. 1942a. Effect of temperature on color change in the rattlesnake. Copeia 1942: 178.

———. 1942b. The reproductive cycle of the prairie rattler. Copeia 1942: 233–240.

Ramirez, G. A., P. L. Fletcher, Jr., and L. D. Possani. 1990. Characterization of the venom from *Crotalus molossus nigrescens* Gloyd (black tail rattlesnake): Isolation of two proteases. Toxicon 28: 285–297.

Ramírez, M. S., E. E. Sánchez, C. García-Prieto, J. C. Pérez, G. R. Chapa, M. R. McKeller, R. Ramírez, and Y. D. Anda. 1999. Screening for fibrinolytic activity in eight viperid venoms. Comp. Biochem. Physiol. 124C: 91–98.

Ramírez-Bautista, A. 1994. Manual y claves illustradas de los amphibios y reptiles de la región de Chamela, Jalisco, México. Universidad Nacional Autónoma de México, México, D. F.

Ramsay, E. C., G. B. Daniel, B. W. Tryon, J. I. Merryman, P. J. Morris, and D. A. Bemis. 2002. Osteomyelitis associated with *Salmonella enterica* ss *arizonae* in a colony of ridgenose rattlesnakes (*Crotalus willardi*). J. Zoo Wildl. Med. 33: 301–310.

Randell, C. J., III, and H. O. Clark, Jr. 2007. Mojave desert sidewinder (*Crotalus cerastes cerastes*) behavior. Sonoran Herpetol. 20: 96.

Rao, R. B., M. Palmer, and M. Touger. 1998. Thrombocytopenia after rattlesnake envenomation. Ann. Emerg. Med. 31: 139–140.

Rathbun, G. A., and L. R. Heim. 1982. The dual nature of complement enhancement by venom from the Mojave rattlesnake (*Crotalus scutulatus scutulatus*). Toxicon 20: 495–499.

Raun, G. G. 1965. A guide to Texas snakes. Texas Mem. Mus., Mus. Notes (9): 1–85.

Raun, G. G., and F. R. Gehlbach. 1972. Amphibians and reptiles in Texas. Dallas Mus. Natur. Hist. Bull. (2): 1–61.

Reed, R. N. 2003a. Interspecific patterns of species richness, geographic range size, and body size among New World venomous snakes. Ecography 26: 107–117.

———. 2003b. Courtship and copulation in the Grand Canyon rattlesnake, *Crotalus viridis abyssus*. Herpetol. Rev. 34: 111–112.

Reed, R. N., and M. E. Douglas. 2002. Ecology of the Grand Canyon rattlesnake (*Crotalus viridis abyssus*) in the Little Colorado River Canyon, Arizona. Southwest. Nat. 47: 30–39.

Rego, A. A. 1980/1981. Sobre a identifacaçao des spécies de *Porocephalus* (Pentastomida) que ocorrem em ofídios da América tropical. Mem. Inst. Butantan 44/45: 219–231.

Rehling, G. C. 2002. Venom expenditure in multiple bites by rattlesnakes and cottonmouths. Master's thesis, Loma Linda Univ., Loma Linda, California.

Reichenbach-Klinke, H., and E. Elkan. 1965. The principal diseases of the lower vertebrates. Academic Press, New York.

Reichling, S. B. 2008. Reptiles and amphibians of the southern pine woods. Univ. Press Florida, Gainesville.

Reid, A. 2005. Craniodichotomy in a *Crotalus*: A two-headed prairie rattlesnake near Leader, SK. Blue Jay 63: 139–143.

Reinert, H. K. 1984a. Habitat separation between sympatric snake populations. Ecology 65: 478–486.

———. 1984b. Habitat variation within sympatric snake populations. Ecology 65: 1673–1682.

———. 1985. Timber rattlesnake, *Crotalus horridus* Linnaeus, pp. 282–285. *In* H. Genoways and F. J. Brenner (eds.), Species of special concern in Pennsylvania. Spec. Publ. Carnegie Mus. Nat. Hist. 11: 1–430.

———. 1990. A profile and impact assessment of organized rattlesnake hunts in Pennsylvania. J. Pennsylvania Acad. Sci. 64: 136–144.

———. 1991. The spatial ecology of timber rattlesnakes (*Crotalus horridus*). Progr. Jt. Ann. Meet. Soc. Stud. Amphib. Rept./Herpetol. League, Pennsylvania State Univ. (Abstract).

————. 1992. Radiotelometric field studies of pitvipers: Data acquisition and analysis, pp. 185–198. *In* J. A. Campbell and E. D. Brodie, Jr. (eds.), Biology of the pitvipers. Selva, Tyler, Texas.

————. 2005. A telemetric study of the survivorship, behavior, and spatial ecology of neonatal timber rattlesnakes, *Crotalus horridus*, p. 45. *In* Biology of the rattlesnakes symposium (program abstracts). Loma Linda Univ., Loma Linda, California.

Reinert, H. K., and R. R. Rupert, Jr. 1999. Impacts of translocation on behavior and survival of timber rattlesnakes, *Crotalus horridus*. J. Herpetol. 33: 45–61.

Reinert, H. K., and R. T. Zappalorti. 1988a. Timber rattlesnakes (*Crotalus horridus*) of the Pine Barrens: Their movement patterns and habitat preference. Copeia 1988: 964–978.

————. 1988b. Field observation of the association of adult and neonatal timber rattlesnakes, *Crotalus horridus*, with possible evidence for conspecific trailing. Copeia 1988: 1057–1059.

Reinert, H. K., D. Cundall, and L. M. Bushar. 1984. Foraging behavior of the timber rattlesnake, *Crotalus horridus*. Copeia 1984: 976–981.

Reiserer, R. S. 2001. Evolution of life histories in rattlesnakes. Ph.D. diss., Univ. California, Berkeley.

————. 2002. Stimulus control of caudal luring and other feeding responses: A program for research on visual perception in vipers, pp. 361–383. *In* G. W. Schuett, M. Höggren, M. E. Douglas, and H. W. Greene (eds.), Biology of the vipers. Eagle Mountain Publ., Eagle Mountain, Utah.

Reiserer, R. S., G. W. Schuett, and R. L. Earley. 2008. Dynamic aggregations of newborn sibling rattlesnakes exhibit thermoregulatory properties. J. Zool. (London) 274: 277–283.

Repp, R. 1998. Wintertime observations on five species of reptiles in the Tucson area: Sheltersite selections/Fidelity to sheltersites/Notes on behavior. Bull. Chicago Herpetol. Soc. 33: 49–56.

————. 1999. Herping Arizona—1998 in review: El Niño, flower shows, media hype, numbers, comparisons, special events and look ahead to La Niña. Sonoran Herpetol. 12: 14–17.

Repp, R. A., and G. W. Schuett. 2003. Herpetofauna of the 100 mile circle: West diamondback rattlesnake (*Crotalus atrox*). Sonoran Herpetol. 16: 8.

————. 2008. Western diamond-backed rattlesnakes, *Crotalus atrox* (Serpentes: Viperidae), gain water by harvesting and drinking rain, sleet, and snow. Southwest. Nat. 53: 108–114.

————. 2009. *Crotalus atrox* (western diamond-backed rattlesnake): Adult predation on lizards. Herpetol. Rev. 40: 353–354.

Retzios, A. D., and F. S. Markland, Jr. 1992. Purification, characterization, and fibrinogen cleavage sites of three fibrinolytic enzymes from the venom of *Crotalus basiliscus basiliscus*. Biochemistry 31: 4547–4557.

————. 1994. Fibrinolytic enzymes from the venoms of *Agkistrodon contortrix contortrix* and *Crotalus basiliscus basiliscus*: Cleavage site specificity towards the α-chain of fibrin. Thromb. Res. 74: 355–367.

Revell, T. K., and W. K. Hayes. 2009. Desert iguanas (*Diposaurus dorsalis*) sleep less when in close proximity to a rattlesnake predator (*Crotalus cerastes*). J. Herpetol. 43: 29–37.

Reyes-Velasco, J., C. I. Grünwald, J. M. Jones, and G. N. Weatherman. 2008. *Crotalus willardi meridionalis* (southern ridge-nosed rattlesnake). Herpetol. Rev. 39: 485.

Reynolds, R. P. 1978. Resource use, habitat selection, and seasonal activity of a Chihuahuan snake community. Ph.D. diss., Univ. New Mexico, Albuquerque.

————. 1982. Seasonal incidence of snakes in northeastern Chihuahua, Mexico. Southwest. Nat. 27: 161–166.

Reynolds, R. P., and N. J. Scott, Jr. 1982. Use of a mammalian resource by a Chihuahuan snake community, pp. 99–118. *In* N. J. Scott, Jr. (ed.), Herpetological communities. U.S. Fish Wildl. Serv. Wildl. Res. Rep. 13.

Rice, A. N., T. L. Roberts IV, and M. E. Dorcas. 2006. Heating and cooling rates of eastern diamondback rattlesnakes, *Crotalus adamanteus*. J. Therm. Biol. 31: 501–505.

Richards, R. L. 1990. Quaternary distribution of the timber rattlesnake (*Crotalus horridus*) in southern Indiana. Proc. Indiana Acad. Sci. 99: 113–122.

Richardson, C. H. 1915. Reptiles of northwestern Nevada and adjacent territory. Proc. U.S. Natl. Mus. 48(2078): 403–435.

Riedle, D. 1994. Distribution of the timber rattlesnake (*Crotalus horridus*) in Chautauqua, Elk, and Montgomery Counties, Kansas. Kansas Herpetol. Soc. Newsl. 95: 11–12.

———. 1996. Some occurrence of the western diamondback rattlesnake (*Crotalus atrox*) in Kansas. Kansas Herpetol. Soc. Newsl. 105: 18–19.

Riffer, E., S. C. Curry, and R. Gerkin. 1987. Successful treatment with antivenin of marked thrombocytopenia without significant coagulopathy following rattlesnake bite. Ann. Emerg. Med. 16: 1297–1299.

Riley, J., and J. T. Self. 1979. On the systematics of the pentastomid genus *Porocephalus* (Humboldt, 1811) with descriptions of two new species. System. Parasitol. 1: 25–42.

Riley, K. B., D. Antoniskis, R. Maris, and J. M. Leedom. 1988. Rattlesnake capsule-associated *Salmonella arizona* infections. Arch. Intern. Med. 148: 1207–1210.

Ritter, W. E. 1921. The rattling of rattle snakes. Copeia 1921(94): 29–31.

Roberts, J., and H. Lillywhite. 1980. Lipid barrier to water exchange in epidermis of reptiles. Science (New York) 207: 1077–1079.

Roddy, H. J. 1928. Reptiles of Lancaster County and the state of Pennsylvania. Science Press, Lancaster, Pennsylvania.

Rodgers, T. L., and W. L. Jellison. 1942. A collection of amphibians and reptiles from western Montana. Copeia 1942: 10–13.

Rodrigo, C., and G. de Souza. 2006. A simple restraining device for venomous snakes. Bull. Chicago Herpetol. Soc. 41: 183–184.

Rodríguez-Acosta, A., S. Magaldi, J. C. Pérez, E. E. Sánchez, M. E. Girón, and I. Aguilar. 2007. A protein homology detection between rattlesnakes (Viperidae: Crotalinae) from South and North America deduced from antigenically related metalloproteases. Anim. Biol. 57: 401–407.

Rodríguez-Robles, J. A. 1994. Are Duvernoy's gland secretions of colubrid snakes venoms? J. Herpetol. 28: 388–390.

Roelke, C. E., and M. J. Childress. 2007. Defensive and infrared reception responses of true vipers, pitvipers, *Azemiops* and colubrids. J. Zool. (London) 273: 421–425.

Rogers, K. L. 1976. Herpetofauna of the Beck Ranch local fauna (upper Pliocene: Blancan) of Texas. Publ. Mus. Michigan St. Univ. Palcontol. Ser. 1: 167–200.

———. 1984. Herpetofauna of the Big Springs and Hornet's Nest quarries (northeastern Nebraska, Pleistocene: late Blancan). Trans. Nebraska Acad. Sci. 12: 81–94.

Rogers, K. L., C. A. Repenning, R. M. Forester, E. E. Larson, S. A. Hall, G. R. Smith, E. Anderson, and T. J. Brown. 1985. Middle Pleistocene (Late Irvingtonian) climatic changes in south-central Colorado. Natl. Geogr. Res. 1: 535–563.

Rome, L. C., D. A. Syme, S. Hollingworth, S. L. Lindstedt, and S. M. Baylor. 1996. The whistle and the rattle: The design of sound producing muscles. Proc. Nat. Acad. Sci. 93: 8095–8100.

Romer, A. S. 1956. Osteology of the reptiles. Univ. Chicago Press, Chicago.

Rorabaugh, J. 2007. Apparent rain harvesting by a Colorado desert sidewinder (*Crotalus cerastes laterorepens*). Sonoran Herpetol. 20: 128–129.

Rose, S. L. 2001. Prairie rattlesnake (*Crotalus viridis viridis*) monitoring in Alberta—preliminary investigations (2000). Alberta Sustainable Resource Development, Fish and Wildlife Division, Alberta Species at Risk Report 28. Edmonton, Alberta, Canada.

Rosen, P. B., J. E. Leiva, and C. P. Ross. 2000. Delayed antivenin treatment for a patient after envenomation with *Crotalus atrox*. Ann. Emerg. Med. 35: 86–88.

Rosen, P. C., and S. R. Goldberg. 2002. Female reproduction in the western diamond-backed rattlesnake, *Crotalus atrox* (Serpentes: Viperidae), from Arizona. Texas J. Sci. 54: 347–356.

Rosen, P. C., and C. H. Lowe. 1994. Highway mortality of snakes in the Sonoran Desert of southern Arizona. Biol. Conserv. 68: 143–148.

Ross, D. A. 1989. Amphibians and reptiles in the diets of North American raptors. Wisconsin Dept. Nat. Resour. Endang. Resour. Rep. 59: 1–33.

Rossi, J. V. 1992. Snakes of the United States and Canada: Keeping them healthy in captivity, vol. I: Eastern area. Krieger, Malabar, Florida.

Rossi, J. V., and J. J. Feldner. 1993. *Crotalus willardi* (Arizona ridgenose rattlesnake) and *Crotalus lepidus klauberi* (banded rock rattlesnake). Arboreal behavior. Herpetol. Rev. 24: 35.

Rossi, J. V., and R. Rossi. 1995. Snakes of the United States and Canada, vol. 2: Western area. Krieger, Malabar, Florida.

Rossman, D. A. 1960. Herpetological survey of the Pine Hills area of southern Illinois. Quart. J. Florida Acad. Sci. 22: 207–225.

Roudabush, R. L., and G. R. Coatney. 1937. On some blood protozoa of reptiles and amphibians. Trans. Am. Micro. Soc. 56: 291–297.

Rowe, M. P., and D. H. Owings. 1978. Meaning of sound rattling by rattlesnakes to California ground squirrels. Behaviour 66: 252–267.

———. 1990. Probing, assessment, and management during interactions between ground squirrels and rattlesnakes, part 1: Risks related to rattlesnake size and body temperature. Ethology 86: 237–249.

———. 1996. Probing, assessment and management during interactions between ground squirrels (Rodentia: Sciuridae) and rattlesnakes (Squamata: Viperidae), part 2: Cues afforded by rattlesnake rattling. Ethology 102: 856–874.

Rowe, M. P., and J. B. Sealy. 2005. Hibernaculum or bust: Species preferences in the scent-trailing behavior of neonatal timber rattlesnakes (Crotalus horridus), p. 45. In Biology of the rattlesnakes symposium (program abstracts). Loma Linda Univ., Loma Linda, California.

Rowe, M. P., R. G. Coss, and D. H. Owings. 1986. Rattlesnake rattles and burrowing owl hisses: A case of acoustic Batesian mimicry. Ethology 72: 53–71.

Rowe, M. P., T. M. Farrell, and P. M. May. 2002. Rattle loss in pygmy rattlesnakes (Sistrurus miliarius): Causes, consequences, and implications for rattle function and evolution, pp. 385–404. In G. W. Schuett, M. Höggren, M. E. Douglas, and H. W. Greene (eds.), Biology of the vipers. Eagle Mountain Publ., Eagle Mountain, Utah.

Ruben, J. A. 1976. Aerobic and anaerobic metabolism during activity in snakes. J. Comp. Physiol. 109B: 147–157.

———. 1979. Blood physiology during activity in the snakes Masticophis flagellum (Colubridae) and Crotalus viridis (Crotalidae). Comp. Biochem. Physiol. 64A: 577–580.

———. 1983. Mineralized tissues and exercise physiology of snakes. Am. Zool. 23: 377–381.

Ruben, J. A., and C. Geddes. 1983. Some morphological correlates of striking in snakes. Copeia 1983: 221–225.

Rubio, M. 1998. Rattlesnake: Portrait of a predator. Smithsonian Inst. Press, Washington, D.C.

Rudolph, D. C., S. J. Burgdork, R. R. Schaefer, R. N. Conner, and R. T. Zappalorti. 1998. Snake mortality associated with late season radio-transmitter implantation. Herpetol. Rev. 29: 155–156.

Rudolph, D. C., R. R. Schaefer, D. Saenz, and R. N. Conner. 2004. Arboreal behavior in the timber rattlesnake, Crotalus horridus, in eastern Texas. Texas J. Sci. 56: 395–404.

Rue, L., Jr. 2005. Red-tailed hawk killing diamondback rattlesnake. Wildl. Soc. Bull. 32(2): 305.

Ruiz, J. M. 1952. Sobre a distinção genérica dos Crotalidae (Ophidia: Crotaloidea) baseada em alguns caracteres osteológicos. (Nota preliminar). Mem. Inst. Butantan 23: 109–113.

Rundquist, E. M. 1981. Longevity records at the Oklahoma City Zoo. Herpetol. Rev. 12: 87.

Rundus, A. R., and D. H. Owings. 2005. Infrared communication: The use of a snake-directed infrared signal by California ground squirrels (Spermophilus beecheyi) during predatory encounters with northern Pacific rattlesnakes (Crotalus viridis oreganus), pp. 45–46. In Biology of the rattlesnakes symposium (program abstracts). Loma Linda Univ., Loma Linda, California.

Russell, F. E. 1960a. Snake venom poisoning in southern California. California Med. 93: 347–350.

———. 1960b. Rattlesnake bites in southern California. Am. J. Med. Sci. 239: 1.

———. 1965. Effects of cortisone during immunization with Crotalus venom. Toxicon 3: 65–67.

———. 1967a. Gel diffusion study of human sera following rattlesnake venom poisoning. Toxicon 5: 147–148.

———. 1967b. Pharmacology of animal venoms. Clin. Pharmacol. Therapeu. 8: 849–873.

———. 1969. Clinical aspects of snake venom poisoning in North America. Toxicon 7: 33–37.

———. 1978. Consecutive bites on three persons by a single rattlesnake. Toxicon 16: 79–80.

———. 1980a. Snake venom poisoning in the United States. Annual Rev. Med. 31: 247–259.

————. 1980b. Snake venom poisoning. Lippincott, Philadelphia.

————. 1983. Snake venom poisoning. Scholium International, Great Neck, New York.

————. 1984. Snake venoms, pp. 469–480. *In* M. W. J. Ferguson (ed.), The structure, development and evolution of reptiles. A Festschrift in honour of Professor A. d'A. Bellairs on occasion of his retirement. Symp. Zool. Soc. London (52), Academic Press, London.

Russell, F. E., and A. F. Brodie. 1974. Venoms of reptiles, pp. 449–478. *In* M. Florkin and B. T. Scheer (eds.), Chemical zoology, vol. 9: Amphibia and Reptilia. Academic Press, New York.

Russell, F. E., and J. A. Emery. 1959. Use of the chick in zootoxicologic studies on venoms. Copeia 1959: 73–74.

Russell, F. E., and L. Lauritzen. 1966. Antivenins. Trans. Roy. Soc. Trop. Med. Hyg. 60: 797–810.

Russell, F. E., and B. A. Michaelis. 1960. Zootoxicologic effects of *Crotalus* venoms. Physiologist 3(3): 135.

Russell, F. E., and A. L. Picchioni. 1983. Snake venom poisoning. Clin. Toxicol. Consultant 5: 73–87.

Russell, F. E., and H. W. Puffer. 1970. Pharmacology of snake venoms. Clin. Toxicol. 3: 433–444.

————. 1971. Pharmacology of snake venoms, pp. 87–98. *In* S. A. Minton (ed.), Snake venoms and envenomation. Marcel Dekker, New York.

Russell, F. E., and R. S. Scharffenberg. 1964. Bibliography of snake venoms and venomous snakes. Bibliographic Associates, West Covina, California.

Russell, F. E., J. A. Emery, and T. E. Long. 1960. Some properties of rattlesnake venom following 26 years storage. Proc. Soc. Exp. Biol. Med. 103: 737–739.

Russell, F. E., N. Ružić, and H. Gonzalez. 1973. Effectiveness of antivenin (Crotalidae) polyvalent following injection of *Crotalus* venom. Toxicon 11: 461–464.

Russell, F. E., R. W. Carlson, W. Wainschel, and J. Osborne. 1975. Snake venom poisoning in the United States: Experiences with 550 cases. J. Am. Med. Assoc. 233: 341–344.

Russell, F. E., T. R. Pattabhiraman, D. C. Buffkin, R. W. Carlson, R. C. Schaeffer, H. Whigham, M. D. Fairchild, N. Tamiya, N. Maeda, and R. Tanz. 1976. Some chemical and physiopharmacologic properties of the venoms of the rattlesnakes *Crotalus viridis helleri* and *Crotalus scutulatus scutulatus*. United States–Republic of China Binational Seminar on Protein Chemistry, Snake Venoms and Hormonal Proteins, Taipei, Taiwan, March, 1976. Toxicon 14: 417–418.

Ruth, S. B. 1974. A kingsnake from Isla Tortuga in the Gulf of California, Mexico. Herpetologica 30: 97–98.

Ruthling, P. D. R. 1916. Observing the feeding habits of the Pacific rattlesnake. Lorquinia 1(4): 26–29.

Ruthven, A. G. 1907. A collection of reptiles and amphibians from southern New Mexico and Arizona. Bull. Am. Mus. Nat. Hist. 23: 483–604.

Ruthven, A. G., C. Thompson, and H. T. Gaige. 1928. The herpetology of Michigan. Univ. Michigan Mus. Handb. Ser. 3: 1–228.

Rutledge, A. 1936. Birds and serpents. Nature Mag. 27: 137–139.

Ryan, K. C., and E. M. Caravati. 1994. Life-threatening anaphylaxis follows envenomation by two different species of Crotalidae. J. Wild. Med. 5: 263–268.

Saenz, D., S. J. Burgdorf, D. C. Rudolph, and C. M. Duran. 1996. *Crotalus horridus* (timber rattlesnake): Climbing. Herpetol. Rev. 27: 145.

Saint Girons, H. 1982. Reproductive cycles of male snakes and their relationships with climate and female reproductive cycles. Herpetologica 38: 5–16.

————. 1986. Les organes thermorécepteurs des serpents: Fossettes loréales et labiales. J. Psychol. Norm. Pathol. 81: 357–367.

Sajdak, R. A., and A. W. Bartz. 2004. *Crotalus horridus* (timber rattlesnake): Arboreality, diet. Herpetol. Rev. 35: 60–61.

Sajdak, R. A., J. M. Kapfer, and C. S. Berg. 2005. Habitat selection and home ranges of timber rattlesnakes, *Crotalus horridus,* in the upper Mississippi River Valley, p. 46. *In* Biology of the rattlesnakes symposium (program abstracts). Loma Linda Univ., Loma Linda, California.

Samejima, Y., Y. Aoki, and D. Mebs. 1991. Amino acid sequence of a myotoxin from venom of the eastern diamondback rattlesnake (*Crotalus adamanteus*). Toxicon 29: 461–468.

Sánchez, E. E., L. A. Soliz, M. S. Ramírez, and J. C. Pérez. 2001. Partial characterization of a basic protein from *Crotalus molossus molossus* (northern blacktail rattlesnake) venom and production of a monoclonal antibody. Toxicon 39: 523–537.

Sánchez, E. E., J. A. Galán, R. Powell, J. G. Soto, W. K. Russell, D. H. Russell, and J. C. Pérez. 2005a. Isolation of two disintegrins from *Crotalus scutulatus scutulatus* (Mojave rattlesnake) venom lacking Mojave toxin, p. 47. *In* Biology of the rattlesnakes symposium (program abstracts). Loma Linda Univ., Loma Linda, California.

Sánchez, E. E., J. A. Galán, R. L. Powell, S. R. Reyes, J. G. Soto, W. K. Russell, D. H. Russell, and J. C. Pérez. 2005b. Disintegrin, hemorrhagic, and proteolytic activities of Mojave rattlesnake, *Crotalus scutulatus scutulatus* venoms lacking Mojave toxin. Comp. Biochem. Physiol. 141C: 124–132.

Sánchez, E. E., J. A. Galán, and J. C. Pérez. 2006. Inhibition of lung tumor formation in BALB/c mice treated with disintegrins isolated from *Crotalus atrox* (western diamondback rattlesnake) venom, pp. 41–42. *In* S. A. Seifert (ed.), Snakebites in the new millennium: A state-of-the-art symposium, University of Nebraska Medical Center, 21–23 October 2005, Omaha, Nebraska. J. Med. Toxicol. 2: 29–45.

Sánchez, E. E., A. M. Salazar, B. Guerrero, A. Rodríguez-Acosta, and J. C. Pérez. 2007. Hemorrhagic, fibrino(geno)lytic, coagulant and lethal activities of venoms of the southern Pacific rattlesnakes. *In* Venom week 2007 (program abstracts). Univ. Arizona Colleges of Medicine, Arizona Health Sci. Center, Off. Continuing Medical Educ., Tucson.

Sanchez, F. C., A. Gonzalez Ruiz, E. Godinez Cano, and J. F. Delgadillo Espinosa. 1999. *Crotalus lepidus morulus* (rock rattlesnake). Reproduction. Herpetol. Rev. 30: 168.

Sánchez-Herrera, O. 1980. Herpetofauna of the Pedregal de San Angel. Bull. Maryland Herpetol. Soc. 16: 9–18.

Sanders, L. 2009. Venom hunters: Scientists probe toxins, revealing the healing powers of biochemical weapons. Science News 176(4): 16–20.

Sanders, R. T. 1951. Effect of venom injections in rattlesnakes. Herpetologica 7: 47–52.

Sant' Ana, C. D., F. K. Ticli, L. L. Oliveira, J. R. Giglio, C. G. V. Rechia, A. L. Fuly, H. S. Selistre de Araúja, J. J. Franco, R. G. Stabeli, A. M. Soares, and S. V. Sampaio. 2008. BjussuSP-I: A new thrombin-like enzyme isolated from *Bothrops jararacussu* snake venom. Comp. Biochem. Physiol. 151A: 443–454.

Sapru, Z. Z., A. T. Tu, and G. S. Bailey. 1983. Purification and characterization of fibrinogenase from the venom of western diamondback rattlesnake (*C. atrox*). Biochem. Biophys. Acta 747: 225–231.

Sasaki, K., and D. Duvall. 2006. Field notes: Negative impacts of her-ping on timber rattlesnakes (*Crotalus horridus*). Bull. Herpetol. Soc. Japan 2006(1): 45–47.

Savage, J. M. 1960. Evolution of a peninsular herpetofauna. Syst. Zool. 9: 184–211.

———. 1966. The origins and history of the central American herpetofauna. Copeia 1966: 719–766.

Savage, J. M., and F. S. Cliff. 1953. A new subspecies of sidewinder, *Crotalus cerastes*, from Arizona. Chicago Acad. Sci. Nat. Hist. Misc. 119: 1–7.

Savage, T. 1967. The diet of rattlesnakes and copperheads in the Great Smoky Mountains National Park. Copeia 1967: 226–227.

Savary, W. 1999. *Crotalus molossus molossus* (northern blacktail rattlesnake): Brood defense. Herpetol. Rev. 30: 45.

Savitsky, A. H. 1992. Embryonic development of the maxillary and prefrontal bones of crotaline snakes, pp. 119–142. *In* J. A. Campbell and E. D. Brodie, Jr. (eds.), Biology of the pitvipers. Selva, Tyler, Texas.

Savitzky, A. H., and B. R. Moon. 2008. Tail morphology in the western diamond-backed rattlesnake, *Crotalus atrox*. J. Morphol. 269: 935–944.

Schaefer, N. 1976. The mechanism of venom transfer from the venom duct to the fang in snakes. Herpetologica 32: 71–76.

Schaefer, W. H. 1934. Diagnosis of sex in snakes. Copeia 1934: 181.

Schaeffer, G. C. 1969. Sex independent ground color in the timber rattlesnake, *Crotalus horridus horridus*. Herpetologica 25: 65–66.

Schaeffer, P. J., K. E. Conley, and S. L. Lindstedt. 1996. Structural correlates of speed and endurance in skeletal muscle: The rattlesnake tailshaker muscle. J. Exp. Biol. 199: 354–358.

Schaeffer, R. C., Jr., S. Bernick, T. H. Rosenquist, and F. E. Russell. 1972a. The histochemistry of the venom glands of the rattlesnake *Crotalus viridis helleri*—I. Lipid and non-specific esterase. Toxicon 10: 183–186.

Schaeffer, R. C., Jr., S. Bernick, T. H. Rosenquist, and F. E. Russell. 1972b. The histochemistry of the venom glands of the rattlesnake *Crotalus viridis helleri*—II. Monoamine oxidase, acid and alkaline phosphatase. Toxicon 10: 295–297.

Schaeffer, R. C., Jr., R. W. Carlson, H. Whigham, F. E. Russell, and M. H. Weil. 1973. Some hemodynamic effects of rattlesnake (*Crotalus viridis helleri*) venom. Proc. West. Pharmacol. Soc. 16: 58–62.

Schaeffer, R. C., Jr., T. R. Pattabhiraman, R. W. Carlson, F. E. Russell, and M. H. Weil. 1979. Cardiovascular failure produced by a peptide from the venom of the southern Pacific rattlesnake, *Crotalus viridis helleri*. Toxicon 17: 447–453.

Schmidt, C. 2002. A demographical analysis of the prairie rattlesnakes collected for the 2000 and 2001 Sharon Springs, Kansas, rattlesnake roundups. J. Kansas Herpetol. 1: 12–18.

Schmidt, D. F., W. K. Hayes, and F. E. Hayes. 1993. Influence of prey movement on the aim of predatory strikes of the western rattlesnake (*Crotalus viridis*). Great Basin Nat. 53: 203–206.

Schmidt, K. P. 1953. A check list of North American amphibians and reptiles. 6th ed. Am. Soc. Ichthyol. Herpetol., Chicago.

Schmidt, K. P., and D. D. Davis. 1941. Field book of snakes of the United States and Canada. Putnam, New York.

Schmidt, K. P., and F. A. Shannon. 1947. Notes on amphibians and reptiles of Michoacan, Mexico. Fieldiana: Zool. 31: 63–85.

Schoettler, W. H. A. 1951. On the stability of desiccated snake venoms. J. Immun. 67: 299–304.

Schofer, J. 2005. Population modeling of Arizona black rattlesnakes (*Crotalus oreganus cerberus*) at a den site near Flagstaff, Arizona, p. 47. *In* Biology of the rattlesnakes symposium (program abstracts). Loma Linda Univ., Loma Linda, California.

Schorger, A. W. 1968. Rattlesnakes in early Wisconsin. Trans. Wisconsin Acad. Sci., Arts, Lett. 56: 29–48.

Schroeder, C. R. 1934. The snake mite (*Ophionyssus serpentium* Hirst.). J. Econ. Entomol. 27: 1004–1014.

Schuett, G. W. 1992. Is long-term sperm storage an important component of the reproductive biology of temperate pitvipers? pp. 169–184. *In* J. A. Campbell and E. D. Brode, Jr. (eds.), Biology of the pit vipers. Selva, Tyler, Texas.

———. 1998. Current research on male aggression and parthenogenesis in snakes. Sonoran Herpetol. 11: 98–101.

Schuett, G. W., and P. A. Buttenhoff. 1993. Corroborative evidence for the lack of spring-mating in certain populations of prairie rattlesnakes, *Crotalus viridis*. Herpetol. Nat. Hist. 1: 101–106.

Schuett, G. W., P. J. Fernandez, W. F. Gergits, N. J. Casna, D. Chiszar, H. M. Smith, J. B. Mitton, S. P. Mackessey, R. A. Odum, and M. J. Demlong. 1997. Production of offspring in the absence of males: Evidence for facultative parthenogenesis in bisexual snakes. Herpetol. Nat. Hist. 5: 1–10.

Schuett, G. W., P. J. Fernandez, D. Chiszar, and H. M. Smith. 1998. Fatherless reproduction: A new type of parthenogenesis in snakes. Fauna 1: 20–25.

Schuett, G. W., S. L. Carlisle, A. T. Holycross, J. K. O'Leile, D. L. Hardy, Sr., E. A Van Kirk, and W. J. Murdoch. 2002a. Mating system of male Mojave rattlesnakes (*Crotalus scutulatus*): Seasonal timing of mating, agonistic behavior, spermatogenesis, sexual segment of kidney, and plasma sex steroids, pp. 515–532. *In* G. W. Schuett, M. Höggren, M. E. Douglas, and H. W. Greene (eds.), Biology of the vipers. Eagle Mountain Publ., Eagle Mountain, Utah.

Schuett, G. W., E. M. Nowak, and R. A. Repp. 2002b. *Crotalus cerberus* (Arizona black rattlesnake): Diet and prey size. Herpetol. Rev. 33: 210–211.

Schuett, G. W., M. Höggren, M. E. Douglas, and H. W. Greene (eds.). 2002c. Biology of the vipers. Eagle Mountain Publ., Eagle Mountain, Utah.

Schuett, G. W., E. N. Taylor, E. A. Van Kirk, and W. J. Murdoch. 2004a. Handling stress and plasma corticosterone levels in captive male western diamond-backed rattlesnakes (*Crotalus atrox*). Herpetol. Rev. 35: 229–233.

Schuett, G. W., M. S. Grober, E. A. Van Kirk, and W. J. Murdoch. 2004b. Long-term sperm storage and plasma steroid profile of pregnancy in a western diamond-backed rattlesnake (*Crotalus atrox*). Herpetol. Rev. 35: 328–333.

Schuett, G. W., D. L. Hardy, Sr., H. W. Greene, R. L. Earley, M. S. Grober, E. A. Van Kirk, and W. J. Murdoch. 2005. Sympatric rattlesnakes with contrasting mating systems show differences in seasonal patterns of plasma sex steroids. Anim. Behav. 70: 257–266.

Schuett, G. W., R. A. Repp, E. N. Taylor, D. F. De Nardo, R. L. Earley, E. A. Van Kirk, and W. J. Murdoch. 2006. Winter profile of plasma sex steroid levels in free-living male western diamond-backed rattlesnakes, *Crotalus atrox* (Serpentes: Viperidae). Gen. Comp. Endocrinol. 149: 72–80.

Schultz, E., A. W. Clark, A. Susuki, and R. G. Cassens. 1980. Rattlesnake shaker muscle: 1. A light microscopic and histochemical study. Tissue Cell 12: 323–334.

Schwab, D. 1988. Growth and rattle development in a captive timber rattlesnake, *Crotalus horridus*. Bull. Chicago Herpetol. Soc. 23: 26–27.

Scudder, K. M., and D. Chiszar. 1977. Effects of six visual stimulus conditions on defensive and exploratory behavior in two species of rattlesnakes. Psychol. Rec. 3: 519–526.

Scudder, K. M., D. Chiszar, and H. M. Smith. 1983. Effects of environmental odors on strike-induced chemosensory searching by rattlesnakes. Copeia 1983: 519–522.

Scudder, K. M., D. Chiszar, H. M. Smith, and T. Melcer. 1988. Response of neonatal prairie rattlesnakes (*Crotalus viridis*) to conspecific and heterospecific chemical cues. Psychol. Rec. 38: 459–471.

Scudder, K. M., D. Chiszar, and H. M. Smith. 1992. Strike-induced chemosensory searching and trailing behaviour in neonatal rattlesnakes. Anim. Behav. 44: 574–576.

Sealy, J. B. 1996. *Crotalus horridus* (timber rattlesnake): Mating. Herpetol. Rev. 27: 23–24.

———. 1997. Short-distance translocations of timber rattlesnakes in a North Carolina state park: A successful conservation and management program. Sonoran Herpetol. 10: 94–99.

———. 2002. Ecology and behavior of the timber rattlesnake (*Crotalus horridus*) in the Upper Piedmont of North Carolina: Identified threats and conservation recommendations, pp. 561–578. *In* G. W. Schuett, M. Höggren, M. E. Douglas, and H. W. Greene (eds.), Biology of the vipers. Eagle Mountain Publ., Eagle Mountain, Utah.

Secor, S. M. 1992. A preliminary analysis of the movement and home range size of the sidewinder, *Crotalus cerastes*, pp. 389–393. *In* J. A. Campbell and E. D. Brodie, Jr. (eds.), Biology of the pitvipers. Selva, Tyler, Texas.

———. 1994. Natural history of the sidewinder, *Crotalus cerastes*, pp. 281–301. *In* P. R. Brown and J. W. Wright (eds.), Herpetology of North American deserts: Proceedings of a symposium. Southwest. Herpetol. Soc. Spec. Publ. 5.

———. 1995. Ecological aspects of foraging mode for the snakes *Crotalus cerastes* and *Masticophis flagellum*. Herpetol. Monogr. 9: 169–186.

Secor, S. M., and J. Diamond. 1998. A vertebrate model of extreme physiological regulation. Nature (London) 395: 659–662.

Secor, S. M., and K. A. Nagy. 1994. Bioenergetic correlates of foraging mode for the snakes *Crotalus cerastes* and *Masticophis flagellum*. Ecology 75: 1600–1614.

Secor, S. M., B. C. Jayne, and A. F. Bennett. 1992. Locomotor performance and energetic cost of sidewinding by the snake *Crotalus cerastes*. J. Exp. Biol. 163: 1–14.

Secor, S. M., E. D. Stein, and J. Diamond. 1994. Rapid regulation of snake intestine in response to feeding: A new model of intestinal adaptation. Am. J. Physiol. 266: G695–G705.

Seelex, S. F. 1963. Interim statement on first-aid therapy for bites by venomous snakes. Toxicon 1: 81–87.

Seifert, S. A. 2006. Snakebites in the new millennium: A state-of-the-art symposium, University of Nebraska Medical Center, 21–32 October 2005, Omaha, Nebraska. J. Med. Toxicol. 2: 29–45.

———. 2007a. NPDS-based characterization of native U.S. elapid and viperid envenomations. *In* Venom week 2007 (program abstracts). Univ. Arizona Colleges of Medicine, Arizona Health Sci. Center, Off. Continuing Medical Educ., Tucson.

———. 2007b. Exotic antivenoms in the United States. 2007. *In* Venom week 2007 (program abstracts). Univ. Arizona Colleges of Medicine, Arizona Health Sci. Center, Off. Continuing Medical Educ., Tucson.

Seifert, S. A., L. V. Boyer, R. C. Dart, R. S. Porter, and L. Sjostrom. 1997. Relationship of venom effects to venom antigen and antivenom serum concentrations in a patient with *Crotalus atrox* envenomation treated with a Fab antivenom. Ann. Emerg. Med. 30: 49–53.

Seigel, R. A., and J. T. Collins (eds.). 1993. Snakes: Ecology & behavior. McGraw-Hill, New York.

——— (eds.) 2002. Snakes: Ecology & behavior. Blackburn Press, Caldwell, New Jersey.

Seigel, R. A., and H. S. Fitch. 1984. Ecological patterns of relative clutch mass in snakes. Oecologia (Berlin) 61: 293–301.

Seigel, R. A., L. E. Hunt, J. L. Knight, L. Malaret, and N. L. Zuschlag. 1984. Vertebrate ecology and systematics: A tribute to Henry S. Fitch. Univ. Kansas Mus. Nat. Hist. Spec. Publ. 10: 1–278.

Seigel, R. A., J. T. Collins, and S. S. Novak. 2001. Snakes: Ecology and evolutionary biology. Blackburn Press, Caldwell, New Jersey.

Self, J. T., and R. E. Kuntz. 1967. Host-parasite relations in some Pentastomida. J. Parasitol. 53: 202–206.

Senior, K. 1999. Taking the bite out of snake venoms. Lancet 353: 1946–1947.

Serrano, S. M. T., J. Kim, D. Wang, B. Dragulev, J. D. Shannon, H. H. Mann, G. Veit, R. Wagener, M. Koch, and J. W. Fox. 2006. The cysteine-rich domain of snake venom metalloproteinases is a ligand for von Willebrand factor A domains: Role in substrate targeting. J. Biolog. Chem. 281(52): 39746–39756.

Setser, K. 2007. Use of anesthesia increases precision of snake length measurements. Herpetol. Rev. 38: 409–411.

Setser, K., E. Mociño-Deloya, and B. G. Fedorko. 2005. *Crotalus willardi obscurus* (New Mexico ridge-nosed rattlesnake): Foraging. Herpetol. Rev. 36: 68–69.

Sexton, O. J., and K. R. Marion. 1981. Experimental analysis of movements by prairie rattlesnakes, *Crotalus viridis*, during hibernation. Oecologia (Berlin) 51: 37–41.

Shams, A. M. S., D. H. Sifford, and B. D. Johnson. 1995. 5 [minute] nucleotidase and thrombin-like activities of selected crotalid venoms. Arkanasa Acad. Sci. Proc. 49: 169–172.

Shaw, C. E. 1948. The male combat "dance" of some crotalid snakes. Herpetologica 4: 137–145.

———. 1951. Male combat in American colubrid snakes with remarks on combat in other colubrid and elapid snakes. Herpetologica 7: 149–168.

———. 1964a. A snake easily ruffled but unrattled. Zoonooz 37: 1–7.

———. 1964b. Note on *Crotalus catalinensis* birthing. Zoonooz 37: 8.

———. 1966. Southern Pacific rattlesnake. Zoonooz 39: 19.

———. 1969. Longevity of snakes in North American collections as of 1 January 1968. Der Zool. Gart. 37(4/5): 193–196.

Shaw, C. E., and S. Campbell. 1974. Snakes of the American West. A. E. Knopf, New York.

Sheldon, D. 1929. Another reputed remedy of the American Indians. Bull. Antiv. Inst. Am. 3: 59.

Sherbrooke, W. C. 2003. Introduction to horned lizards of North America. Univ. California Press, Berkeley.

Sherbrooke, W. C., and C. J. May. 2008a. *Phrynosoma solare* (regal horned lizard): *Crotalus* envenomation. Herpetol. Rev. 39: 90–91.

———. 2008b. Body-flip and immobility behavior in regal horned lizards: A gape-limiting defense selectively displayed toward one of the two snake predators. Herpetol. Rev. 39: 156–162.

Sherbrooke, W. C., and M. F. Westphal. 2006. Responses of greater roadrunners during attacks on sympatric venomous and nonvenomous snakes. Southwest. Nat. 51: 41–47.

Shine, R. 1978. Sexual size dimorphism and male combat in snakes. Oecologia (Berlin) 33: 269–277.

———. 2003. Reproductive strategies in snakes. Proc. Royal Soc. London B 270: 995–1004.

———. 2008. Tracking elusive timber rattlers with molecular genetics. Mol. Ecol. 17: 715–718.

Shine, R., and R. A. Seigel. 1996. A neglected life-history trait: Clutch-size variance in snakes. J. Zool. (London) 239: 209–223.

Shipley, B. K., and R. P. Reading. 2006. A comparison of herpetofauna and small mammal diversity on black-tailed prairie dog (*Cynomys ludovicianus*) colonies and non-colonized grasslands in Colorado. J. Arid Environ. 66: 27–41.

Short, H. L. 1983. Wildlife guilds in Arizona desert habitats. U.S. Dept. Int., Bur. Land Mgmt. Tech. Note (362).

Shu, Y. Y., H. R. Allen, and C. R. Geren. 1988. Isolation and characterization of the weakly acidic hemorrhagins from timber rattlesnake venom, pp. 445–455. *In* H. Pirkle and F. S. Markland, Jr. (eds.), Hemostasis and animal venoms. Marcel Dekker, New York.

Sibley, H. 1951. Snakes are scared of you! Field and Stream 55(9): 46–48.

Sievert, J. 2002a. Erfahrungen bei der Haltung und Nachzucht von *Crotalus viridis cerberus* (Coues, 1875). Sauria (Berlin) 24: 19–25.

———. 2002b. Haltung und Zucht der Hopi-Klapperschlange *Crotalus viridis nuntius* Klauber, 1935. Sauria (Berlin) 24: 25–30.

———. 2002c. Beobachtungen bei der Vergesellschaftung von männlichen *Crotalus cerastes cerastes* Hallowell, 1854 im Terrarium. Sauria (Berlin) 24: 45–46.

———. 2002d. *Crotalus cerastes* Hallowell. Sauria (Berlin) 24 (suppl.): 559–564.

Sifford, C. A., D. H. Sifford, and B. D. Johnson. 1996. Acid phosphatase and proteinase activities of selected crotalid venoms. SAAS Bull. Biochem. Biotech. 9: 9–16.

Siigur, J., and E. Siigur. 1992. The direct acting α-fibrin(ogen)olytic enzymes from snake venoms. J. Toxicol. Toxin Rev. 11: 91–113.

———. 2006. Factor X activating proteases from snake venoms. J. Toxicol. Toxin Rev. 25: 235–255.

Silvani, S. H., S. De Valentine, B. L. Scurran, and J. M. Karlin. 1980. Poisonous snakebites of the extremities: A case history. J. Am. Podiatry Assoc. 70: 172–176.

Simon, T. L., and T. G. Grace. 1981. Envenomation coagulopathy in wounds from pit vipers. New England J. Med. 305(8): 443–447.

Simons, L. H. 1986. *Crotalus atrox* (western diamondback rattlesnake): Pattern. Herpetol. Rev. 17: 20–22.

Simpson, J. W., and L. J. Rider. 1971. Collagenolytic activity from venom of the rattlesnake *Crotalus atrox*. Proc. Soc. Exp. Biol. Med. 137: 893–895.

Simpson, J. W., A. C. Taylor, and B. M. Levy. 1973. Elastolytic activity from venom of the rattlesnake *Crotalus atrox*. Proc. Soc. Exp. Biol. Med. 144: 380–383.

Sinclair, R., W. Hon, and R. B. Ferguson. 1965. Amphibians and reptiles of Tennessee. Tennessee Game Fish Comm., Nashville.

Slavens, F. L., and K. Slavens. 2000. Reptiles and amphibians in captivity: Breeding, longevity, and inventory current January 1, 1999. Slaveware, Seattle, Washington.

Slowinski, J. B., and S. L. Rasmussen. 1985. *Crotalus viridis viridis* (prairie rattlesnake): Coloration. Herpetol. Rev. 16: 29.

Smart, E. W. 1951. Color analysis in the Great Basin rattlesnake. Herpetologica 7: 41–46.

Smith, B. D., and H. M. Smith. 1972. Bipedalism in the lizard *Sceloporus undulatus erythrocheilus*, and other notes on Wyoming herpetozoa. J. Herpetol. 6: 81–82.

Smith, C. 1992. *Crotalus adamanteus* (eastern diamondback rattlesnake). Behavior. Herpetol. Rev. 23: 118.

Smith, H. M. 1943. Summary of the collections of snakes and crocodilians made in Mexico under the Walter Rathbone Bacon Traveling Scholarship. Proc. U.S. Natl. Mus. 93: 393–504.

———. 1946. Preliminary notes and speculations on the *Triseriatus* group of rattlesnakes in Mexico. Univ. Kansas Sci. Bull. 31: 75–101.

———. 1952. A revised arrangement of maxillary fangs of snakes. Turtox News 30: 214–218.

———. 1956. Handbook of amphibians and reptiles of Kansas. 2nd ed. Univ. Kansas Mus. Nat. Hist. Misc. Publ. 9: 1–356.

————. 1990. Signs and symptoms following human envenomation by the Mojave rattlesnake, *Crotalus scutulatus*, treated without use of antivenom. Bull. Maryland Herpetol. Soc. 26: 105–110.

————. 2001. Searching for herps in Mexico in the 1930s—II. Bull. Chicago Herpetol. Soc. 36: 31–42.

Smith, H. M., and E. D. Brodie, Jr. 1982. Reptiles of North America: A guide to field identification. Golden Press, New York.

Smith, H. M., and R. L. Holland. 1971. Noteworthy snakes and lizards from Baja California. J. Herpetol. 5: 56–59.

Smith, H. M., and R. G. Van Gelder. 1955. New and noteworthy amphibians and reptiles from Sinaloa and Puebla, Mexico. Herpetologica 11: 145–149.

Smith, H. M., R. L. Holland, and R. L. Brown. 1971. The prairie rattlesnake in Baja California del Sur. J. Herpetol. 5: 200.

Smith, H. M., G. L. Smith, and D. Chiszar. 1996. A new record and review of partially scaleless snakes. Bull. Maryland Herpetol. Soc. 32: 107–112.

Smith, H. M., J. A. Lemos-Espinal, and P. Heimes. 2005. 2005 Amphibians and reptiles from northwestern Mexico. Bull. Chicago Herpetol. Soc. 40: 206–212.

Smith, J. J., and M. Goode. 2005. A rattler's tale: Inferring a snake's history through its rattle, p. 48. *In* Biology of the rattlesnakes symposium (program abstracts). Loma Linda Univ., Loma Linda, California.

Smith, L. J., A. T. Holycross, C. W. Painter, and M. E. Douglas. 2001. Montane rattlesnakes and prescribed fire. Southwest. Nat. 46: 54–61.

Smith, M. T., and M. L. Collyer. 2008. Regional variation and sexual dimorphism in head form of the prairie rattlesnake (*Crotalus viridis viridis*): Comparisons using new analytical techniques and collection methods, pp. 79–90. *In* W. K. Hayes, K. R. Beaman, M. D. Cardwell, and S. P. Bush (eds.), The biology of rattlesnakes. Loma Linda Univ. Press, Loma Linda, California.

Smith, P. W., and M. M. Hensley. 1958. Notes on a small collection of amphibians and reptiles from the vicinity of the Pinacate Lava Cap in northwestern Sonora, Mexico. Trans. Kansas Acad. Sci. 61: 64–76.

Smith, P. W., and L. M. Page. 1972. Repeated mating of a copperhead and timber rattlesnake. Herpetol. Rev. 4: 196.

Smith, R. M., W. F. Bien, H. W. Avery, and J. R. Spotila. 2008. Coexistence of rattlesnakes and military operations: Occurrence and spatial ecology of the timber rattlesnake (*Crotalus horridus*) on the Warren Grove Gunnery Range in the Pinelands of New Jersey, pp. 317–326. *In* W. K. Hayes, K. R. Beaman, M. D. Cardwell, and S. P. Bush (eds.), The biology of rattlesnakes. Loma Linda Univ. Press, Loma Linda, California.

Smith, T. 1818. On the structure of the poisonous fangs of serpents. Philos. Trans. Royal Soc. London 1818: 471–476.

Smith, T. L., and K. V. Kardong. 2000. Absence of polarity perception by rattlesnakes of envenomated prey trails. J. Herpetol. 34: 621–624.

Smith, T. L., and K. V. Kardong. 2005. Blood is not a cue for poststrike trailing in rattlesnakes, pp. 389–396. *In* R. T. Mason, M. P. LeMaster, and D. Müller-Schwarze (eds.), Chemical signals in vertebrates 10, Springer, New York.

Smith, T. L., K. V. Kardong, and P. A. Lavín-Murcio. 2000. Persistence of trailing behavior: Cues involved in poststrike behavior by the rattlesnake (*Crotalus viridis oreganus*). Behaviour 137: 691–703.

Smith, T. L., G. S. Bevelander, and K. V. Kardong. 2005. Influence of prey odor concentration on the poststrike trailing behavior of the northern Pacific rattlesnake. Herpetologica 61: 111–115.

Smits, A. W., and H. B. Lillywhite. 1985. Maintenance of blood volume in snakes: Transcapillary shifts of extravascular fluids during acute hemorrhage. J. Comp. Physiol. 155B: 305–310.

Smyth, T. 1949. Notes on the timber rattlesnake at Mountain Lake, Virginia. Copeia 1949: 78–79.

Snellings, E., Jr. 1986. The gentleman of snakes. Florida Nat. 59: 6–8.

Snider, A. T., and J. K. Bowler. 1992. Longevity of reptiles and amphibians in North American collections. Soc. Stud. Amphib. Rept. Herpetol. Circ. 21: 1–40.

Snider, A. T., Y. M. Lee, D. Hyde, and R. Christoffel. 2005. The eastern massasauga, *Sistrurus c. catenatus*, in Michigan: Conservation through education, p. 49. *In* Biology of the rattlesnakes symposium (program abstracts). Loma Linda Univ., Loma Linda, California.

Snyder, B. 1949. Diamondbacks and dollar bills. Florida Wildl. 4(5): 3–5, 16.

Snyder, C. C., and R. P. Knowles. 1988. Snakebites: Guidelines for practical management. Postgrad. Med. 83: 52–75.

Snyder, C. C., J. E. Pickins, R. P. Knowles, J. L. Emerson, and W. A. Hines. 1968. A definitive study of snakebite. J. Florida Med. Assoc. 55: 330–337.

Snyder, D. H. 1972. Amphibians and reptiles of Land Between the Lakes. Tennessee Valley Auth., Knoxville.

Soares, A. M., and J. R. Giglio. 2003. Chemical modifications of phospholipases A_2 from snake venoms: Effects on catalytic and pharmacological properties. Toxicon 42: 855–868.

Solomon, G. B. 1974. Probable role of the timber rattlesnake, *Crotalus horridus*, in the release of *Capillaria hepatica* (Nematoda) eggs from small mammals. Virginia J. Sci. 25: 182–184.

Solórzano, A. 2004. Serpientes de Costa Rica. Univ. Costa Rica, San José.

Sonnini, C. S., and P. A. Latreille. 1801. Histoire naturelle des reptiles . . . Vol. 3. Deterville, Paris.

Soto, J. G., J. C. Perez, M. M. Lopez, M. Martinez, T. B. Quintanilla-Hernandez, M. S. Santa-Hernandez, K. Turner, J. L. Glenn, R. C. Straight, and S. A. Minton. 1989. Comparative enzymatic study of HPLC-fractionated *Crotalus* venoms. Comp. Biochem. Physiol. 93B: 847–855.

Soto, J. G., S. A. White, S. R. Reyes, R. Regalado, E. E. Sanchez, and J. C. Perez. 2007. Molecular evolution of *PIII-SVMP* and *RGD disintegrin* genes from the genus *Crotalus*. Gene 389: 66–72.

Soulé, M. E., and A. J. Sloan. 1966. Biogeography and distribution of reptiles and amphibians on the islands in the Gulf of California, Mexico. Trans. San Diego Soc. Nat. Hist. 14: 137–156.

Speake, D. W., and R. H. Mount. 1973. Some possible ecological effects of "rattlesnake roundups" in the southeastern coastal plain. Proc. 27th Ann. Conf. Southeastern Assoc. Game Fish Commissioners 1973: 267–277.

Spencer, C. L. 2003. Geographic variation in the morphology, diet, and reproduction of a widespread pitviper, the western diamondback rattlesnake (*Crotalus atrox*). Ph.D. diss., Univ. Texas, Arlington.

———. 2008. Geographic variation in western diamond-backed rattlesnake (*Crotalus atrox*) morphology, pp. 55–78. *In* W. K. Hayes, K. R. Beaman, M. D. Cardwell, and S. P. Bush (eds.), The biology of rattlesnakes. Loma Linda Univ. Press, Loma Linda, California.

Sperry, C. C. 1941. Food habits of the coyote. Wildlife Res. Bull. 4: 1–70.

Spille, M., and B. Hamilton. 2006. *Crotalus molossus* (black-tailed rattlesnake): Behavior. Herpetol. Rev. 37: 477.

Sprent, J. F. A. 1978. Ascaridoid nematodes of amphibians and reptiles: *Polydelphis, Travassosascaris* n. g. and *Hexametra*. J. Helminthol. 52: 355–384.

Stabler, R. M. 1939. Frequency of skin shedding in snakes. Copeia 1939: 227–229.

———. 1948. Prairie rattlesnake eats spadefoot toad. Herpetologica 4: 168.

Stafford, P. 2000. Snakes. Smithsonian Inst. Press, Washington, D.C.

Stahlecker, D. W. 2004. Grand Canyon rattlesnake preys on juvenile spotted sandpipers. Southwest. Nat. 49: 412–414.

Stahnke, H. L. 1966. The treatment of venomous bites and stings. Revised ed. Arizona St. Univ., Tempe.

Starck, M. 1984. A prairie rattlesnake drinking water! Blue Jay 42: 195–196.

———. 1986. Overwintering by an ambystomid salamander in a prairie rattlesnake hibernaculum. Herpetol. Rev. 17: 7.

———. 1987. An active prairie rattlesnake den taken over by foxes. Blue Jay 45: 53–54.

Stark, C. P., D. Chiszar, and H. M. Smith. 2006. A noninvasive technique for blocking vomeronasal chemoreception in rattlesnakes. Psychol. Rec. 56: 471–487.

Stark, M. A. 1985. A simple technique for trapping prairie rattlesnakes during spring emergence. Herpetol. Rev. 16: 75, 77.

Starrett, B. L., and A. T. Holycross. 2000. *Crotalus lepidus klauberi* (banded rock rattlesnake). Caudal luring. Herpetol. Rev. 31: 245.

———. 2003. *Crotalus atrox* (western diamondback rattlesnake): Behavior. Herpetol. Rev. 34: 249.

Steadman, D. W., and L. J. Craig. 1993. Late Pleistocene and Holocene vertebrates from Jora Lemon's (Fish Club) Cave, Albany County, New York. Bull. New York St. Archaeol. Assoc. 105: 9–15.

Steadman, D. W., L. J. Craig, and J. Bopp. 1993 [1992]. Didly Cave: A new late Quaternary vertebrate fauna from New York state. Current Res. Pleistocene 9: 110–112.

Stebbins, R. C. 1943. Diurnal activity of *Crotalus cerastes*. Copeia 1943: 128–129.

———. 1954. Amphibians and reptiles of western North America. McGraw-Hill, New York.

———. 1985. A field guide to western reptiles and amphibians. Houghton Mifflin, Boston.

———. 2003. A field guide to western reptiles and amphibians. 3rd ed. Houghton Mifflin, Boston.

Stechert, R. 1980. Observations on northern snake dens. HERP: Bull. New York Herpetol. Soc. 15: 7–14.

———. 1981. Historical depletion of timber rattlesnake colonies in New York state. HERP: Bull. New York Herpetol. Soc. 17: 23–24.

Steehouder, T. 1991. Growth of the northern rattlesnakes. Litt. Serpent. Engl. Ed. 11: 74–77.

Steen, D. A., L. L. Smith, L. M. Conner, J. C. Brock, and S. K. Hoss. 2007. Habitat use of sympatric rattlesnake species within the Gulf Coastal Plain. J. Wildl. Manage. 71: 759–764.

Steenhof, K., and M. N. Kochert. 1985. Dietary shifts of sympatric buteos during prey decline. Oecologia (Berlin) 66: 6–16.

Stegall, T. D., D. H. Sifford, and B. D. Johnson. 1994. Proteinase, BAEEase, and TAMEase activities of selected crotalid venoms. SAAS Bull. Biochem. Biotech. 7: 7–13.

Stejneger, L. S. 1893. Annotated list of the reptiles and batrachians collected by the Death Valley Expedition in 1891, with descriptions of new species. North Am. Fauna 7: 159–228.

———. 1895 [1893]. The poisonous snakes of North America, pp. 337–487. *In* Annual Report of the U.S. National Museum, Smithsonian Institution, Washington, D.C.

———. 1898. The poisonous snakes of North America. Smithsonian Institution Report 1898: 338–487.

———. 1902. The reptiles of the Huachuca Mountains, Arizona. Proc. U.S. Natl. Mus. 25(1282): 149–158.

Stephenson, B. R., and G. R. Pisani. 1991. Notes on early spring parasites and pathologies of Oklahoma *Crotalus atrox*. Herpetol. Rev. 22: 88–90.

Stevenson, D. J. 2003. *Crotalus adamanteus* (eastern diamondback rattlesnake): Winter feeding. Herpetol. Rev. 34: 372–373.

Stewart, M. M., G. E. Larson, and T. H. Mathews. 1960. Morphological variation in a litter of timber rattlesnakes. Copeia 1960: 366–367.

Stewart, S. G., and D. J. Morafka. 1989. Karyotypes of Gulf of California insular rattlesnakes (Viperidae: *Crotalus*) compared to those of peninsular sister taxa (abstract). First World Congress of Herpetology, Canterbury, England.

Stewart, S. G., D. J. Morafka, and A. D. Stock. 1990. Karyotypes of Gulf of California insular rattlesnakes (Viperidae: *Crotalus*) compared to those of peninsular sister taxa, pp. 261–266. *In* E. Olmo (ed.), Advances in life sciences: Cytogenetics of amphibians and reptiles. Birkhäuser Verlag, Basel.

Stickel, W. H., and J. B. Cope. 1947. The home ranges and wanderings of snakes. Copeia 1947: 127 136.

Stille, B. 1987. Dorsal scale microdermatoglyphics and rattlesnake (*Crotalus* and *Sistrurus*) phylogeny (Reptilia: Viperidae: Crotalinae). Herpetologica 43: 98–104.

Stimson, A. C., and H. T. Engelhardt. 1960. The treatment of snakebite. J. Occup. Med. 2: 163–168.

Stinner, J. N., L. K. Hartzler, M. R. Grguric, and D. L. Newlon. 1998. A protein titration hypothesis for the temperature-dependence of tissue CO_2 content in reptiles and amphibians. J. Exp. Biol. 201: 415–424.

St. John, A. D. 1980. Knowing Oregon reptiles. Salem Audubon Soc., Salem, Oregon.

Stocker, K. F. 1990a. Medical use of snake venom proteins. CRC Press, Boca Raton, Florida.

———. 1990b. Snake venom proteins affecting hemostasis and fibrinolysis, pp. 97–160. *In* K. F. Stocker. Medical use of snake venom proteins. CRC Press, Boca Raton, Florida.

Stoddard, H. L. 1942. The bobwhite quail: Its habits, preservation, and increase. Charles Scribner's Sons, New York.

Storer, D. H. 1839. Reptiles of Massachusetts. Rep. Comm. Zool. Surv. Massachusetts: 203–253.

Storer, T. I., and B. M. Wilson. 1932. Feeding habits and molt of *Crotalus confluentus oreganus* in captivity. Copeia 1932: 169–173.

Storment, D. 1990. Field observations of sexual dimorphism in head pattern (markings) in timber rattlesnakes (*Crotalus horridus*). Bull. Chicago Herpetol. Soc. 25: 160–162.

Straight, R. C., and J. L. Glenn. 1993. Human fatalities caused by venomous animals in Utah, 1900–90. Great Basin Nat. 53: 390–394.

Straight, R. C., J. L. Glenn, T. B. Wolt, and M. C. Wolfe. 1991. Regional differences in content of small basic peptide toxins in the venoms of *Crotalus adamanteus* and *Crotalus horridus*. Comp. Biochem. Physiol. 100B: 51–58.

Straight, R. C., J. L. Glenn, T. B. Wolt, and M. C. Wolfe. 1992. North-south regional variation in phospholipase A activity in the venom of *Crotalus ruber*. Comp. Biochem. Physiol. 103B: 635–639.

Strimple, P. 1992a. *Crotalus mitchelli*, the speckled rattlesnake. Litt. Serpent. Engl. Ed. 12: 26–31.

———. 1992b. Report on the maintenance and growth of a juvenile eastern diamondback rattlesnake, *Crotalus adamanteus*, during its first year in captivity. Litt. Serpent. Engl. Ed. 12: 85–88.

———. 1993a. Report on the feeding and growth of a juvenile mottled rock rattlesnake, *Crotalus lepidus lepidus*, during three years in captivity. Litt. Serpent. Engl. Ed. 13: 89–94.

———. 1993b. Captive birth of Mojave rattlesnakes, *Crotalus scutulatus scutulatus*. Litt. Serpent. Engl. Ed. 13: 166–168.

———. 1996. *Crotalus scutulatus* (Kennicott), the Mojave rattlesnake. Litt. Serpent. Engl. Ed. 16: 36–38.

Stroupe, D. A., and M. E. Dorcas. 2001. The apparent persistence of *Crotalus horridus* in the western Piedmont of North Carolina. Herpetol. Rev. 32: 287–288.

Surface, H. A. 1906. The serpents of Pennsylvania. Bull. Pennsylvania St. Dept. Agric. Div. Zool. 4: 133–208.

Sutcliffe, R. 1952. Notes made by Dr. Edward Hallowell. Copeia 1952: 113–114.

Sutherland, I. D. W. 1958. The "combat dance" of the timber rattlesnake. Herpetologica 14: 23–24.

Sutherland, S. K., and A. R. Coulter. 1981. Early management of bites by the eastern diamondback rattlesnake (*Crotalus adamanteus*): Studies in monkeys (*Macaca fascicularis*). Am. J. Trop. Med. Hyg. 30: 497–500.

Suzuki, T., and S. Iwanaga. 1958a. Studies on snake venom: II. Some observations on the alkaline phosphatases of Japanese and Formosan snake venoms. J. Pharm. Soc. Japan 78: 354–361.

———. 1958b. Studies on snake venom: IV. Purification of the alkaline phosphatases in cobra venom. J Pharm. Soc. Japan 78: 368–375.

Svoboda, P., C. H. Adler, and J. Meier. 1992. Antimicrobial activity of several Elapidae and Viperidae snake venoms. *In* P. Gopalakrishnakone and C. D. Tan (eds.), Proceedings 10th world congress on animal, plant and microbial toxins (3–8 November 1991), Singapore.

Svoboda, P., J. Meier, and T. A. Freyvogel. 1995. Purification and characterization of three [alpha] 2-antiplasmin and [alpha] 2-macroglobulin inactivating enzymes from the venom of the Mexican west coast rattlesnake (*Crotalus basiliscus*). Toxicon 33: 1331–1346.

Swaisgood, R. R., D. H. Owings, and M. P. Rowe. 1999a. Conflict and assessment in a predator-prey system: Ground squirrels versus rattlesnakes. Anim. Behav. 57: 1033–1044.

Swaisgood, R. R., M. P. Rowe, and D. H. Owings. 1999b. Assessment of rattlesnake dangerousness by California ground squirrels: Exploitation of cues from rattling sounds. Anim. Behav. 57: 1301–1310.

————. 2003. Antipredator responses of California ground squirrels to rattlesnakes and rattling sounds: The roles of sex, reproductive parity, and offspring age in assessment and decision-making rules. Behav. Ecol. Sociobiol. 55: 22–31.

Swanson, P. 1952. The reptiles of Venango County, Pennsylvania. Am. Midl. Nat. 47: 161–182.

Swanson, P. L. 1946. Effects of snake venoms on snakes. Copeia 1946: 242–249.

Swaroop, W. H., and B. Grab. 1954. Snakebite mortality in the World. Bull. World Health Org. 10: 35–76.

Sweet, S. S. 1985. Geographic variation, convergent crypsis and mimicry in gopher snakes (*Pituophis melanoleucus*) and western rattlesnakes (*Crotalus viridis*). J. Herpetol. 19: 55–67.

Swift, L. W. 1933. Death of a rattlesnake from continued exposure to direct sunlight. Copeia 1933: 150.

Swinford, G. W. 1989. Captive reproduction of the banded rock rattlesnake *Crotalus lepidus klauberi*, pp. 99–110. *In* M. J. Uricheck (ed.), 13th International Herpetological Symposium on Captive Propagation and Husbandry. International Herpetological Symposiums, Stanford, California.

Szyndlar, Z., and J. C. Rage. 2002. Fossil record of the true vipers, pp. 419–444. *In* G. W. Schuett, M. Höggren, M. E. Douglas, and H. W. Greene (eds.), Biology of the vipers. Eagle Mountain Publ., Eagle Mountain, Utah.

Tabor, S. P., and D. J. Germano. 1997. *Masticophis flagellum* (coachwhip). Prey. Herpetol. Rev. 28: 90.

Taggart, T. W., and C. J. Schmidt. 2004. *Crotalus viridis* (prairie rattlesnake): New maximum size for entire range. J. Kansas Herpetol. 12: 18.

————. 2005. *Crotalus horridus* (timber rattlesnake). J. Kansas Herpetol. 14: 11.

Tai-A-Pin, J. 2008. *Crotalus cerastes*: De sidewinder of hoornratelslang. Lacerta 66 (1–3): 22–29.

Takahashi, H., and H. Mashiko. 1998. Haemorrhagic factors from snake venoms: I. Properties of haemorrhagic factors and antihaemorrhagic factors. J. Toxicol. Toxin Rev. 17: 315–335.

Takeya, H., A. Onikura, T. Nikai, H. Sugihara, and S. Iwanaga. 1990. Primary structure of a hemorrhagic metalloproteinase, HT-2, isolated from the venom of *Crotalus ruber ruber*. J. Biochem. (Tokyo) 108: 711–719.

Takeya, H., S. Nishida, N. Nishino, Y. Makinori, T. Omori-satoh, T. Nikai, H. Sugihara, and S. Iwanaga. 1993. Primary structures of platelet aggregation inhibitors (Disintegrins) auto-proteolytically released from snake venom hemorrhagic metalloproteinases and new flavogenic peptide substrates for these enzymes. J. Biochem. (Tokyo) 113: 473–483.

Talan, D. A., D. M. Citron, G. D. Overturf, B. Singer, P. Froman, and E. J. C. Goldstein. 1991. Antibacterial activity of crotalid venoms against oral snake flora and other clinical bacteria. J. Infect. Dis. 164: 195–198.

Tan, N.-H., and G. Ponnudurai. 1991. A comparative study of the biological activities of rattlesnake (genera *Crotalus* and *Sistrurus*) venom. Comp. Biochem. Physiol. 98C: 455–461.

Tanner, V. M. 1930. The amphibians and reptiles of Bryce Canyon National Park, Utah. Copeia 1930: 41–43.

Tanner, W. W. 1958. Herpetology of Glen Canyon of the Upper Colorado River Basin. Herpetologica 14: 193–195.

————. 1960. *Crotalus mitchelli pyrrhus* Cope in Utah. Herpetologica 16: 140.

————. 1966. A systematic review of the Great Basin reptiles in the collections of Brigham Young University and the University of Utah. Gr. Basin Nat. 26: 87–135.

————. 1975. Checklist of Utah amphibians and reptiles. Proc. Utah Acad. Sci. Arts Lett. 52: 4–8.

————. 1985. Snakes of western Chihuahua. Great Basin Nat. 45: 615–676.

Tanner, W. W., J. R. Dixon, and H. S. Harris, Jr. 1972. A new subspecies of *Crotalus lepidus* from western Mexico. Great Basin Nat. 32: 16–24.

Taylor, E. H. 1936. Notes on the herpetological fauna of the Mexican state of Sonora. Univ. Kansas Sci. Bull. 24: 475–503.

————. 1950. Second contribution to the herpetology of San Luis Potosí. Univ. Kansas Sci. Bull. 33: 441–457.

Taylor, E. N. 2001. Dict of the Baja California rattlesnake, *Crotalus enyo* (Viperidae). Copeia 2001: 553–555.

Taylor, E. N., and D. F. De Nardo. 2004. Reproductive ecology of western diamond-backed rattlesnakes in the Sonoran Desert, p. 26. *In* Proceedings of the Snake Ecology Group 2004 Conference, Jackson County, Illinois.

———. 2005a. Reproductive ecology of western diamond-backed rattlesnakes (*Crotalus atrox*) in the Sonoran Desert. Copeia 2005: 152–158.

———. 2005b. Sexual size dimorphism and growth plasticity in snakes: An experiment on the western diamond-backed rattlesnake (*Crotalus atrox*). J. Exp. Zool. 303A: 598–607.

———. 2008. Proximate determinants of sexual size dimorphism in the western diamond-backed rattlesnake (*Crotalus atrox*), pp. 91–100. *In* W. K. Hayes, K. R. Beaman, M. D. Cardwell, and S. P. Bush (eds.), The biology of rattlesnakes. Loma Linda Univ. Press, Loma Linda, California.

Taylor, E. N., and G. W. Schuett. 2004. Effect of temperature and storage duration on the stability of steroid hormones in blood samples from western diamond-backed rattlesnakes (*Crotalus atrox*). Herpetol. Rev. 35: 14–17.

Taylor, E. N., D. F. De Nardo, and D. H. Jennings. 2004a. Seasonal steroid hormone levels and their relation to reproduction in the western diamond-backed rattlesnake, *Crotalus atrox* (Serpentes: Viperidae). Gen. Comp. Endrocinol. 136: 328–337.

Taylor, E. N., D. F. De Nardo, and M. A. Malawy. 2004b. A comparison between point- and semi-continuous sampling for assessing body temperature in a free-ranging ectotherm. J. Thermal Biol. 29: 91–96.

Taylor, E. N., M. A. Malawy, D. M. Browning, S. V. Lemar, and D. F. De Nardo. 2005. Effects of food supplementation on the physiological ecology of female Western diamond-backed rattlesnakes (*Crotalus atrox*). Oecologia (Berlin) 144: 206–213.

Taylor, W. P. 1935. Notes on *Crotalus atrox* near Tucson, Arizona, with a special reference to its breeding habits. Copeia 1935: 154–155.

Telford, S. R., Jr. 1952. A herpetological survey in the vicinity of Lake Shipp, Polk County, Florida. Quart. J. Florida Acad. Sci. 15: 175–185.

———. 1965. A study of filariasis in Mexican snakes. Japan. J. Exp. Med. 35(6): 565–586.

Telford, S. R., Jr., P. E. Moler, and J. F. Butler. 2008. *Hepatozoon* species of the timber rattlesnake in northern Florida: Description of a new species, evidence of salivary gland oocysts, and a natural cross-familial transmission of an *Hepatozoon* species. J. Parasitol. 94: 520–523.

Tennant, A. 1984. The snakes of Texas. Texas Monthly Press, Austin.

———. 1985. A field guide to Texas snakes. Texas Monthly Press, Austin.

———. 1997. A field guide to snakes of Florida. Gulf Publ. Co., Houston, Texas.

———. 1998. A field guide to Texas snakes. 2nd ed. Gulf Publ. Co., Houston, Texas.

———. 2003. Snakes of North America: Eastern and central regions. Lone Star Books, Lantham, Maryland.

Tevis, L., Jr. 1943. Field notes on a red rattlesnake in Lower California. Copeia 1943: 242–245.

Thayer, F. D., Jr. 1988. *Crotalus atrox* (western diamondback rattlesnake). Hunting behavior. Herpetol. Rev. 19: 35.

Theakston, R. D. G., and H. A. Reid. 1978. Changes in the biological properties of venom from *Crotalus atrox* with aging. Period. Biol. 80 (suppl. 1): 123–133.

Theodoratus, D. H., and D. Chiszar. 2000. Habitat selection and prey odor in the foraging behavior of western rattlesnakes (*Crotalus viridis*). Behaviour 137: 119–135.

Thireau, M. 1991. Types and historically important specimens of rattlesnakes in the Muséum National d'Histoire Naturelle (Paris). Smithsonian Herpetol. Inform. Serv. (87): 1–10.

Thomas, R. G., and F. H. Pough. 1979. The effects of rattlesnake venom on digestion of prey. Toxicon 17: 221–228.

Thompson, L. S., and P. S. Nichols. 1982. Circle West wildlife monitoring study: 4th annual report for period March 1, 1981–May 31, 1982. Montana Dept. Nat. Res. Conserv., Circle West Tech. Rep. 10: 19–20.

Thompson, S. W. 1982. Snakes of South Dakota. South Dakota Conserv. Dig. 49(4): 12–18.

Thorne, E. T. 1977. Sybille Creek snake dance. Wyoming Wildl. 41(6): 14.

Thorpe, R. S., W. Wüster, and A. Malhotra. 1997. Venomous snakes: Ecology, evolution and snakebite. Clarendon Press, Oxford.

Tihen, J. A. 1962. A review of New World fossil bufonids. Am. Midl. Nat. 68: 1–50.

Timmerman, W. W. 1995. Home range, habitat use, and behavior of the eastern diamondback rattlesnake (*Crotalus adamanteus*) on the Ordway Preserve. Bull. Florida Mus. Nat. Hist. 38: 127–158.

Timmerman, W. W., and W. H. Martin. 2003. Conservation guide to the eastern diamondback rattlesnake, *Crotalus adamanteus*. Soc. Stud. Amphib. Rept. Herpetol. Circ. 32.

Tinkle, D. W. 1962. Reproductive potential and cycles in female *Crotalus atrox* from northwestern Texas. Copeia 1962: 306–313.

———. 1967. The life and demography of the side-blotched lizard, *Uta stansburiana*. Misc. Publ. Mus. Zool. Univ. Michigan 132: 1–182.

Tinkle, D. W., and J. W. Gibbons. 1977. The distribution and evolution of viviparity in reptiles. Misc. Publ. Mus. Zool. Univ. Michigan 154: 1–55.

Tipton, B. L. 2005. Snakes of the Americas: Checklist and lexicon. Krieger, Melbourne, Florida.

Tobey, F. J. 1985. Virginia's amphibians and reptiles: A distributional survey. Virginia Herpetol. Surv., Purcellville.

Tomes, C. S. 1875. On the structure and development of the teeth of Ophidia. Philos. Trans. Royal Soc. London 165: 297–302.

———. 1877. On the development and succession of the poison-fangs of snakes. Philos. Trans. Royal Soc. London 166: 377–385.

Tomko, D. S. 1975. The reptiles and amphibians of the Grand Canyon. Plateau 47: 161–166.

Toom, P. M., P. G. Squire, and A. T. Tu. 1969. Characterization of the enzymatic and biological activities of snake venoms by isoelectric focusing. Biochim. Biophys. Acta 181: 339–341.

Toweill, D. E. 1982. Winter foods of eastern Oregon bobcats. Northwest. Sci. 56: 310–315.

Tracey, J. A. 2000. Movement of red diamondback rattlesnakes (*Crotalus ruber*) in heterogeneous landscapes in coastal southern California. Master's thesis, Univ. California, San Diego.

Tracey, J. A., J. Zhu, and K. Crooks. 2005. A set of nonlinear regression models for animal movement in response to a single landscape feature. J. Agri., Biol., and Environ. Stat. 10(1): 1–18.

Trapido, H. 1937. The snakes of New Jersey: A guide. Newark Mus.

———. 1939. Parturition in the timber rattlesnake, *Crotalus horridus horridus* Linné. Copeia 1939: 230.

Trauth, S. E., H. W. Robison, and M. V. Plummer. 2004. The amphibians and reptiles of Arkansas. Univ. Arkansas Press, Fayetteville.

Triplehorn, C. A. 1955. Notes on the young of some North American reptiles. Copeia 1955: 248–249.

Trutnau, L. 1998. Schlangen im Terrarium, vol. 2: Giftschlangen. Verlag Eugen Ulmer, Stuttgart, Germany.

———. 2002. *Crotalus willardi* Meek, 1905: Eine kleine Berg-Klapperschlange. Draco 3(4): 50–53.

Tryon, B. 1978. Reproduction in a pair of captive Arizona ridge-nosed rattlesnakes, *Crotalus willardi willardi* (Reptilia, Serpentes, Crotalidae). Bull. Maryland Herpetol. Soc. 14: 83–88.

———. 1985. Snake hibernation and breeding: In and out of the zoo, pp. 19–31. *In* S. Townson and K. Lawrence (eds.), Reptiles: Breeding, behaviour, and veterinary aspects. British Herpetological Society, London.

Tryon, B. W., and C. W. Radcliffe. 1977. Reproduction in captive lower California rattlesnakes, *Crotalus enyo enyo* (Cope). Herpetol. Rev. 8: 34–36.

Tsai, I. H., Y. H. Chen, Y. M. Wang, M. C. Tu, and A. T. Tu. 2001. Purification, sequencing, and phylogenetic analyses of novel Lys-49 phospholipases A$_2$ from the venoms of rattlesnakes and other pit vipers. Arch. Biochem. Biophys. 394: 236–244.

Tsai, M. C., C. Y. Lee, and A. Bdolah. 1983. Mode of neuromuscular blocking action of a toxic phospholipase A$_2$ from *Pseudocerastes fieldi* (Field's horned viper) snake venom. Toxicon 21: 527–534.

Tu, A. T. 1977. Venoms: Chemistry and molecular biology. John Wiley and Sons, New York.

——— (ed.). 1982a. Rattlesnake venoms: Their actions and treatment. Marcel Dekker, New York.

——— (ed.). 1982b. Chemistry of rattlesnake venoms, pp. 247–312. *In* A. T. Tu (ed.), Rattlesnake venoms: Their actions and treatment. Marcel Dekker, New York.

———. 1988. Overview of snake venom chemistry. Adv. Exp. Med. Biol. 391: 37–62.

Tubbs, K. A., R. W. Nelson, J. R. Krone, and A. L. Bieber. 2000. Mass spectral studies of snake venoms and some of their toxins. J. Toxicol. 19: 1–22.

Turner, F. B. 1955. Reptiles and amphibians of Yellowstone National Park. Yellowstone Interp. Ser. 5: 1–40.

Tyning, T. 1987. In the path of progress. Sanctuary (Lincoln) 26(9): 3–5.

Uhler, F. M., C. Cottam, and T. E. Clarke. 1939. Food of snakes of the George Washington National Forest, Virginia. Trans. N. Am. Wildl. Conf. 4: 605–622.

Underwood, G. 1967. A contribution to the classification of snakes. Publ. British Mus. Nat. Hist. 653: i–x, 1–179.

U.S. Fish and Wildlife Service (USFWS). 1985. New Mexico ridgenose rattlesnake recovery plan. U.S. Fish Wildl. Serv., Albuquerque, New Mexico.

Van Bourgondien, T. M., and R. C. Bothner. 1969. A comparative study of the arterial systems of some New World Crotalinae (Reptilia: Ophidia). Am. Midl. Nat. 81: 107–147.

Van Dam, G. H. 1978. Amphibians and reptiles, pp. 19–25. In J. E. Guilday (ed.), The Baker Bluff Cave deposit, Tennessee, and the Late Pleistocene faunal gradient. Bull. Carnegie Mus. Nat. Hist. 11.

van Dam, R., and M. K. Hecht. 1954. Fossil rattlesnakes of the genus *Crotalus* from northern Massachusetts. Copeia 1954: 158–159.

Van Denburgh, J. 1895. Description of a new rattlesnake (*Crotalus pricei*) from Arizona. Proc. California Acad. Sci. Ser. 2, 5: 856–857.

———. 1920. Description of a new species of rattlesnake (*Crotalus lucasensis*) from lower California. Proc. California Acad. Sci. Ser. 4, 10: 29–30.

———. 1922. The reptiles of western North America, vol. 2: Snakes and turtles. Occ. Pap. California Acad. Sci. 10: 617–1028.

———. 1978. Herpetology of lower California: Herpetology of Baja California, Mexico (collected papers). Facsimile reprints in herpetology, Soc. Stud. Amphib. Rept.

Van Denburgh, J., and J. R. Slevin. 1921a. A list of the amphibians and reptiles of the peninsula of Lower California, with notes on the species in the collection of the Academy. Proc. California Acad. Sci. Ser. 4, 11(4): 49–72.

———. 1921b. Preliminary diagnoses of more new species of reptiles from islands in the Gulf of California, Mexico. Proc. California Acad. Sci. Ser. 4, 11(17): 395–398.

Vanderpool, R., J. Malcolm, and M. Hill. 2005. *Crotalus atrox* (western diamondback rattlesnake): Predation. Herpetol. Rev. 36: 191–192.

Van Devender, T. R., and G. L. Bradley. 1994. Late Quaternary amphibians and reptiles from Maravillas Canyon Cave, Texas, with discussion of the biogeography and evolution of the Chihuahuan Desert herpetofauna, pp. 23–53. In P. R. Brown and J. W. Wright (eds.), Herpetology of the North American deserts: Proceedings of a symposium. Southwest. Herpetol. Soc. Spec. Publ. 5.

Van Devender, T. R., and C. H. Lowe, Jr. 1977. Amphibians and reptiles of Yepómera, Chihuahua, Mexico. J. Herpetol. 11: 41–50.

Van Devender, T. R., and J. I. Mead. 1978. Early Holocene and late Pleistocene amphibians and reptiles in Sonoran Desert packrat middens. Copeia 1978: 464–475.

Van Devender, T. R., K. B. Moodie, and A. H. Harris. 1976. The desert tortoise (*Gopherus agassizii*) in the Pleistocene of the northern Chihuahuan Desert. Herpetologica 32: 298–304.

Van Devender, T. R., A. M. Phillips III, and J. I. Mead. 1977. Late Pleistocene reptiles and small mammals from the Lower Grand Canyon of Arizona. Southwest. Nat. 22: 49–66.

Van Devender, T. R., A. M. Rea, and M. L. Smith. 1985. The Sangamon interglacial vertebrate fauna from Rancho la Brisca, Sonora, Mexico. Trans. San Diego Soc. Nat. Hist. 21: 23–55.

Van Devender, T. R., A. M. Rea, and W. E. Hall. 1991a. Faunal analysis of late Quaternary vertebrates from Organ Pipe Cactus National Monument, southwestern Arizona. Southwest. Nat. 36: 94–106.

Van Devender, T. R., J. I. Mead, and A. M. Rea. 1991b. Late Quaternary plants and vertebrates from Picacho Peak, Arizona. Southwest. Nat. 36: 302–314.

Vandeventer, T. L. 1977. Report of a double pre-molt period in a mottled rock rattlesnake. Bull. Chicago Herpetol. Soc. 12: 60.

Van Frank, R., and M. K. Hecht. 1954. Fossil rattlesnakes of the genus *Crotalus* from northern Massachusetts. Copeia 1954: 158–159.

Van Hyning, O. C. 1931. Reproduction of some Florida snakes. Copeia 1931: 59–60.

Van Mierop, L. H. S. 1976a. Poisonous snakebite: A review. 1. Snakes and their venom. J. Florida Med. Assoc. 63: 191–200.

Van Mierop, L. H., and C. S. Kitchens. 1980. Defibrination syndrome following bites by the eastern diamondback rattlesnake. J. Florida Med. Assoc. 67: 21–27.

Van Riper, W. 1954. Measuring the speed of a rattlesnake's strike. Anim. Kingd. 57: 50–53.

———. 1955. How a rattlesnake strikes. Nat. Hist. 64: 308–311.

Vaughan, T. A. 1961. Vertebrates inhabiting pocket gopher burrows in Colorado. J. Mamm. 42: 171–174.

Vaughan, T. A., and S. T. Schwartz. 1980. Behavioral ecology of an insular woodrat. J. Mamm. 61: 205–218.

v. d. Velde, H. 1995a. *Crotalus mitchelli mitchelli*: Speckled rattlesnake. Litt. Serpent. Engl. Ed. 15: 78–79.

———. 1995b. *Crotalus lepidus klauberi*. Litt. Serpent. Engl. Ed. 15: 134.

Vermersch, T. G., and R. E. Kuntz. 1986. Snakes of south-central Texas. Eakin Press, Austin, Texas.

Vetas, B. 1951. Temperatures of entrance and emergence. Herpetologica 7: 15–20.

Vick, J. A. 1971. Symptomatology of experimental and clinical crotalid envenomation, pp. 71–86. *In* L. L. Simpson (ed.), Neuropoisons: Their pathophysiological actions, vol. 1. Plenum Press, New York.

Vidal Breard, J. J. 1950. Iron in serpent venoms. Arch. Farm. Bioquím. (Tucuman) 4: 321.

Villa, R. A., P. T. Condon, T. A. Hare, S. Avila-Villegas, and D. G. Barker. 2007. *Crotalus willardi willardi* (Arizona ridge-nosed rattlesnake). Herpetol. Rev. 38: 220.

Villarreal, X., J. Bricker, H. K. Reinert, L. Gelbert, and L. M. Bushar. 1996. Isolation and characterization of microsatellite loci for use in population genetic analysis in the timber rattlesnake, *Crotalus horridus*. J. Heredity 87(2): 152–155.

Vincent, J. W. 1982a. Phenotypic variation in *Crotalus lepidus lepidus* (Kennicott). J. Herpetol. 16: 189–191.

———. 1982b. Color pattern variation in *Crotalus lepidus lepidus* (Viperidae) in southwestern Texas. Southwest. Nat. 27: 263–272.

Vitt, L. J. 1974. Body temperatures of high latitude reptiles. Copeia 1974: 225–256.

Vitt, L. J., and R. D. Ohmart. 1978. Herpetofauna of the Lower Colorado River: Davis Dam to the Mexican Border. Proc. West. Found. Vert. Zool. 2: 33–72.

Voge, M. 1953. New host records for *Mesocestoides* (Cestoda: Cyclophyllidea) in California. Am. Midl. Nat. 49: 249–251.

Vogel, C. W. 1991. Cobra venom factor: The complement-activating protein of cobra venom, pp. 147–188. *In* A. T. Tu (ed.), Handbook of natural toxins, vol. 5: Reptile venoms and toxins. Marcel Dekker, New York.

Vogler, J. 1973. Teach respect—not fear: Getting along with rattlesnakes. Wyoming Wildl. 35(11): 22–25.

Vogt, R. C. 1981. Natural history of amphibians and reptiles of Wisconsin. Milwaukee Public Museum, Milwaukee, Wisconsin.

Vogt, W. 1990. Snake venom constituents affecting the complement system, pp. 79–96. *In* K. F. Stocker, Medical use of snake venom proteins. CRC Press, Boca Raton, Florida.

von Bloeker, J. C., Jr. 1942. Amphibians and reptiles of the dunes. Bull. So. California Acad. Sci. 41: 29–38.

Vorhies, C. T.. 1948. Food items of rattlesnakes. Copeia 1948: 302–303.

Vorhies, C. T., and W. P. Taylor. 1940. Life history and ecology of the white-throated wood rat, *Neotoma albigula albigula*, in relation to grazing in Arizona. Univ. Arizona, Agric. Tech. Bull. 86: 453–529.

Voris, H. K. 1951. Miscellaneous notes on the eggs and young of Texan and Mexican reptiles. Zoologica 36: 37–48.

Wacha, R. S., and J. L. Christiansen. 1982a. Development of *Caryospora bigenica*, new species (Apicomplexa, Eimeriidae) in rattlesnakes and laboratory mice. J. Protozool. 29: 272–278.

————. 1982b. Life cycle pattern of *Caryospora* sp. (Coccidia). J. Protozool. 29: 289.

Wadsworth, J. R. 1954. Neoplasms of snakes. Vet. Exten. Quart. 133: 65–72.

————. 1956. Serpentine tumors. Vet. Med. 51: 326–328.

Wagner, F. W., and J. M. Prescott. 1966. A comparative study of proteolytic activities in the venoms of some North American snakes. Comp. Biochem. Physiol. 17: 191–201.

Wagner, R. T. 1962. Notes on the combat dance in *Crotalus adamanteus*. Bull. Philadelphia Herpetol. Soc. 10(1): 7–8.

Waldron, J. L., S. M. Welch, S. Bennett, W. Kalinowsky, M. E. Dorcas, and D. Lanham. 2005. A multiscale examination of habitat use by sympatric populations of eastern diamondback rattlesnakes and canebrake rattlesnakes in the South Carolina coastal plain, p. 51. *In* Biology of the rattlesnakes symposium (program abstracts). Loma Linda Univ., Loma Linda, California.

Waldron, J. L., S. H. Bennett, S. M. Welch, M. E. Dorcas, J. D. Lanham, and W. Kalinowsky. 2006a. Habitat specificity and home-range size as attributes of species vulnerability to extinction: A case study using sympatric rattlesnakes. Anim. Conserv. 9: 414–420.

Waldron, J. L., J. D. Lanham, and S. H. Bennett. 2006b. Using behaviorally-based seasons to investigate canebrake rattlesnake (*Crotalus horridus*) movement patterns and habitat selection. Herpetologica 62: 389–398.

Walker, J. M. 1963. Amphibians and reptiles of Jackson Parish, Louisiana. Proc. Louisiana Acad. Sci. 26: 91–101.

Walker, K. J. 2003. An illustrated guide to trunk vertebrae of cottonmouth (*Agkistrodon piscivorus*) and diamondback rattlesnake (*Crotalus adamanteus*) in Florida. Bull. Florida Mus. Nat. Hist. 44: 91–100.

Wallace, R. L., and L. V. Diller. 1990. Feeding ecology of the rattlesnake, *Crotalus viridis oreganus*, in northern Idaho. J. Herpetol. 24: 246–253.

————. 2001. Variation in emergence, egress, and ingress among life-history stages and sexes of *Crotalus viridis oreganus* in northern Idaho. J. Herpetol. 35: 583–589.

Wallach, V. 1998. The lungs of snakes, pp. 93–295. *In* C. Gans and A. S. Gaunt (eds.), Biology of the Reptilia, vol. 19: Morphology G. Soc. Stud. Amphib. Rept., Ithaca, New York.

Wallach, J. D., and C. Hoessle. 1967. Visceral gout in captive reptiles. J. Am. Vet. Med. Assoc. 151: 897–899.

Wallach, V. 2007. Axial bifurcation and duplication in snakes: Part I. A synopsis of authentic and anecdotal cases. Bull. Maryland Herpetol. Soc. 43: 57–95.

Walley, H. D. 1963. The rattlesnake, *Crotalus horridus horridus*, in north-central Illinois. Herpetologica 19: 216.

Walls, G. L. 1932. Pupil shapes in reptilian eyes. Bull. Antiv. Inst. Am. 5: 68–70.

Warkentin, K. M. 1995. Adaptive plasticity in hatching age: A response to predation risk trade-offs. Proc. Natl. Acad. Sci. USA 92: 3507 3510.

Warrell, D. A. 2007. World antivenom shortage. *In* Venom week 2007 (program abstracts). Univ. Arizona Colleges of Medicine, Arizona Health Sci. Center, Off. Continuing Medical Educ., Tucson, Arizona.

Warwick, C. 1990. Disturbance of natural habitats arising from rattlesnake round-ups. Environ. Conserv. 17(2): 172–174.

Watt, C. H., Jr. 1985. Treatment of poisonous snakebite with emphasis on digit dermotomy. South. Med. J. 78: 694–699.

Weaver, R. E., and M. E. Lahti. 2005. Diet of the northern Pacific rattlesnake (*Crotalus viridis oreganus*) in the Yakima River Canyon of central Washington state, p. 51. *In* Biology of the Rattlesnakes Symposium (program abstracts). Loma Linda Univ., Loma Linda, California.

Webb, R. G. 1970. Reptiles of Oklahoma. Univ. Oklahoma Press, Norman.

Weigel, R. D. 1962. Fossil vertebrates of Vero, Florida. Florida Geol. Surv. Spec. Publ. 10: 1–59.

Weinstein, S. A., and L. A. Smith. 1990. Preliminary fractionation of tiger rattlesnake (*Crotalus tigris*) venom. Toxicon 28: 1447–1455.

Weinstein, S. A., S. A. Minton, and C. E. Wilde. 1985. The distribution among ophidian venoms of a toxin isolated from the venom of the Mojave rattlesnake (*Crotalus scutulatus scutulatus*). Toxicon 23: 825–844.

Weinstein, S. A., C. F. DeWitt, and L. A. Smith. 1992. Variability of venom-neutralizing properties of serum from snakes of the colubrid genus *Lampropeltis*. J. Herpetol. 26: 452–461.

Weir, J. 1992. The Sweetwater rattlesnake round-up: A case study in environmental ethics. Conserv. Biol. 6(1): 116–127.

Weldon, P. J., and G. M. Burghardt. 1979. The ophiophage defensive response in crotaline snakes: Extension to new taxa. J. Chem. Ecol. 5: 141–151.

Weldon, P. J., and D. B. Fagre. 1989. Responses by canids to scent gland secretions of the western diamondback rattlesnake (*Crotalus atrox*). J. Chem. Ecol. 15: 1589–1604.

Weldon, P. J., and F. M. Schell. 1984. Responses of king snakes (*Lampropeltis getulus getulus*) to chemicals from colubrid and crotaline snakes. J. Chem. Ecol. 10: 1509–1520.

Weldon, P. J., N. B. Ford, and J. J. Perry-Richardson. 1990a. Responses by corn snakes (*Elaphe guttata*) to chemicals from heterospecific snakes. J. Chem. Ecol. 16: 37–43.

Weldon, P. J., H. A. Lloyd, and M. S. Blum. 1990b. Glycerol monoethers in the scent gland secretions of the western diamondback rattlesnake (*Crotalus atrox*). Experentia 46: 774–775.

Weldon, P. J., R. Ortiz, and T. R. Sharp. 1992. The chemical ecology of crotaline snakes, pp. 309–319. *In* J. A. Campbell and E. D. Brodie, Jr. (eds.), Biology of the pitvipers. Selva, Tyler, Texas.

Wells, M. A., and D. J. Hanahan. 1969. Studies on phospholipase A: I. Isolation and characterization of two enzymes from *Crotalus adamanteus* venom. Biochemistry, N.Y. 8: 414–424.

Werler, J. E. 1950. The poisonous snakes of Texas and the first aid treatment of their bites. Texas Game Fish Bull. 31: 1–40.

Werler, J. E., and J. R. Dixon. 2000. Texas snakes: Identification, distribution, and natural history. Univ. Texas Press, Austin.

Werman, S. D. 2008. Phylogeny and the evolution of β-neurotoxic phospholipases A$_2$ (PLA$_2$) in the venoms of rattlesnakes, *Crotalus* and *Sistrurus* (Serpentes: Viperidae), pp. 511–536. *In* W. K. Hayes, K. R. Beaman, M. D. Cardwell, and S. P. Bush (eds.), The biology of rattlesnakes. Loma Linda Univ. Press, Loma Linda, California.

Wermelinger, L. S., D. L. S. Dutra, A. L. Oliveira-Carvalho, M. R. Soares, C. Bloch, Jr., and R. B. Zingali. 2005. Fast analysis of low molecular mass compounds present in snake venom: Identification of ten new pyroglutamate-containing peptides. Rapid Commun. Mass Spectom. 19: 1703–1708.

Werner, J. K., B. A. Maxwell, P. Hendricks, and D. L. Flath. 2004. Amphibians and reptiles of Montana. Montana Press Publ. Co., Missoula.

West, G. S. 1895. On the buccal glands and teeth of certain poisonous snakes. Proc. Zool. Soc. London 1895: 812–826.

Wharton, M. E., and R. W. Barbour. 1991. Bluegrass land & life. Univ. Press Kentucky, Lexington.

Wheeler, D. G. 1994. An unusual incidence of venom squirting by a captive western diamondback rattlesnake, *Crotalus atrox*. Bull. Chicago Herpetol. Soc. 29: 199.

Wheeler, G. C. 1947. The amphibians and reptiles of North Dakota. Am. Midl. Nat. 38: 162–190.

Wheeler, G. C., and J. Wheeler. 1966. The amphibians and reptiles of North Dakota. Univ. North Dakota Press, Grand Forks.

Whisenhunt, M. H., Jr. 1949. An account of copulation of the western diamondback rattlesnake. Nat. Hist. Misc. 49: 1–2.

White, F. H. 1963. Leptospiral agglutinins in snake serums. Am. J. Vet. Res. 24: 179–182.

White, F. N., and R. C. Lasiewski. 1971. Rattlesnake denning: Theoretical considerations on winter temperatures. J. Theor. Biol. 30: 553–557.

White, J. 2005. Snake venoms and coagulopathy. Toxicon 45: 951–967.

Whitlow, K. S., J. T. Ascue, R. N. Patel, and S. R. Rose. 2007. Arizona black rattlesnake envenomation. *In* Venom week 2007 (program abstracts). Univ. Arizona Colleges of Medicine, Arizona Health Sci. Center, Off. Continuing Medical Educ., Tucson, Arizona.

Whitt, A. L., Jr. 1970. Some mechanisms with which *Crotalus horridus horridus* responds to stimuli. Trans. Kentucky Acad. Sci. 31: 45–48.

Whorley, J. R. 2000. Keys to partial mammals: A method for identifying prey items from snakes. Herpetol. Rev. 31: 227–229.

Widmer, E. A. 1966. Synonomy of *Oochoristica crotalicola* Alexander and Alexander, 1957. Bull. Wildl. Dis. Assoc. 2: 82–84.

———. 1967. Helminth parasites of the prairie rattlesnake, *Crotalus viridis* Rafinesque, 1818, in Weld County, Colorado. J. Parasitol. 53: 362–363.

Widmer, E. A., and G. B. Hanson. 1983. Site selection and penetration studies of *Mesocestoides corti* (Eucestoda) tetrathyridia in the southern Pacific rattlesnake, *Crotalus viridis helleri*. J. Parasitol. 69: 788–789.

Widmer, E. A., and O. W. Olsen. 1967. The life history of *Oochoristica osheroffi* Meggitt, 1934 (Cyclophyllidea: Anoplocephalidae). J. Parasitol. 53: 343–349.

Widmer, E. A., and H. D. Specht. 1991. Asynchronous capsule formation in the gastrointestinal tract of a prairie rattlesnake (*Crotalus viridis viridis*) induced by *Mesocestoides* sp. tetrathyridea. J. Wildl. Dis. 27: 161–163.

———. 1992. Isolation of asexually proliferative tetrathyridia of *Mesocestoides* sp. from the southern Pacific rattlesnake (*Crotalus viridia helleri*) with additional data from two previous isolates from the Great Basin fence lizard (*Sceloporus occidentalis longipes*). J. Parasitol. 78: 921–923.

Widmer, E. A., P. C. Engen, and G. L. Bradley. 1995. Intracapsular asexual proliferation of *Mesocestoides* sp. tetrathyridia in the gastrointestinal tract and mesenteries of the prairie rattlesnake (*Crotalus viridis viridis*). J. Parasitol. 81: 493–496.

Wilcox, J. T. 2005. *Rana catesbeiana* (American bullfrog): Diet. Herpetol. Rev. 36: 306.

Wilcox, J. T., and D. H. Van Vuren. 2009. Wild pigs as predators in oak woodlands of California. J. Mammal. 90: 114–118.

Wiley, E. O. 1972. The Pleistocene herpetofauna of Dark Canyon Cave, New Mexico. Herpetol. Rev. 4: 128.

Wiley, G. O. 1929. Notes on the Texas rattlesnake in captivity with special reference to the birth of a litter of young. Bull. Antiv. Inst. Am. 3: 8–14.

———. 1930. Notes on the neotropical rattlesnake (*Crotalus terrificus basiliscus*) in captivity. Bull. Antiv. Inst. Am. 3: 100–103.

Wilkinson, J. A., J. L. Glenn, R. C. Straight, and J. W. Sites, Jr. 1991. Distribution and genetic variation in venom A and B populations of the Mojave rattlesnake (*Crotalus scutulatus scutulatus*) in Arizona. Herpetologica 47: 54–68.

Williams, E. J., S. C. Sung, and M. Laskowski, Sr. 1961. Action of venom phosphodiesterase on deoxyribonucleic acid. J. Biol. Chem. 236: 1130–1134.

Williamson, M. A. 1971. An instance of cannibalism in *Crotalus lepidus* (Serpentes: Crotalidae). Herpetol. Rev. 3: 18.

Willis, T. W., and A. T. Tu. 1988. Purification and biochemical characterization of atroxase, a nonhemorrhagic fibrinolytic protease from western diamondback rattlesnake venom. Biochemistry 27: 4769–4777.

Willis, T. W., A. T. Tu, and C. W. Miller. 1989. Thrombolysis with a snake venom protease in a rat model of venous thrombosis. Thromb. Res. 53: 19–29.

Williston, S. W. 1878. The prairie dog, owl and rattlesnake. Am. Nat. 12: 203–208.

Wills, C. A., and S. J. Beaupre. 2000. An application of randomization for detecting evidence of thermoregulation in timber rattlesnakes (*Crotalus horridus*) from Northwest Arkansas. Physiol. Biochem. Zool. 73: 325–334.

Wilson, L. D., and L. Porras. 1983. The ecological impact of man on the south Florida herpetofauna. Univ. Kansas Mus. Nat. Hist. Spec. Publ. 9: i–vi, 1–89.

Wilson, S. C. 1954. Snake fight. Texas Game Fish. 12(5): 16–17.

Wilson, T. P. 2000. Notes and observations on body size in an unusually large western diamondback rattlesnake, *Crotalus atrox* (Baird and Girard, 1853). Bull. Maryland Herpetol. Soc. 36: 65–67.

Wingert, W. A., and L. Chan. 1988. Rattlesnake bites in southern California and rationale for recommended treatment. West. J. Med. 148: 37–44.

Wingert, W. A., T. R. Pattabhiraman, D. Powers, and F. E. Russell. 1981. Effect of a rattlesnake venom (*Crotalus viridis helleri*) on bone marrow. Toxicon 19: 181–183.

Winkel, K. 2007. U.S. snakebite mortality, 1979–2005. *In* Venom week 2007 (program abstracts).

Univ. Arizona Colleges of Medicine, Arizona Health Sci. Center, Off. Continuing Medical Educ., Tucson, Arizona.

Wittner, D. 1978. A discussion of venomous snakes of North America: Distribution and occurrence of North American snakes of the genera *Crotalus, Sistrurus,* and *Agkistrodon.* HERP: Bull. New York Herpetol. Soc. 14: 12–17.

Wolfenbarger, K. A. 1952. Systematic and biological studies on North American chiggers of the genus *Trombicula,* subgenus *Eutrombicula.* Ann. Entomol. Soc. Am. 45: 645–672.

Wong, H. 1997. Comments on the snake records of *Chilomeniscus cinctus, C. exsul,* and *C. mitchellii* from Islas Magdalena and Santa Margarita, Baja California, México. Herpetol. Rev. 28: 188–189.

Wood, F. D. 1933. Mating of the prairie rattlesnake, *Crotalus confluentus confluentus* Say. Copeia 1933: 84–87.

Wood, S. F. 1944. The reptile associates of wood rats and cone-nosed bugs. Bull. So. California Acad. Sci. 43: 44–48.

Wood, S. F., and F. D. Wood. 1936. Occurrence of Haematozoa in some California cold-blooded vertebrates. J. Parasitol. 22(5): 518–520.

Woodbury, A. M. 1929. A new rattlesnake from Utah. Bull. Univ. Utah (Biol. Ser.) 20: 106.

———. 1930. *Crotalus confluentus concolor* (Woodbury). Bull. Antiv. Inst. Am. 4: 23.

———. 1931. A descriptive catalog of the reptiles of Utah. Bull. Univ. Utah 21(5): x, 1–129.

———. 1942. Status of the name *Crotalus concolor.* Copeia 1942: 258.

———. 1945. My rattlesnake bite. Proc. Utah Acad. Sci. Arts Lett. 19–20: 179–184.

———. 1948. Marking reptiles with an electric tattooing outfit. Copeia 1948: 127–128.

———. 1951. Symposium: A snake den in Toole County, Utah. Introduction—a ten year study. Herpetologica 7: 4–14.

———. 1952. Amphibians and reptiles of the Great Salt Lake Valley. Herpetologica 8: 42–50.

———. 1958. The name *Crotalus viridis concolor* Woodbury. Copeia 1958: 151.

Woodbury, A. M., and J. D. Anderson. 1945. Report of rattlesnake bite of J. Dwain Anderson. Proc. Utah Acad. Sci. Arts Lett. 19–20: 185–188.

Woodbury, A. M., and R. M. Hansen. 1950. A snake den in Tintic Mountains, Utah. Herpetologica 6: 66–70.

Woodbury, A. M., and R. Hardy. 1947a. The Mojave rattlesnake in Utah. Copeia 1947: 66.

———. 1947b. The speckled rattlesnake in NW Arizona. Herpetologica 3: 169.

———. 1948. Studies of the desert tortoise *Gopherus agassizii.* Ecol. Monogr. 18: 145–200.

Woodbury, A. M., and D. D. Parker. 1956. A snake den in Cedar Mountains and notes on snakes and parasitic mites. Herpetologica 12: 261–268.

Woodbury, A. M., and E. W. Smart. 1950. Unusual snake records from Utah and Nevada. Herpetologica 6: 45–47.

Woodbury, A. M., and D. M. Woodbury. 1944. Notes on Mexican snakes from Oaxaca. J. Washington Acad. Sci. 34: 360–373.

Woodin, W. H., III. 1953. Notes on some reptiles from the Huachuca area of southeastern Arizona. Bull. Chicago Acad. Sci. 9: 285–296.

Wooldridge, B. J., G. Pineda, J. J. Banuelas-Ornelas, R. K. Dagda, S. E. Gasanov, E. D. Rael, and C. S. Lieb. 2001. Mojave rattlesnakes (*Crotalus scutulatus scutulatus*) lacking the acidic subunit DNA sequence lack Mojave toxin in their venom. Comp. Biochem. Physiol. 130B: 169–179.

World Health Organization. 2007. Rabies and envenomations: A neglected public health issue. Report of a Consultative Meeting, World Health Organization, Geneva, 10 January 2007. WHO Press, Geneva, Switzerland.

Wozniak, E. J., G. L. McLaughlin, and S. R. Telford, Jr. 1994. Description of the vertebrate states of haemogregarine species naturally infecting Mojave Desert sidewinders (*Crotalus cerastes cerastes*). J. Zoo Wildl. Med. 25: 103–110.

Wozniak, E. J., J. Wisser, and M. Schwartz. 2006. Venomous adversaries: A reference to snake identification, field safety, and bite-victim first aid for disaster-response personnel deploying into the hurricane-prone regions of North America. Wild. Environ. Med. 17: 246–266.

Wright, A. H., and A. A. Wright. 1957. Handbook of snakes of the United States and Canada, vols. 1 and 2. Comstock Publ. Associates, Cornell Univ. Press, Ithaca, New York.

―――. 1962. Handbook of snakes of the United States and Canada, vol. 3: Bibliography. Edwards Brothers, Ann Arbor, Michigan.

Wright, R. A. S. 1987. Natural history observations on venomous snakes near the Peaks of Otter, Bedford County, Virginia. Catesbeiana 7(2): 2–9.

Wüster, W., J. E. Ferguson, J. A. Quijada-Mascareñas, C. E. Pook, M. da Graça Salomão, and R. S. Thorpe. 2005. Tracing an invasion: Landbridges, refugia, and the phylogeography of the Neotropical rattlesnake (Serpentes: Viperidae: *Crotalus durissus*). Mol. Ecol. 14(4): 1095–1108.

Wyeth-Ayerst Laboratories. 1998. Antivenin (Crotalidae) polyvalent, pp. 3009–3010. *In* Physicians' desk reference, 52nd ed. Medical Economics, Montvale, New Jersey.

Wynn, D. E., and S. M. Moody. 2006. Ohio turtle, lizard, and snake atlas. Ohio Biolog. Surv. Misc. Contr. 10: iv–82.

Xiong, Y., W. Wand, X. Pu, and J. Song. 1992. Preliminary study on the mechanism of using snake venoms to substitute for morphine. Toxicon 30: 567.

Yamaguti, S. 1961. Systema Helminthum: The nematodes of vertebrates, part I. Interscience Publ., New York.

Yamazaki, Y., and T. Morita. 2004. Structure and function of snake venom cysteine-rich secretory proteins. Toxicon 44: 227–231.

Yancey, F. D., II, W. Meinzer, and C. Jones. 1997. Aberrant morphology in western diamondback rattlesnakes (*Crotalus atrox*). Occ. Paps. Mus. Texas Tech Univ. 164: 1–3.

Yarrow, H. C. 1875. Report upon the collections of batrachians and reptiles made in portions of Nevada, Utah, California, Colorado, New Mexico, and Arizona, during the years 1871, 1872, 1873 and 1874. Rep. Geog. Geol. Expl. Surv. W. 100th Merid. Geo. M. Wheeler 5: 509–584.

Young, B. A. 2008. Perspectives on the regulation of venom expulsion in snakes, pp. 181–190. *In* W. K. Hayes, K. R. Beaman, M. D. Cardwell, and S. P. Bush (eds.), The biology of rattlesnakes. Loma Linda Univ. Press, Loma Linda, California.

Young, B. A., and A. Aguiar. 2002. Response of western diamondback rattlesnakes *Crotalus atrox* to airborne sounds. J. Exp. Biol. 205: 3087–3092.

Young, B. A., and I. P. Brown. 1993. On the acoustic profile of the rattlesnake rattle. Amphibia-Reptilia 14: 373–380.

Young, B. A., and J. Harris. 2006. Auditory sensitivity of the northern Pacific rattlesnake, *Crotalus viridis oreganus*: Do behavioral responses conform to physiological performance? Northwest Sci. 80: 218–223.

Young, B. A., and K. Jackson. 2008. Functional specialization of the extrinsic venom gland musculature within the Crotaline snakes (Reptilia: Serpentes) and the role of the M. Pterygoideus Glandulae, pp. 47–54. *In* W. K. Hayes, K. R. Beaman, M. D. Cardwell, and S. P. Bush (eds.), The biology of rattlesnakes. Loma Linda Univ. Press, Loma Linda, California.

Young, B. A., and K. V. Kardong. 2007. Mechanisms controlling venom expulsion in the western diamondback rattlesnake, *Crotalus atrox*. J. Exp. Zool. 307A: 18–27.

Young, B. A., and K. Zahn. 2001. Venom flow in rattlesnakes: Mechanics and metering. J. Exp. Biol. 204: 4345–4351.

Young, B. A., M. Phelan, J. Jaggers, and N. Nejman. 2001. Kinematic modulation of the strike of the western diamondback rattlesnake (*Crotalus atrox*). Hamadryad 26(2): 316–349.

Young, B. A., M. Phelan, M. Morain, M. Ommundsen, and R. Kurt. 2003. Venom injection by rattlesnakes (*Crotalus atrox*): Peripheral resistance and the pressure-balance hypothesis. Can. J. Zool. 81: 313–320.

Young, N. 1940. Snakebite: Treatment and nursing care. Am. J. Nurs. 40: 657–660.

Young, R. A., and D. M. Miller. 1980. Notes on the natural history of the Grand Canyon rattlesnake, *Crotalus viridis abyssus* Klauber. Bull. Chicago Herpetol. Soc. 15: 1–5.

Young, R. A., D. M. Miller, and D. C. Ochsner. 1980. The Grand Canyon rattlesnake (*Crotalus viridis abyssus*): Comparison of venom protein profiles with other *viridis* subspecies. Comp. Biochem. Physiol. 66B: 601–603.

Yousef, G. M., M. B. Elliott, A. D. Kopolovic, E. Serry, and E. P. Diamandis. 1995. Families and clans of serine peptidases. Arch. Biochem. Biophys. 318: 247–250.

———. 2004. Sequence and evolutionary analysis of the human trypsin subfamily of serine peptidases. Biochim. Biophys. Acta 1698: 77–86.

Zaidan, F., III, and S. J. Beaupre. 2003. Effects of body mass, meal size, fast length, and temperature on specific dynamic action in the timber rattlesnake (*Crotalus horridus*). Physiol. Biochem. Zool. 76(4): 447–458.

Zamudio, K. R., D. L. Hardy Sr., M. Martins, and H. W. Greene. 2000. Fang tip spread, puncture distance, and suction for snake bite. Toxicon 38: 723–728.

Zappalorti, R. T., and H. K. Reinert. 1986. Final report on habitat utilization by the rattlesnake, *Crotalus horridus* (Linnaeus) in southern New Jersey with notes on hibernation. New Jersey Endang. Nongame Spec. Progr., Div. Fish Game Wildl., Trenton.

———. 2005. Habitat use, foraging behavior, and overwintering sites of timber rattlesnakes (*Crotalus horridus*) in the New Jersey Pine Barrens, p. 52. *In* Biology of the rattlesnakes symposium (program abstracts). Loma Linda Univ., Loma Linda, California.

Zarafonetis, C. J. D., and J. P. Kalas. 1960. Serotonin, catechol amines, and amine oxidase activity in the venoms of certain reptiles. Am. J. Med. Sci. 240: 764–768.

Zeller, E. A. 1948. Enzymes of snake venoms and their biological significance, pp. 459–495. *In* F. F. Nora (ed.), Advances in enzymology, vol. 8. Interscience, New York.

———. 1950a. Über Phosphatasen II. Über eine neue Adenosintriphosphatase. Helv. Chim. Acta 33: 821–833.

———. 1950b. The formation of pyrophosphate from adenosine triphosphate in the presence of snake venom. Arch. Biochem. 28: 138–139.

Zimmermann, A. A., and C. H. Pope. 1948. Development and growth of the rattle of rattlesnakes. Fieldiana: Zool. 32: 355–413.

Zimmerman, E. G., and C. W. Kilpatrick. 1973. Karyology of North American crotaline snakes (family Viperidae) of the genera *Agkistrodon*, *Sistrurus*, and *Crotalus*. Can. J. Genet. Cytol. 15(3): 389–395.

Ziolkowski, C., and A. L. Bieber. 1992. Mojave toxin affects fusion of myoblasts and viability of myotubes in cell cultures. Toxicon 30: 733–744.

Ziolkowski, C., H. A. Murchison, and A. L. Bieber. 1992. Effects of myotoxin alpha on fusion and contractile activity in myoblast-myotube cell cultures. Toxicon 30: 397–409.

Zozaya, J., and R. E. Stadelman. 1930. Hypersensitiveness to snake venom proteins: A case report. Bull. Antiv. Inst. Am. 3: 93–95.

Zug, G. R., and C. H. Ernst. 2004. Smithsonian answer book: Snakes. Smithsonian Books, Washington, D.C.

Zug, G. R., L. J. Vitt, and J. P. Caldwell. 2001. Herpetology: An introductory biology of amphibians and reptiles. 2nd ed. Academic Press, New York.

Zwart, P., F. G. Poelma, and W. J. Strik. 1970. The distribution of various types of Salmonellae and Arizonas in reptiles. Zentralb. Bak. Parasitenk. Infek. Hyg. 213A: 201–212.

Zweifel, R. G. 1952. Notes on the lizards of the Coronados Islands, Baja California, Mexico. Herpetologica 8: 9–11.

———. 1959. Additions to the herpetofauna of Nayarit, Mexico. Am. Mus. Novit. 1953: 1–16.

Zweifel, R. G., and K. S. Norris. 1955. Contribution to the herpetology of Sonora, Mexico: Descriptions of new subspecies of snakes (*Micruroides euryxanthus* and *Lampropeltis getulus*) and miscellaneous collecting notes. Am. Midl. Nat. 54: 230–249.

Index to Common and Scientific Names